Human Retrotransposons in Health and Disease

Gael Cristofari
Editor

Human Retrotransposons in Health and Disease

Springer

Editor
Gael Cristofari, Ph.D.
Institute for Research on Cancer and Aging of Nice (IRCAN)
French National Institute of Health and Medical Research (INSERM)
Centre National de la Recherche Scientifique (CNRS)
Faculty of Medicine
University of Nice-Sophia Antipolis (UNS)
Université Côte d'Azur (UCA)
Nice, France

ISBN 978-3-319-48343-6 ISBN 978-3-319-48344-3 (eBook)
DOI 10.1007/978-3-319-48344-3

Library of Congress Control Number: 2016961233

Printed on acid-free paper

This Springer imprint is published by Springer Nature
The registered company is Springer International Publishing AG
The registered company address is: Gewerbestrasse 11, 6330 Cham, Switzerland

To Callisto and Lou

Preface

Retrotransposons are highly repetitive and dispersed sequences. These transposable elements have the ability to proliferate via an RNA-mediated copy-and-paste mechanism, called retrotransposition, and belong to several distant subclasses in humans.

The Long INterspersed Element-1 (L1 or LINE-1) is the only autonomous transposable element able to generate new copies in the modern human genome. The role of this process as a source of genetic diversity and diseases in humans has been recognized since the late 1980s. However, the advances of deep-sequencing technologies have recently shed new light on the extent of L1-mediated genome variations. They have also led to the discovery that L1 is not only able to mobilize in the germline—resulting in inheritable genetic variations—but can also jump in somatic tissues, such as embryonic stem cells, neuronal progenitor cells, and in many cancers. L1 is also able to mobilize *in trans* other sequences, leading to the expansion of Alu elements, which belong to another class of repeats, the Short INterspersed Elements (SINEs); or to the formation of processed pseudogenes, which also contribute to genome plasticity. Understanding the link between retrotransposon-mediated structural genomic variation and human phenotypes or diseases has become an intense field of research.

Although insertional mutagenesis is one of the mechanisms by which retrotransposons reshape our genome, retrotransposition-*in*dependent mechanisms also impact genome stability and cellular physiology (e.g., recombination, exaptation, creation of novel or alternative transcripts or proteins, DNA damage). These processes implicate a much broader set of retrotransposon sequences, some being very ancient and totally unable to mobilize, at least in humans. For example, the remodeling of the epigenetic landscape during early embryonic development activates the transcription of defective retrotransposons. The latter are unable to jump, but their transcription allows the synthesis of long noncoding RNA (lncRNA) with essential roles in stemness.

One of the originalities of this book is to explore not only insertion-based effects and their related consequences on germline and somatic genome dynamics but also the role and impact of retrotransposon sequences in a broader context, including a number of novel topics that emerged recently (lncRNA, neuronal disorders, exaptation, aging). We hope that it will illustrate how much retrotransposon biology is tightly connected to a myriad of cellular or physiological processes and will stimulate the next generation of young scientists to join a fascinating field and a highly dynamic research community.

Nice, France Gael Cristofari

Contents

Retrotransposons and the Mammalian Germline

Ian R. Adams

1 Retrotransposons in Mammalian Genomes

Retrotransposons are a class of mobile genetic elements that make up around 40 % of the sequenced mammalian genome (Chinwalla et al. 2002; Lander et al. 2001). Retrotransposons contribute to genomic instability in mammalian genomes by providing interspersed repeats of homologous sequences that can act as substrates for recombination causing deletions, duplications and structural rearrangements in the genome (Romanish et al. 2010). Retrotransposons are thought to be the only active class of mobile genetic element in most mammalian genomes, and can also cause genome instability through jumping to new locations in the genome. These de novo retrotransposon insertions have been reported as the causal mutation in various human genetic diseases (Crichton et al. 2014; Hancks and Kazazian 2012). The copy-and-paste mechanism that retrotransposons use to jump to new locations in the genome involves reverse-transcription of retrotransposon RNA, and integration of the resulting cDNA into new locations in the genome. There are typically a few hundred different types of retrotransposon annotated in each mammalian genome, with each type of retrotransposon being present in up to 10,000 copies. However, the types of retrotransposon, their copy numbers and their genomic locations vary significantly between species.

Mammalian retrotransposons are classified into LINE (long interspersed nuclear elements), SINE (short interspersed nuclear elements) and LTR (long terminal repeat) retrotransposon classes (Chinwalla et al. 2002; Lander et al. 2001). Each class of retrotransposons can be further subdivided into families, and each family into individual types. The LINE class of retrotransposons in mammals is primarily

I.R. Adams (✉)
MRC Human Genetics Unit, MRC Institute of Genetics and Molecular Medicine,
University of Edinburgh, Western General Hospital, Crewe Road, Edinburgh EH4 2XU, UK
e-mail: Ian.Adams@igmm.ed.ac.uk

G. Cristofari (ed.), *Human Retrotransposons in Health and Disease*,
DOI 10.1007/978-3-319-48344-3_1

represented by the LINE-1 family. Only the human-specific L1HS-Ta subfamily of LINE-1 is still active in the human genome, whereas the L1MdA, L1MdTf and L1MdGf types of LINE-1 are all active in the mice (Beck et al. 2011; Hancks and Kazazian 2012; Sookdeo et al. 2013). Full-length LINE-1s are typically 6–7 kb long and are transcribed from an internal promoter located in the 5′ untranslated region (UTR) of these elements (Beck et al. 2011; Hancks and Kazazian 2012; Swergold 1990). The LINE-1 promoter also generates an antisense transcript that can extend into the adjacent flanking cellular DNA (Cruickshanks et al. 2013; Li et al. 2014; Macia et al. 2011; Speek 2001). The LINE-1 5′ UTR varies significantly between species, and even between individual types of LINE-1 within species (Khan et al. 2006; Lee et al. 2010; Sookdeo et al. 2013). LINE-1 encodes two open reading frames in mice and rodents, and three open reading frames in humans and primates (Beck et al. 2011; Denli et al. 2015; Hancks and Kazazian 2012). The recently discovered primate-specific LINE-1 ORF0 protein is localised to nuclear bodies and enhances LINE-1 retrotransposition activity, although its mechanism of action is not currently understood (Denli et al. 2015). LINE-1 ORF1 protein encodes an RNA-binding protein that forms a particle with LINE-1 RNA and is required for LINE-1 retrotransposition activity (Khazina and Weichenrieder 2009; Martin and Branciforte 1993; Moran et al. 1996). The LINE-1-encoded ORF2 protein is also required for LINE-1 retrotransposition and encodes an endonuclease and reverse transcriptase that nicks the host genomic DNA and catalyses target-primed reverse transcription of LINE-1 RNA into DNA at the site of genomic integration (Cost et al. 2002; Feng et al. 1996; Mathias et al. 1991; Moran et al. 1996). LINE-1 elements display a strong *cis*-preference where LINE-1 ORF1p and ORF2p tend to associate with the same mRNA molecule from which they are translated (Esnault et al. 2000; Kulpa and Moran 2006; Wei et al. 2001).

The SINE class of retrotransposons includes a group of elements that are typically 100–300 bp long and are derived from small non-coding cellular RNAs including 7SL RNA, 5S rRNA and tRNAs (Kramerov and Vassetzky 2011). SINE retrotransposons are transcribed from internal RNA polymerase III promoters, and include the Alu family in humans and primates. The human genome contains hundreds of active Alu elements, particularly those belonging to AluY and AluS subfamilies (Bennett et al. 2008). Alu elements have been proposed to utilise a ‘stealth’ mode of amplification where elements mobilise at low frequencies for a long time, occasionally generating hyperactive copies that expand aggressively but rapidly become extinct (Han et al. 2005). Alu elements, and SINEs in general, are non-autonomous retrotransposons that rely on LINE-1-encoded proteins to catalyse their retrotransposition (Dewannieux et al. 2003; Hancks et al. 2011; Raiz et al. 2012). The crystal structure of the Alu ribonucleoprotein particle suggests that Alu elements hijack LINE-1 reverse transcriptase by binding to ribosomes that are stalled when LINE-1 ORF2p reverse transcribes its encoding mRNA (Ahl et al. 2015). Other active SINEs in the human genome include SVA elements which can be up to 2 kb in length and contain regions derived from SINE-R and Alu retrotransposons (Hancks and Kazazian 2010; Wang et al. 2005).

The LTR retrotransposon class, whose members are also known as endogenous retroviruses (ERVs), typically encodes the Gag, Pol, Pro and sometimes also Env proteins that are found in retrovirus genomes (Bannert and Kurth 2006). These retroviral proteins function in the production of retroviral capsid proteins, retroviral DNA synthesis and integration into the host genome, processing of retroviral proteins, and forming the surface envelope on the retroviral capsid, respectively. The 5′ LTR acts as a promoter for these elements. LTR retrotransposons typically reverse-transcribe their RNA in the cytoplasm of the host, then translocate the cDNA into the nucleus and use a LTR retrotransposon-encoded integrase to insert the DNA into the genome. Some LTR retrotransposons are autonomous and encode the proteins required to catalyse their own retrotransposition, whereas others use proteins encoded by other types of LTR retrotransposon to retrotranspose *in trans* (Dewannieux et al. 2004; Ribet et al. 2004). LTR retrotransposons actively mobilise in rodents, and de novo retrotransposition of LTR retrotransposons accounts for around 10 % of spontaneously occurring mutations in mice (Maksakova et al. 2006). However, the vast majority of LTR retrotransposons in human genomes are probably extinct, and mobilisation of LTR retrotransposons in humans is extremely limited (Wildschutte et al. 2016).

Despite the diversity in retrotransposon structure and life cycle, the reason why each and every one of these elements has been able to accumulate multiple copies in the genome during evolution is because they are able to retrotranspose in the cells that can transmit those new genomic copies to the next generation. These crucial cells that play a key role in the life cycle and biology of retrotransposons in mammals belong to the germline.

2 The Mammalian Germline

In mammals, genetic information is transmitted from generation to generation by germ cells. Germ cells are conceptually distinct from the somatic cells that populate organs like the liver, brain, kidney, lungs and heart in that any genetic change that arises in germ cells can potentially be transmitted to subsequent generations whereas those that arise in somatic cells cannot (Fig. 1). This defining distinction between germ cells and somatic cells, as first proposed by the evolutionary biologist August Weismann (1834–1914) (Weismann 1889), means that it is events and activities that occur within germ cells that shape the landscape of mammalian genomes during evolution. Thus, while any individual retrotransposon might retrotranspose in any specific somatic tissue, all successful retrotransposons must retrotranspose in the germline.

Weismann's distinction between germline and soma means that genetic information and mutations that arise in somatic cells cannot be transmitted to the germ cells and subsequent generations, but this barrier between germ cells and soma is unidirectional (Fig. 1). Weismann realised that the soma must originate from the germline during early development, and therefore genetic information and mutations that

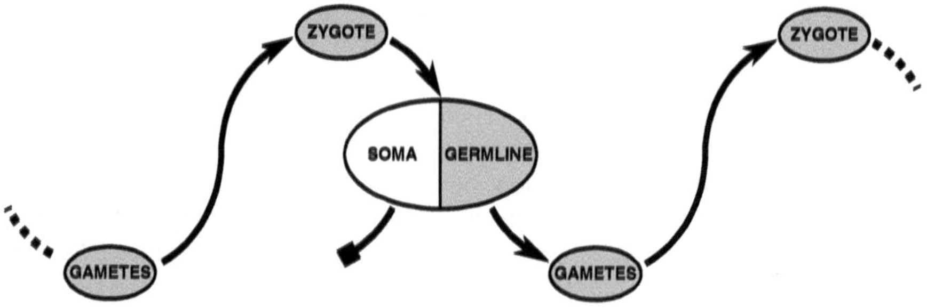

Fig. 1 The germline cycle. A schematic diagram showing the mammalian germline cycle. Genetic information, variation and mutations are inherited in the zygote from the parental gametes. The zygote gives rise to an individual containing soma (*white*) and germline (*grey*) tissues. Genetic information, variation and mutation in germline tissues can be incorporated into the gametes and transmitted to the next generation, whereas that in the soma cannot

arise in the germline can be propagated into the soma. These early embryonic cells that can give rise to both the germline and all somatic tissues would correspond to totipotent and pluripotent cells in modern terminology (Nichols and Smith 2009), while the term ‘germ cell’ is now typically reserved for lineage-restricted cells that have the capacity to differentiate into the mature gametes, but that do not normally contribute to somatic tissues in that individual (McLaren 2003). Weismann's germline encompasses totipotent cells, pluripotent cells and germ cells as they are all able to transmit genetic information and mutations to the next generation. Similarly, I will include the totipotent and pluripotent cells that are present in early mammalian development as part of the germline for the purposes of this review.

The major stages in mammalian germline development are outlined in Fig. 2. As many of these events are best characterised in mice, the timings and stages of germline development will be described for this species although there are likely to be broad similarities with other mammalian species. Mouse development initiates when mature germ cells, that is, female eggs and male sperm, fuse during fertilisation to generate a single-celled zygote. This zygote is totipotent in that it has the potential to differentiate into all extra-embryonic and embryonic cell types in the conceptus, which includes the germ cells. As pre-implantation development proceeds to blastocyst stage, some cells in the embryo differentiate into trophectoderm and primitive endoderm lineages that contribute only to extra-embryonic tissues. The remaining cells differentiate into pluripotent epiblast cells which retain the capacity to differentiate into all cell types in the embryo proper, including the germ cells (Magnúsdóttir and Surani 2014; Nichols and Smith 2009). After the blastocyst implants into the uterus, some pluripotent epiblast cells located close to the extra-embryonic ectoderm are induced by extracellular signals to differentiate into primordial germ cells. This germ cell specification event occurs around 6.25–7.25 days *post coitum* (dpc) in mouse embryos, and generates a founding population of around 40 primordial germ cells (Lawson and Hage 1994; Ohinata et al. 2005). The nascent primordial germ cells then embark on a phase of proliferation as they migrate

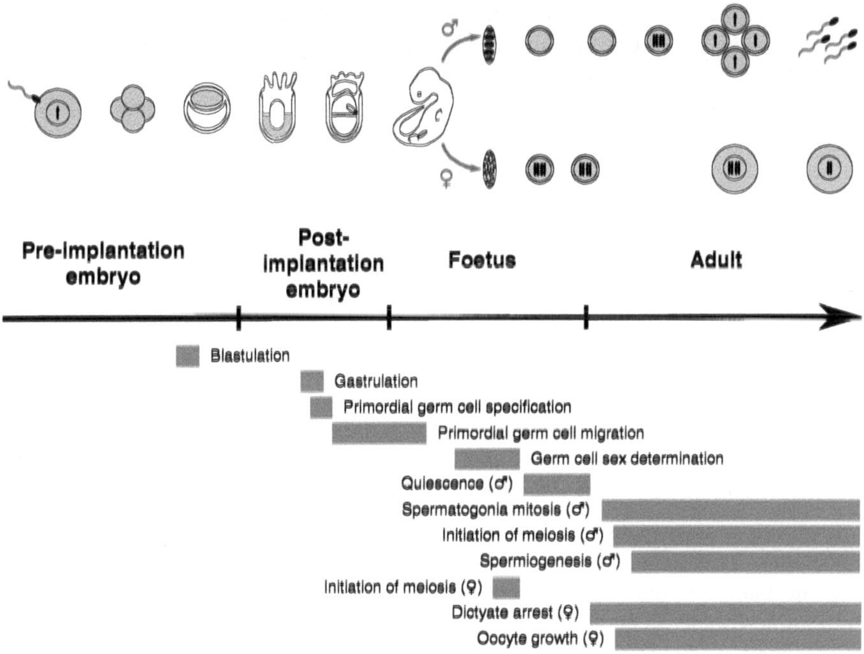

Fig. 2 Germline development in mice. A schematic diagram summarising the main stages of germline development in mice. Germ cells and their developmental precursors (totipotent and pluripotent cells) are coloured *grey*. For pre-implantation stages, a totipotent zygote and cleavage stage embryo, and a blastocyst containing pluripotent epiblast cells are shown. Post-implantation embryos before and after gastrulation are also shown, with pluripotent epiblast cells and primordial germ cells coloured *grey*. For foetal stages, primordial germ cells are shown within the gonads, and meiotic cells are indicated by a pair of homologous chromosomes in their nuclei. Pachytene stage of meiosis is depicted by the ladder-like synaptonemal complex between homologous chromosomes, which is absent in dictyate oocytes. Products of the first and second meiotic divisions are depicted by nuclei with a single replicated chromosome and an individual chromatid, respectively. Note that the second meiotic division in oocytes is typically only completed at fertilisation. Chromosomes are not shown in diploid mitotic cells for clarity, and the distinctive morphologies of fully grown oocytes and mature sperm are also indicated

through the embryo to reach the emerging gonads around 10.5 dpc. The germ cells are typically referred to as primordial germ cells during the early stages of their development until they reach the gonad, and primordial germ cell development at this point occurs similarly in male and female embryos. Once the primordial germ cells colonise the gonad, they continue to proliferate for a few more days, then initiate sex-specific differentiation into either male prospermatogonia or female oocytes (Kocer et al. 2009).

The germ cells' decision to differentiate along a male or a female pathway depends on sex-determining cues present in the gonadal environment rather than the sex chromosome constitution of the germ cells themselves (Kocer et al. 2009). In mice, germ cells in a foetal ovary become committed to develop down a female pathway between 12.5 dpc and 13.5 dpc (Adams and McLaren 2002). These oocytes initiate meiosis around 13.5 dpc and progress through most of the first meiotic prophase during late foetal development, then arrest as dictyate oocytes a few days after birth. The dictyate oocytes will eventually be stimulated to grow and resume meiosis in response to hormonal cues in adult animals. Fully grown oocytes then arrest meiosis during the second meiotic division before being ovulated and potentially fertilised (MacLennan et al. 2015).

In contrast, germ cells in a foetal mouse testis become committed to male development between 11.5 dpc and 12.5 dpc (Adams and McLaren 2002). The resulting prospermatogonia, also termed gonocytes, enter a period of quiescence and differentiation during late foetal development, then resume mitotic proliferation a few days after birth (Rossitto et al. 2015). The prospermatogonia give rise to mitotic spermatogonia and to a pool of spermatogonial stem cells which will self-renew and differentiate into more mitotic spermatogonia, thereby maintaining spermatogenesis throughout adulthood (Yang and Oatley 2014). The spermatogonia undergo a number of mitotic divisions before initiating meiosis which, in contrast to oocytes, typically proceeds without interruption. During spermiogenesis, the post-meiotic spermatids undergo a series of morphological changes that include condensation of the chromatin, elongation of the nucleus, specialisation of the Golgi membranes into an acrosome, generation of a flagella and elimination of residual cytoplasm (O'Donnell 2015). The sperm chromatin is delivered, along with a limited amount of cytoplasmic material, into the oocyte at fertilisation. The highly condensed and specialised sperm chromatin is predominantly associated with protamines rather than histones, and is reprogrammed with oocyte-derived histones shortly after fertilisation (Hogg and Western 2015).

3 Retrotransposon Expression in the Mammalian Germline

For a retrotransposon to accumulate new genomic integrations during evolution it needs to be active in the mammalian germline. While each retrotransposon does not need to be expressed and active at all stages in the germline cycle, each retrotransposon needs to be expressed and active at least at one point in the germline cycle. As outlined in the previous section, the germline cycle involves multiple distinct phases of development, and germline cells appear to use distinct transcription factor networks during these phases. Thus, it is rare to find germline-specific genes or transcription factors that are expressed throughout pre-implantation development, primordial germ cell development, foetal gametogenesis, oogenesis and spermatogenesis. One might expect individual retrotransposon expression profiles to behave similarly.

LINE-1 element transcripts, for example, are reported to be present in pre-implantation embryos, and in primordial germ cells in foetal gonads from 11.5 onwards (Fadloun et al. 2013; Hayashi et al. 2008; Molaro et al. 2014; Seisenberger et al. 2012). Although LINE-1 transcripts are present in foetal germ cells from

11.5 dpc, LINE-1 ORF1 protein is not detected in these cells until 15.5 dpc (Trelogan and Martin 1995). In female germ cells, LINE-1 ORF1p protein is expressed during early meiotic prophase but does not appear to be as abundant during post-natal oocyte development (Malki et al. 2014; Trelogan and Martin 1995). In male germ cells, LINE-1 ORF1 protein levels decrease after birth and, somewhat analogously to females, increase transiently during early meiotic prophase, then decrease as spermatogenesis proceeds (Branciforte and Martin 1994). Thus, early meiotic prophase appears to be a point in the germline cycle that LINE-1 retrotransposons target for expression. Interestingly, different types of LINE-1 retrotransposon have distinct RNA expression profiles during mouse spermatogenesis (Zamudio et al. 2015), which presumably reflects differences in the transcription factors binding to the distinct 5′ UTRs that each type of LINE-1 possesses.

LTR retrotransposons have a rich diversity of expression patterns during the germline cycle. Multiple types of LTR retrotransposon are highly expressed in pre-implantation and early post-implantation embryos, and retroviral-like particles are abundant in the totipotent and pluripotent cells present in these stages, although each LTR retrotransposon has quite specific expression profiles within these developmental stages (Brûlet et al. 1985; Dupressoir and Heidmann 1996; Fadloun et al. 2013; Macfarlan et al. 2012; Peaston et al. 2004; Piko et al. 1984; Reichmann et al. 2012; Ribet et al. 2008; Yotsuyanagi and Szöllösi 1981). The complete germline expression patterns of many types of LTR retrotransposon have not been reported, and additional intricacies are likely to emerge from next generation sequencing of RNA isolated from germline cells. As the retrotransposon LTRs will contain binding sites for transcription factors that are expressed in the germline, understanding how retrotransposons are expressed at specific stages in the germline cycle may help decipher some aspects of the transcriptional regulatory networks operating in these cells and can potentially help identify developmentally distinct subpopulations in the germline cycle (Macfarlan et al. 2012).

4 Retrotransposon Activity in the Mammalian Germline

The rate of de novo retrotransposition in the germline is presumably subject to evolutionary constraint. Although retrotransposons and de novo retrotransposition provide a rich source of genetic material and genetic variation, the insertional mutations that arise from these events contribute to genome instability and high de novo retrotransposition rates could prove to be deleterious for both the host and the retrotransposon. Rates of de novo retrotransposition in the germline have been estimated from sequencing genomic DNA, and from identification of disease and phenotype-causing mutations in mice and humans (Hancks and Kazazian 2012). In humans, de novo LINE-1 insertions are estimated to arise in 1 in every 100 births, with de novo Alu insertions estimated to occur five times more frequently. SVA elements have a somewhat lower de novo retrotransposition rate of approximately 1 in every 1000 births. LTR retrotransposons are not thought to be retrotranspositionally active in humans, but in mice de novo LTR retrotransposition events account for around 10–15 % of sequenced spontaneous mutant alleles (Maksakova et al. 2006). However, in general the stages of germline development during which the retrotransposition can occur are not known.

Retrotransposition at different stages of germline development can have distinct consequences for the host. Retrotransposition in the one cell zygote immediately after fertilisation and before the first round of DNA replication would theoretically result in the de novo retrotransposition event being present in a heterozygous state in all germ cells and somatic cells in that individual. However, retrotransposition at later times during pre-implantation development or during early post-implantation development would likely result in a mosaic conceptus containing some cells that have new heterozygous retrotransposition events, and some that do not. The characterisation of a mutagenic LINE-1 insertion in humans suggests that LINE-1 retrotransposition can generate extensive somatic mosaicism consistent with retrotransposition occurring early in development (van den Hurk et al. 2007). Importantly, if a de novo retrotransposition event has phenotypic consequences, the genetically distinct cells in the conceptus could potentially compete with each other, and select for or against cells carrying the de novo retrotransposition event. Lastly, de novo retrotransposition within the germ cells themselves once they are specified will generate mosaicism and potential competition and selection in the germ cell population, but these events will not be present in the soma.

The timing of de novo retrotransposition is probably best studied for LINE-1 elements. Experiments using transgenic mice and rats carrying human or mouse LINE-1 retrotransposition reporter cassettes suggest that de novo LINE-1 retrotransposition occurs infrequently in the germ cells themselves, and is more readily detectable in pre-implantation embryos (Kano et al. 2009). It is not clear if the inner cell mass, trophectoderm and primitive endoderm layers present in mouse blastocysts have differential susceptibilities to LINE-1 retrotransposition. Even though expression of these LINE-1 reporter transgenes was significantly higher in spermatogenic cells than somatic cells, the relative abundance of cells carrying de novo retrotransposition events in sperm was an order of magnitude lower than that in somatic tissues (Kano et al. 2009). Thus germ cells may possess host defence mechanisms that inhibit LINE-1 retrotransposition at a post-transcriptional level. Intriguingly, this study also indicates that pre-implantation embryos can inherit LINE-1 RNA from both the mature parental sperm and egg, and that this parentally transcribed RNA can retrotranspose in pre-implantation embryos (Kano et al. 2009). The finding that transgenic LINE-1 reporters can retrotranspose in early pre-implantation embryos when pluripotent cells are present in mice is consistent with the observation that transgenic LINE-1 reporter and endogenous LINE-1 and SINE retrotransposition occurs in human induced pluripotent stem cells and human embryonic stem (ES) cells in culture (Garcia-Perez et al. 2010; Klawitter et al. 2016; Wissing et al. 2011). However, data on retrotransposition rates of endogenous LINE-1 elements at different stages of the germline cycle is still lacking. Similarly, the rates of de novo retrotransposition of LTR retrotransposons at different stages of the mouse germline cycle also remain poorly understood. The application and adaptation of new methodologies to identify de novo retrotransposition events in genomic DNA (Baillie et al. 2011; Evrony et al. 2012; Ewing and Kazazian 2010; Upton et al. 2015) should help characterise the natural retrotransposition rates of individual retrotransposons during different stages of the germline cycle.

Although viable retrotransposons must be able to retrotranspose in the germline cycle, some retrotransposons may also be active in somatic tissues. In recent years, evidence has accumulated for de novo LINE-1 retrotransposition in human and mouse brain tissue, and de novo retrotransposition has been shown to be a mutational mechanism that can inactivate tumour suppressor genes in cancer (Baillie et al. 2011; Coufal et al. 2009; Evrony et al. 2012; Muotri et al. 2005; Shukla et al. 2013; Solyom et al. 2012; Upton et al. 2015). This aspect is detailed in chapters 'Retrotransposon Contribution to Genomic Plasticity', 'The Mobilisation of Processed Transcripts in Germline and Somatic Tissues', 'Neuronal Genome Plasticity: Retrotransposons, Environment and Disease' and 'Activity of Retrotransposons in Stem Cells and Differentiated Cells' of this book. Similarly, various types of LTR retrotransposon are also expressed in specific somatic tissues (Chuong et al. 2013; Faulkner et al. 2009; Gimenez et al. 2010; Seifarth et al. 2005). Some retrotransposon expression in somatic tissues may represent additional somatic roles for the transcription factors that individual retrotransposons are using to drive their expression in the germline cycle. For example, SOX2, which associates with the LINE-1 5′UTR in human cells (Coufal et al. 2009), is an integral component of the transcription factor network in pluripotent cells, but it also maintains the identity of neural progenitor cells (Avilion et al. 2003; Boyer et al. 2005; Graham et al. 2003). For other retrotransposons, expression in somatic tissues may represent the transcription of a small number of specific integration events that have occurred at genomic loci that promote their expression in somatic tissues. The liver-specific expression of a subset of IAP elements in mice appears to fall into this category (Puech et al. 1997), as does expression of the active L1HS-Ta LINE-1 subfamily in human cell lines (Philippe et al. 2016).

It is often not clear whether retrotransposon expression in somatic tissues has a functional role or if it is evolutionarily neutral. Retrotransposition itself can generate mosaicism in an individual, and it is possible that this provides some evolutionary advantage in some somatic tissues (Muotri et al. 2007). Some retrotransposons can influence expression of nearby host genes, for example, the LTRs of some types of LTR retrotransposon appears to act as enhancers in the placenta, and the activity of these elements in the placenta could be being selected for (Chuong et al. 2013). In some cases, specific copies of retrotransposon-encoded proteins appear to have been co-opted into the host genome to provide key functions in somatic cells, with the repeated independent co-option of LTR retrotransposon proteins to promote cell–cell fusions in the placenta during mammalian evolution being a good example of this (Dupressoir et al. 2011; Mi et al. 2000) (see chapter 'Roles of Endogenous Retrovirus-Encoded Syncytins in Human Placentation'). The domestication of an ancient LINE-like retrotransposon to generate telomerase, the enzyme that maintains telomeres at the ends of chromosomes in eukaryotes, suggests that retrotransposon-derived sequences can evolve to have functions in somatic tissues (Belfort et al. 2011;

Curcio and Belfort 2007; Eickbush 1997). Thus, although viable retrotransposons must be expressed and active in the germline cycle, this requirement is not incompatible with potential roles for these elements in somatic tissues.

5 The Impact of Retrotransposons on the Mammalian Germline

Retrotransposons are able to impact on the germline as a source of trans-generational genomic instability that can cause insertional mutations due to jumping to new locations in the genome, and due to recombination between homologous retrotransposon loci causing genetic deletions, segmental duplications and other chromosomal rearrangements (Fig. 3). However, retrotransposons also influence the biology of germline cells in other ways (Fig. 3). Retrotransposons use germline transcription factors to drive their expression, therefore each retrotransposon locus provides clusters of binding sites for transcription factors that are active in the germline that can provide a useful source of DNA sequence modules for evolution (Bourque et al. 2008; Rebollo et al. 2012). For example, retrotransposon sequences, particularly those from the ERV1 family of LTR retrotransposons, account for around 15–20 % of the genomic locations occupied by either OCT4 or NANOG pluripotency-associated transcription factors in human ES cells, and can drive expression of nearby genes in human ES cells (Kunarso et al. 2010). The ERVL family of LTR

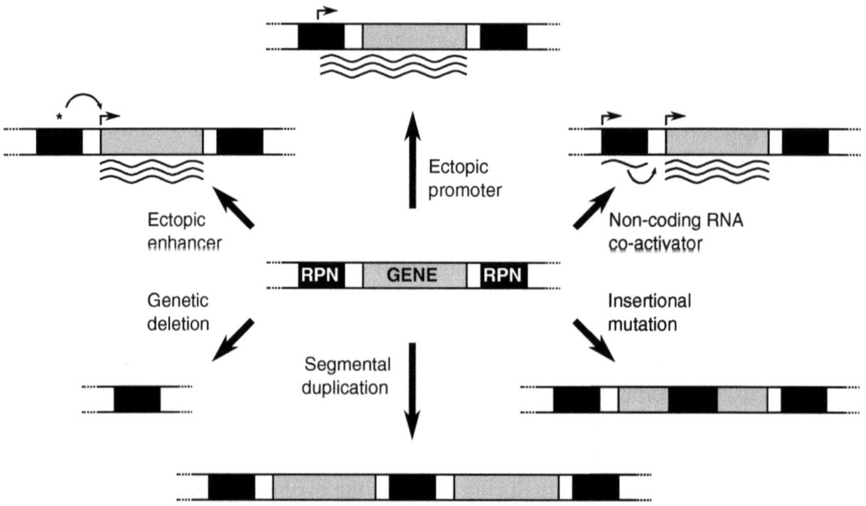

Fig. 3 Impact of retrotransposons on germline biology. A schematic diagram summarising some of the ways that retrotransposons impact on mammalian germ cells. A region of a chromosome containing a gene (*grey*) flanked by retrotransposons (RPNs, *black*) is indicated. Transcription is indicated by *corner arrows*, and active enhancers by *asterisks*. RNA transcripts are indicated by *wavy lines*. For details see main text

retrotransposons similarly appears to strongly influence the oocyte transcriptome and a subset of oocyte transcripts are chimaeras between ERVL retrotransposons and host genes (Peaston et al. 2004). One gene that has an oocyte-specific isoform driven by an ERVL retrotransposon promoter in oocytes is *DICER1*, which encodes an RNA endonuclease that is involved in the production of endogenous short interfering RNAs (Flemr et al. 2013). This oocyte-specific isoform of *DICER1* influences the abundance of retrotransposon transcripts in mouse oocytes, presumably through endogenous siRNAs targeting retrotransposon transcripts (Flemr et al. 2013). Thus, retrotransposon sequences in the genome are even promoting expression of retrotransposon defence mechanisms in the germline. Chimaeric oocyte transcripts originating from ERVL-derived promoters can encode retrotransposon-host gene fusion proteins that are translated in the oocytes (Peaston et al. 2004). Similarly, the recently discovered primate LINE-1 ORF0 transcript can run through to adjacent host genes and generate fusions between the ORF0 polypeptide and host-encoded proteins in human ES cells (Denli et al. 2015). Non-coding RNAs derived from LTR retrotransposons are also reported to be essential to maintain human ES cells in a pluripotent state, possibly through facilitating the binding of some transcriptional co-activator proteins to chromatin (Lu et al. 2014; Wang et al. 2014). Thus retrotransposons appear to be a rich source of genetic material that is contributing to the evolution of the transcriptome and the proteome of germline cells.

Mutations in retrotransposon defence mechanisms can result in high levels of retrotransposon expression during the germline cycle (Crichton et al. 2014; Ollinger et al. 2010; Zamudio and Bourc'his 2010). Mutations in the PIWI-piRNA pathway (Fu and Wang 2014), or in the accessory de novo methyltransferase DNMT3L, can result in high levels of LINE-1 expression in male germ cells, particularly during meiotic prophase (De Fazio et al. 2011; Di Giacomo et al. 2013; Soper et al. 2008; Zamudio et al. 2015). In general, these mutants arrest spermatogenesis during meiosis, typically around the pachytene stage (Crichton et al. 2014; Ollinger et al. 2010; Zamudio and Bourc'his 2010). Mice that have mutations in *MAEL*, a component of the PIWI-piRNA pathway, are reported to arrest during meiosis with high levels of meiosis-independent DNA damage that could potentially be caused by high levels of de novo retrotransposition of the de-repressed retrotransposons (Soper et al. 2008). In contrast, *DNMT3L*$^{-/-}$ mice are reported to have no detectable increase in meiosis-independent DNA damage, and have been proposed to arrest during meiosis due to histone modifications at transcriptionally active LINE-1 retrotransposon loci recruiting the meiotic recombination machinery, and disrupting the pairing of homologous chromosomes that characterise meiotic prophase (Zamudio et al. 2015). However, mice that de-repress LINE-1 post-transcriptionally also exhibit chromosome asynapsis and pachytene arrest (Di Giacomo et al. 2013), suggesting that there may be additional aspects to this phenotype that are not currently understood. Importantly, although de-repression of retrotransposons has been reported in various germline genome defence mutants, it remains to be determined whether de novo retrotransposition events are accumulating in the mutant germ cells.

De-repression of retrotransposons in oocytes is also associated with defects in progression through meiotic prophase. Mutations in *LSH*, a gene implicated in the establishment or maintenance of DNA methylation at retrotransposons and some single copy genes, result in loss of methylation at IAP retrotransposon sequences in oocytes, a failure of oocytes to progress through pachytene, and foetal oocyte death (De La Fuente et al. 2006). Mutating *LSH* in somatic cells also results in loss of DNA methylation and de-repression of retrotransposons (Dunican et al. 2013). This loss of DNA methylation appears to have indirect effects on other chromatin modifications in the genome as polycomb repressive complexes re-localise to sites normally occupied by DNA methylation, and sequestering polycomb repressive complexes away from their normal targets (Dunican et al. 2013). A similar phenomenon is seen in somatic cells with mutations in the maintenance DNA methyltransferase DNMT1 (Reddington et al. 2013), and in hypomethylated embryonic stem cells (Lynch et al. 2012). It is not clear if relocalisation of polycomb repressive complexes also happens in LSH mutant oocytes or in PIWI-piRNA mutant spermatocytes, but DNA hypomethylation at abundant retrotransposon sequences does have the potential to cause significant effects on the genome-wide distribution of other histone modifications that could be contributing more to the mutant phenotypes than retrotransposon de-repression itself.

Mutations in *MAEL* result in de-repression of retrotransposons, meiotic abnormalities and foetal oocyte death (Malki et al. 2014). The $MAEL^{-/-}$ oocyte phenotype appears to be related to de-repression of LINE-1 retrotransposons, and differences in the level of LINE-1 expression between individual oocytes in wild-type mice has been proposed to influence foetal oocyte attrition and the number of oocytes present in the ovarian pool at birth (Malki et al. 2014). The rate of foetal oocyte attrition is accelerated in transgenic mice carrying an active LINE-1 transgene, and delayed by treating pregnant mice with an anti-retroviral drug, although these manipulations do not change the final number of oocytes in the ovarian pool (Malki et al. 2014). There are fundamental differences in the way that the oocyte pool influences fertility and menopause between humans and mice, and it will be of interest to determine if manipulating LINE-1 activity can influence the size of the oocyte pool in humans. Human oocytes do, however, contain the host factors to support LINE-1 retrotransposition, at least using engineered LINE-1 reporter constructs (Georgiou et al. 2009).

6 Genome Defence Mechanisms Operating in the Mammalian Germline

The mammalian germline possesses a number of defence mechanisms that suppress the potentially mutagenic activity of retrotransposons in these cells (Crichton et al. 2014; Friedli and Trono 2015; Zamudio and Bourc'his 2010). Histone modification appears to play an important role in repressing retrotransposons in mouse ES cells, with H3K9me3 and H3K27me3 chromatin marks frequently associating with silenced retrotransposons in these cells (Day et al. 2010). Canonical and alternative polycomb repressive complexes, which catalyse trimethylation of H3K27, are involved in repressing multiple families of LTR retrotransposons in mouse ES cells

(Hisada et al. 2012; Leeb and Wutz 2007; Reichmann et al. 2012). SETDB1 and SUV39H1/SUV39H2, which trimethylate H3K9, are similarly implicated in repressing LINE-1 and multiple families of LTR retrotransposons on mouse ES cells (Bulut-Karslioglu et al. 2014; Karimi et al. 2011; Matsui et al. 2010; Reichmann et al. 2012). The lysine demethylase KDM1A also plays a role in repressing LINE-1 elements and ERVL LTR retrotransposons (Macfarlan et al. 2012), and histone deacetylases are implicated in repression of ERVK LTR retrotransposons in mouse ES cells (Reichmann et al. 2012), and in suppressing de novo but not pre-existing LINE-1 integrations in human embryonal carcinoma cells (Garcia-Perez et al. 2010). The variety of different mechanisms operating in ES cells may reflect the diversity of retrotransposons in the genome and the multiple strategies that these elements use to drive their transcription in the germline.

As retrotransposons hijack the transcriptional networks present in the host to drive their transcription, it is possible that some of the silencing mechanisms operating in ES cells reflect the mechanisms normally used to regulate the developmental timing of host genes whose expression is driven by the same transcription factors. For example, endogenous gene transcripts that have a role in the zygote might be downregulated as pre-implantation development proceeds such that they are not expressed in pluripotent epiblast cells or ES cells. Similarly, retrotransposons that are using this transcription factor network to drive their expression in zygotes would be expected to be downregulated in ES cells. Some of the mechanisms involved in repressing retrotransposons in ES cells will presumably reflect the normal mechanisms that mediate developmental changes in host gene transcription in these cells, with retrotransposon sequences being repressed 'by association' due to their co-regulation with host genes that are downregulated at this stage of development.

In contrast, the specific targeting of repressive chromatin marks to retrotransposons that is mediated by Krüppel-associated box zinc finger proteins (KRAB-ZFPs) appears to represent a more active and directed defence mechanism against these elements. Specific KRAB-ZFPs bind to specific retrotransposon sequences, recruiting the co-repressor KAP1 (also known as TRIM28) and H3K9me3 chromatin modifications to these sites (Friedman et al. 1996; Rowe et al. 2010; Wolf and Goff 2009, 2007). KAP1 interacts with the histone H3K9 methyltransferase SETDB1, which appears to be the major H3K9 histone methyltransferase involved in silencing LTR retrotransposons in mouse ES cells (Karimi et al. 2011; Matsui et al. 2010; Sripathy et al. 2006). Some LINE-1 elements are silenced by SUV39H1/SUV39H2 H3K9 histone methyltransferases rather than SETDB1 (Bulut-Karslioglu et al. 2014; Matsui et al. 2010), but it is not clear why different histone methyltransferases are being used to silence different retrotransposons. For KAP1 to target a retrotransposon for silencing, a KRAB-ZFP must evolve to bind to that retrotransposon sequence, and KRAB-ZFPs appear to be evolving rapidly for this purpose (Jacobs et al. 2014). Furthermore, retrotransposon loci that mutate or delete their KRAB-ZFP binding sites can escape from KAP1-dependent repression and evolve into new retrotransposon sub-types (Jacobs et al. 2014). Thus while some LINE-1 elements recruit and are repressed by KAP1 in mouse and human ES cells, the youngest types of LINE-1 are repressed by alternative mechanisms (Castro-Diaz et al. 2014).

DNA methylation plays an important role in repressing retrotransposons in germ cells and somatic cells (Bourc'his and Bestor 2004; Davis et al. 1989; De La Fuente et al. 2006; Dunican et al. 2013; Jackson-Grusby et al. 2001; Walsh et al. 1998), although its role in repressing retrotransposons in mouse ES cells may be more restricted than either KAP1 or SETDB1 (Karimi et al. 2011; Matsui et al. 2010; Rowe et al. 2010). The ability of mouse ES cells to induce compensatory histone modifications at some retrotransposons in response to DNA hypomethylation (Walter et al. 2016) may contribute to this. Pluripotent epiblast cells in mouse embryos undergo a wave of de novo DNA methylation during post-implantation development (Borgel et al. 2010) and it is possible that DNA methylation is recruited to retrotransposons that are already silenced by histone modifications at this stage to reinforce and stabilise the repressed state (Rowe et al. 2013). This mechanism bears some resemblance to the observation that bulk of the de novo DNA methylation that occurs during tumourigenesis is located at genes that are already silenced by other mechanisms (Sproul et al. 2011). However, at least for some retrotransposons, persistent KAP1/KRAB-ZFP activity is required to maintain repression of some retrotransposon loci in differentiating and adult somatic cells (Ecco et al. 2016), and the interplay between histone modifications and DNA methylation in establishing and maintaining silencing at different retrotransposon sequences in different cell types is likely to be complex and requires further study.

In the developing germ cells, DNA methylation is globally lost from the genome from 8.5 dpc through to 11.5 dpc as the primordial germ cells migrate to and colonise the genital ridge (Hajkova et al. 2008, 2002). This global loss of DNA methylation includes retrotransposon sequences, although IAP elements are amongst the most resistant to this phenomenon (Popp et al. 2010; Seisenberger et al. 2012). This loss of DNA methylation is part of a more extensive epigenetic reprogramming event that also involves genome-wide loss of various repressive histone modifications including H3K9me1 and H3K9me2 (Hajkova et al. 2008). It is not clear how retrotransposons are transcriptionally repressed during this stage of germ cell development, but the histone arginine methyltransferase PRMT5 localises to the nucleus and symmetric dimethylation of histones H2AR3 and/or H4R3 is upregulated during the early part of this reprogramming event and are present at LINE-1 and IAP retrotransposon sequences (Ancelin et al. 2006; Kim et al. 2014). *PRMT5*$^{-/-}$ primordial germ cells have undetectable levels of H2A/H4R3me2s in their nuclei, and de-repress LINE-1, IAP and other retrotransposons suggesting that PRMT5-dependent symmetric dimethylation of histones H2AR3 and H4R3 may contribute directly to the repression of retrotransposons in hypomethylated primordial germ cells (Kim et al. 2014). There may of course be additional, presently uncharacterised, histone modifications associated with specific types of retrotransposon that are helping to repress transcription of these elements in hypomethylated germ cells.

At around 11.5 dpc, PRMT5 is re-localised from the primordial germ cell nucleus into the cytoplasm and levels of H2A/H4R3me2s concomitantly decrease (Ancelin et al. 2006). Additional mechanisms likely become important to limit retrotransposon activity in hypomethylated germ cells at this stage. Analysis of retrotransposon transcript abundance in the transcriptome of 13.5 dpc primordial germ cell from wild-type mice supports widespread low level transcriptional de-repression of retrotransposons in hypomethylated germ cells at this stage (Molaro et al. 2014). The DNA hypomethylation that occurs in the developing primordial germ cells is not restricted to retrotransposons and extends to most genomic features including endogenous gene promoters (Popp et al. 2010; Seisenberger et al. 2012). Thus this global DNA hypomethylation event can influence expression of host genes in the developing germline (Hackett et al. 2012). Interestingly, many genes that are primarily and causally regulated by DNA methylation in mice are germline-specific genes that are involved in suppressing retrotransposon activity (Hackett et al. 2012). DNA hypomethylation in the developing germline therefore appears to induce expression of a group of retrotransposon-suppressing genome defence genes to compensate for the increase in potential retrotransposon activity caused by loss of this repressive epigenetic mark.

One of the genome defence genes that is most sensitive to DNA hypomethylation is *TEX19.1*. *TEX19.1* is required to repress MMERVK10C LTR retrotransposons in spermatocytes, and also to repress LINE-1 elements and LTR retrotransposons in the hypomethylated somatic cells present in the placenta (Öllinger et al. 2008; Reichmann et al. 2013). Most of the other germline genome defence genes induced in response to global DNA hypomethylation are components of the PIWI-piRNA pathway for repressing retrotransposons. The PIWI-piRNA pathway uses small RNAs encoded in the genome to target retrotransposons for suppression by epigenetic and post-transcriptional mechanisms in the germline (Fu and Wang 2014; Iwasaki et al. 2015). There are three PIWI proteins in mouse and rat genomes, but four PIWI proteins in many other mammals including humans. The three mouse PIWI proteins have distinct expression profile during spermatogenesis. These PIWI proteins physically interact with small single-stranded PIWI-interacting RNAs (piRNAs) whose sequence is thought to target the PIWI proteins to retrotransposon sequences (Aravin et al. 2006, 2008, 2007; Carmell et al. 2007; Kuramochi-Miyagawa et al. 2001). Genomic piRNAs are derived from long RNA precursors that undergo a number of processing events to generate mature piRNAs. These primary piRNAs can facilitate processing of complementary precursor sequences, such as retrotransposon transcripts, into secondary piRNAs which in turn can promote processing of genomic precursors into primary piRNAs. This ping-pong cycle can amplify groups of piRNAs and is important for generating an effective piRNA response against LINE-1 elements in male mouse germ cells (De Fazio et al. 2011). The slicer RNA endonuclease activity of PIWI proteins plays an important role in processing piRNA precursors, and PIWI-piRNA-directed slicing of retrotransposon RNAs by PIWIL1 and PIWIL2 contribute to the PIWI-piRNA defence against retrotransposons (De Fazio et al. 2011; Di Giacomo et al. 2013; Reuter et al. 2011). PIWIL2 and PIWIL4 are required for male germ cells to establish de novo DNA methylation at retrotransposon sequences (Aravin et al. 2008; Kuramochi-Miyagawa et al. 2008). De novo methylation of retrotransposons occurs from 16.5 dpc onwards in foetal male germ cells, and it is possible that sequence information present in the PIWI-piRNA pathway is being used to direct the de novo DNA methylation machinery to these sequences. It is not clear if PIWI-piRNA complexes regulate DNA methylation directly, or indirectly through other chromatin modifications. H3K9 methylation has been implicated in PIWI-dependent retrotransposon silencing in *Drosophila*, and in silencing retrotransposons in post-natal male germ cells in mice (Di Giacomo et al. 2014, 2013; Huang et al. 2013; Pezic et al. 2014). Perhaps one of the reasons that genome-wide loss of DNA methylation occurs in the developing germ cells is to expose retrotransposon loci to the PIWI-piRNA pathway so that retrotransposons can be identified and epigenetic repression of these elements established de novo in preparation for transmission of the genome to the next generation. Removing and resetting epigenetic marks on these sequences may be preferable to propagating existing marks in order to prevent epimutations from being transmitted across multiple generations. There may also be some analogies to the mechanisms operating in *Arabidopsis*, where programmed loss of DNA demethylation in the pollen's vegetative nucleus results in de-repression of retrotransposons whose transcripts are processed into small RNAs, transported to the pollen's germline nucleus, and used to direct epigenetic silencing of retrotransposons in the germline DNA (Calarco et al. 2012; Slotkin et al. 2009). The DNA methylation-sensitive coupling of expression of post-transcriptional genome defence mechanisms and components of the PIWI-piRNA pathway to transcriptional de-repression of retrotransposons in mouse germ cells may similarly allow mouse germ cells to generate the retrotransposon RNA transcripts needed to direct de novo identification and silencing of retrotransposon loci in the mammalian germline (Fig. 4).

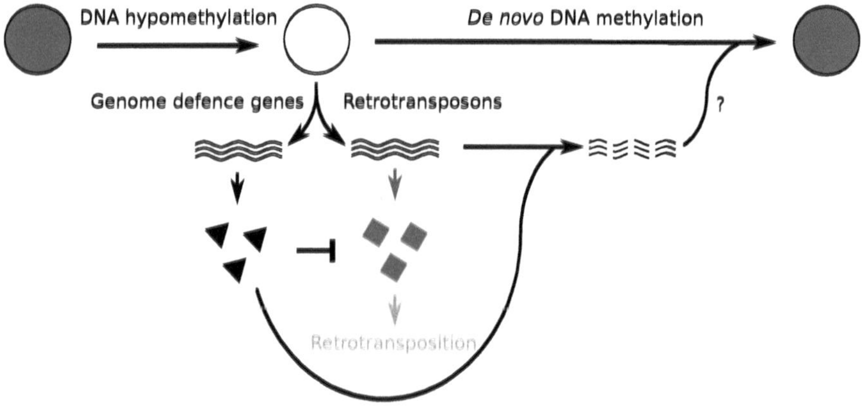

Fig. 4 Potential role of genome defence genes in the male germline. A schematic diagram outlining the potential role of genome defence genes during epigenetic reprogramming in the male germline. Germ cells losing DNA methylation (*filled grey circles → filled white circles*) transcribe RNA (*wavy lines*) encoding genome defence genes and retrotransposons, which can be translated into protein (*filled triangles* and *squares*, respectively). Genome defence proteins, including components of the PIWI-piRNA pathway, that inhibit any post-transcriptional stages of the retrotransposon life cycle (*grey*) can limit mutations caused by retrotransposition, while allowing retrotransposon RNA transcripts to prime the PIWI-piRNA pathway (*broken wavy lines*). The PIWI-piRNA pathway slices retrotransposon RNA transcripts to generate piRNAs and can potentially use sequence information in the piRNAs to direct de novo DNA methylation onto retrotransposon sequences (indicated by *question mark*)

Components of the PIWI-piRNA system also appear to play roles in post-transcriptional suppression of retrotransposons in oocytes, and PIWIL2 suppresses LINE-1 mobility in human induced pluripotent cells (Lim et al. 2013; Malki et al. 2014; Marchetto et al. 2013; Watanabe et al. 2008). In mice, PIWI function in oocytes is primarily provided by PIWIL2 although additional members of the PIWI family may also contribute to PIWI function in oocytes in other mammalian species including human (Roovers et al. 2015). Mutating PIWIL2 in mice does not have the severe consequences for fertility in females that it does in males (Kuramochi-Miyagawa et al. 2004; Lim et al. 2013), which may in part reflect differences in the way that de novo DNA methylation is regulated between spermatogenesis and oogenesis (Smallwood and Kelsey 2012). De novo DNA methylation occurs post-natally during oocyte growth in the female germline, and oocytes are therefore in a DNA hypomethylated state throughout their prolonged dictyate arrest and for much of their adult life. In the absence of PIWIL2, the abundance of some retrotransposon transcripts is elevated in oocytes, potentially reflecting PIWIL2-dependent post-transcriptional suppression of these elements during oogenesis (Lim et al. 2013; Watanabe et al. 2008). However, the PIWI-piRNA system is not the only mechanism that operates in oocytes to post-transcriptionally suppress retrotransposons, and DICER1-dependent endogenous siRNAs also make a significant contribution (Flemr et al. 2013; Stein et al. 2015; Tam et al. 2008; Watanabe et al. 2008). MARF1 may represent another mechanism regulating retrotransposons at a post-transcriptional level in these cells (Su et al. 2012a, b).

The distinct post-transcriptional suppression mechanisms operating in mouse oocytes appear to complement each other to target different types of retrotransposon (Watanabe et al. 2008). Pools of endogenous siRNA and piRNA present in fully grown oocytes can potentially be transmitted to the next generation to provide some protection against retrotransposons in pre-implantation embryos. However, retrotransposon-encoded transcripts, proteins, and ribonucleoprotein particles that are expressed during the oocyte's prolonged dictyate arrest or post-natal growth can similarly be transmitted in the oocyte cytoplasm and can cause retrotransposition in the next generation (Kano et al. 2009). The presence of multiple overlapping and complementary genome defence mechanisms in oocytes may therefore provide some protection against retrotransposon mobilisation during oogenesis, and also help to limit maternal transmission of retrotransposon-derived ribonucleoprotein particles than can retrotranspose in the next generation.

7 Concluding Remarks

As described in this chapter, the mammalian germline and retrotransposons are intrinsically linked in multiple ways. All retrotransposons need to be expressed and active in the mammalian germline in order to accumulate in the genome during evolution, and while the germline appears to have evolved multiple defence mechanisms to limit the mutagenic activity of these elements, these mechanisms are helping to drive evolution of retrotransposons to escape suppression. This situation is analogous to the Red Queen hypothesis (van Valen 1973) as both retrotransposons and germline defence mechanisms need to continue to evolve simply to keep up with each other. However, in addition to this antagonistic relationship, retrotransposons appear to be participating in the transcriptional and proteomic networks of germline cells and providing regulatory modules for gene expression that are being repurposed by the germline to help it evolve. Additional intricacies will likely emerge in the coming years as the interplay between retrotransposons and the germline becomes better understood, but it would appear that retrotransposons can be viewed as having both beneficial and deleterious effects on their germline hosts.

Acknowledgements I thank the MRC for funding research in this area in my laboratory through an intramural programme grant from MRC Human Genetics Unit. I am extremely grateful to Chris Playfoot and Jose Luis Garcia-Perez (both MRC Human Genetics Unit, Edinburgh) for their critical comments and excellent suggestions on this chapter.

References

Adams IR, McLaren A (2002) Sexually dimorphic development of mouse primordial germ cells: switching from oogenesis to spermatogenesis. Development 129:1155–1164

Ahl V, Keller H, Schmidt S, Weichenrieder O (2015) Retrotransposition and crystal structure of an Alu RNP in the ribosome-stalling conformation. Mol Cell 60:715–727. doi:10.1016/j.molcel.2015.10.003

Ancelin K, Lange UC, Hajkova P, Schneider R, Bannister AJ, Kouzarides T, Surani MA (2006) Blimp1 associates with Prmt5 and directs histone arginine methylation in mouse germ cells. Nat Cell Biol 8:623–630. doi:10.1038/ncb1413

Aravin AA, Sachidanandam R, Bourc'his D, Schaefer C, Pezic D, Toth KF, Bestor T, Hannon GJ (2008) A piRNA pathway primed by individual transposons is linked to de novo DNA methylation in mice. Mol Cell 31:785–799. doi:10.1016/j.molcel.2008.09.003

Aravin AA, Sachidanandam R, Girard A, Fejes-Toth K, Hannon GJ (2007) Developmentally regulated piRNA clusters implicate MILI in transposon control. Science 316:744–747. doi:10.1126/science.1142612

Aravin A, Gaidatzis D, Pfeffer S, Lagos-Quintana M, Landgraf P, Iovino N, Morris P, Brownstein MJ, Kuramochi-Miyagawa S, Nakano T, Chien M, Russo JJ, Ju J, Sheridan R, Sander C, Zavolan M, Tuschl T (2006) A novel class of small RNAs bind to MILI protein in mouse testes. Nature 442:203–207. doi:10.1038/nature04916

Avilion AA, Nicolis SK, Pevny LH, Perez L, Vivian N, Lovell-Badge R (2003) Multipotent cell lineages in early mouse development depend on SOX2 function. Genes Dev 17:126–140. doi:10.1101/gad.224503

Baillie JK, Barnett MW, Upton KR, Gerhardt DJ, Richmond TA, De Sapio F, Brennan P, Rizzu P, Smith S, Fell M, Talbot RT, Gustincich S, Freeman TC, Mattick JS, Hume DA, Heutink P, Carninci P, Jeddeloh JA, Faulkner GJ (2011) Somatic retrotransposition alters the genetic landscape of the human brain. Nature 479:534–537. doi:10.1038/nature10531

Bannert N, Kurth R (2006) The evolutionary dynamics of human endogenous retroviral families. Annu Rev Genomics Hum Genet 7:149–173. doi:10.1146/annurev.genom.7.080505.115700

Beck CR, Garcia-Perez JL, Badge RM, Moran JV (2011) LINE-1 elements in structural variation and disease. Annu Rev Genomics Hum Genet 12:187–215. doi:10.1146/annurev-genom-082509-141802

Belfort M, Curcio MJ, Lue NF (2011) Telomerase and retrotransposons: reverse transcriptases that shaped genomes. Proc Natl Acad Sci U S A 108:20304–20310. doi:10.1073/pnas.1100269109
Bennett EA, Keller H, Mills RE, Schmidt S, Moran JV, Weichenrieder O, Devine SE (2008) Active Alu retrotransposons in the human genome. Genome Res 18:1875–1883. doi:10.1101/gr.081737.108
Borgel J, Guibert S, Li Y, Chiba H, Schübeler D, Sasaki H, Forné T, Weber M (2010) Targets and dynamics of promoter DNA methylation during early mouse development. Nat Genet 42:1093–1100. doi:10.1038/ng.708
Bourc'his D, Bestor TH (2004) Meiotic catastrophe and retrotransposon reactivation in male germ cells lacking Dnmt3L. Nature 431:96–99. doi:10.1038/nature02886
Bourque G, Leong B, Vega VB, Chen X, Lee YL, Srinivasan KG, Chew J-L, Ruan Y, Wei C-L, Ng HH, Liu ET (2008) Evolution of the mammalian transcription factor binding repertoire via transposable elements. Genome Res 18:1752–1762. doi:10.1101/gr.080663.108
Boyer LA, Lee TI, Cole MF, Johnstone SE, Levine SS, Zucker JP, Guenther MG, Kumar RM, Murray HL, Jenner RG, Gifford DK, Melton DA, Jaenisch R, Young RA (2005) Core transcriptional regulatory circuitry in human embryonic stem cells. Cell 122:947–956. doi:10.1016/j.cell.2005.08.020
Branciforte D, Martin SL (1994) Developmental and cell type specificity of LINE-1 expression in mouse testis: implications for transposition. Mol Cell Biol 14:2584–2592
Brûlet P, Condamine H, Jacob F (1985) Spatial distribution of transcripts of the long repeated ETn sequence during early mouse embryogenesis. Proc Natl Acad Sci U S A 82:2054–2058
Bulut-Karslioglu A, De La Rosa-Velázquez IA, Ramirez F, Barenboim M, Onishi-Seebacher M, Arand J, Galán C, Winter GE, Engist B, Gerle B, O'Sullivan RJ, Martens JHA, Walter J, Manke T, Lachner M, Jenuwein T (2014) Suv39h-dependent H3K9me3 marks intact retrotransposons and silences LINE elements in mouse embryonic stem cells. Mol Cell 55:277–290. doi:10.1016/j.molcel.2014.05.029
Calarco JP, Borges F, Donoghue MTA, Van Ex F, Jullien PE, Lopes T, Gardner R, Berger F, Feijó JA, Becker JD, Martienssen RA (2012) Reprogramming of DNA methylation in pollen guides epigenetic inheritance via small RNA. Cell 151:194–205. doi:10.1016/j.cell.2012.09.001
Carmell MA, Girard A, van de Kant HJG, Bourc'his D, Bestor TH, de Rooij DG, Hannon GJ (2007) MIWI2 is essential for spermatogenesis and repression of transposons in the mouse male germline. Dev Cell 12:503–511. doi:10.1016/j.devcel.2007.03.001
Castro-Diaz N, Ecco G, Coluccio A, Kapopoulou A, Yazdanpanah B, Friedli M, Duc J, Jang SM, Turelli P, Trono D (2014) Evolutionally dynamic L1 regulation in embryonic stem cells. Genes Dev 28:1397–1409. doi:10.1101/gad.241661.114
Chinwalla AT, Cook LL, Delehaunty KD, Fewell GA, Fulton LA, Fulton RS, Graves TA, Hillier LW, Mardis ER, McPherson JD, Miner TL, Nash WE, Nelson JO, Nhan MN, Pepin KH, Pohl CS, Ponce TC, Schultz B, Thompson J, Trevaskis E, Waterston RH, Wendl MC, Wilson RK, Yang S-P, An P, Berry E, Birren B, Bloom T, Brown DG, Butler J, Daly M, David R, Deri J, Dodge S, Foley K, Gage D, Gnerre S, Holzer T, Jaffe DB, Kamal M, Karlsson EK, Kells C, Kirby A, Kulbokas EJ, Lander ES, Landers T, Leger JP, Levine R, Lindblad-Toh K, Mauceli E, Mayer JH, McCarthy M, Meldrim J, Mesirov JP, Nicol R, Nusbaum C, Seaman S, Sharpe T, Sheridan A, Singer JB, Santos R, Spencer B, Stange-Thomann N, Vinson JP, Wade CM, Wierzbowski J, Wyman D, Zody MC, Birney E, Goldman N, Kasprzyk A, Mongin E, Rust AG, Slater G, Stabenau A, Ureta-Vidal A, Whelan S, Ainscough R, Attwood J, Bailey J, Barlow K, Beck S, Burton J, Clamp M, Clee C, Coulson A, Cuff J, Curwen V, Cutts T, Davies J, Eyras E, Grafham D, Gregory S, Hubbard T, Hunt A, Jones M, Joy A, Leonard S, Lloyd C, Matthews L, McLaren S, McLay K, Meredith B, Mullikin JC, Ning Z, Oliver K, Overton-Larty E, Plumb R, Potter S, Quail M, Rogers J, Scott C, Searle S, Shownkeen R, Sims S, Wall M, West AP, Willey D, Williams S, Abril JF, Guigó R, Parra G, Agarwal P, Agarwala R, Church DM, Hlavina W, Maglott DR, Sapojnikov V, Alexandersson M, Pachter L, Antonarakis SE, Dermitzakis ET, Reymond A, Ucla C, Baertsch R, Diekhans M, Furey TS, Hinrichs A, Hsu F, Karolchik D, Kent WJ, Roskin KM, Schwartz MS, Sugnet C, Weber RJ, Bork P, Letunic I, Suyama M, Torrents D,

Zdobnov EM, Botcherby M, Brown SD, Campbell RD, Jackson I, Bray N, Couronne O, Dubchak I, Poliakov A, Rubin EM, Brent MR, Flicek P, Keibler E, Korf I, Batalov S, Bult C, Frankel WN, Carninci P, Hayashizaki Y, Kawai J, Okazaki Y, Cawley S, Kulp D, Wheeler R, Chiaromonte F, Collins FS, Felsenfeld A, Guyer M, Peterson J, Wetterstrand K, Copley RR, Mott R, Dewey C, Dickens NJ, Emes RD, Goodstadt L, Ponting CP, Winter E, Dunn DM, von Niederhausern AC, Weiss RB, Eddy SR, Johnson LS, Jones TA, Elnitski L, Kolbe DL, Eswara P, Miller W, O'Connor MJ, Schwartz S, Gibbs RA, Muzny DM, Glusman G, Smit A, Green ED, Hardison RC, Yang S, Haussler D, Hua A, Roe BA, Kucherlapati RS, Montgomery KT, Li J, Li M, Lucas S, Ma B, McCombie WR, Morgan M, Pevzner P, Tesler G, Schultz J, Smith DR, Tromp J, Worley KC, Green ED (2002) Initial sequencing and comparative analysis of the mouse genome. Nature 420:520–562. doi:10.1038/nature01262

Chuong EB, Rumi MAK, Soares MJ, Baker JC (2013) Endogenous retroviruses function as species-specific enhancer elements in the placenta. Nat Genet 45:325–329. doi:10.1038/ng.2553

Cost GJ, Feng Q, Jacquier A, Boeke JD (2002) Human L1 element target-primed reverse transcription in vitro. EMBO J 21:5899–5910

Coufal NG, Garcia-Perez JL, Peng GE, Yeo GW, Mu Y, Lovci MT, Morell M, O'Shea KS, Moran JV, Gage FH (2009) L1 retrotransposition in human neural progenitor cells. Nature 460:1127–1131. doi:10.1038/nature08248

Crichton JH, Dunican DS, Maclennan M, Meehan RR, Adams IR (2014) Defending the genome from the enemy within: mechanisms of retrotransposon suppression in the mouse germline. Cell Mol Life Sci 71:1581–1605. doi:10.1007/s00018-013-1468-0

Cruickshanks HA, Vafadar-Isfahani N, Dunican DS, Lee A, Sproul D, Lund JN, Meehan RR, Tufarelli C (2013) Expression of a large LINE-1-driven antisense RNA is linked to epigenetic silencing of the metastasis suppressor gene TFPI-2 in cancer. Nucleic Acids Res 41:6857–6869. doi:10.1093/nar/gkt438

Curcio MJ, Belfort M (2007) The beginning of the end: links between ancient retroelements and modern telomerases. Proc Natl Acad Sci U S A 104:9107–9108. doi:10.1073/pnas.0703224104

Davis CM, Constantinides PG, van der Riet F, van Schalkwyk L, Gevers W, Parker MI (1989) Activation and demethylation of the intracisternal A particle genes by 5-azacytidine. Cell Differ Dev 27:83–93

Day DS, Luquette LJ, Park PJ, Kharchenko PV (2010) Estimating enrichment of repetitive elements from high-throughput sequence data. Genome Biol 11:R69–R69. doi:10.1186/gb-2010-11-6-r69

De Fazio S, Bartonicek N, Di Giacomo M, Abreu-Goodger C, Sankar A, Funaya C, Antony C, Moreira PN, Enright AJ, O'Carroll D (2011) The endonuclease activity of Mili fuels piRNA amplification that silences LINE1 elements. Nature 480:259–263. doi:10.1038/nature10547

De La Fuente R, Baumann C, Fan T, Schmidtmann A, Dobrinski I, Muegge K (2006) Lsh is required for meiotic chromosome synapsis and retrotransposon silencing in female germ cells. Nat Cell Biol 8:1448–1454. doi:10.1038/ncb1513

Denli AM, Narvaiza I, Kerman BE, Pena M, Benner C, Marchetto MCN, Diedrich JK, Aslanian A, Ma J, Moresco JJ, Moore L, Hunter T, Saghatelian A, Gage FH (2015) Primate-specific ORF0 contributes to retrotransposon-mediated diversity. Cell 163:583–593. doi:10.1016/j.cell.2015.09.025

Dewannieux M, Dupressoir A, Harper F, Pierron G, Heidmann T (2004) Identification of autonomous IAP LTR retrotransposons mobile in mammalian cells. Nat Genet 36:534–539. doi:10.1038/ng1353

Dewannieux M, Esnault C, Heidmann T (2003) LINE-mediated retrotransposition of marked Alu sequences. Nat Genet 35:41–48. doi:10.1038/ng1223

Di Giacomo M, Comazzetto S, Saini H, De Fazio S, Carrieri C, Morgan M, Vasiliauskaite L, Benes V, Enright AJ, O'Carroll D (2013) Multiple epigenetic mechanisms and the piRNA pathway enforce LINE1 silencing during adult spermatogenesis. Mol Cell 50:601–608. doi:10.1016/j.molcel.2013.04.026

Di Giacomo M, Comazzetto S, Sampath SC, Sampath SC, O'Carroll D (2014) G9a co-suppresses LINE1 elements in spermatogonia. Epigenetics Chromatin 7:24. doi:10.1186/1756-8935-7-24

Dunican DS, Cruickshanks HA, Suzuki M, Semple CA, Davey T, Arceci RJ, Greally J, Adams IR, Meehan RR (2013) Lsh regulates LTR retrotransposon repression independently of Dnmt3b function. Genome Biol 14:R146. doi:10.1186/gb-2013-14-12-r146
Dupressoir A, Heidmann T (1996) Germ line-specific expression of intracisternal A-particle retrotransposons in transgenic mice. Mol Cell Biol 16:4495–4503
Dupressoir A, Vernochet C, Harper F, Guégan J, Dessen P, Pierron G, Heidmann T (2011) A pair of co-opted retroviral envelope syncytin genes is required for formation of the two-layered murine placental syncytiotrophoblast. Proc Natl Acad Sci U S A 108:E1164–E1173. doi:10.1073/pnas.1112304108
Ecco G, Cassano M, Kauzlaric A, Duc J, Coluccio A, Offner S, Imbeault M, Rowe HM, Turelli P, Trono D (2016) Transposable elements and their KRAB-ZFP controllers regulate gene expression in adult tissues. Dev Cell 36:611–623. doi:10.1016/j.devcel.2016.02.024
Eickbush TH (1997) Telomerase and retrotransposons: which came first? Science 277:911–912
Esnault C, Maestre J, Heidmann T (2000) Human LINE retrotransposons generate processed pseudogenes. Nat Genet 24:363–367. doi:10.1038/74184
Evrony GD, Cai X, Lee E, Hills LB, Elhosary PC, Lehmann HS, Parker JJ, Atabay KD, Gilmore EC, Poduri A, Park PJ, Walsh CA (2012) Single-neuron sequencing analysis of L1 retrotransposition and somatic mutation in the human brain. Cell 151:483–496. doi:10.1016/j.cell.2012.09.035
Ewing AD, Kazazian HH (2010) High-throughput sequencing reveals extensive variation in human-specific L1 content in individual human genomes. Genome Res 20:1262–1270. doi:10.1101/gr.106419.110
Fadloun A, Le Gras S, Jost B, Ziegler-Birling C, Takahashi H, Gorab E, Carninci P, Torres-Padilla M-E (2013) Chromatin signatures and retrotransposon profiling in mouse embryos reveal regulation of LINE-1 by RNA. Nat Struct Mol Biol 20:332–338. doi:10.1038/nsmb.2495
Faulkner GJ, Kimura Y, Daub CO, Wani S, Plessy C, Irvine KM, Schroder K, Cloonan N, Steptoe AL, Lassmann T, Waki K, Hornig N, Arakawa T, Takahashi H, Kawai J, Forrest ARR, Suzuki H, Hayashizaki Y, Hume DA, Orlando V, Grimmond SM, Carninci P (2009) The regulated retrotransposon transcriptome of mammalian cells. Nat Genet 41:563–571. doi:10.1038/ng.368
Feng Q, Moran JV, Kazazian HH Jr, Boeke JD (1996) Human L1 retrotransposon encodes a conserved endonuclease required for retrotransposition. Cell 87:905–916
Flemr M, Malik R, Franke V, Nejepinska J, Sedlacek R, Vlahovicek K, Svoboda P (2013) A retrotransposon-driven dicer isoform directs endogenous small interfering RNA production in mouse oocytes. Cell 155:807–816. doi:10.1016/j.cell.2013.10.001
Friedli M, Trono D (2015) The developmental control of transposable elements and the evolution of higher species. Annu Rev Cell Dev Biol 31:429–451. doi:10.1146/annurev-cellbio-100814-125514
Friedman JR, Fredericks WJ, Jensen DE, Speicher DW, Huang XP, Neilson EG, Rauscher FJ (1996) KAP-1, a novel corepressor for the highly conserved KRAB repression domain. Genes Dev 10:2067–2078
Fu Q, Wang PJ (2014) Mammalian piRNAs: biogenesis, function, and mysteries. Spermatogenesis 4, e27889. doi:10.4161/spmg.27889
Garcia-Perez JL, Morell M, Scheys JO, Kulpa DA, Morell S, Carter CC, Hammer GD, Collins KL, O'Shea KS, Menendez P, Moran JV (2010) Epigenetic silencing of engineered L1 retrotransposition events in human embryonic carcinoma cells. Nature 466:769–773. doi:10.1038/nature09209
Georgiou I, Noutsopoulos D, Dimitriadou E, Markopoulos G, Apergi A, Lazaros L, Vaxevanoglou T, Pantos K, Syrrou M, Tzavaras T (2009) Retrotransposon RNA expression and evidence for retrotransposition events in human oocytes. Hum Mol Genet 18:1221–1228. doi:10.1093/hmg/ddp022
Gimenez J, Montgiraud C, Pichon J-P, Bonnaud B, Arsac M, Ruel K, Bouton O, Mallet F (2010) Custom human endogenous retroviruses dedicated microarray identifies self-induced HERV-W family elements reactivated in testicular cancer upon methylation control. Nucleic Acids Res 38:2229–2246. doi:10.1093/nar/gkp1214

Graham V, Khudyakov J, Ellis P, Pevny L (2003) SOX2 functions to maintain neural progenitor identity. Neuron 39:749–765

Hackett JA, Reddington JP, Nestor CE, Dunican DS, Branco MR, Reichmann J, Reik W, Surani MA, Adams IR, Meehan RR (2012) Promoter DNA methylation couples genome-defence mechanisms to epigenetic reprogramming in the mouse germline. Development 139:3623–3632. doi:10.1242/dev.081661

Hajkova P, Ancelin K, Waldmann T, Lacoste N, Lange UC, Cesari F, Lee C, Almouzni G, Schneider R, Surani MA (2008) Chromatin dynamics during epigenetic reprogramming in the mouse germ line. Nature 452:877–881. doi:10.1038/nature06714

Hajkova P, Erhardt S, Lane N, Haaf T, El-Maarri O, Reik W, Walter J, Surani MA (2002) Epigenetic reprogramming in mouse primordial germ cells. Mech Dev 117:15–23. doi:10.1016/S0925-4773(02)00181-8

Hancks DC, Goodier JL, Mandal PK, Cheung LE, Kazazian HH (2011) Retrotransposition of marked SVA elements by human L1s in cultured cells. Hum Mol Genet 20:3386–3400. doi:10.1093/hmg/ddr245

Hancks DC, Kazazian H (2010) SVA retrotransposons: evolution and genetic instability. Semin Cancer Biol 20:234–245. doi:10.1016/j.semcancer.2010.04.001

Hancks DC, Kazazian HH Jr (2012) Active human retrotransposons: variation and disease. Curr Opin Genet Dev 22:191–203. doi:10.1016/j.gde.2012.02.006

Han K, Xing J, Wang H, Hedges DJ, Garber RK, Cordaux R, Batzer MA (2005) Under the genomic radar: the stealth model of Alu amplification. Genome Res 15:655–664. doi:10.1101/gr.3492605

Hayashi K, Chuva de Sousa Lopes SM, Kaneda M, Tang F, Hajkova P, Lao K, O'Carroll D, Das PP, Tarakhovsky A, Miska EA, Surani MA (2008) MicroRNA biogenesis is required for mouse primordial germ cell development and spermatogenesis. PLoS One 3:e1738. doi:10.1371/journal.pone.0001738

Hisada K, Sánchez C, Endo TA, Endoh M, Román-Trufero M, Sharif J, Koseki H, Vidal M (2012) RYBP represses endogenous retroviruses and preimplantation- and germ line-specific genes in mouse embryonic stem cells. Mol Cell Biol 32:1139–1149. doi:10.1128/MCB.06441-11

Hogg K, Western PS (2015) Refurbishing the germline epigenome: out with the old, in with the new. Semin Cell Dev Biol 45:104–113. doi:10.1016/j.semcdb.2015.09.012

Huang XA, Yin H, Sweeney S, Raha D, Snyder M, Lin H (2013) A major epigenetic programming mechanism guided by piRNAs. Dev Cell 24:502–516. doi:10.1016/j.devcel.2013.01.023

Iwasaki YW, Siomi MC, Siomi H (2015) PIWI-interacting RNA: its biogenesis and functions. Annu Rev Biochem 84:405–433. doi:10.1146/annurev-biochem-060614-034258

Jackson-Grusby L, Beard C, Possemato R, Tudor M, Fambrough D, Csankovszki G, Dausman J, Lee P, Wilson C, Lander E, Jaenisch R (2001) Loss of genomic methylation causes p53-dependent apoptosis and epigenetic deregulation. Nat Genet 27:31–39. doi:10.1038/83730

Jacobs FMJ, Greenberg D, Nguyen N, Haeussler M, Ewing AD, Katzman S, Paten B, Salama SR, Haussler D (2014) An evolutionary arms race between KRAB zinc-finger genes ZNF91/93 and SVA/L1 retrotransposons. Nature 516:242–245. doi:10.1038/nature13760

Kano H, Godoy I, Courtney C, Vetter MR, Gerton GL, Ostertag EM, Kazazian HH Jr (2009) L1 retrotransposition occurs mainly in embryogenesis and creates somatic mosaicism. Genes Dev 23:1303–1312. doi:10.1101/gad.1803909

Karimi MM, Goyal P, Maksakova IA, Bilenky M, Leung D, Tang JX, Shinkai Y, Mager DL, Jones S, Hirst M, Lorincz MC (2011) DNA methylation and SETDB1/H3K9me3 regulate predominantly distinct sets of genes, retroelements, and chimeric transcripts in mESCs. Cell Stem Cell 8:676–687. doi:10.1016/j.stem.2011.04.004

Khan H, Smit A, Boissinot S (2006) Molecular evolution and tempo of amplification of human LINE-1 retrotransposons since the origin of primates. Genome Res 16:78–87. doi:10.1101/gr.4001406

Khazina E, Weichenrieder O (2009) Non-LTR retrotransposons encode noncanonical RRM domains in their first open reading frame. Proc Natl Acad Sci U S A 106:731–736. doi:10.1073/pnas.0809964106

Kim S, Günesdogan U, Zylicz JJ, Hackett JA, Cougot D, Bao S, Lee C, Dietmann S, Allen GE, Sengupta R, Surani MA (2014) PRMT5 protects genomic integrity during global DNA demethylation in primordial germ cells and preimplantation embryos. Mol Cell 56:564–579. doi:10.1016/j.molcel.2014.10.003

Klawitter S, Fuchs NV, Upton KR, Muñoz-Lopez M, Shukla R, Wang J, Garcia-Cañadas M, Lopez-Ruiz C, Gerhardt DJ, Sebe A, Grabundzija I, Merkert S, Gerdes P, Pulgarin JA, Bock A, Held U, Witthuhn A, Haase A, Sarkadi B, Löwer J, Wolvetang EJ, Martin U, Ivics Z, Izsvák Z, Garcia-Perez JL, Faulkner GJ, Schumann GG (2016) Reprogramming triggers endogenous L1 and Alu retrotransposition in human induced pluripotent stem cells. Nat Commun 7:10286. doi:10.1038/ncomms10286

Kocer A, Reichmann J, Best D, Adams IR (2009) Germ cell sex determination in mammals. Mol Hum Reprod 15:205–213. doi:10.1093/molehr/gap008

Kramerov DA, Vassetzky NS (2011) SINEs. Wiley Interdiscip Rev RNA 2:772–786. doi:10.1002/wrna.91

Kulpa DA, Moran JV (2006) Cis-preferential LINE-1 reverse transcriptase activity in ribonucleoprotein particles. Nat Struct Mol Biol 13:655–660. doi:10.1038/nsmb1107

Kunarso G, Chia N-Y, Jeyakani J, Hwang C, Lu X, Chan Y-S, Ng H-H, Bourque G (2010) Transposable elements have rewired the core regulatory network of human embryonic stem cells. Nat Genet 42:631–634. doi:10.1038/ng.600

Kuramochi-Miyagawa S, Kimura T, Ijiri TW, Isobe T, Asada N, Fujita Y, Ikawa M, Iwai N, Okabe M, Deng W, Lin H, Matsuda Y, Nakano T (2004) Mili, a mammalian member of piwi family gene, is essential for spermatogenesis. Development 131:839–849. doi:10.1242/dev.00973

Kuramochi-Miyagawa S, Kimura T, Yomogida K, Kuroiwa A, Tadokoro Y, Fujita Y, Sato M, Matsuda Y, Nakano T (2001) Two mouse piwi-related genes: miwi and mili. Mech Dev 108:121–133

Kuramochi-Miyagawa S, Watanabe T, Gotoh K, Totoki Y, Toyoda A, Ikawa M, Asada N, Kojima K, Yamaguchi Y, Ijiri TW, Hata K, Li E, Matsuda Y, Kimura T, Okabe M, Sakaki Y, Sasaki H, Nakano T (2008) DNA methylation of retrotransposon genes is regulated by Piwi family members MILI and MIWI2 in murine fetal testes. Genes Dev 22:908–917. doi:10.1101/gad.1640708

Lander ES, Linton LM, Birren B, Nusbaum C, Zody MC, Baldwin J, Devon K, Dewar K, Doyle M, FitzHugh W, Funke R, Gage D, Harris K, Heaford A, Howland J, Kann L, Lehoczky J, LeVine R, McEwan P, McKernan K, Meldrim J, Mesirov JP, Miranda C, Morris W, Naylor J, Raymond C, Rosetti M, Santos R, Sheridan A, Sougnez C, Stange-Thomann N, Stojanovic N, Subramanian A, Wyman D, Rogers J, Sulston J, Ainscough R, Beck S, Bentley D, Burton J, Clee C, Carter N, Coulson A, Deadman R, Deloukas P, Dunham A, Dunham I, Durbin R, French L, Grafham D, Gregory S, Hubbard T, Humphray S, Hunt A, Jones M, Lloyd C, McMurray A, Matthews L, Mercer S, Milne S, Mullikin JC, Mungall A, Plumb R, Ross M, Shownkeen R, Sims S, Waterston RH, Wilson RK, Hillier LW, McPherson JD, Marra MA, Mardis ER, Fulton LA, Chinwalla AT, Pepin KH, Gish WR, Chissoe SL, Wendl MC, Delehaunty KD, Miner TL, Delehaunty A, Kramer JB, Cook LL, Fulton RS, Johnson DL, Minx PJ, Clifton SW, Hawkins T, Branscomb E, Predki P, Richardson P, Wenning S, Slezak T, Doggett N, Cheng JF, Olsen A, Lucas S, Elkin C, Uberbacher E, Frazier M, Gibbs RA, Muzny DM, Scherer SE, Bouck JB, Sodergren EJ, Worley KC, Rives CM, Gorrell JH, Metzker ML, Naylor SL, Kucherlapati RS, Nelson DL, Weinstock GM, Sakaki Y, Fujiyama A, Hattori M, Yada T, Toyoda A, Itoh T, Kawagoe C, Watanabe H, Totoki Y, Taylor T, Weissenbach J, Heilig R, Saurin W, Artiguenave F, Brottier P, Bruls T, Pelletier E, Robert C, Wincker P, Smith DR, Doucette-Stamm L, Rubenfield M, Weinstock K, Lee HM, Dubois J, Rosenthal A, Platzer M, Nyakatura G, Taudien S, Rump A, Yang H, Yu J, Wang J, Huang G, Gu J, Hood L, Rowen L, Madan A, Qin S, Davis RW, Federspiel NA, Abola AP, Proctor MJ, Myers RM, Schmutz J, Dickson M, Grimwood J, Cox DR, Olson MV, Kaul R, Raymond C, Shimizu N, Kawasaki K, Minoshima S, Evans GA, Athanasiou M, Schultz R, Roe BA, Chen F, Pan H, Ramser J, Lehrach H, Reinhardt R, McCombie WR, de la Bastide M, Dedhia N, Blöcker H, Hornischer K, Nordsiek G, Agarwala R, Aravind L, Bailey JA, Bateman A, Batzoglou S, Birney E, Bork P, Brown DG,

Burge CB, Cerutti L, Chen HC, Church D, Clamp M, Copley RR, Doerks T, Eddy SR, Eichler EE, Furey TS, Galagan J, Gilbert JG, Harmon C, Hayashizaki Y, Haussler D, Hermjakob H, Hokamp K, Jang W, Johnson LS, Jones TA, Kasif S, Kaspryzk A, Kennedy S, Kent WJ, Kitts P, Koonin EV, Korf I, Kulp D, Lancet D, Lowe TM, McLysaght A, Mikkelsen T, Moran JV, Mulder N, Pollara VJ, Ponting CP, Schuler G, Schultz J, Slater G, Smit AF, Stupka E, Szustakowski J, Thierry-Mieg D, Thierry-Mieg J, Wagner L, Wallis J, Wheeler R, Williams A, Wolf YI, Wolfe KH, Yang SP, Yeh RF, Collins F, Guyer MS, Peterson J, Felsenfeld A, Wetterstrand KA, Patrinos A, Morgan MJ, de Jong P, Catanese JJ, Osoegawa K, Shizuya H, Choi S, Chen YJ, Szustakowki J, International Human Genome Sequencing Consortium (2001) Initial sequencing and analysis of the human genome. Nature 409:860–921. doi:10.1038/35057062

Lawson KA, Hage WJ (1994) Clonal analysis of the origin of primordial germ cells in the mouse. Ciba Found Symp 182:68–84, discussion 84–91

Leeb M, Wutz A (2007) Ring1B is crucial for the regulation of developmental control genes and PRC1 proteins but not X inactivation in embryonic cells. J Cell Biol 178:219–229. doi:10.1083/jcb.200612127

Lee S-H, Cho S-Y, Shannon MF, Fan J, Rangasamy D (2010) The impact of CpG island on defining transcriptional activation of the mouse L1 retrotransposable elements. PLoS One 5, e11353. doi:10.1371/journal.pone.0011353

Li J, Kannan M, Trivett AL, Liao H, Wu X, Akagi K, Symer DE (2014) An antisense promoter in mouse L1 retrotransposon open reading frame-1 initiates expression of diverse fusion transcripts and limits retrotransposition. Nucleic Acids Res 42:4546–4562. doi:10.1093/nar/gku091

Lim AK, Lorthongpanich C, Chew TG, Tan CWG, Shue YT, Balu S, Gounko N, Kuramochi-Miyagawa S, Matzuk MM, Chuma S, Messerschmidt DM, Solter D, Knowles BB (2013) The nuage mediates retrotransposon silencing in mouse primordial ovarian follicles. Development 140:3819–3825. doi:10.1242/dev.099184

Lu X, Sachs F, Ramsay L, Jacques P-É, Göke J, Bourque G, Ng H-H (2014) The retrovirus HERVH is a long noncoding RNA required for human embryonic stem cell identity. Nat Struct Mol Biol 21:423–425. doi:10.1038/nsmb.2799

Lynch MD, Smith AJH, De Gobbi M, Flenley M, Hughes JR, Vernimmen D, Ayyub H, Sharpe JA, Sloane-Stanley JA, Sutherland L, Meek S, Burdon T, Gibbons RJ, Garrick D, Higgs DR (2012) An interspecies analysis reveals a key role for unmethylated CpG dinucleotides in vertebrate Polycomb complex recruitment. EMBO J 31:317–329. doi:10.1038/emboj.2011.399

Macfarlan TS, Gifford WD, Driscoll S, Lettieri K, Rowe HM, Bonanomi D, Firth A, Singer O, Trono D, Pfaff SL (2012) Embryonic stem cell potency fluctuates with endogenous retrovirus activity. Nature 487:57–63. doi:10.1038/nature11244

Macia A, Muñoz-Lopez M, Cortes JL, Hastings RK, Morell S, Lucena-Aguilar G, Marchal JA, Badge RM, Garcia-Perez JL (2011) Epigenetic control of retrotransposon expression in human embryonic stem cells. Mol Cell Biol 31:300–316. doi:10.1128/MCB.00561-10

MacLennan M, Crichton JH, Playfoot CJ, Adams IR (2015) Oocyte development, meiosis and aneuploidy. Semin Cell Dev Biol 45:68–76. doi:10.1016/j.semcdb.2015.10.005

Magnúsdóttir E, Surani MA (2014) How to make a primordial germ cell. Development 141:245–252. doi:10.1242/dev.098269

Maksakova IA, Romanish MT, Gagnier L, Dunn CA, van de Lagemaat LN, Mager DL (2006) Retroviral elements and their hosts: insertional mutagenesis in the mouse germ line. PLoS Genet 2:e2. doi:10.1371/journal.pgen.0020002

Malki S, van der Heijden GW, O'Donnell KA, Martin SL, Bortvin A (2014) A role for retrotransposon LINE-1 in fetal oocyte attrition in mice. Dev Cell 29:521–533. doi:10.1016/j.devcel.2014.04.027

Marchetto MCN, Narvaiza I, Denli AM, Benner C, Lazzarini TA, Nathanson JL, Paquola ACM, Desai KN, Herai RH, Weitzman MD, Yeo GW, Muotri AR, Gage FH (2013) Differential L1 regulation in pluripotent stem cells of humans and apes. Nature 503:525–529. doi:10.1038/nature12686

Martin SL, Branciforte D (1993) Synchronous expression of LINE-1 RNA and protein in mouse embryonal carcinoma cells. Mol Cell Biol 13:5383–5392

Mathias SL, Scott AF, Kazazian HH Jr, Boeke JD, Gabriel A (1991) Reverse transcriptase encoded by a human transposable element. Science 254:1808–1810
Matsui T, Leung D, Miyashita H, Maksakova IA, Miyachi H, Kimura H, Tachibana M, Lorincz MC, Shinkai Y (2010) Proviral silencing in embryonic stem cells requires the histone methyltransferase ESET. Nature 464:927–931. doi:10.1038/nature08858
McLaren A (2003) Primordial germ cells in the mouse. Dev Biol 262:1–15
Mi S, Lee X, Li X, Veldman GM, Finnerty H, Racie L, LaVallie E, Tang X-Y, Edouard P, Howes S, Keith JC, McCoy JM (2000) Syncytin is a captive retroviral envelope protein involved in human placental morphogenesis. Nature 403:785–789. doi:10.1038/35001608
Molaro A, Falciatori I, Hodges E, Aravin AA, Marran K, Rafii S, McCombie WR, Smith AD, Hannon GJ (2014) Two waves of de novo methylation during mouse germ cell development. Genes Dev 28:1544–1549. doi:10.1101/gad.244350.114
Moran JV, Holmes SE, Naas TP, DeBerardinis RJ, Boeke JD, Kazazian HH Jr (1996) High frequency retrotransposition in cultured mammalian cells. Cell 87:917–927
Muotri AR, Chu VT, Marchetto MCN, Deng W, Moran JV, Gage FH (2005) Somatic mosaicism in neuronal precursor cells mediated by L1 retrotransposition. Nature 435:903–910. doi:10.1038/nature03663
Muotri AR, Marchetto MCN, Coufal NG, Gage FH (2007) The necessary junk: new functions for transposable elements. Hum Mol Genet 16:R159–R167. doi:10.1093/hmg/ddm196
Nichols J, Smith A (2009) Naive and primed pluripotent states. Cell Stem Cell 4:487–492. doi:10.1016/j.stem.2009.05.015
O'Donnell L (2015) Mechanisms of spermiogenesis and spermiation and how they are disturbed. Spermatogenesis 4:e979623. doi:10.4161/21565562.2014.979623
Ohinata Y, Payer B, O'Carroll D, Ancelin K, Ono Y, Sano M, Barton SC, Obukhanych T, Nussenzweig M, Tarakhovsky A, Saitou M, Surani MA (2005) Blimp1 is a critical determinant of the germ cell lineage in mice. Nature 436:207–213. doi:10.1038/nature03813
Öllinger R, Childs AJ, Burgess HM, Speed RM, Lundegaard PR, Reynolds N, Gray NK, Cooke HJ, Adams IR (2008) Deletion of the pluripotency-associated Tex19.1 gene causes activation of endogenous retroviruses and defective spermatogenesis in mice. PLoS Genet 4:e1000199. doi:10.1371/journal.pgen.1000199
Ollinger R, Reichmann J, Adams IR (2010) Meiosis and retrotransposon silencing during germ cell development in mice. Differentiation 79:147–158. doi:10.1016/j.diff.2009.10.004
Peaston AE, Evsikov AV, Graber JH, de Vries WN, Holbrook AE, Solter D, Knowles BB (2004) Retrotransposons regulate host genes in mouse oocytes and preimplantation embryos. Dev Cell 7:597–606. doi:10.1016/j.devcel.2004.09.004
Pezic D, Manakov SA, Sachidanandam R, Aravin AA (2014) piRNA pathway targets active LINE1 elements to establish the repressive H3K9me3 mark in germ cells. Genes Dev 28:1410–1428. doi:10.1101/gad.240895.114
Philippe C, Vargas-Landin DB, Doucet AJ, van Essen D, Vera-Otarola J, Kuciak M, Corbin A, Nigumann P, Cristofari G (2016) Activation of individual L1 retrotransposon instances is restricted to cell-type dependent permissive loci. Elife 5:e13926. doi:10.7554/eLife.13926
Piko L, Hammons MD, Taylor KD (1984) Amounts, synthesis, and some properties of intracisternal A particle-related RNA in early mouse embryos. Proc Natl Acad Sci U S A 81:488–492
Popp C, Dean W, Feng S, Cokus SJ, Andrews S, Pellegrini M, Jacobsen SE, Reik W (2010) Genome-wide erasure of DNA methylation in mouse primordial germ cells is affected by AID deficiency. Nature 463:1101–1105. doi:10.1038/nature08829
Puech A, Dupressoir A, Loireau MP, Mattei MG, Heidmann T (1997) Characterization of two age-induced intracisternal A-particle-related transcripts in the mouse liver. Transcriptional read-through into an open reading frame with similarities to the yeast ccr4 transcription factor. J Biol Chem 272:5995–6003
Raiz J, Damert A, Chira S, Held U, Klawitter S, Hamdorf M, Löwer J, Strätling WH, Löwer R, Schumann GG (2012) The non-autonomous retrotransposon SVA is trans-mobilized by the human LINE-1 protein machinery. Nucleic Acids Res 40:1666–1683. doi:10.1093/nar/gkr863

Rebollo R, Romanish MT, Mager DL (2012) Transposable elements: an abundant and natural source of regulatory sequences for host genes. Annu Rev Genet 46:21–42. doi:10.1146/annurev-genet-110711-155621

Reddington JP, Perricone SM, Nestor CE, Reichmann J, Youngson NA, Suzuki M, Reinhardt D, Dunican DS, Prendergast JG, Mjoseng H, Ramsahoye BH, Whitelaw E, Greally JM, Adams IR, Bickmore WA, Meehan RR (2013) Redistribution of H3K27me3 upon DNA hypomethylation results in de-repression of Polycomb target genes. Genome Biol 14:R25. doi:10.1186/gb-2013-14-3-r25

Reichmann J, Crichton JH, Madej MJ, Taggart M, Gautier P, Garcia-Perez JL, Meehan RR, Adams IR (2012) Microarray analysis of LTR retrotransposon silencing identifies Hdac1 as a regulator of retrotransposon expression in mouse embryonic stem cells. PLoS Comput Biol 8:e1002486. doi:10.1371/journal.pcbi.1002486

Reichmann J, Reddington JP, Best D, Read D, Öllinger R, Meehan RR, Adams IR (2013) The genome-defence gene Tex19.1 suppresses LINE-1 retrotransposons in the placenta and prevents intra-uterine growth retardation in mice. Hum Mol Genet 22:1791–1806. doi:10.1093/hmg/ddt029

Reuter M, Berninger P, Chuma S, Shah H, Hosokawa M, Funaya C, Antony C, Sachidanandam R, Pillai RS (2011) Miwi catalysis is required for piRNA amplification-independent LINE1 transposon silencing. Nature 480:264–267. doi:10.1038/nature10672

Ribet D, Dewannieux M, Heidmann T (2004) An active murine transposon family pair: retrotransposition of "master" MusD copies and ETn trans-mobilization. Genome Res 14:2261–2267. doi:10.1101/gr.2924904

Ribet D, Louvet-Vallée S, Harper F, de Parseval N, Dewannieux M, Heidmann O, Pierron G, Maro B, Heidmann T (2008) Murine endogenous retrovirus MuERV-L is the progenitor of the "orphan" epsilon viruslike particles of the early mouse embryo. J Virol 82:1622–1625. doi:10.1128/JVI.02097-07

Romanish MT, Cohen CJ, Mager DL (2010) Potential mechanisms of endogenous retroviral-mediated genomic instability in human cancer. Semin Cancer Biol 20:246–253. doi:10.1016/j.semcancer.2010.05.005

Roovers EF, Rosenkranz D, Mahdipour M, Han C-T, He N, Chuva de Sousa Lopes SM, van der Westerlaken LAJ, Zischler H, Butter F, Roelen BAJ, Ketting RF (2015) Piwi proteins and piRNAs in mammalian oocytes and early embryos. Cell Rep 10:2069–2082. doi:10.1016/j.celrep.2015.02.062

Rossitto M, Philibert P, Poulat F, Boizet-Bonhoure B (2015) Molecular events and signalling pathways of male germ cell differentiation in mouse. Semin Cell Dev Biol 45:84–93. doi:10.1016/j.semcdb.2015.09.014

Rowe HM, Friedli M, Offner S, Verp S, Mesnard D, Marquis J, Aktas T, Trono D (2013) De novo DNA methylation of endogenous retroviruses is shaped by KRAB-ZFPs/KAP1 and ESET. Development 140:519–529. doi:10.1242/dev.087585

Rowe HM, Jakobsson J, Mesnard D, Rougemont J, Reynard S, Aktas T, Maillard PV, Layard-Liesching H, Verp S, Marquis J, Spitz F, Constam DB, Trono D (2010) KAP1 controls endogenous retroviruses in embryonic stem cells. Nature 463:237–240. doi:10.1038/nature08674

Seifarth W, Frank O, Zeilfelder U, Spiess B, Greenwood AD, Hehlmann R, Leib-Mösch C (2005) Comprehensive analysis of human endogenous retrovirus transcriptional activity in human tissues with a retrovirus-specific microarray. J Virol 79:341–352. doi:10.1128/JVI.79.1.341-352.2005

Seisenberger S, Andrews S, Krueger F, Arand J, Walter J, Santos F, Popp C, Thienpont B, Dean W, Reik W (2012) The dynamics of genome-wide DNA methylation reprogramming in mouse primordial germ cells. Mol Cell 48:849–862. doi:10.1016/j.molcel.2012.11.001

Shukla R, Upton KR, Muñoz-Lopez M, Gerhardt DJ, Fisher ME, Nguyen T, Brennan PM, Baillie JK, Collino A, Ghisletti S, Sinha S, Iannelli F, Radaelli E, Dos Santos A, Rapoud D, Guettier C, Samuel D, Natoli G, Carninci P, Ciccarelli FD, Garcia-Perez JL, Faivre J, Faulkner GJ (2013) Endogenous retrotransposition activates oncogenic pathways in hepatocellular carcinoma. Cell 153:101–111. doi:10.1016/j.cell.2013.02.032

Slotkin RK, Vaughn M, Borges F, Tanurdzić M, Becker JD, Feijó JA, Martienssen RA (2009) Epigenetic reprogramming and small RNA silencing of transposable elements in pollen. Cell 136:461–472. doi:10.1016/j.cell.2008.12.038

Smallwood SA, Kelsey G (2012) De novo DNA methylation: a germ cell perspective. Trends Genet 28:33–42. doi:10.1016/j.tig.2011.09.004

Solyom S, Ewing AD, Rahrmann EP, Doucet T, Nelson HH, Burns MB, Harris RS, Sigmon DF, Casella A, Erlanger B, Wheelan S, Upton KR, Shukla R, Faulkner GJ, Largaespada DA, Kazazian HH Jr (2012) Extensive somatic L1 retrotransposition in colorectal tumors. Genome Res 22:2328–2338. doi:10.1101/gr.145235.112

Sookdeo A, Hepp CM, McClure MA, Boissinot S (2013) Revisiting the evolution of mouse LINE-1 in the genomic era. Mob DNA 4:3. doi:10.1186/1759-8753-4-3

Soper SFC, van der Heijden GW, Hardiman TC, Goodheart M, Martin SL, de Boer P, Bortvin A (2008) Mouse maelstrom, a component of nuage, is essential for spermatogenesis and transposon repression in meiosis. Dev Cell 15:285–297. doi:10.1016/j.devcel.2008.05.015

Speek M (2001) Antisense promoter of human L1 retrotransposon drives transcription of adjacent cellular genes. Mol Cell Biol 21:1973–1985. doi:10.1128/MCB.21.6.1973-1985.2001

Sproul D, Nestor C, Culley J, Dickson JH, Dixon JM, Harrison DJ, Meehan RR, Sims AH, Ramsahoye BH (2011) Transcriptionally repressed genes become aberrantly methylated and distinguish tumors of different lineages in breast cancer. Proc Natl Acad Sci U S A 108:4364–4369. doi:10.1073/pnas.1013224108

Sripathy SP, Stevens J, Schultz DC (2006) The KAP1 corepressor functions to coordinate the assembly of de novo HP1-demarcated microenvironments of heterochromatin required for KRAB zinc finger protein-mediated transcriptional repression. Mol Cell Biol 26:8623–8638. doi:10.1128/MCB.00487-06

Stein P, Rozhkov NV, Li F, Cárdenas FL, Davydenko O, Davydenk O, Vandivier LE, Gregory BD, Hannon GJ, Schultz RM (2015) Essential role for endogenous siRNAs during meiosis in mouse oocytes. PLoS Genet 11:e1005013. doi:10.1371/journal.pgen.1005013

Su Y-Q, Sugiura K, Sun F, Pendola JK, Cox GA, Handel MA, Schimenti JC, Eppig JJ (2012a) MARF1 regulates essential oogenic processes in mice. Science 335:1496–1499. doi:10.1126/science.1214680

Su Y-Q, Sun F, Handel MA, Schimenti JC, Eppig JJ (2012b) Meiosis arrest female 1 (MARF1) has nuage-like function in mammalian oocytes. Proc Natl Acad Sci U S A 109:18653–18660. doi:10.1073/pnas.1216904109

Swergold GD (1990) Identification, characterization, and cell specificity of a human LINE-1 promoter. Mol Cell Biol 10:6718–6729

Tam OH, Aravin AA, Stein P, Girard A, Murchison EP, Cheloufi S, Hodges E, Anger M, Sachidanandam R, Schultz RM, Hannon GJ (2008) Pseudogene-derived small interfering RNAs regulate gene expression in mouse oocytes. Nature 453:534–538. doi:10.1038/nature06904

Trelogan SA, Martin SL (1995) Tightly regulated, developmentally specific expression of the first open reading frame from LINE-1 during mouse embryogenesis. Proc Natl Acad Sci U S A 92:1520–1524

Upton KR, Gerhardt DJ, Jesuadian JS, Richardson SR, Sánchez-Luque FJ, Bodea GO, Ewing AD, Salvador-Palomeque C, van der Knaap MS, Brennan PM, Vanderver A, Faulkner GJ (2015) Ubiquitous L1 mosaicism in hippocampal neurons. Cell 161:228–239. doi:10.1016/j.cell.2015.03.026

van den Hurk JAJM, Meij IC, Seleme M d C, Kano H, Nikopoulos K, Hoefsloot LH, Sistermans EA, de Wijs IJ, Mukhopadhyay A, Plomp AS, de Jong PTVM, Kazazian HH, Cremers FPM (2007) L1 retrotransposition can occur early in human embryonic development. Hum Mol Genet 16:1587–1592. doi:10.1093/hmg/ddm108

van Valen L (1973) A new evolutionary law. Evol Theory 1:1–30

Walsh CP, Chaillet JR, Bestor TH (1998) Transcription of IAP endogenous retroviruses is constrained by cytosine methylation. Nat Genet 20:116–117. doi:10.1038/2413

Walter M, Teissandier A, Pérez-Palacios R, Bourc'his D (2016) An epigenetic switch ensures transposon repression upon dynamic loss of DNA methylation in embryonic stem cells. Elife 5:e11418. doi:10.7554/eLife.11418
Wang H, Xing J, Grover D, Hedges DJ, Han K, Walker JA, Batzer MA (2005) SVA elements: a hominid-specific retroposon family. J Mol Biol 354:994–1007. doi:10.1016/j.jmb.2005.09.085
Wang J, Xie G, Singh M, Ghanbarian AT, Raskó T, Szvetnik A, Cai H, Besser D, Prigione A, Fuchs NV, Schumann GG, Chen W, Lorincz MC, Ivics Z, Hurst LD, Izsvák Z (2014) Primate-specific endogenous retrovirus-driven transcription defines naive-like stem cells. Nature 516:405–409. doi:10.1038/nature13804
Watanabe T, Totoki Y, Toyoda A, Kaneda M, Kuramochi-Miyagawa S, Obata Y, Chiba H, Kohara Y, Kono T, Nakano T, Surani MA, Sakaki Y, Sasaki H (2008) Endogenous siRNAs from naturally formed dsRNAs regulate transcripts in mouse oocytes. Nature 453:539–543. doi:10.1038/nature06908
Weismann A (1889) Essays upon heredity. Clarendon Press, Oxford
Wei W, Gilbert N, Ooi SL, Lawler JF, Ostertag EM, Kazazian HH, Boeke JD, Moran JV (2001) Human L1 retrotransposition: cis preference versus trans complementation. Mol Cell Biol 21:1429–1439. doi:10.1128/MCB.21.4.1429-1439.2001
Wildschutte JH, Williams ZH, Montesion M, Subramanian RP, Kidd JM, Coffin JM (2016) Discovery of unfixed endogenous retrovirus insertions in diverse human populations. Proc Natl Acad Sci U S A 113:E2326–E2334. doi:10.1073/pnas.1602336113
Wissing S, Montano M, Garcia-Perez JL, Moran JV, Greene WC (2011) Endogenous APOBEC3B restricts LINE-1 retrotransposition in transformed cells and human embryonic stem cells. J Biol Chem 286:36427–36437. doi:10.1074/jbc.M111.251058
Wolf D, Goff SP (2009) Embryonic stem cells use ZFP809 to silence retroviral DNAs. Nature 458:1201–1204. doi:10.1038/nature07844
Wolf D, Goff SP (2007) TRIM28 mediates primer binding site-targeted silencing of murine leukemia virus in embryonic cells. Cell 131:46–57. doi:10.1016/j.cell.2007.07.026
Yang Q-E, Oatley JM (2014) Spermatogonial stem cell functions in physiological and pathological conditions. Curr Top Dev Biol 107:235–267. doi:10.1016/B978-0-12-416022-4.00009-3
Yotsuyanagi Y, Szöllösi D (1981) Early mouse embryo intracisternal particle: Fourth type of retrovirus-like particle associated with the mouse. J Natl Cancer Inst 67:677–685
Zamudio N, Barau J, Teissandier A, Walter M, Borsos M, Servant N, Bourc'his D (2015) DNA methylation restrains transposons from adopting a chromatin signature permissive for meiotic recombination. Genes Dev 29:1256–1270. doi:10.1101/gad.257840.114
Zamudio N, Bourc'his D (2010) Transposable elements in the mammalian germline: a comfortable niche or a deadly trap? Heredity 105:92–104. doi:10.1038/hdy.2010.53

L1 Regulation in Mouse and Human Germ Cells

Simon J. Newkirk and Wenfeng An

1 Introduction

Transposable elements (TEs) represent one of the most dynamic components of mammalian genomes. Genome sequencing has illuminated the abundance, complexity, and diversity of TEs. TEs constitute approximately 50 % of human and mouse genomes (Lander et al. 2001; Waterston et al. 2002). Based on the mode of mobilization, they are classified as either DNA transposons, which mobilize through a cut-paste mechanism, or retrotransposons, which mobilize through a copy-paste mechanism (i.e., via an RNA intermediate). While DNA transposons have lost their activities in the human and mouse genomes during evolution, retrotransposons have been far more successful in colonizing these genomes (Lander et al. 2001; Waterston et al. 2002). Retrotransposons are further categorized into LTR (long terminal repeat) retrotransposons and non-LTR retrotransposons. LTR retrotransposons are also known as endogenous retroviruses (ERVs) and comprise 9 % and 10 % of the human and mouse genomes, respectively (Lander et al. 2001; Waterston et al. 2002). ERVs remain active in rodents (Maksakova et al. 2006), but do not show any recent insertional activity within the human genome (Magiorkinis et al. 2015). Non-LTR retrotransposons include long interspersed elements (LINEs) and short interspersed elements (SINEs). LINE-1s (L1s) constitute 17 % and 19 % of the human and mouse

S.J. Newkirk
School of Molecular Biosciences, Washington State University, Pullman, WA 99164, USA

Department of Pharmaceutical Sciences, South Dakota State University, Brookings, SD 57007, USA
e-mail: simon.newkirk@sdstate.edu

W. An (✉)
Department of Pharmaceutical Sciences, South Dakota State University, Brookings, SD 57007, USA
e-mail: wenfeng.an@sdstate.edu

G. Cristofari (ed.), *Human Retrotransposons in Health and Disease*,
DOI 10.1007/978-3-319-48344-3_2

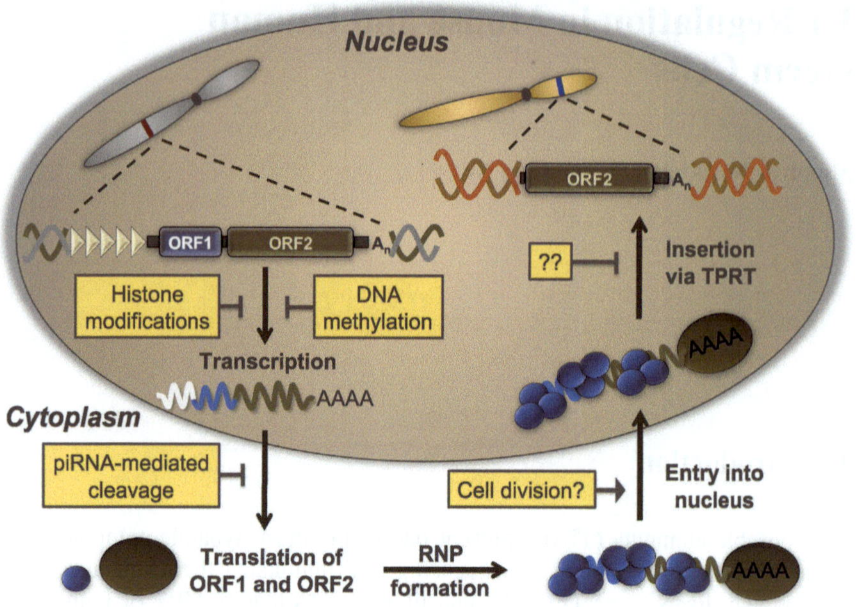

Fig. 1 L1 is regulated at distinct stages of replication. L1 mobilizes through a bicistronic mRNA intermediate and inserts into the genome through target primed reverse transcription (TPRT). L1s must overcome transcriptional silencing mechanisms. Both piRNA-guided DNA methylation and several histone marks (H3K9me3/2 and H2A/H4R3me2) suppress the expression of L1s in the germline. Posttranscriptionally L1s face (both pachytene and fetal) piRNA-mediated endonucleolytic cleavage. It remains unclear what barriers or host-factors are involved in nuclear import and TRPT

genomes, respectively (Lander et al. 2001; Waterston et al. 2002). A full-length L1 is ~6 kb in length and contains an internal promoter in its 5′ untranslated region (5′UTR), two open reading frames (ORF1 and ORF2), and a poly(A) signal in its 3′UTR (Furano 2000) (Fig. 1). SINEs, on the other hand, do not encode any proteins and rely on L1 proteins for mobilization.

Both mouse and human genomes have over 500,000 copies of L1, most of which are incapable of further mobilization. The lack of activity is primarily due to "structural defects." In other words, mutations concurrent or subsequent to retrotransposition (i.e., 5′ truncations, inversions, substitutions, INDELs, and any other types of debilitating mutations) render the elements incompetent (Szak et al. 2002). Within these inactive elements lie the remnants of many L1 families and subfamilies, highlighting the evolving battles between retrotransposons and the host (Khan et al. 2006). Only full-length intact L1s retain the ability to mobilize autonomously and amplify within the genome, and have the potential to generate additional full-length copies. It is estimated that the diploid human genome harbors an average of 80–100 retrotransposition competent L1 elements (Brouha et al. 2003). The number of potentially active elements is much higher in the mouse genome (Goodier et al. 2001). It is worth noting that retrotransposition-competent L1s usually belong to younger L1 families and many are polymorphic among individuals. Accordingly, there can be large interindividual variation in genomic L1 retrotransposition potential (Beck et al. 2010).

Retrotransposons impact and modify the genome in a multitude of ways (Goodier and Kazazian 2008). The most obvious is insertional mutagenesis, which is highlighted by the observed >100 cases of human diseases caused by L1-mediated retrotransposition events (Hancks and Kazazian 2016). Thus, in the short term, retrotransposition activity represents a formidable threat to the host genome and must be properly regulated in both somatic and germ cells. However, in the long term, TEs may benefit a species through expansion of genomic material and alteration of expression patterns. Since L1s mobilize through an RNA intermediate, it expands the repertoire of alterations that L1s can engender to the genome. Due to its weak polyadenylation signal, donor L1s frequently utilize downstream polyadenylation signals and thus mobilize 3′ flanking genetic information (3′ transduction) (Moran et al. 1999). Intronic L1s can be exonized through cryptic splicing events (Zemojtel et al. 2007). Cellular mRNAs can also be mobilized, resulting in the expansion of processed pseudogenes (Esnault et al. 2000). In fact, through a process termed exaptation, TE-derived genes have adopted critical functions in many physiological processes, including immune response, centromere assembly, and placental formation (Alzohairy et al. 2013). Additionally, there is evidence that some TE sequences have been exapted into novel regulatory elements, potentially serving as promoters, enhancers, silencers, or insulators (Rebollo et al. 2012; de Souza et al. 2013). TEs may also spread repressive epigenetic marks to neighboring sequences and affect the activity of nearby promoters (Rebollo et al. 2011; Macfarlan et al. 2011; Grandi et al. 2015). Finally, L1 sequences may serve important roles during X chromosome inactivation through participating in the assembly and propagation of heterochromatin on the inactive X chromosome (Chow et al. 2010).

In order to survive or remain active in the genome, an L1 family needs to consistently generate full-length insertions that can be passed through the germline to the next generation, albeit expectedly at rates that do not significantly endanger the reproductive fitness of the host (Bestor 2003). Such insertions may occur by *bona fide* retrotransposition in developing germ cells, or by retrotransposition during early embryonic development in cells fated to become primordial germ cells (PGCs). The nature of TE regulation in the germline has been expounded in several recent reviews (Bao and Yan 2012; Zamudio and Bourc'his 2010). However, our understanding of just how complex the dance is between the two unwilling partners—host and L1s—has been expanded by new insights into the transcriptional and posttranscriptional coordination and the subtleties of regulatory mechanisms. This review will focus on retrotransposons, especially L1s, and how they are regulated in the mouse and human germlines. As LTR retrotransposons remain active in the mouse genome, differential regulation of L1 and LTR retrotransposons will be compared as appropriate. Our goal is to accent recent regulatory insights and provide a synopsis of the interplay between the host and L1.

2 L1 Activities Can Be Controlled at Different Stages of L1 Replication

L1 retrotransposition involves multiple steps, which can be schematically summarized as transcription, translation, ribonucleoprotein (RNP) formation, nuclear entry, and target-primed reverse transcription (Goodier and Kazazian 2008) (Fig. 1). It begins when a full-length donor L1 is transcribed into a bicistronic mRNA by RNA polymerase II. L1 mRNA is then exported to the cytoplasm, where the two L1 proteins are synthesized and form an RNP complex with the template L1 mRNA. The ORF1 protein (ORF1p) has nucleic acid binding and chaperone activities. Both activities are essential for retrotransposition (Kolosha and Martin 2003; Martin et al. 2005). The ORF2 protein (ORF2p) possesses two enzymatic activities: endonuclease and reverse transcriptase, both playing vital roles in target-primed reverse transcription (TPRT) of L1 mRNA (Feng et al. 1996; Mathias et al. 1991; Moran et al. 1996). To generate a new copy, the full-length donor L1 has to negotiate with the host cell through each of these steps. On the other hand, each step provides an opportunity for the host cell to control L1 activity.

From the L1's perspective, host regulatory mechanisms can be generally categorized as transcriptional regulation or posttranscriptional regulation. Inarguably, transcriptional regulation can be the most effective means of controlling L1 activities, simply because without transcription retrotransposition is impossible. In vivo, this has been illustrated by conditional activation of L1 transgenes in mice either through Cre-mediated excision of a transcriptional stop cassette (An et al. 2008) or through tetracycline induction of an inducible promoter (O'Donnell et al. 2013). Indeed, multiple mechanisms have been adopted to transcriptionally repress full-length L1s in the germline, including DNA methylation, histone modifications, and the fetal Piwi-interacting RNA (piRNA) pathway (Fig. 1). Posttranscriptional regulation includes both fetal and pachytene piRNA pathways. Additionally, entry into the nucleus and access to open chromatin seems to be an inherent obstacle for mobilization. While it may seem arbitrary to group regulatory methods into these categories, it highlights the multiple layers of regulation observed within the germ cells. Each of these regulatory mechanisms will be discussed in the context of germ cell development. Most of the mechanisms have been principally characterized in the mouse male germline (Sects. 4 and 5), but emerging factors involved in L1 regulation in the female mouse germline will be discussed (Sect. 6). L1 expression and regulation in human germ cells will also be summarized (Sect. 8). As this discussion occurs, keep in mind that essentially two antagonistic evolutionary processes are at play: the host needs to suppress mobile DNA from irreparable damage and TEs need to adapt and "reproduce" in the germline to survive.

3 L1 Expression During Male Germ Cell Development

Male germ cell development involves distinct cellular stages (Fig. 2). In the mouse, it starts from a founder population of PGCs at around embryonic day (E) 7.25 (Saitou and Yamaji 2012). Shortly after, they begin migration toward the future gonadal site and colonize the genital ridge at ~E10.5. PGCs exit the mitotic cell cycle and arrest in G0 phase between E12.5 and 14.5 (McLaren 2001; Western et al. 2008), becoming prospermatogonia. The mitotically arrested prospermatogonia resume proliferation after birth at around postnatal day (P) 2, and give rise to undifferentiated and differentiating spermatogonia via mitotic divisions (Yoshida et al. 2006; de Rooij and Russell 2000; Drumond et al. 2011). The most primitive spermatogonia may undergo up to ten successive mitotic divisions before differentiating into preleptotene spermatocytes, starting at P8–P10, signaling the first wave of meiosis. Leptotene, zygotene, pachytene, and diplotene spermatocytes start to appear approximately at P10, P12, P14, and P17, respectively, during the prophase of meiosis I. Haploid spermatids first appear at P20 after two meiotic divisions and differentiate into mature spermatozoa at P30 (de Rooij and Russell 2000; Nebel et al. 1961; Bellve et al. 1977). Spermatogenesis continues in the adult testis in an

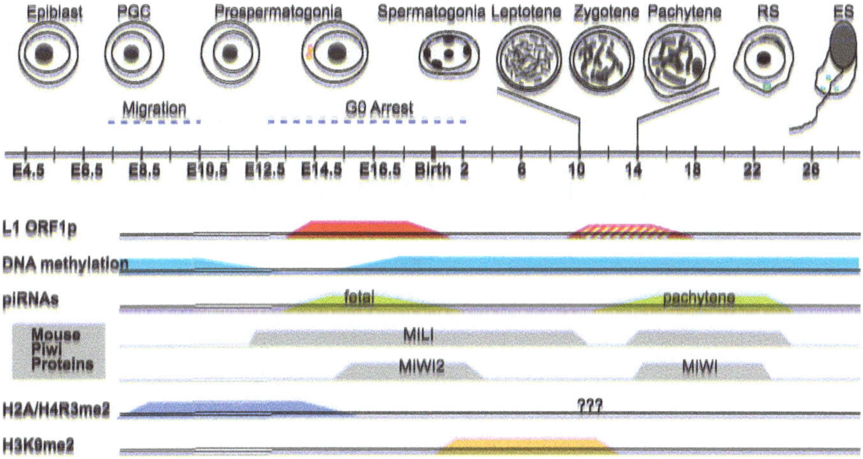

Fig. 2 L1 expression and important regulatory factors during male germ cell development. The timeline depicts male germ cell development in mice from PGC specification through the first wave of spermatogenesis (*RS* round spermatid, *ES* elongating spermatid). Normally suppressed through multiple transcriptional and posttranscriptional mechanisms, there are two periods of well-documented L1 expression. The first results from the loss of DNA methylation during fetal DNA reprogramming around E13.5–E14.5. DNA methylation is quickly restored, presumably guided to active TE elements through MIWI2-bound fetal piRNAs. H2A/H4R3me2 acts to suppress excessive L1 expression at this stage. This histone modification has also been found in postnatal germ cells but a stage-specific profile remains unavailable (indicated by *question marks*). The other period of L1 expression (*striped area*) seems to occur sporadically in a subset of spermatocytes. H3K9me2 is diminished at L1 loci during the transition from spermatogonia to spermatocytes. Additionally, there is extensive chromatin remodeling during meiosis, both of which likely contribute to the sporadic L1 expression observed in spermatocytes. Pachytene piRNAs associated with MIWI posttranscriptionally degrade TEs in pachytene spermatocytes and are known to contribute to the absence of L1 expression for the remainder of spermatogenesis. A role in the germline has not been established for zinc finger proteins (*box in dotted line*)

asynchronous yet precisely timed fashion: spermatogonia stem cells divide mitotically every 8.5 days and form spermatozoa 34.5 days later (Oakberg 1957).

In normal human and mouse individuals, L1 expression is predominantly found in germ cells (Rosser and An 2012). In the male germline, it is restricted to two specific developmental time windows (Fig. 2). The first time window occurs in the mitotically arrested prospermatogonia. Abundant ORF1p signals are detected in E14.5–E17.5 fetal testes (Trelogan and Martin 1995; Aravin et al. 2009). The second time window is in spermatocytes. Prominent ORF1p signals are observed in primary spermatocytes (leptotene and zygotene stages) of the prepuberal mouse testis (P10, P14, P18, and P25) (Branciforte and Martin 1994). In the adult testis, L1 ORF1p is detected sporadically (i.e., not in all cells or seminiferous tubules) by immunofluorescence in leptotene, zygotene, and pachytene spermatocytes (Soper et al. 2008), although positive signal was also reported in round spermatids, elongating spermatids and residual bodies when detected by immunohistochemistry in the initial study (Branciforte and Martin 1994). The same polyclonal antibody was used for both studies. The variation in ORF1p expression pattern might be due to methodological differences. Note in the following sections, when the term upregulation or derepression is used, it refers to a significant increase in L1 expression (i.e., above the normal levels) or unexpected expression in otherwise non-expressing cellular stages.

4 Transcriptional Silencing of L1s During Male Germ Cell Development

4.1 Transcriptional Silencing by DNA Methylation

DNA methylation is a critical mechanism for transcriptional silencing of retrotransposons in the germline. During animal development, there are two critical periods during which CpG methylation is programmatically removed from TEs: first in preimplantation embryos and then in migrating/post-migratory PGCs (Messerschmidt et al. 2014). In particular, L1 5′UTR sequences undergo rapid demethylation in mouse PGCs between E10.5 and E13.5 (from 65 to 17 %) (Lane et al. 2003; Hajkova et al. 2002; Ohno et al. 2013), and later become fully remethylated, through de novo methylation, in prospermatogonia by E17.5 (Lees-Murdock et al. 2003). This demethylation correlates with modest L1 expression in normal prospermatogonia (Fig. 2). As will be discussed in later sections, now we know that other regulatory mechanisms exist to prevent excessive L1 activities when L1s are hypomethylated.

The role of DNA methylation is clearly supported by TE activation in mice deficient in DNA methyltransferases (DNMTs). The mammalian genome encodes three enzymatically active cytosine methyltransferases (DNMT1, DNMT3A, and DNMT3B). The earliest in vivo evidence is from functional ablation of DNMT1. DNMT1 prefers hemi-methylated DNA as its substrate and mediates the maintenance of genomic methylation patterns. Deletion of *Dnmt1* causes embryonic lethality in mice (Li et al. 1992). In somatic tissues of *Dnmt1*-deficient mouse embryos, there is widespread hypomethylation and derepression of a major class of mouse LTR retrotransposons, intracisternal A particles (IAPs) (Walsh et al. 1998). However, a change in methylation of other types of retrotransposons, including L1s, has not been reported in *Dnmt1*-deficient embryos. It is possible that L1 methylation may be unaffected because in other cellular environments, such as mouse embryonic stem cells (ESCs), DNMT1 appears to play a major role in regulating IAPs but not in regulating L1 expression (Li et al. 2015a). Indeed, unlike IAPs, L1 expression is undetectable in *Dnmt1*-deficient embryos (Bourc'his and Bestor 2004). UHRF1 (also known as Np95 and ICBP90) has an essential role in DNA methylation maintenance by recruiting DNMT1 to hemi-methylated DNA sites (Sharif et al. 2007; Bostick et al. 2007). Similar to *Dnmt1* knockouts, deletion of *Uhrf1* causes embryonic lethality. Interestingly, *Uhrf1* deletion is accompanied by demethylation and derepression of IAPs, L1s, and SINE-1 (Sharif et al. 2007).

DNMT3A and DNMT3B are required for *de novo* DNA methylation during animal development. Mice deficient for *Dnmt3a* die at ~4 weeks of age and those deficient for *Dnmt3b* die in utero (Okano et al. 1999). Embryos deficient for both *Dnmt3a* and *Dnmt3b*, but not for either enzyme alone, display reduced methylation at C-type retroviruses, IAP, and major satellite repeats, suggesting DNMT3A and DNMT3B have redundant functions at these sequences. In contrast, minor satellite repeats appear to be specific targets of DNMT3B (Okano et al. 1999). Detailed methylation analyses of conditional *Dnmt3a* and *Dnmt3b* knockouts revealed the timing and target specificity of DNMT3A and DNMT3B in the male germline (Kaneda et al. 2004; Kato et al. 2007). In terms of timing, the methylation of repetitive sequences and paternally differentially methylated regions (DMRs) occurs progressively in fetal prospermatogonia and is completed by the newborn stage (Kato et al. 2007). In terms of target specificity, DNMT3A mainly methylates B1 SINEs and some paternal DMRs, such as the H19 and Dlk1/Gtl2 loci. DNMT3B predominantly methylates minor and major satellite repeats. Both DNMT3A and DNMT3B are required for *de novo* methylation of L1 and IAPs as well as the paternal DMR at the Rasgrf1 locus (Kaneda et al. 2004; Kato et al. 2007).

DNMT3L, a homologue of DNMT3A/3B, lacks cytosine methyltransferase activity but is specifically expressed in growing oocytes of adult females and prospermatogonia of perinatal males (Bourc'his et al. 2001). Consistent with its germ cell specific expression pattern, deletion of *Dnmt3l* does not affect animal viability but causes infertility in both sexes (Bourc'his et al. 2001). In knockout males, spermatogenesis is arrested at the zygotene/early pachytene spermatocyte stages (Bourc'his and Bestor 2004; Webster et al. 2005). Despite the lack of enzymatic activity, DNMT3L plays a pivotal role in de novo methylation of all repetitive sequences and paternal/maternal DMRs examined in the germline (Bourc'his et al. 2001; Hata et al. 2002; Bourc'his and Bestor 2004; Webster et al. 2005; Kato et al. 2007). For example, deletion of *Dnmt3l* results in the global loss of de novo methylation at L1s and IAPs (Bourc'his and Bestor 2004; Kato et al. 2007) as well as B1 SINEs (Kato et al. 2007) in male germ cells. More significantly, hypomethylation leads to TE derepression, as evidenced by high levels of L1 and IAP transcripts in neonatal and juvenile testes (Bourc'his and Bestor 2004). As compared to some piRNA pathway mutants discussed later, the spermatogenic defect manifested in *Dnmt3l* mutants appears to be more pronounced. Germ cells were completely absent in some seminiferous tubules by 4 weeks and in almost all tubules by 8–10 weeks (Hata et al. 2006). This Sertoli-cell-only phenotype suggests a functional deficit in spermatogonial stem cells. Indeed, careful analysis of neonatal animals revealed a mitotic defect in spermatogonia and reduced germ cell numbers as early as P6 (La Salle et al. 2007). In contrast, the female infertility has been attributed to the loss of maternal imprinting and inappropriate expression of maternally imprinted genes in fetuses (Bourc'his et al. 2001; Hata et al. 2002). Accordingly, IAPs and major satellite repeats show little change in methylation in fetuses carried by *Dnmt3l* knockout females (Bourc'his et al. 2001).

Barring a disruption of the de novo methylation process, the vast majority of the full-length L1 elements are remethylated by birth and remain hypermethylated as the germ cells progress through spermatogenesis (Fig. 2). So far there is no evidence for global or dynamic demethylation of L1s at the onset of meiosis (Oakes et al. 2007). However, not all copies of retrotransposons are methylated in spermatocytes (Molaro et al. 2014). A small fraction of L1s, SINEs, and LTR retrotransposons remain hypomethylated in adult spermatocytes as compared to E13.5 PGCs. These hypomethylated loci comprise both old and young subfamilies but are enriched for "middle-aged" subfamilies, such as the F subfamily of L1s and the MT subfamily of LTRs (Molaro et al. 2014). Such constitutively hypomethylated loci may underlie the sporadic L1 ORF1p expression seen in prepuberal and adult spermatocytes (Branciforte and Martin 1994; Soper et al. 2008) (Fig. 2), especially when considering the extensive chromatin remodeling during meiosis (Crichton et al. 2014).

4.2 The Role of Fetal piRNA Pathway in Transcriptional Silencing

piRNAs play critical roles in regulating retrotransposon expression. piRNAs are small noncoding RNAs abundant in germ cells. Unlike miRNAs (typically 19–22 nucleotides), piRNAs are longer and range from 24 to 31 nucleotides. First described in mouse, rat, and human adult testes in 2006 (Aravin et al. 2006; Grivna et al. 2006a; Girard et al. 2006; Lau et al. 2006), piRNAs are expressed mainly in two separate phases of male germ cell development. Accordingly, they are loosely categorized into fetal piRNAs and pachytene piRNAs (Fig. 2). The mouse genome encodes three Piwi proteins (MIWI/PIWIL1, MILI/PIWIL2, and MIWI2/PIWIL4). Fetal piRNAs are associated with MILI and MIWI2, highly enriched for TE sequences (Aravin et al. 2007; Kuramochi-Miyagawa et al. 2008), and function in both transcriptional and posttranscriptional silencing of retrotransposons (this section and Sect. 5.1). In contrast, pachytene piRNAs are associated with MILI and

MIWI, relatively depleted of TE sequences, expressed from genomic clusters (Aravin et al. 2006; Grivna et al. 2006a; Girard et al. 2006; Grivna et al. 2006b), and have an important role in posttranscriptional silencing (Sect. 5.2). More details on piRNA production and processing can be found in recent reviews (Iwasaki et al. 2015; Czech and Hannon 2016). Here we will focus our discussion of piRNAs as to what is applicable to TE regulation.

Transcriptional silencing by the fetal piRNA pathway is primarily mediated through DNA methylation. This putative mechanism is supported by molecular phenotypes in *Mili* and *Miwi2* knockout mice. Both *Mili* and *Miwi2* are expressed in embryonic testes when the genome is being remethylated (Kuramochi-Miyagawa et al. 2001) (Fig. 2). MILI expression starts at E12.5 in both sexes and stops after birth in female but continues until the round spermatid stage in males. In contrast, MIWI2 expression is restricted from E15.5 to P3 in male and absent in female germ cells (Aravin et al. 2008; Kuramochi-Miyagawa et al. 2001). Deletion of either *Mili* or *Miwi2* leads to male-specific infertility, mimicking the *Dnmt3l* knockout phenotype in spermatogenesis. Spermatogenesis arrests at the zygotene stage in *Miwi2* mutants (Carmell et al. 2007) and at the early pachytene stage in *Mili* mutants (Kuramochi-Miyagawa et al. 2004). The spermatogenic defect is accompanied by the concomitant loss of piRNAs and TE methylation (Aravin et al. 2008; Kuramochi-Miyagawa et al. 2008; Carmell et al. 2007). The role of the fetal piRNA pathway on DNA remethylation is highly specific towards repetitive sequences, as whole genome bisulfite sequencing shows that global methylation is not affected in *Mili* and *Miwi2* mutants (Manakov et al. 2015; Molaro et al. 2014). In agreement with the proposed pathway, deletion of *Dnmt3l* has no effect on piRNA production, confirming that the fetal piRNA pathway acts upstream of *de novo* DNA methylation (Aravin et al. 2008).

The fetal piRNA-mediated methylation appears to selectively target active full-length retrotransposons (Molaro et al. 2014). In spermatocytes, where genome-wide remethylation has already occurred, overall TE methylation levels are very similar between wild-type and *Mili*-deficient mice. This observation suggests that most TEs are remethylated nonselectively along with the rest of the genome by a default wave of de novo DNA methylation (Molaro et al. 2014). In contrast, the youngest L1 subfamilies (e.g., A and T) manifest the most pronounced reduction in methylation and corresponding upregulation in *Mili*-deficient spermatocytes (Molaro et al. 2014). A remarkable feature of this phenomenon is that it selectively targets transcriptional regulatory regions, i.e., L1 and IAP promoters are differentially methylated between wild-type and *Mili*-deficient backgrounds (Molaro et al. 2014). This provides a mechanism to effectively silence full-length, transcriptionally active elements. In conclusion, fetal piRNAs partner with the DNA methylation machinery to re-suppress L1s through de novo DNA methylation, specifically homing in on active full-length elements (Fig. 3).

The two principal components of the fetal piRNA pathway, MILI and MIWI2, cooperate but carry distinct roles in fetal germ cells. Besides temporal difference in expression, they differ in intracellular location (Aravin et al. 2008; Aravin et al. 2009). In male prospermatogonia, MILI is localized primarily in the cytoplasm, in

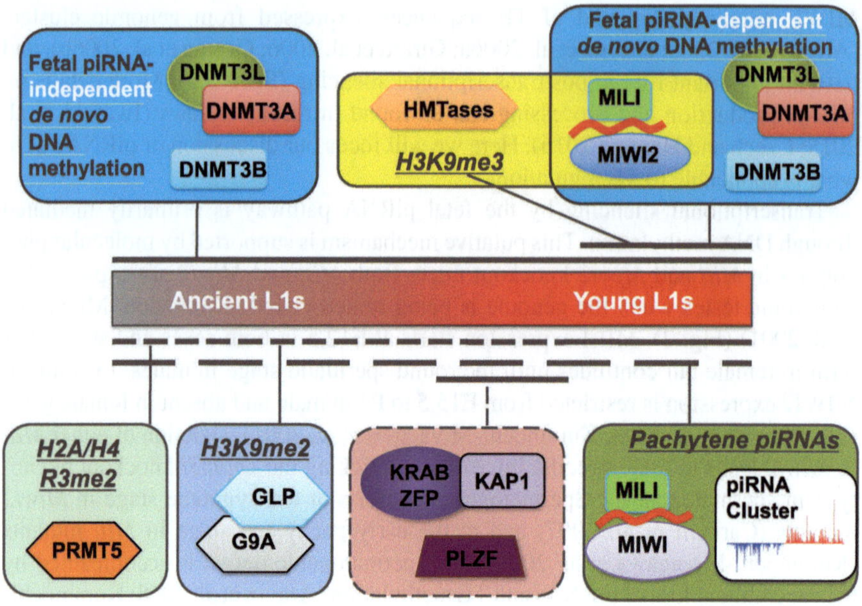

Fig. 3 L1 subfamily-specific regulation in the male germline. Next-generation sequencing-based approaches have revealed that L1 regulatory mechanisms differ in their subfamily specificity. MIWI2-guided de novo DNA methylation and H3K9me3 primarily target young active L1s through specificity granted by fetal piRNAs. In vitro data in ESCs suggest that KAP1, presumably guided by KRAB-ZFPs, recruits repressive histone marks to "middle-aged" L1s but not the youngest subfamilies. More general suppression occurs through piRNA-independent de novo methylation and posttranscriptional regulation by pachytene piRNAs. Pachytene piRNAs are generated from piRNA clusters and enrichment for specific subfamilies seems stochastic. Loss of H2A/H4R3me2 or H3K9me2 is associated with L1 activation in PGCs or spermatogonia. However, the distribution of either histone mark has not been defined for L1 subfamilies

perinuclear cytoplasmic granules that represent intramitochondrial cement. MIWI2 proteins, on the other hand, are both cytoplasmic and nuclear. In the cytoplasm, MIWI2 is localized to distinct granules sharing many components of the processing body. In the nucleus, MIWI2 distribution is homogenous. An early model built on observations from global *Mili* and *Miwi2* deletion mutants suggested that TE-derived fetal piRNAs are amplified through a heterotypic Ping-Pong loop, involving slicer activities of both MILI and MIWI2 (Aravin et al. 2008). However, when the slicer activity of MIWI2 is abolished by a point mutation, piRNA amplification is unaffected, both L1 and IAP remain repressed, and mutant males stay fertile (De Fazio et al. 2011). These results suggest that MIWI2 slicer activity is not required for piRNA amplification. In a revised model, MILI alone was proposed to drive the production of secondary piRNAs through an intra-MILI Ping-Pong cycle, which feeds anti-sense piRNAs to MIWI2 for subsequent targeted DNA methylation in the nucleus (De Fazio et al. 2011). Interestingly, in *Mili* slicer mutants, L1 is derepressed but

IAP remains silenced, indicating that L1 but not IAP silencing is dependent on piRNA amplification (De Fazio et al. 2011).

Transcriptional silencing by fetal piRNAs may also involve the repressive histone modification H3K9me3. SUV39H1/2 methyltransferases are known to specifically deposit H3K9me3 marks at pericentric and intergenic major satellite repeats. Using genome-wide ChIP-Seq approaches, two recent studies have provided novel insights into a functional role of H3K9me3 in transcriptional silencing of L1 retrotransposons (Bulut-Karslioglu et al. 2014; Pezic et al. 2014). In both somatic cells and premeiotic male germ cells (i.e., spermatogonia), H3K9me3 is enriched not only in major satellite repeats but also in L1s and LTR retrotransposons. However, its enrichment on L1s is much more pronounced in germ cells. Overall, transcriptionally active L1 subfamilies (A, F, Gf, T), which contain the majority of full-length elements, display higher levels of H3K9me3. The distribution of H3K9me3 over full-length L1s is asymmetric, with significant concentration toward the 5′UTR promoter (Pezic et al. 2014). In spermatogonia from *Miwi2*-deficient mice, levels of H3K9me3 are reduced from L1 A, Gf, and T elements, especially from the promoter region, but relatively constant at F elements. Coincidently, L1 A, Gf, and T transcripts are elevated (Pezic et al. 2014). This study did not identify the histone methyltransferases responsible for H3K9me3 modification on L1 loci in the spermatogonia. In mouse ESCs, by comparing H3K9me3 distribution in wild-type and *Suv39h1/2*-deficient ESCs, L1s and ERVs are identified as the major targets of SUV39H1/2 outside constitutive heterochromatin. The 5′UTR of full-length L1s are specifically targeted for H3K9me3 modification. Accordingly, these L1s are derepressed in *Suv39h1/2*-deficient ESCs (Bulut-Karslioglu et al. 2014). Additionally, it has been previously shown that SUV39H1/2 are responsible for H3K9me3 in type B spermatogonia and preleptotene spermatocytes (Peters et al. 2001). *Suv39h1/2*-deficient male mice are infertile with most spermatocytes becoming apoptotic between mid- and late-pachytene stages (Peters et al. 2001). However, the effect of *Suv39h1/2* deletion on L1 expression in germ cells has not been reported. H3K9me3 also marks L1s and IAPs in E13.5 PGCs (Liu et al. 2014). However, conditional ablation of *Setdb1*, another H3K9 methyltransferase, in PGCs results in IAP upregulation only (Liu et al. 2014).

How fetal piRNAs target retrotransposons for de novo DNA methylation and/or H3K9me3 modification remains elusive. The observations that these marks are enriched at transcriptionally active L1 loci suggest a model in which fetal piRNAs guide the chromatin modifying apparatus to specific loci through interacting with nascent L1 transcripts (Molaro et al. 2014; Pezic et al. 2014). GTSF1 is a mouse protein with two tandem copies of the putative RNA-binding CHHC Zn-finger domain. Deletion of *Gtsf1* leads to phenotypes similar to those found in *Mili* or *Miwi2* mutants (Yoshimura et al. 2009). Additional insights into the function of GTSF1 in the piRNA pathway came from flies. In *Drosophila*, Gtsf1 is required for Piwi-guided transcriptional silencing and deposition of H3K9me3 on retrotransposons, potentially mediating piRNA/target RNA interaction through its zinc finger domains (Donertas et al. 2013). Another *Drosophila* protein, CG3893 (Asterix), belongs to the same protein family characterized by the CHHC Zn-finger domain. Coincidently, CG3893 is also essential for TE silencing and selected deposition of H3K9me3 on certain transposons (Muerdter et al. 2013). Therefore, mouse GTSF1 may play roles analogous to Gtsf1 or Asterix in *Drosophila*, bridging the interaction between fetal piRNAs and nascent transcripts of targeted retrotransposons (Yu et al. 2015). It is unknown how piRNAs coordinate target-specific DNA methylation and histone modification. However, preliminary analysis of selected L1 loci indicates that H3K9me3 modification is uncoupled from DNA methylation (Pezic et al. 2014).

4.3 Transcriptional Silencing by Histone Modifications

Histone modifications have been well established as critical mechanisms in transcriptional silencing of ERVs in mouse ESCs and preimplantation embryos (Leung and Lorincz 2012), but little was known about their roles in L1 regulation during germ cell development until very recently. Compelling evidence indicates that two histone marks (H2A/H4R3me2 and H3K9me2; discussed below) repress L1 retrotransposons at different stages in the germline. In addition, as discussed in the previous section, H3K9me3 may function in fetal piRNA-mediated transcriptional silencing of L1s.

The symmetric methylation of arginine 3 of histones H2A and H4 (H2A/H4R3me2) is essential for transcriptional silencing of retrotransposons in PGCs (Kim et al. 2014). In PGCs, H2A/H4R3me2 is deposited by protein arginine methyltransferase 5 (PRMT5) (Ancelin et al. 2006). PRMT5 is expressed in both PGCs and somatic cells as early as E7.5 (Ancelin et al. 2006). It is transiently enriched in the nucleus of PGCs between E8.0 and E11.0, and later relocates back to the cytoplasm at E11.5 (Kim et al. 2014). The nuclear presence of PRMT5 is accompanied by progressive enrichment of H2A/H4R3me2 in PGCs from E8.5 to E10.5 and this modification persists at least until E12.5 (Kim et al. 2014) (Fig. 2). The function of H2A/H4R3me2 accumulation in PGCs was revealed by conditional ablation of *Prmt5* specifically in fetal germ cells (Kim et al. 2014; Li et al. 2015b). Both mutant males and females are sterile because of complete loss of PGCs by E16.5 (Kim et al. 2014; Li et al. 2015b). This early germ cell developmental defect is associated with an upregulation of L1 and IAP expression at both RNA and protein levels (Kim et al. 2014). In wild-type animals, both L1s and IAPs are enriched for H2A/H4R3me2 in PGCs at E10.5-E11.5. In mutant PGCs, H2A/H4R3me2 is globally lost with no significant changes in other chromatin markers, such as H3K9me2, H3K9me3, and H3K27me3. There is also no reduction of DNA methylation at L1 and IAP sequences. These results suggest PRMT5-mediated H2A/H4R3me2 modification is a dominant silencing mechanism for retrotransposons during PGC growth (Kim et al. 2014). This time period coincides with global DNA demethylation and loss of H3K9me2, when the germ cell genome is most vulnerable for retrotransposon activation, highlighting the importance of H2A/H4R3me2 modification in transcriptionally controlling retrotransposon expression in fetal gonads (Fig. 2).

The function of PRMT5 is not limited to its modification of histones. PRMT5 has many nonhistone protein substrates, including spliceosomal Sm proteins and Piwi proteins (Vagin et al. 2009; Kirino et al. 2009). Therefore, its function in PGC development may partially depend on its role in the methylation of Sm proteins. Indeed, conditional loss of *Prmt5* leads to aberrant splicing in PGCs (Li et al. 2015b). The PRMT5 protein also functions in the piRNA pathway by symmetrically dimethylating specific arginine residues of all three mouse Piwi proteins (Vagin et al. 2009). These modifications are required for proper interaction with specific Tudor family members (Vagin et al. 2009). In addition, PRMT5 is expressed postnatally during spermatogenesis. It is predominantly cytoplasmic in spermatogonia and nuclear in spermatocytes (Wang et al. 2015). Conditional deletion of *Prmt5* in postnatal male germ cells leads to loss of H2A/H4R3me2 and a meiotic defect. However, there is no significant change in L1 and IAP transcription (Wang et al. 2015). This observation does not necessarily refute a role in transcriptional silencing of retrotransposons by H2A/H4R3me2 modification in spermatocytes because other silencing mechanisms, such as DNA methylation, likely remain intact.

H3K9me2 marks silenced domains in euchromatin. It is catalyzed by the histone-lysine N-methyltransferase G9a/GLP heteromeric complex. The level of H3K9me2 is reduced in PGCs at E8.0 (Seki et al. 2005) and this low level is maintained in fetal prospermatogonia as a result of posttranscriptional downregulation of GLP (Deguchi et al. 2013). In postnatal mice, the H3K9me2 mark is detected in spermatogonia through the zygotene stage of spermatocytes (Fig. 2). The eventual loss of H3K9me2 is due to diminished G9a expression from the preleptotene stage (Di Giacomo et al. 2013, 2014). Conditional ablation of *G9a* in germ cells leads to the loss of H3K9me2 signal in postnatal germ cells, resulting in meiotic arrest at early pachytene stage, but has no impact on L1 and IAP expression (Tachibana et al. 2007). This unperturbed expression is not an indication that H3K9me2 is not involved in controlling L1 and IAP transcription. Instead it reflects remarkable functional redundancy among multiple retrotransposon silencing pathways (Di Giacomo et al. 2013). This multilayered regulation was revealed by comparing retrotransposon expression in different genetic backgrounds (Di Giacomo et al. 2014). Like many other piRNA pathway mutants, in *Mili*-deficient mice, L1s are derepressed only in spermatocytes but not in spermatogonia despite impaired piRNA biogenesis and the consequent loss of DNA methylation at L1 promoters. However, conditional inactivation of *G9a* in *Mili*-deficient adult mice aggravates the disruption of spermatogenesis: spermatogonia are the only germ cells remaining in seminiferous tubules of such animals (Di Giacomo et al. 2014). Unlike mice deficient in either *G9a* or *Mili* alone, L1 expression is derepressed in spermatogonia of double-deficient mice, indicating H3K9me2 is required to co-suppress L1 expression in spermatogonia (Di Giacomo et al. 2014). However, H3K9me2 alone is not sufficient in silencing IAP in spermatogonia, as evidenced by IAP upregulation in mice deficient for *Mili* alone (Di Giacomo et al. 2014). Thus, the programmed loss of H3K9me2 repression is the most likely cause of sudden TE expression at the onset of meiosis seen in many of the DNA methylation/piRNA pathway knockouts.

4.4 Transcriptional Silencing Mediated by Zinc Finger Proteins

Both human and mouse genomes encode a highly diverse family of transcription factors that can recognize specific target sequences through the tandem zinc finger DNA binding domains (Collins et al. 2001). The most abundant subgroup is known collectively as KRAB-zinc finger proteins (KRAB-ZFPs), which are characterized by the repressive Kruppel-associated box (KRAB) domain. A smaller subgroup comprises POZ-ZFPs, which is defined by the poxvirus and zinc finger (POZ) domain. Although a role in the germline has not been established, several members of KRAB-ZFP and POZ-ZFP families have been shown to mediate sequence-specific targeting and transcriptional silencing of L1 elements in other cellular contexts (Castro-Diaz et al. 2014; Jacobs et al. 2014; Puszyk et al. 2013).

KRAB-ZFPs are able to tether KRAB-associated protein 1 (KAP1, also known as tripartite motif-containing 28 or TRIM28) to the target site, which further recruits chromatin modifying enzymes to transcriptionally repress the target genes (Wolf et al. 2015). First demonstrated for exogenous retroviruses, KRAB-ZFPs also repress endogenous retrotransposons in both human and mouse ESCs. The depletion of KAP1 results in upregulation of specific subfamilies of LTR retrotransposons (e.g., IAP) (Rowe et al. 2010; Turelli et al. 2014) and L1s (Castro-Diaz et al. 2014). Interestingly, only "middle-aged" L1 subfamilies (e.g., full-length human L1PA5 and L1PA4 and mouse L1MdF2 elements) are subjected to KAP1-mediated suppression. Evolutionarily older and younger L1s are not affected in KAP1-depleted ESCs (Castro-Diaz et al. 2014) (Fig. 3). This phenomenon is consistent with evolutionary dynamics of KRAB-ZFPs and their retrotransposon targets. Simply put, a substantial amount of time is required for the host to evolve specific KRAB-ZFPs targeting active L1 lineages. Meanwhile, older L1s have accumulated sufficient mutations to obscure the original interaction with cognate KRAB-ZFPs. KRAB-ZFPs that are responsible for sequence-specific L1 targeting have started to emerge. Mouse L1MdF2 elements are specifically bound by Gm6871, a mouse-specific KRAB-ZFP, and derepressed in Gm6871-depleted mouse ESCs (Castro-Diaz et al. 2014). Primate-specific ZNF93 recognizes L1PA3-6 in the human genome (Jacobs et al. 2014). Most extraordinarily, a subgroup of L1PA3 elements, as well as the evolutionarily younger L1PA2 and L1PA1 subfamilies, are able to evade ZNF93-mediated transcriptional silencing due to a deletion at the ZNF93 binding site (Jacobs et al. 2014), highlighting an evolutionary arms race between retrotransposons and host factors.

Promyelocytic leukemia zinc finger (PLZF), encoded by *Zfp145*, is a member of the POZ-ZFP family (Collins et al. 2001). Acting as a transcription repressor, PLZF is an important factor in regulating cell growth and differentiation, typically acting through binding to specific DNA motifs and instigating repressive chromatin state locally by recruiting DNMTs and histone deacetylases. In the male germline, PLZF is first seen in prospermatogonia at E17.5 and subsequently restricted to undifferentiated spermatogonia after birth. Functional ablation of *Plzf* causes progressive loss of spermatogonial stem cells and male infertility (Costoya et al. 2004). Interestingly, in a mouse model that is defective in PLZF DNA binding, methylated DNA immunoprecipitation (MeDIP) identified many hypomethylated L1 loci in the bone marrow and testis, and RT-PCR detected increased L1 RNA levels in these tissues (Puszyk et al. 2013). Follow-up experiments in cultured cells confirmed that many L1s contain a PLZF binding site in a conserved region of ORF2. PLZF binding appeared to establish and propagate DNA methylation and repressive histone marks to the 5′UTR of a full-length L1 element. Lastly, L1 expression was negatively correlated with PLZF rich tissues and those tissues had a closed chromatin conformation around the L1 sequences (Puszyk et al. 2013).

5 Posttranscriptional Silencing of L1s During Male Germ Cell Development

5.1 *Posttranscriptional Silencing by the Fetal piRNA Pathway*

As discussed earlier, much emphasis has been placed on the role of fetal piRNA in transcriptional silencing thorough targeting retrotransposons for DNA methylation. Its function in posttranscriptional silencing should not be ignored. Obviously, one way the fetal piRNA pathway can act posttranscriptionally is by cleaving TE transcripts into short piRNAs, which reduces the abundance of full-length TE transcripts in this process. The challenge is that most fetal piRNA pathway mutations have a deficit in de novo methylation of retrotransposons. So it is almost impossible to differentiate transcriptional from posttranscriptional mechanisms in these mutants. However, there are at least two cases where piRNA genesis was disrupted without a disruption of de novo methylation, revealing posttranscriptional silencing of TEs by the fetal piRNA pathway (Aravin et al. 2009; Ichiyanagi et al. 2014).

Maelstrom (MAEL) contributes to posttranscriptional silencing of L1/IAP in fetal prospermatogonia (Aravin et al. 2009). Dynamics of L1 ORF1p expression has been examined in wild-type and *Mael* mutant germ cells from E14.5 to P10. In wild-type germ cells, L1 expression peaks at E16.5, when the L1 promoter has not been fully remethylated (61 % of CpGs methylated), but significantly declined by E18.5 and became almost undetectable by P2 (Aravin et al. 2009). This expression pattern is consistent with DNA methylation dynamics as L1 promoters are mostly remethylated by E18.5 (Fig. 2). In *Mael* mutant mice, L1 expression is similar to wild type at E16.5, but is further elevated in E18.5 germ cells. The high levels of ORF1p signals persist beyond P6. The increase in ORF1p is not due to loss of methylation as methylation levels are comparable to the wild type at E18.5 and P2. These results suggest that MAEL participates in posttranscriptional silencing of retrotransposons in fetal germ cells. The authors also reported that a similar mechanism targets IAPs (Aravin et al. 2009).

HSP90AA1 is one of the two isoforms of heat-shock protein 90 in mammals. HSP90AA1 protein is specifically expressed in germ cells of the mouse testis (Gruppi and Wolgemuth 1993). Deletion of *Hsp90aa1* causes a threefold reduction of fetal piRNA in E16.5 testes (Ichiyanagi et al. 2014). The levels of L1 ORF1p are increased in both E16.5 and E18.5 germ cells. However, L1 mRNA levels remain largely unchanged in E16.5 and P10. There is little change in L1 promoter methylation in P0 prospermatogonia between wild-type and mutant mice across multiple subfamilies, suggesting the presence of posttranscriptional regulation (Ichiyanagi et al. 2014).

5.2 Pachytene piRNA Pathway

Pachytene piRNAs were the first piRNA species reported in mammals (Aravin et al. 2006; Grivna et al. 2006a; Girard et al. 2006; Lau et al. 2006). Two mouse PIWI proteins, MIWI and MILI, are involved in pachytene piRNA biogenesis. MIWI expression is specific to meiotic and post-meiotic male germ cells (Deng and Lin 2002). It begins in mid-pachytene spermatocytes, peaks in diplotene spermatocytes, and persists in steps 1–3 round spermatids (Deng and Lin 2002). In adult animals, pachytene piRNAs are close to background level in spermatogonia, expressed at the highest level in spermatocytes, modesty expressed in round spermatids, but absent in elongating spermatids (Beyret and Lin 2011) (Fig. 2). In spermatocytes, MIWI is cytoplasmic; in spermatids, MIWI is cytoplasmic but enriched in chromatoid body (Grivna et al. 2006b). Deletion of *Miwi* causes spermatogenic arrest at step 4 in round spermatids (Deng and Lin 2002). The spermatogenic phenotype is accompanied by increased L1 RNA and ORF1p in purified round spermatids as well as in spermatocytes by immunofluorescence analysis (Reuter et al. 2011).

MIWI silences L1 retrotransposons posttranscriptionally. This was demonstrated by using a *Miwi* mutant that carries a point mutation in its conserved slicer domain DDH (termed *MiwiADH*) (Reuter et al. 2011). Similar to global *Miwi* knockout, spermatogenesis is arrested at the round spermatid stage in *MiwiADH* mutants. Unlike *Miwi* knockout, which does not express MIWI, *MiwiADH* mutant shows no change in MIWI protein abundance, subcellular localization, piRNA abundance, or genomic annotation. There is also no change in DNA methylation at L1 promoters in both *Miwi* and *MiwiADH* mutants. Nevertheless, L1 RNA and ORF1p are upregulated in both mutants. Together, these results suggest that MIWI silences L1s posttranscriptionally by cleaving L1 mRNAs (Reuter et al. 2011).

MILI's role in pachytene piRNAs cannot be interrogated by a global deletion of *Mili* because MILI is also involved in fetal piRNA biogenesis (see Sect. 4.2). Conditional deletion of *Mili* or its slicer activity (i.e., via a point mutation in the slicer domain) in spermatocytes, by crossing to Stra8-Cre mice, leads to a spermatogenic arrest at stages ranging from pachytene spermatocytes to elongating spermatids (Di Giacomo et al. 2013). Interestingly, L1, but not IAP, is upregulated only in late zygotene and pachytene spermatocytes, but not in surviving round and elongating spermatids. Detailed analysis of MILI expression pattern reveals that, contrary to a common belief, which holds that MILI is continuously expressed in all stages of spermatogenesis, *Mili* ceases its expression from preleptotene to late zygotene stage (Di Giacomo et al. 2013) (Fig. 2). Because L1 methylation is unaffected in these mutant animals, these results suggest that MILI is required for posttranscriptional silencing of L1 at the onset of the zygotene/pachytene transition and in pachytene spermatocytes (Di Giacomo et al. 2013). MIWI expression and piRNA loading remain intact in these mutants, which could explain the lack of L1 derepression in spermatids.

Whether or not pachytene piRNAs themselves are necessary for L1 suppression is still unclear. MOV10L1 is an RNA helicase (Vourekas et al. 2015) that is required for both fetal (Frost et al. 2010; Zheng et al. 2010) and pachytene piRNA biogenesis (Zheng and Wang 2012). A conditional knockout of *Mov10l1* in spermatocytes displays limited pachytene piRNAs with no effect on fetal piRNA production. However, no derepression of L1 and IAP is observed in these mutants. These results suggest that pachytene piRNAs per se are not required for retrotransposon silencing (Zheng and Wang 2012). In these mutant mice, L1 transcripts may have been degraded by the slicer activity of MIWI through a pachytene piRNA-independent mechanism (Zheng and Wang 2012).

6 L1 Regulation During Female Germ Cell Development

The timing of mouse female germ cell development is significantly different from the male counterpart (Pepling 2006). The difference manifests shortly after PGCs have colonized the gonadal ridge at E10.5. Unlike male germ cells, which enter mitotic arrest between E12.5 and E14.5, the majority of female germ cells begin meiosis at E13.5–E15.5, becoming primary oocytes (Peters 1970; Baltus et al. 2006) (see also chapter "Retrotransposons and the Mammalian Germline" in this book). Primary oocytes progress through leptotene, zygotene, and pachytene stages, and become arrested at the diplotene stage of meiosis I between E17.5 and P5 (Peters 1970). Shortly after birth, oocytes become individually surrounded by somatic granulosa cells, forming primordial follicles (Peters 1969; Pepling and Spradling 2001). The majority of the primordial follicles will remain dormant for months or years until being recruited to the growing follicle pool. After the initial recruitment, primordial follicles must progress through primary, secondary, and antral stages before oocytes resume meiosis and ovulate as secondary oocytes resting at meiotic metaphase II (McGee and Hsueh 2000).

Male and female germ cells also differ in genome-wide DNA methylation dynamics. Unlike male germ cells, which have regained high levels of DNA methylation by E17.5, demethylation at the L1 promoter is protracted in fetal oocytes. There is no apparent remethylation even at E17.5 (Lees-Murdock et al. 2003; Seisenberger et al. 2012). In postnatal mice, primary oocytes arrested at the diplotene stage still display minimal methylation at L1 sequences (Sanford et al. 1987;

Howlett and Reik 1991). Methylation increases during meiosis, but L1 sequences remain modestly methylated in ovulated secondary oocytes (Howlett and Reik 1991), including active L1 A, Gf, and T subfamilies as evidenced by recent methylomic studies (Smith et al. 2012).

Consistent with relaxed DNA methylation, L1 expression is readily detected in primary oocytes during the meiotic I prophase. L1 ORF1p is first detected at E15.5 in primary follicles and throughout all stages of prophase I in fetal and neonatal ovaries (Trelogan and Martin 1995; Malki et al. 2014). ORF1p signals were not found in adult ovaries (Trelogan and Martin 1995). Interestingly, ORF1p signals in the nucleus are highly variable among primary oocytes (Malki et al. 2014). Detailed analyses of nuclear ORF1p signals in E15.5, E18.5, and P5 oocytes suggest that differential L1 expression may play a role in fetal oocyte attrition, a programmed loss of developing oocytes in humans and rodents (Malki et al. 2014).

Among the three mouse Piwi genes, only *Mili* is expressed during female germ cell development (Deng and Lin 2002; Kuramochi-Miyagawa et al. 2001; Aravin et al. 2008; Watanabe et al. 2008). MILI protein can be detected as early as E12.5 (Unhavaithaya et al. 2009; Aravin et al. 2008). The signal increases at E13.5 with the onset of meiosis but is significantly diminished at E18.5 (Unhavaithaya et al. 2009). Using transgenic animals that express GFP-tagged MILI under the endogenous *Mili* promoter, distinct MILI signals are detected in postnatal primary follicles and localized to cytoplasmic granules in both diplotene-arrested and growing primary oocytes (Aravin et al. 2008). Protein quantification indicates that MILI is more abundant in earlier stages of growing oocytes (Watanabe et al. 2008).

Although MILI is the only mouse Piwi gene expressed in female germ cells, piRNAs are produced in growing oocytes (Watanabe et al. 2008; Tam et al. 2008). The majority of these piRNAs are 25-26 nucleotides long and bound to MILI (Watanabe et al. 2008). Similar to fetal piRNA in prospermatogonia, oocyte piRNAs are enriched for retrotransposon sequences, including L1s and IAPs (Watanabe et al. 2008; Tam et al. 2008). Deletion of *Mili* leads to increased expression of L1s as well as IAPs (Watanabe et al. 2008; Lim et al. 2013). In addition to piRNAs, growing oocytes produce 21–22 nucleotides long small interfering RNAs (siRNAs) (Watanabe et al. 2006; Watanabe et al. 2008; Tam et al. 2008). Most of such siRNAs are derived from retrotransposons. In *Dicer*-deficient oocytes, neither IAP nor L1 transcription increases although other LTR retrotransposons are upregulated (Watanabe et al. 2008). Therefore, piRNAs and siRNAs complement each other to regulate distinct retrotransposon families in growing oocytes.

7 Other Factors Implicated in L1 Regulation

The male germline is a potential hub for retrotransposition. A recent study suggests that, for all three active classes of retrotransposons (L1, SINE, and LTR), new insertions occur almost exclusively through the male germline genome (Nellaker et al. 2012). Using whole-genome sequencing data, the authors catalogued polymorphic TE variants among 18 inbred mouse strains (Nellaker et al. 2012). Based on the evolutionary divergence among the mouse strains analyzed, these polymorphic TE variants represent "recent" insertion events within the past two million years. Interestingly, when chromosomal distribution was analyzed, all classes of polymorphic variants (L1, SINE, and LTR) were relatively depleted on the X chromosome. The deficit of polymorphic TEs on the X chromosome contrasts sharply with fixed TEs, which are distributed evenly among all chromosomes (an exception is fixed L1s, which are enriched on the X chromosome due to positive selection). The relative depletion of TE variants on the X chromosome versus autosomes is indicative of a male-driven insertional mechanism. Simply put, if retrotransposition occurs at equivalent rates in males and females, there should be no chromosomal bias. However, the calculated ratio of male-to-female insertion rates is 7.3, 151.8, and 7.8 for polymorphic L1, SINE, and LTR variants (Nellaker et al. 2012).

What might be responsible for this apparent gender bias in retrotransposition? A prime candidate is the vast difference in the number of germ cell divisions between male and females (Drost and Lee 1995). Mouse PGCs have committed to sex-specific trajectories by E13.5. From this point on, female germ cells enter meiosis and meiotically divide twice to become eggs, whereas male germ cells undergo a minimum of 9 mitotic divisions plus two meiotic divisions (i.e., 11 divisions) to become sperm at the end of the first wave of spermatogenesis (de Rooij and Russell 2000). The number of cell divisions can be substantially higher in adult animals as spermatogonia stem cells divide every 8.5 days to sustain continued spermatogenesis (Oakberg 1957). For example, the corresponding number of cell divisions would be around 17 for a 3-month-old male mouse. Thus, the ratio of male-to-female cell divisions would be ~8.5 (e.g., 17 divided by 2). This back-of-the-envelope calculation suggests that the bulk of retrotransposition in the male germline must occur before male meiosis, either in prospermatogonia or dividing spermatogonia.

Mitotically dividing spermatogonia might be the most important cellular niche for retrotransposition. In vitro experiments demonstrate that cell division promotes L1 retrotransposition (Shi et al. 2007; Xie et al. 2013). In these experiments, retrotransposition of engineered L1 constructs is diminished in cell-cycle arrested cells (Shi et al. 2007; Xie et al. 2013) and non-dividing cells (Kubo et al. 2006). Even in dividing but cell-cycle synchronized cells, retrotransposition frequency is lower if cells just go through one fewer cell division (Xie et al. 2013). These results suggest that the intact nuclear membrane may act as a barrier for L1 RNP to access the nuclear genome (Fig. 1). A recent analysis of polymorphic processed pseudogenes in the 1000 Human Genomes Project supports the importance of cell division in L1-mediated retrotransposition (Abyzov et al. 2013). Processed pseudogenes are copies of cellular mRNAs that have been reversed transcribed and inserted into the genome by the L1 machinery (Esnault et al. 2000). The first piece of evidence is that polymorphic processed pseudogenes are enriched for parent genes being expressed during M or M/G1 transition. The second piece of evidence is that M/G1 expressed genes possess higher copies of fixed processed pseudogenes. Both are consistent with a model in which retrotransposition is coupled to cell division (Abyzov et al. 2013). The potential requirement for active cell division may help to explain a peculiar observation from L1 transgenic mice (Kano et al. 2009). Despite abundant RNA signals in male meiotic and post-meiotic germ cells, no heritable L1 insertions were detected in these L1 transgenic mice. Part of this "posttranscriptional block" to retrotransposition may be the expectedly low number of cell divisions by spermatocytes (e.g., twice for primary spermatocytes, once for secondary spermatocytes, and no further division for spermatids and sperm), although the unique properties of meiotic chromatin and spermatid DNA compaction may present a formidable challenge to the retrotransposition machinery (Fig. 1). However, it has not been formally shown that a cell cycle arrest in vivo blocks L1 retrotransposition.

8 L1 Expression and Regulation in Human Germ Cells

The vast majority of the findings discussed above were obtained from mouse models. How do they translate into humans? Due to tissue accessibility issues, time points of L1 expression in human germ cells are rather limited but the overall expression pattern is similar to what has been found in mice (Rosser and An 2012). Both human L1 ORF1p and ORF2p are detectable in fetal and adult testes (Ergun et al. 2004). L1 protein expression has not been reported in human ovaries. As for DNA methylation, multiple recent methylomic analyses indicate that global demethylation is conserved in human PGCs (von Meyenn and Reik 2015; Guo et al. 2015; Tang et al. 2015; Gkountela et al. 2015). However, global demethylation is not accompanied by significant derepression of retrotransposon transcripts in human PGCs, suggesting other silencing mechanisms are in place. H3K9me3 has been suggested as a candidate. It is consistently present during this stage of germ cell development (Gkountela et al. 2015; Guo et al. 2015; Tang et al. 2015), but whether H3K9me3 is enriched on retrotransposons has not been investigated.

Human germ cells also have an intact piRNA pathway. The human genome encodes four Piwi proteins: HILI (MILI for mouse), HIWI1 (MIWI for mouse), HIWI2 (MIWI2 for mouse), and HIWI3 (Sasaki et al. 2003). At the RNA level, HILI and HIWI are expressed specifically in testes and ovaries (Sasaki et al. 2003; Williams et al. 2015). HILI and HIWI proteins have also been confirmed in adult ovaries (Roovers et al. 2015). Unlike mouse MIWI2, which is restricted to embryonic/neonatal male germ cells, HIWI2 is ubiquitously expressed in normal human tissues (Keam et al. 2014; Williams et al. 2015). HIWI3, which has no ortholog in mouse, is also detected specifically in human testes and ovaries, but its transcript abundance is at least tenfold lower than its three other paralogs (Williams et al. 2015). Pachytene piRNAs are abundant in adult human testes (Aravin et al. 2006; Girard et al. 2006; Ha et al. 2014; Yang et al. 2013; Williams et al. 2015). They can be enriched by antibodies against HILI and HIWI, suggesting these piRNAs are HILI and HIWI-bound (Williams et al. 2015). piRNAs are also present in human ovaries, and more abundant in fetal ovaries than in adult ovaries (Williams et al. 2015; Roovers et al. 2015). The role of the piRNA pathway on human L1 regulation has not been determined. There is a general lack of enrichment of retrotransposon sequences in piRNA libraries isolated from human adult testes and fetal ovaries (Williams et al. 2015; Roovers et al. 2015). However, in parallel to male infertility found in *Miwi2*-deficient mice, increased L1 ORF1p expression is correlated with reduced HIWI2 expression in the spermatogonia of boys with cryptorchidism who are at a high risk of infertility (Hadziselimovic et al. 2015).

9 Functional Consequences of L1 (De)regulation in the Germline

9.1 Implication on the Timing of L1 Retrotransposition

Both the frequency and timing of new L1 insertions remain unexplored during germline development. Nevertheless, recent studies have improved and refined our knowledge about L1 expression during normal germ cell development (Sect. 3). L1 is expressed at both RNA and protein levels in both male and female germ cells. For males, there are two developmental time windows (Fig. 2); for females, only in meiotic germ cells. How do these restricted expression windows impact L1 retrotransposition? As transcriptional activation is a prerequisite for L1 retrotransposition, it is expected that the bulk of retrotransposition in germ cells occurs during these time windows. As discussed in Sect. 7, retrotransposition may also favor actively dividing cellular stages. One caveat is that, for the great majority of studies, ORF1p is used as a surrogate for L1 expression at the protein level. This is because endogenous ORF2p has been notoriously difficult to detect, partially due to its low expression levels (Dai et al. 2014). As ORF2p function is required for L1 retrotransposition (Moran et al. 1996), the control of ORF2p expression may be a key mechanism. L1 ORF2p has been detected in human fetal and adult testes (Ergun et al. 2004), but a careful analysis of ORF2p during germ cell development is lacking in mouse models.

The timing of retrotransposition in the male germ line will have a huge impact on the allelic frequency of new L1 insertions in the resulting spermatozoa. If an insertion occurs at the prospermatogonia/spermatogonia stage, it may be represented in thousands of gametes. In contrast, if retrotransposition occurs at the spermatid stage, this specific insertion can only be found in a single gamete. Thus, early retrotransposition timing increases the allelic frequency for a specific insertion by several thousand-fold as compared to late retrotransposition events. The above interpretation is based on the premise that all insertions are selectively neutral and do not affect the fitness of afflicted germ cells. However, the orientation bias of intronic L1 and LTR elements in the human genome suggests that some TE insertions are subjected to purifying selection (Smit 1999; Medstrand et al. 2002). Even if a small fraction of the insertions are deleterious or adaptive mutations, the difference in retrotransposition timing will significantly impact the overall fitness landscape. This is because late insertions, by definition, occur only in terminal stage of germ cell development and the selection pressure will act on a limited number of developmental pathways (i.e., post-meiotic development for spermatids). In contrast, early-stage germ cells carrying insertions have to survive a myriad of additional developmental processes and eventually emerge as de novo mutations in the population (i.e., the journey for prospermatogonia to become spermatozoa). A corollary of germ cell selection is that the actual frequency of retrotransposition in developing germ cells can be substantially higher than what would be observed in live births. Future studies in animal models are required to address these important questions.

9.2 Impact on Germ Cell Development

What is the functional consequence of retrotransposon deregulation? As discussed above, genetic ablation of various host factors leads to retrotransposon derepression, evidenced by increased RNA and protein expression. Indeed, some of these mechanisms preferentially target transcriptionally and transpositionally active retrotransposon lineages (Fig. 3). In mice, these include several evolutionarily young non-LTR and LTR-retrotransposon subfamilies, such as L1s and IAPs, which can potentially wreak havoc to the germline genome due to insertional mutagenesis. A common theme emerging from these mouse knockout studies is a disruption of the spermatogenic program and subsequent male infertility. An unanswered question is whether spermatogenic failure is caused by the derepression of retrotransposons itself. One of the biggest obstacles is how to quantitatively determine the number of new insertions while facing an extremely high baseline signal from preexisting retrotransposon copies. Real-time PCR has been used to quantify endogenous human and mouse L1 copies (Coufal et al. 2009; Muotri et al. 2010). A similar strategy failed to detect any significant changes in L1 copy number in *Dnmt3l*-deficient mice (Zamudio et al. 2015). Therefore, there is an urgent need for novel, sensitive detection methods so that the relationship between L1 expression and mobilization in vivo can be established. It is also important to emphasize that TE activation can potentially act on germ cells at three different stages of the replication cycle: transcription, translation, and insertion (Fig. 1). Besides insertional mutagenesis, overexpression of L1 ORF2p is known to be cytotoxic (Belgnaoui et al. 2006; Gasior et al. 2006; Malki et al. 2014; Wallace et al. 2008). Additionally, transcriptional activation of normally silenced L1 loci genome-wide has the potential to interfere with chromosome pairing, synapses, and recombination (Zamudio et al. 2015).

10 Concluding Remarks

It has been a decade since the initial discovery of mammalian piRNAs in 2006. Follow-up studies spotlight critical roles of fetal and pachytene piRNA pathways for retrotransposon regulation in the male germline. Remarkable progress has also been made in delineating the intricate relationships between piRNAs, DNA methylation, and histone modifications. It is clear that the host has evolved multiple layers of mechanisms to prevent uncontrolled retrotransposon activities at each stage of the male germ cell development (Fig. 2). Retrotransposons are demethylated in post-migratory PGCs, which rely on H2A/H4R3me2 modification to transcriptionally silence L1 and IAP (Sect. 4.3). The mitotically arrested prospermatogonia are protected by the fetal piRNA pathway, which establishes DNA methylation (and potentially H3K9me3) by birth (Sect. 4.2). After birth, retrotransposons are silenced in spermatogonia by H3K9me2, in addition to DNA methylation (Sect. 4.3). In meiotic germ cells, retrotransposons are transcriptionally silenced by DNA methylation and posttranscriptionally suppressed by pachytene piRNAs (Sect. 5.2). In contrast, our understanding of how retrotransposons are regulated in the female germline is much less complete (Sect. 6). Even less is our knowledge about L1 regulation in human germ cells (Sect. 8). It is imperative to understand how different repression systems cooperate to suppress individual retrotransposons. Future studies can no longer view all L1s as equivalent and subfamilies should be separately analyzed (Fig. 3). Furthermore, analysis must be germ cell specific and stage specific. Tools to quantify retrotransposition must be improved and new methods be developed.

Acknowledgments Research in our laboratory is supported by the National Institutes of Health (NIH) grants R21OD017965, R21HD080143, and P50GM107632. SN was supported, in part, by NIH award T32GM008336. The content is solely the responsibility of the authors and does not necessarily represent the official views of the NIH.

References

Abyzov A, Iskow R, Gokcumen O, Radke DW, Balasubramanian S, Pei B, Habegger L, Genomes Project C, Lee C, Gerstein M (2013) Analysis of variable retroduplications in human populations suggests coupling of retrotransposition to cell division. Genome Res 23(12):2042–2052. doi:10.1101/gr.154625.113

Alzohairy AM, Gyulai G, Jansen RK, Bahieldin A (2013) Transposable elements domesticated and neofunctionalized by eukaryotic genomes. Plasmid 69(1):1–15. doi:10.1016/j.plasmid.2012.08.001

An W, Han JS, Schrum CM, Maitra A, Koentgen F, Boeke JD (2008) Conditional activation of a single copy L1 transgene in mice by Cre. Genesis 46(7):373–383. doi:10.1002/dvg.20407

Ancelin K, Lange UC, Hajkova P, Schneider R, Bannister AJ, Kouzarides T, Surani MA (2006) Blimp1 associates with Prmt5 and directs histone arginine methylation in mouse germ cells. Nat Cell Biol 8(6):623–630. doi:10.1038/ncb1413

Aravin A, Gaidatzis D, Pfeffer S, Lagos-Quintana M, Landgraf P, Iovino N, Morris P, Brownstein MJ, Kuramochi-Miyagawa S, Nakano T, Chien M, Russo JJ, Ju J, Sheridan R, Sander C, Zavolan M, Tuschl T (2006) A novel class of small RNAs bind to MILI protein in mouse testes. Nature 442(7099):203–207. doi:10.1038/nature04916

Aravin AA, Sachidanandam R, Bourc'his D, Schaefer C, Pezic D, Toth KF, Bestor T, Hannon GJ (2008) A piRNA pathway primed by individual transposons is linked to de novo DNA methylation in mice. Mol Cell 31(6):785–799. doi:10.1016/j.molcel.2008.09.003

Aravin AA, Sachidanandam R, Girard A, Fejes-Toth K, Hannon GJ (2007) Developmentally regulated piRNA clusters implicate MILI in transposon control. Science 316(5825):744–747. doi:10.1126/science.1142612

Aravin AA, van der Heijden GW, Castaneda J, Vagin VV, Hannon GJ, Bortvin A (2009) Cytoplasmic compartmentalization of the fetal piRNA pathway in mice. PLoS Genet 5(12):e1000764. doi:10.1371/journal.pgen.1000764

Baltus AE, Menke DB, Hu YC, Goodheart ML, Carpenter AE, de Rooij DG, Page DC (2006) In germ cells of mouse embryonic ovaries, the decision to enter meiosis precedes premeiotic DNA replication. Nat Genet 38(12):1430–1434. doi:10.1038/ng1919

Bao J, Yan W (2012) Male germline control of transposable elements. Biol Reprod 86(5):162. doi:10.1095/biolreprod.111.095463

Beck CR, Collier P, Macfarlane C, Malig M, Kidd JM, Eichler EE, Badge RM, Moran JV (2010) LINE-1 retrotransposition activity in human genomes. Cell 141(7):1159–1170. doi:10.1016/j.cell.2010.05.021

Belgnaoui SM, Gosden RG, Semmes OJ, Haoudi A (2006) Human LINE-1 retrotransposon induces DNA damage and apoptosis in cancer cells. Cancer Cell Int 6:13. doi:10.1186/1475-2867-6-13

Bellve AR, Cavicchia JC, Millette CF, O'Brien DA, Bhatnagar YM, Dym M (1977) Spermatogenic cells of the prepuberal mouse. Isolation and morphological characterization. J Cell Biol 74(1):68–85

Bestor TH (2003) Cytosine methylation mediates sexual conflict. Trends Genet 19(4):185–190. doi:10.1016/S0168-9525(03)00049-0

Beyret E, Lin H (2011) Pinpointing the expression of piRNAs and function of the PIWI protein subfamily during spermatogenesis in the mouse. Dev Biol 355(2):215–226. doi:10.1016/j.ydbio.2011.04.021

Bostick M, Kim JK, Esteve PO, Clark A, Pradhan S, Jacobsen SE (2007) UHRF1 plays a role in maintaining DNA methylation in mammalian cells. Science 317(5845):1760–1764. doi:10.1126/science.1147939

Bourc'his D, Bestor TH (2004) Meiotic catastrophe and retrotransposon reactivation in male germ cells lacking Dnmt3L. Nature 431(7004):96–99. doi:10.1038/nature02886

Bourc'his D, Xu GL, Lin CS, Bollman B, Bestor TH (2001) Dnmt3L and the establishment of maternal genomic imprints. Science 294(5551):2536–2539. doi:10.1126/science.1065848

Branciforte D, Martin SL (1994) Developmental and cell type specificity of LINE-1 expression in mouse testis: implications for transposition. Mol Cell Biol 14(4):2584–2592

Brouha B, Schustak J, Badge RM, Lutz-Prigge S, Farley AH, Moran JV, Kazazian HH Jr (2003) Hot L1s account for the bulk of retrotransposition in the human population. Proc Natl Acad Sci U S A 100(9):5280–5285. doi:10.1073/pnas.0831042100

Bulut-Karslioglu A, De La Rosa-Velazquez IA, Ramirez F, Barenboim M, Onishi-Seebacher M, Arand J, Galan C, Winter GE, Engist B, Gerle B, O'Sullivan RJ, Martens JH, Walter J, Manke T, Lachner M, Jenuwein T (2014) Suv39h-dependent H3K9me3 marks intact retrotransposons and silences LINE elements in mouse embryonic stem cells. Mol Cell 55(2):277–290. doi:10.1016/j.molcel.2014.05.029

Carmell MA, Girard A, van de Kant HJ, Bourc'his D, Bestor TH, de Rooij DG, Hannon GJ (2007) MIWI2 is essential for spermatogenesis and repression of transposons in the mouse male germline. Dev Cell 12(4):503–514. doi:10.1016/j.devcel.2007.03.001

Castro-Diaz N, Ecco G, Coluccio A, Kapopoulou A, Yazdanpanah B, Friedli M, Duc J, Jang SM, Turelli P, Trono D (2014) Evolutionally dynamic L1 regulation in embryonic stem cells. Genes Dev 28(13):1397–1409. doi:10.1101/gad.241661.114

Chow JC, Ciaudo C, Fazzari MJ, Mise N, Servant N, Glass JL, Attreed M, Avner P, Wutz A, Barillot E, Greally JM, Voinnet O, Heard E (2010) LINE-1 activity in facultative heterochromatin formation during X chromosome inactivation. Cell 141(6):956–969. doi:10.1016/j.cell.2010.04.042

Collins T, Stone JR, Williams AJ (2001) All in the family: the BTB/POZ, KRAB, and SCAN domains. Mol Cell Biol 21(11):3609–3615. doi:10.1128/MCB.21.11.3609-3615.2001

Costoya JA, Hobbs RM, Barna M, Cattoretti G, Manova K, Sukhwani M, Orwig KE, Wolgemuth DJ, Pandolfi PP (2004) Essential role of Plzf in maintenance of spermatogonial stem cells. Nat Genet 36(6):653–659. doi:10.1038/ng1367

Coufal NG, Garcia-Perez JL, Peng GE, Yeo GW, Mu Y, Lovci MT, Morell M, O'Shea KS, Moran JV, Gage FH (2009) L1 retrotransposition in human neural progenitor cells. Nature 460(7259):1127–1131. doi:10.1038/nature08248

Crichton JH, Playfoot CJ, Adams IR (2014) The role of chromatin modifications in progression through mouse meiotic prophase. J Genet Genomics 41(3):97–106. doi:10.1016/j.jgg.2014.01.003

Czech B, Hannon GJ (2016) One loop to rule them all: the ping-pong cycle and piRNA-guided silencing. Trends Biochem Sci 41(4):324–337. doi:10.1016/j.tibs.2015.12.008

Dai L, LaCava J, Taylor MS, Boeke JD (2014) Expression and detection of LINE-1 ORF-encoded proteins. Mob Genet elements 4:e29319. doi:10.4161/mge.29319

De Fazio S, Bartonicek N, Di Giacomo M, Abreu-Goodger C, Sankar A, Funaya C, Antony C, Moreira PN, Enright AJ, O'Carroll D (2011) The endonuclease activity of Mili fuels piRNA amplification that silences LINE1 elements. Nature 480(7376):259–263. doi:10.1038/nature10547

de Rooij DG, Russell LD (2000) All you wanted to know about spermatogonia but were afraid to ask. J Androl 21(6):776–798

de Souza FS, Franchini LF, Rubinstein M (2013) Exaptation of transposable elements into novel cis-regulatory elements: is the evidence always strong? Mol Biol Evol 30(6):1239–1251. doi:10.1093/molbev/mst045

Deguchi K, Nagamatsu G, Miyachi H, Kato Y, Morita S, Kimura H, Kitano S, Hatada I, Saga Y, Tachibana M, Shinkai Y (2013) Posttranscriptional regulation of histone lysine methyltransferase GLP in embryonic male mouse germ cells. Biol Reprod 88(2):36. doi:10.1095/biolreprod.112.103572

Deng W, Lin H (2002) miwi, a murine homolog of piwi, encodes a cytoplasmic protein essential for spermatogenesis. Dev Cell 2(6):819–830

Di Giacomo M, Comazzetto S, Saini H, De Fazio S, Carrieri C, Morgan M, Vasiliauskaite L, Benes V, Enright AJ, O'Carroll D (2013) Multiple epigenetic mechanisms and the piRNA pathway enforce LINE1 silencing during adult spermatogenesis. Mol Cell 50(4):601–608. doi:10.1016/j.molcel.2013.04.026

Di Giacomo M, Comazzetto S, Sampath SC, Sampath SC, O'Carroll D (2014) G9a co-suppresses LINE1 elements in spermatogonia. Epigenetics Chromatin 7:24. doi:10.1186/1756-8935-7-24

Donertas D, Sienski G, Brennecke J (2013) Drosophila Gtsf1 is an essential component of the Piwi-mediated transcriptional silencing complex. Genes Dev 27(15):1693–1705. doi:10.1101/gad.221150.113

Drost JB, Lee WR (1995) Biological basis of germline mutation: comparisons of spontaneous germline mutation rates among drosophila, mouse, and human. Environ Mol Mutagen 25(Suppl 26):48–64

Drumond AL, Meistrich ML, Chiarini-Garcia H (2011) Spermatogonial morphology and kinetics during testis development in mice: a high-resolution light microscopy approach. Reproduction 142(1):145–155. doi:10.1530/REP-10-0431

Ergun S, Buschmann C, Heukeshoven J, Dammann K, Schnieders F, Lauke H, Chalajour F, Kilic N, Stratling WH, Schumann GG (2004) Cell type-specific expression of LINE-1 open reading frames 1 and 2 in fetal and adult human tissues. J Biol Chem 279(26):27753–27763

Esnault C, Maestre J, Heidmann T (2000) Human LINE retrotransposons generate processed pseudogenes. Nat Genet 24(4):363–367. doi:10.1038/74184

Feng Q, Moran JV, Kazazian HH Jr, Boeke JD (1996) Human L1 retrotransposon encodes a conserved endonuclease required for retrotransposition. Cell 87(5):905–916. doi:10.1016/S0092-8674(00)81997-2

Frost RJ, Hamra FK, Richardson JA, Qi X, Bassel-Duby R, Olson EN (2010) MOV10L1 is necessary for protection of spermatocytes against retrotransposons by Piwi-interacting RNAs. Proc Natl Acad Sci U S A 107(26):11847–11852. doi:10.1073/pnas.1007158107

Furano AV (2000) The biological properties and evolutionary dynamics of mammalian LINE-1 retrotransposons. Prog Nucleic Acid Res Mol Biol 64:255–294

Gasior SL, Wakeman TP, Xu B, Deininger PL (2006) The human LINE-1 retrotransposon creates DNA double-strand breaks. J Mol Biol 357(5):1383–1393. doi:10.1016/j.jmb.2006.01.089

Girard A, Sachidanandam R, Hannon GJ, Carmell MA (2006) A germline-specific class of small RNAs binds mammalian Piwi proteins. Nature 442(7099):199–202. doi:10.1038/nature04917

Gkountela S, Zhang KX, Shafiq TA, Liao WW, Hargan-Calvopina J, Chen PY, Clark AT (2015) DNA demethylation dynamics in the human prenatal germline. Cell 161(6):1425–1436. doi:10.1016/j.cell.2015.05.012

Goodier JL, Kazazian HH Jr (2008) Retrotransposons revisited: the restraint and rehabilitation of parasites. Cell 135(1):23–35. doi:10.1016/j.cell.2008.09.022

Goodier JL, Ostertag EM, Du K, Kazazian HH Jr (2001) A novel active L1 retrotransposon subfamily in the mouse. Genome Res 11(10):1677–1685

Grandi FC, Rosser JM, Newkirk SJ, Yin J, Jiang X, Xing Z, Whitmore L, Bashir S, Ivics Z, Izsvak Z, Ye P, Yu YE, An W (2015) Retrotransposition creates sloping shores: a graded influence of hypomethylated CpG islands on flanking CpG sites. Genome Res 25(8):1135–1146. doi:10.1101/gr.185132.114

Grivna ST, Beyret E, Wang Z, Lin H (2006a) A novel class of small RNAs in mouse spermatogenic cells. Genes Dev 20(13):1709–1714. doi:10.1101/gad.1434406

Grivna ST, Pyhtila B, Lin H (2006b) MIWI associates with translational machinery and PIWI-interacting RNAs (piRNAs) in regulating spermatogenesis. Proc Natl Acad Sci U S A 103(36):13415–13420. doi:10.1073/pnas.0605506103

Gruppi CM, Wolgemuth DJ (1993) HSP86 and HSP84 exhibit cellular specificity of expression and co-precipitate with an HSP70 family member in the murine testis. Dev Genet 14(2):119–126. doi:10.1002/dvg.1020140206

Guo F, Yan L, Guo H, Li L, Hu B, Zhao Y, Yong J, Hu Y, Wang X, Wei Y, Wang W, Li R, Yan J, Zhi X, Zhang Y, Jin H, Zhang W, Hou Y, Zhu P, Li J, Zhang L, Liu S, Ren Y, Zhu X, Wen L, Gao YQ, Tang F, Qiao J (2015) The Transcriptome and DNA Methylome Landscapes of Human Primordial Germ Cells. Cell 161(6):1437–1452. doi:10.1016/j.cell.2015.05.015

Ha H, Song J, Wang S, Kapusta A, Feschotte C, Chen KC, Xing J (2014) A comprehensive analysis of piRNAs from adult human testis and their relationship with genes and mobile elements. BMC Genomics 15:545. doi:10.1186/1471-2164-15-545

Hadziselimovic F, Hadziselimovic NO, Demougin P, Krey G, Oakeley E (2015) Piwi-pathway alteration induces LINE-1 transposon derepression and infertility development in cryptorchidism. Sex Dev 9(2):98–104. doi:10.1159/000375351

Hajkova P, Erhardt S, Lane N, Haaf T, El-Maarri O, Reik W, Walter J, Surani MA (2002) Epigenetic reprogramming in mouse primordial germ cells. Mech Dev 117(1-2):15–23, doi:S0925477302001818

Hancks DC, Kazazian HH Jr (2016) Roles for retrotransposon insertions in human disease. Mob DNA 7:9. doi:10.1186/s13100-016-0065-9

Hata K, Kusumi M, Yokomine T, Li E, Sasaki H (2006) Meiotic and epigenetic aberrations in Dnmt3L-deficient male germ cells. Mol Reprod Dev 73(1):116–122. doi:10.1002/mrd.20387

Hata K, Okano M, Lei H, Li E (2002) Dnmt3L cooperates with the Dnmt3 family of de novo DNA methyltransferases to establish maternal imprints in mice. Development 129(8):1983–1993

Howlett SK, Reik W (1991) Methylation levels of maternal and paternal genomes during preimplantation development. Development 113(1):119–127

Ichiyanagi T, Ichiyanagi K, Ogawa A, Kuramochi-Miyagawa S, Nakano T, Chuma S, Sasaki H, Udono H (2014) HSP90alpha plays an important role in piRNA biogenesis and retrotransposon repression in mouse. Nucleic Acids Res 42(19):11903–11911. doi:10.1093/nar/gku881

Iwasaki YW, Siomi MC, Siomi H (2015) PIWI-interacting RNA: its biogenesis and functions. Annu Rev Biochem 84:405–433. doi:10.1146/annurev-biochem-060614-034258

Jacobs FM, Greenberg D, Nguyen N, Haeussler M, Ewing AD, Katzman S, Paten B, Salama SR, Haussler D (2014) An evolutionary arms race between KRAB zinc-finger genes ZNF91/93 and SVA/L1 retrotransposons. Nature 516(7530):242–245. doi:10.1038/nature13760

Kaneda M, Okano M, Hata K, Sado T, Tsujimoto N, Li E, Sasaki H (2004) Essential role for de novo DNA methyltransferase Dnmt3a in paternal and maternal imprinting. Nature 429(6994):900–903. doi:10.1038/nature02633

Kano H, Godoy I, Courtney C, Vetter MR, Gerton GL, Ostertag EM, Kazazian HH Jr (2009) L1 retrotransposition occurs mainly in embryogenesis and creates somatic mosaicism. Genes Dev 23(11):1303–1312. doi:10.1101/gad.1803909

Kato Y, Kaneda M, Hata K, Kumaki K, Hisano M, Kohara Y, Okano M, Li E, Nozaki M, Sasaki H (2007) Role of the Dnmt3 family in de novo methylation of imprinted and repetitive sequences during male germ cell development in the mouse. Hum Mol Genet 16(19):2272–2280. doi:10.1093/hmg/ddm179

Keam SP, Young PE, McCorkindale AL, Dang TH, Clancy JL, Humphreys DT, Preiss T, Hutvagner G, Martin DI, Cropley JE, Suter CM (2014) The human Piwi protein Hiwi2 associates with tRNA-derived piRNAs in somatic cells. Nucleic Acids Res 42(14):8984–8995. doi:10.1093/nar/gku620

Khan H, Smit A, Boissinot S (2006) Molecular evolution and tempo of amplification of human LINE-1 retrotransposons since the origin of primates. Genome Res 16(1):78–87. doi:10.1101/gr.4001406

Kim S, Gunesdogan U, Zylicz JJ, Hackett JA, Cougot D, Bao S, Lee C, Dietmann S, Allen GE, Sengupta R, Surani MA (2014) PRMT5 protects genomic integrity during global DNA demethylation in primordial germ cells and preimplantation embryos. Mol Cell 56(4):564–579. doi:10.1016/j.molcel.2014.10.003

Kirino Y, Kim N, de Planell-Saguer M, Khandros E, Chiorean S, Klein PS, Rigoutsos I, Jongens TA, Mourelatos Z (2009) Arginine methylation of Piwi proteins catalysed by dPRMT5 is required for Ago3 and Aub stability. Nat Cell Biol 11(5):652–658. doi:10.1038/ncb1872

Kolosha VO, Martin SL (2003) High-affinity, non-sequence-specific RNA binding by the open reading frame 1 (ORF1) protein from long interspersed nuclear element 1 (LINE-1). J Biol Chem 278(10):8112–8117. doi:10.1074/jbc.M210487200

Kubo S, Seleme MC, Soifer HS, Perez JL, Moran JV, Kazazian HH Jr, Kasahara N (2006) L1 retrotransposition in nondividing and primary human somatic cells. Proc Natl Acad Sci U S A 103(21):8036–8041. doi:10.1073/pnas.0601954103

Kuramochi-Miyagawa S, Kimura T, Ijiri TW, Isobe T, Asada N, Fujita Y, Ikawa M, Iwai N, Okabe M, Deng W, Lin H, Matsuda Y, Nakano T (2004) Mili, a mammalian member of piwi family gene, is essential for spermatogenesis. Development 131(4):839–849. doi:10.1242/dev.00973

Kuramochi-Miyagawa S, Kimura T, Yomogida K, Kuroiwa A, Tadokoro Y, Fujita Y, Sato M, Matsuda Y, Nakano T (2001) Two mouse piwi-related genes: miwi and mili. Mech Dev 108(1-2):121–133

Kuramochi-Miyagawa S, Watanabe T, Gotoh K, Totoki Y, Toyoda A, Ikawa M, Asada N, Kojima K, Yamaguchi Y, Ijiri TW, Hata K, Li E, Matsuda Y, Kimura T, Okabe M, Sakaki Y, Sasaki H, Nakano T (2008) DNA methylation of retrotransposon genes is regulated by Piwi family members MILI and MIWI2 in murine fetal testes. Genes Dev 22(7):908–917. doi:10.1101/gad.1640708

La Salle S, Oakes CC, Neaga OR, Bourc'his D, Bestor TH, Trasler JM (2007) Loss of spermatogonia and wide-spread DNA methylation defects in newborn male mice deficient in DNMT3L. BMC Dev Biol 7:104. doi:10.1186/1471-213X-7-104

Lander ES, Linton LM, Birren B, Nusbaum C, Zody MC, Baldwin J, Devon K, Dewar K, Doyle M, FitzHugh W, Funke R, Gage D, Harris K, Heaford A, Howland J, Kann L, Lehoczky J, LeVine R, McEwan P, McKernan K, Meldrim J, Mesirov JP, Miranda C, Morris W, Naylor J, Raymond C, Rosetti M, Santos R, Sheridan A, Sougnez C, Stange-Thomann N, Stojanovic N, Subramanian A, Wyman D, Rogers J, Sulston J, Ainscough R, Beck S, Bentley D, Burton J, Clee C, Carter N, Coulson A, Deadman R, Deloukas P, Dunham A, Dunham I, Durbin R, French L, Grafham D, Gregory S, Hubbard T, Humphray S, Hunt A, Jones M, Lloyd C, McMurray A, Matthews L, Mercer S, Milne S, Mullikin JC, Mungall A, Plumb R, Ross M, Shownkeen R, Sims S, Waterston RH, Wilson RK, Hillier LW, McPherson JD, Marra MA,

Mardis ER, Fulton LA, Chinwalla AT, Pepin KH, Gish WR, Chissoe SL, Wendl MC, Delehaunty KD, Miner TL, Delehaunty A, Kramer JB, Cook LL, Fulton RS, Johnson DL, Minx PJ, Clifton SW, Hawkins T, Branscomb E, Predki P, Richardson P, Wenning S, Slezak T, Doggett N, Cheng JF, Olsen A, Lucas S, Elkin C, Uberbacher E, Frazier M, Gibbs RA, Muzny DM, Scherer SE, Bouck JB, Sodergren EJ, Worley KC, Rives CM, Gorrell JH, Metzker ML, Naylor SL, Kucherlapati RS, Nelson DL, Weinstock GM, Sakaki Y, Fujiyama A, Hattori M, Yada T, Toyoda A, Itoh T, Kawagoe C, Watanabe H, Totoki Y, Taylor T, Weissenbach J, Heilig R, Saurin W, Artiguenave F, Brottier P, Bruls T, Pelletier E, Robert C, Wincker P, Smith DR, Doucette-Stamm L, Rubenfield M, Weinstock K, Lee HM, Dubois J, Rosenthal A, Platzer M, Nyakatura G, Taudien S, Rump A, Yang H, Yu J, Wang J, Huang G, Gu J, Hood L, Rowen L, Madan A, Qin S, Davis RW, Federspiel NA, Abola AP, Proctor MJ, Myers RM, Schmutz J, Dickson M, Grimwood J, Cox DR, Olson MV, Kaul R, Shimizu N, Kawasaki K, Minoshima S, Evans GA, Athanasiou M, Schultz R, Roe BA, Chen F, Pan H, Ramser J, Lehrach H, Reinhardt R, McCombie WR, de la Bastide M, Dedhia N, Blocker H, Hornischer K, Nordsiek G, Agarwala R, Aravind L, Bailey JA, Bateman A, Batzoglou S, Birney E, Bork P, Brown DG, Burge CB, Cerutti L, Chen HC, Church D, Clamp M, Copley RR, Doerks T, Eddy SR, Eichler EE, Furey TS, Galagan J, Gilbert JG, Harmon C, Hayashizaki Y, Haussler D, Hermjakob H, Hokamp K, Jang W, Johnson LS, Jones TA, Kasif S, Kaspryzk A, Kennedy S, Kent WJ, Kitts P, Koonin EV, Korf I, Kulp D, Lancet D, Lowe TM, McLysaght A, Mikkelsen T, Moran JV, Mulder N, Pollara VJ, Ponting CP, Schuler G, Schultz J, Slater G, Smit AF, Stupka E, Szustakowski J, Thierry-Mieg D, Thierry-Mieg J, Wagner L, Wallis J, Wheeler R, Williams A, Wolf YI, Wolfe KH, Yang SP, Yeh RF, Collins F, Guyer MS, Peterson J, Felsenfeld A, Wetterstrand KA, Patrinos A, Morgan MJ, de Jong P, Catanese JJ, Osoegawa K, Shizuya H, Choi S, Chen YJ (2001) Initial sequencing and analysis of the human genome. Nature 409(6822):860–921. doi:10.1038/35057062

Lane N, Dean W, Erhardt S, Hajkova P, Surani A, Walter J, Reik W (2003) Resistance of IAPs to methylation reprogramming may provide a mechanism for epigenetic inheritance in the mouse. Genesis 35(2):88–93. doi:10.1002/gene.10168

Lau NC, Seto AG, Kim J, Kuramochi-Miyagawa S, Nakano T, Bartel DP, Kingston RE (2006) Characterization of the piRNA complex from rat testes. Science 313(5785):363–367. doi:10.1126/science.1130164

Lees-Murdock DJ, De Felici M, Walsh CP (2003) Methylation dynamics of repetitive DNA elements in the mouse germ cell lineage. Genomics 82(2):230–237, doi:S0888754303001058

Leung DC, Lorincz MC (2012) Silencing of endogenous retroviruses: when and why do histone marks predominate? Trends Biochem Sci 37(4):127–133. doi:10.1016/j.tibs.2011.11.006

Li E, Bestor TH, Jaenisch R (1992) Targeted mutation of the DNA methyltransferase gene results in embryonic lethality. Cell 69(6):915–926

Li Z, Dai H, Martos SN, Xu B, Gao Y, Li T, Zhu G, Schones DE, Wang Z (2015a) Distinct roles of DNMT1-dependent and DNMT1-independent methylation patterns in the genome of mouse embryonic stem cells. Genome Biol 16:115. doi:10.1186/s13059-015-0685-2

Li Z, Yu J, Hosohama L, Nee K, Gkountela S, Chaudhari S, Cass AA, Xiao X, Clark AT (2015b) The Sm protein methyltransferase PRMT5 is not required for primordial germ cell specification in mice. EMBO J 34(6):748–758. doi:10.15252/embj.201489319

Lim AK, Lorthongpanich C, Chew TG, Tan CW, Shue YT, Balu S, Gounko N, Kuramochi-Miyagawa S, Matzuk MM, Chuma S, Messerschmidt DM, Solter D, Knowles BB (2013) The nuage mediates retrotransposon silencing in mouse primordial ovarian follicles. Development 140(18):3819–3825. doi:10.1242/dev.099184

Liu S, Brind'Amour J, Karimi MM, Shirane K, Bogutz A, Lefebvre L, Sasaki H, Shinkai Y, Lorincz MC (2014) Setdb1 is required for germline development and silencing of H3K9me3-marked endogenous retroviruses in primordial germ cells. Genes Dev 28(18):2041–2055. doi:10.1101/gad.244848.114

Macfarlan TS, Gifford WD, Agarwal S, Driscoll S, Lettieri K, Wang J, Andrews SE, Franco L, Rosenfeld MG, Ren B, Pfaff SL (2011) Endogenous retroviruses and neighboring genes are coordinately repressed by LSD1/KDM1A. Genes Dev 25(6):594–607. doi:10.1101/gad.2008511

Magiorkinis G, Blanco-Melo D, Belshaw R (2015) The decline of human endogenous retroviruses: extinction and survival. Retrovirology 12:8. doi:10.1186/s12977-015-0136-x

Maksakova IA, Romanish MT, Gagnier L, Dunn CA, van de Lagemaat LN, Mager DL (2006) Retroviral elements and their hosts: insertional mutagenesis in the mouse germ line. PLoS Genet 2(1):e2. doi:10.1371/journal.pgen.0020002

Malki S, van der Heijden GW, O'Donnell KA, Martin SL, Bortvin A (2014) A role for retrotransposon LINE-1 in fetal oocyte attrition in mice. Dev Cell 29(5):521–533. doi:10.1016/j.devcel.2014.04.027

Manakov SA, Pezic D, Marinov GK, Pastor WA, Sachidanandam R, Aravin AA (2015) MIWI2 and MILI have differential effects on piRNA biogenesis and DNA methylation. Cell Rep 12(8):1234–1243. doi:10.1016/j.celrep.2015.07.036

Martin SL, Cruceanu M, Branciforte D, Wai-Lun Li P, Kwok SC, Hodges RS, Williams MC (2005) LINE-1 retrotransposition requires the nucleic acid chaperone activity of the ORF1 protein. J Mol Biol 348(3):549–561. doi:10.1016/j.jmb.2005.03.003

Mathias SL, Scott AF, Kazazian HH Jr, Boeke JD, Gabriel A (1991) Reverse transcriptase encoded by a human transposable element. Science 254(5039):1808–1810

McGee EA, Hsueh AJ (2000) Initial and cyclic recruitment of ovarian follicles. Endocr Rev 21(2):200–214. doi:10.1210/edrv.21.2.0394

McLaren A (2001) Mammalian germ cells: birth, sex, and immortality. Cell Struct Funct 26(3):119–122

Medstrand P, van de Lagemaat LN, Mager DL (2002) Retroelement distributions in the human genome: variations associated with age and proximity to genes. Genome Res 12(10):1483–1495. doi:10.1101/gr.388902

Messerschmidt DM, Knowles BB, Solter D (2014) DNA methylation dynamics during epigenetic reprogramming in the germline and preimplantation embryos. Genes Dev 28(8):812–828. doi:10.1101/gad.234294.113

Molaro A, Falciatori I, Hodges E, Aravin AA, Marran K, Rafii S, McCombie WR, Smith AD, Hannon GJ (2014) Two waves of de novo methylation during mouse germ cell development. Genes Dev 28(14):1544–1549. doi:10.1101/gad.244350.114

Moran JV, DeBerardinis RJ, Kazazian HH Jr (1999) Exon shuffling by L1 retrotransposition. Science 283(5407):1530–1534

Moran JV, Holmes SE, Naas TP, DeBerardinis RJ, Boeke JD, Kazazian HH Jr (1996) High frequency retrotransposition in cultured mammalian cells. Cell 87(5):917–927. doi:10.1016/S0092-8674(00)81998-4

Muerdter F, Guzzardo PM, Gillis J, Luo Y, Yu Y, Chen C, Fekete R, Hannon GJ (2013) A genome-wide RNAi screen draws a genetic framework for transposon control and primary piRNA biogenesis in Drosophila. Mol Cell 50(5):736–748. doi:10.1016/j.molcel.2013.04.006

Muotri AR, Marchetto MC, Coufal NG, Oefner R, Yeo G, Nakashima K, Gage FH (2010) L1 retrotransposition in neurons is modulated by MeCP2. Nature 468(7322):443–446. doi:10.1038/nature09544

Nebel BR, Amarose AP, Hacket EM (1961) Calendar of gametogenic development in the prepuberal male mouse. Science 134:832–833

Nellaker C, Keane TM, Yalcin B, Wong K, Agam A, Belgard TG, Flint J, Adams DJ, Frankel WN, Ponting CP (2012) The genomic landscape shaped by selection on transposable elements across 18 mouse strains. Genome Biol 13(6):R45. doi:10.1186/gb-2012-13-6-r45

O'Donnell KA, An W, Schrum CT, Wheelan SJ, Boeke JD (2013) Controlled insertional mutagenesis using a LINE-1 (ORFeus) gene-trap mouse model. Proc Natl Acad Sci U S A 110(29):E2706–E2713. doi:10.1073/pnas.1302504110

Oakberg EF (1957) Duration of spermatogenesis in the mouse. Nature 180(4595):1137–1138

Oakes CC, La Salle S, Smiraglia DJ, Robaire B, Trasler JM (2007) Developmental acquisition of genome-wide DNA methylation occurs prior to meiosis in male germ cells. Dev Biol 307(2):368–379. doi:10.1016/j.ydbio.2007.05.002

Ohno R, Nakayama M, Naruse C, Okashita N, Takano O, Tachibana M, Asano M, Saitou M, Seki Y (2013) A replication-dependent passive mechanism modulates DNA demethylation in mouse primordial germ cells. Development 140(14):2892–2903. doi:10.1242/dev.093229

Okano M, Bell DW, Haber DA, Li E (1999) DNA methyltransferases Dnmt3a and Dnmt3b are essential for de novo methylation and mammalian development. Cell 99(3):247–257

Pepling ME (2006) From primordial germ cell to primordial follicle: mammalian female germ cell development. Genesis 44(12):622–632. doi:10.1002/dvg.20258

Pepling ME, Spradling AC (2001) Mouse ovarian germ cell cysts undergo programmed breakdown to form primordial follicles. Dev Biol 234(2):339–351. doi:10.1006/dbio.2001.0269

Peters AH, O'Carroll D, Scherthan H, Mechtler K, Sauer S, Schofer C, Weipoltshammer K, Pagani M, Lachner M, Kohlmaier A, Opravil S, Doyle M, Sibilia M, Jenuwein T (2001) Loss of the Suv39h histone methyltransferases impairs mammalian heterochromatin and genome stability. Cell 107(3):323–337, doi:S0092-8674(01)00542-6

Peters H (1969) The development of the mouse ovary from birth to maturity. Acta Endocrinol 62(1):98–116

Peters H (1970) Migration of gonocytes into the mammalian gonad and their differentiation. Philos Trans R Soc Lond B Biol Sci 259(828):91–101

Pezic D, Manakov SA, Sachidanandam R, Aravin AA (2014) piRNA pathway targets active LINE1 elements to establish the repressive H3K9me3 mark in germ cells. Genes Dev 28(13):1410–1428. doi:10.1101/gad.240895.114

Puszyk W, Down T, Grimwade D, Chomienne C, Oakey RJ, Solomon E, Guidez F (2013) The epigenetic regulator PLZF represses L1 retrotransposition in germ and progenitor cells. EMBO J 32(13):1941–1952. doi:10.1038/emboj.2013.118

Rebollo R, Karimi MM, Bilenky M, Gagnier L, Miceli-Royer K, Zhang Y, Goyal P, Keane TM, Jones S, Hirst M, Lorincz MC, Mager DL (2011) Retrotransposon-Induced Heterochromatin Spreading in the Mouse Revealed by Insertional Polymorphisms. PLoS Genet 7(9):e1002301

Rebollo R, Romanish MT, Mager DL (2012) Transposable elements: an abundant and natural source of regulatory sequences for host genes. Annu Rev Genet 46:21–42. doi:10.1146/annurev-genet-110711-155621

Reuter M, Berninger P, Chuma S, Shah H, Hosokawa M, Funaya C, Antony C, Sachidanandam R, Pillai RS (2011) Miwi catalysis is required for piRNA amplification-independent LINE1 transposon silencing. Nature 480(7376):264–267. doi:10.1038/nature10672

Roovers EF, Rosenkranz D, Mahdipour M, Han CT, He N, Chuva de Sousa Lopes SM, Van der Westerlaken LA, Zischler H, Butter F, Roelen BA, Ketting RF (2015) Piwi proteins and piRNAs in mammalian oocytes and early embryos. Cell Rep 10(12):2069–2082. doi:10.1016/j.celrep.2015.02.062

Rosser JM, An W (2012) L1 expression and regulation in humans and rodents. Front Biosci (Elite Ed) 4:2203–2225

Rowe HM, Jakobsson J, Mesnard D, Rougemont J, Reynard S, Aktas T, Maillard PV, Layard-Liesching H, Verp S, Marquis J, Spitz F, Constam DB, Trono D (2010) KAP1 controls endogenous retroviruses in embryonic stem cells. Nature 463(7278):237–240. doi:10.1038/nature08674

Saitou M, Yamaji M (2012) Primordial germ cells in mice. Cold Spring Harb Perspect Biol 4(11):pii: a008375. doi:10.1101/cshperspect.a008375

Sanford JP, Clark HJ, Chapman VM, Rossant J (1987) Differences in DNA methylation during oogenesis and spermatogenesis and their persistence during early embryogenesis in the mouse. Genes Dev 1(10):1039–1046

Sasaki T, Shiohama A, Minoshima S, Shimizu N (2003) Identification of eight members of the Argonaute family in the human genome. Genomics 82(3):323–330

Seisenberger S, Andrews S, Krueger F, Arand J, Walter J, Santos F, Popp C, Thienpont B, Dean W, Reik W (2012) The dynamics of genome-wide DNA methylation reprogramming in mouse primordial germ cells. Mol Cell 48(6):849–862. doi:10.1016/j.molcel.2012.11.001

Seki Y, Hayashi K, Itoh K, Mizugaki M, Saitou M, Matsui Y (2005) Extensive and orderly reprogramming of genome-wide chromatin modifications associated with specification and early development of germ cells in mice. Dev Biol 278(2):440–458. doi:10.1016/j.ydbio.2004.11.025

Sharif J, Muto M, Takebayashi S, Suetake I, Iwamatsu A, Endo TA, Shinga J, Mizutani-Koseki Y, Toyoda T, Okamura K, Tajima S, Mitsuya K, Okano M, Koseki H (2007) The SRA protein Np95 mediates epigenetic inheritance by recruiting Dnmt1 to methylated DNA. Nature 450(7171):908–912. doi:10.1038/nature06397

Shi X, Seluanov A, Gorbunova V (2007) Cell divisions are required for L1 retrotransposition. Mol Cell Biol 27(4):1264–1270. doi:10.1128/MCB.01888-06

Smit AF (1999) Interspersed repeats and other mementos of transposable elements in mammalian genomes. Curr Opin Genet Dev 9(6):657–663, doi:S0959-437X(99)00031-3

Smith ZD, Chan MM, Mikkelsen TS, Gu H, Gnirke A, Regev A, Meissner A (2012) A unique regulatory phase of DNA methylation in the early mammalian embryo. Nature 484(7394):339–344. doi:10.1038/nature10960

Soper SF, van der Heijden GW, Hardiman TC, Goodheart M, Martin SL, de Boer P, Bortvin A (2008) Mouse maelstrom, a component of nuage, is essential for spermatogenesis and transposon repression in meiosis. Dev Cell 15(2):285–297. doi:10.1016/j.devcel.2008.05.015

Szak ST, Pickeral OK, Makalowski W, Boguski MS, Landsman D, Boeke JD (2002) Molecular archeology of L1 insertions in the human genome. Genome Biol 3(10):research0052

Tachibana M, Nozaki M, Takeda N, Shinkai Y (2007) Functional dynamics of H3K9 methylation during meiotic prophase progression. EMBO J 26(14):3346–3359. doi:10.1038/sj.emboj.7601767

Tam OH, Aravin AA, Stein P, Girard A, Murchison EP, Cheloufi S, Hodges E, Anger M, Sachidanandam R, Schultz RM, Hannon GJ (2008) Pseudogene-derived small interfering RNAs regulate gene expression in mouse oocytes. Nature 453(7194):534–538. doi:10.1038/nature06904

Tang WW, Dietmann S, Irie N, Leitch HG, Floros VI, Bradshaw CR, Hackett JA, Chinnery PF, Surani MA (2015) A unique gene regulatory network resets the human germline epigenome for development. Cell 161(6):1453–1467. doi:10.1016/j.cell.2015.04.053

Trelogan SA, Martin SL (1995) Tightly regulated, developmentally specific expression of the first open reading frame from LINE-1 during mouse embryogenesis. Proc Natl Acad Sci U S A 92(5):1520–1524

Turelli P, Castro-Diaz N, Marzetta F, Kapopoulou A, Raclot C, Duc J, Tieng V, Quenneville S, Trono D (2014) Interplay of TRIM28 and DNA methylation in controlling human endogenous retroelements. Genome Res 24(8):1260–1270. doi:10.1101/gr.172833.114

Unhavaithaya Y, Hao Y, Beyret E, Yin H, Kuramochi-Miyagawa S, Nakano T, Lin H (2009) MILI, a PIWI-interacting RNA-binding protein, is required for germ line stem cell self-renewal and appears to positively regulate translation. J Biol Chem 284(10):6507–6519. doi:10.1074/jbc.M809104200

Vagin VV, Wohlschlegel J, Qu J, Jonsson Z, Huang X, Chuma S, Girard A, Sachidanandam R, Hannon GJ, Aravin AA (2009) Proteomic analysis of murine Piwi proteins reveals a role for arginine methylation in specifying interaction with Tudor family members. Genes Dev 23(15):1749–1762. doi:10.1101/gad.1814809

von Meyenn F, Reik W (2015) Forget the parents: epigenetic reprogramming in human germ cells. Cell 161(6):1248–1251. doi:10.1016/j.cell.2015.05.039

Vourekas A, Zheng K, Fu Q, Maragkakis M, Alexiou P, Ma J, Pillai RS, Mourelatos Z, Wang PJ (2015) The RNA helicase MOV10L1 binds piRNA precursors to initiate piRNA processing. Genes Dev 29(6):617–629. doi:10.1101/gad.254631.114

Wallace NA, Belancio VP, Deininger PL (2008) L1 mobile element expression causes multiple types of toxicity. Gene 419(1-2):75–81. doi:10.1016/j.gene.2008.04.013

Walsh CP, Chaillet JR, Bestor TH (1998) Transcription of IAP endogenous retroviruses is constrained by cytosine methylation. Nat Genet 20(2):116–117. doi:10.1038/2413

Wang Y, Zhu T, Li Q, Liu C, Han F, Chen M, Zhang L, Cui X, Qin Y, Bao S, Gao F (2015) Prmt5 is required for germ cell survival during spermatogenesis in mice. Sci Rep 5:11031. doi:10.1038/srep11031

Watanabe T, Takeda A, Tsukiyama T, Mise K, Okuno T, Sasaki H, Minami N, Imai H (2006) Identification and characterization of two novel classes of small RNAs in the mouse germline: retrotransposon-derived siRNAs in oocytes and germline small RNAs in testes. Genes Dev 20(13):1732–1743. doi:10.1101/gad.1425706

Watanabe T, Totoki Y, Toyoda A, Kaneda M, Kuramochi-Miyagawa S, Obata Y, Chiba H, Kohara Y, Kono T, Nakano T, Surani MA, Sakaki Y, Sasaki H (2008) Endogenous siRNAs from naturally formed dsRNAs regulate transcripts in mouse oocytes. Nature 453(7194):539–543. doi:10.1038/nature06908

Waterston RH, Lindblad-Toh K, Birney E, Rogers J, Abril JF, Agarwal P, Agarwala R, Ainscough R, Alexandersson M, An P, Antonarakis SE, Attwood J, Baertsch R, Bailey J, Barlow K, Beck S, Berry E, Birren B, Bloom T, Bork P, Botcherby M, Bray N, Brent MR, Brown DG, Brown SD, Bult C, Burton J, Butler J, Campbell RD, Carninci P, Cawley S, Chiaromonte F, Chinwalla AT, Church DM, Clamp M, Clee C, Collins FS, Cook LL, Copley RR, Coulson A, Couronne O, Cuff J, Curwen V, Cutts T, Daly M, David R, Davies J, Delehaunty KD, Deri J, Dermitzakis ET, Dewey C, Dickens NJ, Diekhans M, Dodge S, Dubchak I, Dunn DM, Eddy SR, Elnitski L, Emes RD, Eswara P, Eyras E, Felsenfeld A, Fewell GA, Flicek P, Foley K, Frankel WN, Fulton LA, Fulton RS, Furey TS, Gage D, Gibbs RA, Glusman G, Gnerre S, Goldman N, Goodstadt L, Grafham D, Graves TA, Green ED, Gregory S, Guigo R, Guyer M, Hardison RC, Haussler D, Hayashizaki Y, Hillier LW, Hinrichs A, Hlavina W, Holzer T, Hsu F, Hua A, Hubbard T, Hunt A, Jackson I, Jaffe DB, Johnson LS, Jones M, Jones TA, Joy A, Kamal M, Karlsson EK, Karolchik D, Kasprzyk A, Kawai J, Keibler E, Kells C, Kent WJ, Kirby A, Kolbe DL, Korf I, Kucherlapati RS, Kulbokas EJ, Kulp D, Landers T, Leger JP, Leonard S, Letunic I, Levine R, Li J, Li M, Lloyd C, Lucas S, Ma B, Maglott DR, Mardis ER, Matthews L, Mauceli E, Mayer JH, McCarthy M, McCombie WR, McLaren S, McLay K, McPherson JD, Meldrim J, Meredith B, Mesirov JP, Miller W, Miner TL, Mongin E, Montgomery KT, Morgan M, Mott R, Mullikin JC, Muzny DM, Nash WE, Nelson JO, Nhan MN, Nicol R, Ning Z, Nusbaum C, O'Connor MJ, Okazaki Y, Oliver K, Overton-Larty E, Pachter L, Parra G, Pepin KH, Peterson J, Pevzner P, Plumb R, Pohl CS, Poliakov A, Ponce TC, Ponting CP, Potter S, Quail M, Reymond A, Roe BA, Roskin KM, Rubin EM, Rust AG, Santos R, Sapojnikov V, Schultz B, Schultz J, Schwartz MS, Schwartz S, Scott C, Seaman S, Searle S, Sharpe T, Sheridan A, Shownkeen R, Sims S, Singer JB, Slater G, Smit A, Smith DR, Spencer B, Stabenau A, Stange-Thomann N, Sugnet C, Suyama M, Tesler G, Thompson J, Torrents D, Trevaskis E, Tromp J, Ucla C, Ureta-Vidal A, Vinson JP, Von Niederhausern AC, Wade CM, Wall M, Weber RJ, Weiss RB, Wendl MC, West AP, Wetterstrand K, Wheeler R, Whelan S, Wierzbowski J, Willey D, Williams S, Wilson RK, Winter E, Worley KC, Wyman D, Yang S, Yang SP, Zdobnov EM, Zody MC, Lander ES (2002) Initial sequencing and comparative analysis of the mouse genome. Nature 420(6915):520–562. doi:10.1038/nature01262

Webster KE, O'Bryan MK, Fletcher S, Crewther PE, Aapola U, Craig J, Harrison DK, Aung H, Phutikanit N, Lyle R, Meachem SJ, Antonarakis SE, de Kretser DM, Hedger MP, Peterson P, Carroll BJ, Scott HS (2005) Meiotic and epigenetic defects in Dnmt3L-knockout mouse spermatogenesis. Proc Natl Acad Sci U S A 102(11):4068–4073. doi:10.1073/pnas.0500702102

Western PS, Miles DC, van den Bergen JA, Burton M, Sinclair AH (2008) Dynamic regulation of mitotic arrest in fetal male germ cells. Stem Cells 26(2):339–347. doi:10.1634/stemcells.2007-0622

Williams Z, Morozov P, Mihailovic A, Lin C, Puvvula PK, Juranek S, Rosenwaks Z, Tuschl T (2015) Discovery and characterization of piRNAs in the human fetal ovary. Cell Rep 13(4):854–863. doi:10.1016/j.celrep.2015.09.030

Wolf G, Greenberg D, Macfarlan TS (2015) Spotting the enemy within: targeted silencing of foreign DNA in mammalian genomes by the Kruppel-associated box zinc finger protein family. Mob DNA 6:17. doi:10.1186/s13100-015-0050-8

Xie Y, Mates L, Ivics Z, Izsvak Z, Martin SL, An W (2013) Cell division promotes efficient retrotransposition in a stable L1 reporter cell line. Mob DNA 4(1):10. doi:10.1186/1759-8753-4-10

Yang Q, Hua J, Wang L, Xu B, Zhang H, Ye N, Zhang Z, Yu D, Cooke HJ, Zhang Y, Shi Q (2013) MicroRNA and piRNA profiles in normal human testis detected by next generation sequencing. PLoS One 8(6):e66809. doi:10.1371/journal.pone.0066809

Yoshida S, Sukeno M, Nakagawa T, Ohbo K, Nagamatsu G, Suda T, Nabeshima Y (2006) The first round of mouse spermatogenesis is a distinctive program that lacks the self-renewing spermatogonia stage. Development 133(8):1495–1505. doi:10.1242/dev.02316

Yoshimura T, Toyoda S, Kuramochi-Miyagawa S, Miyazaki T, Miyazaki S, Tashiro F, Yamato E, Nakano T, Miyazaki J (2009) Gtsf1/Cue110, a gene encoding a protein with two copies of a CHHC Zn-finger motif, is involved in spermatogenesis and retrotransposon suppression in murine testes. Dev Biol 335(1):216–227. doi:10.1016/j.ydbio.2009.09.003

Yu Y, Gu J, Jin Y, Luo Y, Preall JB, Ma J, Czech B, Hannon GJ (2015) Panoramix enforces piRNA-dependent cotranscriptional silencing. Science 350(6258):339–342. doi:10.1126/science.aab0700

Zamudio N, Barau J, Teissandier A, Walter M, Borsos M, Servant N, Bourc'his D (2015) DNA methylation restrains transposons from adopting a chromatin signature permissive for meiotic recombination. Genes Dev 29(12):1256–1270. doi:10.1101/gad.257840.114

Zamudio N, Bourc'his D (2010) Transposable elements in the mammalian germline: a comfortable niche or a deadly trap? Heredity 105(1):92–104. doi:10.1038/hdy.2010.53

Zemojtel T, Penzkofer T, Schultz J, Dandekar T, Badge R, Vingron M (2007) Exonization of active mouse L1s: a driver of transcriptome evolution? BMC Genomics 8:392. doi:10.1186/1471-2164-8-392

Zheng K, Wang PJ (2012) Blockade of pachytene piRNA biogenesis reveals a novel requirement for maintaining post-meiotic germline genome integrity. PLoS Genet 8(11), e1003038. doi:10.1371/journal.pgen.1003038

Zheng K, Xiol J, Reuter M, Eckardt S, Leu NA, McLaughlin KJ, Stark A, Sachidanandam R, Pillai RS, Wang PJ (2010) Mouse MOV10L1 associates with Piwi proteins and is an essential component of the Piwi-interacting RNA (piRNA) pathway. Proc Natl Acad Sci U S A 107(26):11841–11846. doi:10.1073/pnas.1003953107

Retrotransposon Contribution to Genomic Plasticity

Tara T. Doucet-O'Hare and Haig H. Kazazian Jr.

1 Introduction

L1 sequences comprise approximately 17 % of the genome; however, taking into account the mobilized *Alu*, SVA, and pseudogene insertions which occurred due to L1 protein expression, L1 is responsible for ~33 % of the sequence of the genome (Lander et al. 2001). Most of the transposable elements in the human genome are no longer identifiable as TEs due to the age of their sequences and their sequence divergence is so great that they cannot be assigned to a single TE family (Lander et al. 2001; Smit et al. 1995). The majority of the L1 elements in the human genome are not actively transposing because only one subfamily of each type of elements is active at a time (Smit et al. 1995; Khan et al. 2006). Subfamilies are determined by differences in sequence content which occur as mutations accrue over evolutionary time in the respective elements.

The emergence of new active subfamilies of elements is largely due to a situation referred to as an "arms race," a part of the "Red Queen" hypothesis (Van Valen 1973). The arms race serves the purpose of evading the host defenses and the race is waged between both retrotransposons and the mechanisms for controlling their activity, such as APOBEC3 proteins (Chen et al. 2006; Bogerd et al. 2006a; Muckenfuss et al. 2006; Stenglein and Harris 2006). To limit the activity of potentially mutagenic TE insertion events, eukaryotic cells have acquired several defense mechanisms to derail the process of L1 mobilization at various stages. The fossils of older inactive elements are a testament to the fact that continued evolution of transposable elements occurs nearly constantly although the rate of retrotransposition has not been constant (Khan et al. 2006).

T.T. Doucet-O'Hare • H.H. Kazazian Jr. (✉)
Department of Human Genetics, Johns Hopkins School of Medicine,
733 North Broadway, Baltimore, MD 21205, USA
e-mail: hkazazi1@jhmi.edu

G. Cristofari (ed.), *Human Retrotransposons in Health and Disease*,
DOI 10.1007/978-3-319-48344-3_3

One of the main tools utilized to limit retrotransposition is the methylation of retrotransposon promoters to restrict the transcription of the elements. Varied epigenetic modifiers are active in retrotransposition silencing, including the DNA methyltransferase-like protein Dnmt3L (Bourc'his and Bestor 2004; Kato et al. 2007). Dnmt3L is essential for Dnmt3A mediated methylation of retrotransposons in primordial germ cells (Bourc'his and Bestor 2004; Kato et al. 2007). In addition there is a mechanism for controlling retrotransposon activity in the germline via a piRNA-specific pathway which mediates genome-wide CpG methylation of TEs and restricts their activity (Bourc'his and Bestor 2004; Aravin et al. 2008; Woodcock et al. 1997; Yoder et al. 1997; Walsh et al. 1998). It has been demonstrated that L1 expression levels are inversely correlated with the methylation of the canonical promoter in the 5′ UTR of the element (Baba et al. 2010; Muotri et al. 2010; Perng et al. 2013), and numerous epigenetic modifiers contribute to establishing and maintaining the methylation status of L1 elements in the genome. To reinforce the suppression of TE activity, eukaryotic cells have also developed a Piwi-interacting RNA silencing pathway (Aravin et al. 2008; Aravin et al. 2006). Piwi-interacting RNAs, repeat-associated small interfering RNAs, and microRNAs all act to degrade retrotransposon transcripts via RNA interference (Heras et al. 2013; Macias et al. 2012; Sijen and Plasterk 2003; Watanabe et al. 2006; Chow et al. 2010; Levin and Moran 2011).

RNA interference is yet another mechanism by which the host can control TE expression using repeat-associated small interfering RNAs and microRNAs to degrade TE transcripts (Bogerd et al. 2006a; Heras et al. 2013; Macias et al. 2012; Sijen and Plasterk 2003; Chow et al. 2010; Levin and Moran 2011; Watanabe et al. 2008; Yang and Kazazian 2006; Czech et al. 2008; Malone and Hannon 2009; Slotkin et al. 2009). In addition to epigenetic and posttranscriptional regulation of L1, there are numerous host factor proteins which target the process by which L1s and other retrotransposons integrate into the genome (Pizarro and Cristofari 2016). Oftentimes, these host factors are also used to control retroviral infection in the host. MOV10, a host factor, is a potential RNA helicase and has the ability to restrict L1 retrotransposition in cell culture (Arjan-Odedra et al. 2012; Goodier et al. 2012; Goodier et al. 2013). MOV10 restricts TEs by associating with the key RNA-induced silencing complex component AGO2 and the L1 ribonucleoprotein particle (RNP). After association with AGO2 and the RNP, MOV10 is theorized to degrade or block the translation of L1 RNA (Liu et al. 2012). The exonuclease Trex1 metabolizes reverse transcribed retrotransposon DNA to stunt the process of retrotransposition (Stetson et al. 2008). In addition to Trex1 and MOV10, many studies have reported members of the APOBEC3 (A3) family of cytidine deaminases having a role in restricting the activity of L1 elements in cultured cells (Chen et al. 2006; Muckenfuss et al. 2006; Stenglein and Harris 2006; Bogerd et al. 2006b).

For TE families, especially L1, to continue to propagate in the human genome, elements must evolve to circumvent cellular host control mechanisms. In addition to the competition between the host and the TE, the TEs themselves must compete among other elements to retain their activity as well. The competition among L1 elements has been demonstrated in vivo in studies of rodent L1s (Casavant and Hardies 1994; Cabot et al. 1997) and further supported by evidence that L1 subfamilies seem to only coexist when the elements contain differing 5′ UTR sequences (Khan et al. 2006). In the mouse genome there are three subfamilies of active L1 elements and all have sequence differences in their 5′ UTR (Goodier et al. 2001).

The currently active L1 elements, L1PA1 and L1Hs, are the products of a long succession of L1 element evolution. The active elements can be subclassified based on certain "diagnostic" nucleotides. Elements in the *t*ranscribed group *a* subfamilies are referred to as "Ta" elements (Skowronski et al. 1988) and appeared approximately 2–3 million years ago (Boissinot et al. 2000; Myers et al. 2002) and have "ACA" at positions 5924-5926 and a "G" at 6010 relative to the active $L1_{RP}$ element. Older and inactive elements instead have a "GAG" and an "A" at the same positions. A family which is slightly older, yet still active, known as the pre-Ta elements, have the sequence "ACG" in place of "ACA."

Only a subset of the preTa and Ta L1 elements are capable of transposing to new locations in the genome and there are several requirements which must be met. Because many L1s are 5′ truncated upon insertion, a large number of the elements are unable to promote their own transcription because the 5′ promoter is absent from the insertion site (Grimaldi et al. 1984). The 5′ truncations may be due to poor processivity during the reverse transcriptase reaction or potentially because of degradation of the L1 RNA after translation and prior to the reverse transcription. During or after the insertion process, oftentimes L1 DNA accumulates mutations, frame shifts, or other inactivating alterations in either of the ORFs which causes them to be potentially inactive. It has been determined that approximately 80–100 L1 elements are active in a diploid genome and are therefore able to mobilize themselves and other TEs in *trans* (Brouha et al. 2003). Both copies of the L1 present in a diploid genome can generate new insertions (Brouha et al. 2003). Additionally, there are approximately 2000–3000 *Alu* elements and less than 100 SVA elements capable of retrotransposition in the genome (Mills et al. 2007; Bennett et al. 2008). There is allelic variability between L1 elements (Seleme et al. 2006; Lutz et al. 2003). In a mechanism not dissimilar to single nucleotide polymorphisms (SNPs) (Cheung et al. 2005), SNPs in active L1 elements can change the activity up to 16-fold (Lutz et al. 2003). A study comparing three active L1s across ~200 haploid genomes from six geographic regions resulted in 0–390 % activity when compared to a reference (Seleme et al. 2006). In this study comparing a trio of L1s, it was also noted that one new L1 allele (i.e., the same L1 with a different nucleotide sequence variant) existed for every 3–5 L1s sequenced in the study. Because the active elements are mobilizing to novel insertion sites in the genome, it is logical that individuals will differ with respect to the presence or absence of L1 insertions at loci throughout their genomes. These retrotransposons insertion polymorphisms (RIPs) segregate with populations in much the same way that SNPs do. Because many insertions derived from retrotransposons which are active occurred recently, they are polymorphic with regard to the presence or absence of the insertion in different human populations (Boissinot et al. 2000, 2004; Myers et al. 2002; Sheen et al. 2000; Batzer and Deininger 2002; Wang et al.

2006). In a study using fosmid end resequencing and mapping to identify 6000 nucleotide and greater structural variants, 68 non-reference L1 RIPs were identified (Beck et al. 2010). Of the 68 RIPs identified, 37 were found to be "hot" or highly active when assayed in cell culture using the retrotransposition assay (Moran et al. 1996). In addition to the new RIPs discovered, the authors noted that each of the six individuals studied possessed 2/6 insertions present in the reference genome that were classified as "hot" in previous work (Brouha et al. 2003). In addition to the two aforementioned "hot" L1 elements, each individual possessed between three and nine additional "hot" elements which were not in the reference genome (Beck et al. 2010). Altogether, these studies demonstrate L1 is active and mobilizing in the genome.

2 Retrotransposition Can Cause Disease

Spontaneous and inherited occurrences of disease-causing mutations have been observed in greater than 100 cases, including diseases such as hemophilia, cancer, and diabetes (Kazazian et al. 1988; Beck et al. 2011; Hancks and Kazazian 2012; Kaer and Speek 2013; Kutsche et al. 2002; Gu et al. 2007; Conley et al. 2005; Apoil et al. 2007; Claverie-Martin et al. 2003; Masson et al. 2013; Sukarova et al. 2001; Ganguly et al. 2003; Green et al. 2008; Vidaud et al. 1993; Wulff et al. 2000; Li et al. 2001; Zhang et al. 2000; den Hollander et al. 1999; Beauchamp et al. 2000; Kloor et al. 2004; Muratani et al. 1991; Janicic et al. 1995; Sobrier et al. 2005; Gallus et al. 2010; Anagnou et al. 1989; Halling et al. 1999; Su et al. 2000; Tucker et al. 2011; Manco et al. 2006; Chen et al. 2008; Abdelhak et al. 1997; Udaka et al. 2007; Bouchet et al. 2007; Oldridge et al. 1999; Bochukova et al. 2009; Tighe et al. 2002; Stoppa-Lyonnet et al. 1990; Mustajoki et al. 1999; Tappino et al. 2008; Miki et al. 1996; Teugels et al. 2005; Schollen et al. 2007; Peixoto et al. 2013; Wallace et al. 1991; Wimmer et al. 2011; Meischl et al. 2000; Brouha et al. 2002; Musova et al. 2006; Narita et al. 1993; Holmes et al. 1994; Mukherjee et al. 2004; Morisada et al. 2010; Kondo-Iida et al. 1999; Bernard et al. 2009; Lanikova et al. 2013; Miné et al. 2007; Kagawa et al. 2015; Nakamura et al. 2015; Makino et al. 2007; Arca et al. 2002; Takasu et al. 2007; Kobayashi et al. 1998; Taniguchi-Ikeda et al. 2011; Akman et al. 2010; Segal et al. 1999; Wang et al. 2001; Qian et al. 2015) (see Table 1). Previously, it has been suggested that ~0.27 % of human genetic disease is caused by TE insertions (Callinan et al. 2006). An example of a somatic insertion causing disease would be a processed pseudogene which inserted into the CYBB gene and caused primary immunodeficiency (de Boer et al. 2014). Yet another example of a somatic event causing disease occurred when SVA mediated deletions in the NF1 gene caused disease (Vogt et al. 2014). A somatic L1 insertion caused Choroideremia in a patient when it was inserted into the coding region of the gene (Van Den Hurk et al. 2003). There are many mechanisms by which TEs could disrupt normal gene expression or affect genome structure. TEs can disrupt genomic sequences when they insert; however, they can also cause deletions and rearrangements in the

Table 1 Disease-causing retrotransposon insertions

Element	Subfamily	Gene	Disease	Chr	Reference	Size (bp)	PolyA tail length (bp)
Alu	AluYb9	*ABCD1*	Adrenoleukodystrophy (ALD)	X	Kutsche et al. (2002)	98	20
Alu	AluYa5a2	*ATP7A*	Menkes disease	X	Gu et al. (2007)	282	89
Alu	AluY	*BTK*	X-linked agammaglobulinemia (XLA)	X	Hancks and Kazazian (2012)	N/A	N/A
Alu	AluY	*BTK*	X-linked agammaglobulinemia (XLA)	X	Conley et al. (2005)	281	74
Alu	AluYb8	*CD40LG*	Hyper-immunoglobulin M syndrome (HIGM)	X	Apoil et al. (2007)	292	8
Alu	AluYa5	*CLCN5*	Dent's disease	X	Hancks and Kazazian (2012), Claverie-Martin et al. (2003)	281	50
Alu	Alu	*CTRC*	Chronic pancreatitis	1	Masson et al. (2013)	31	11
Alu	AluYb8	*FVIII*	Hemophilia A	X	Sukarova et al. (2001)	290	47
Alu	AluYb9	*FVIII*	Hemophilia A	X	Ganguly et al. (2003)	288	37
Alu	AluYb8	*FVIII*	Hemophilia A	X	Green et al. (2008)	FL	N/A
Alu	AluYa5a2	*FIX*	Hemophilia B	X	Vidaud et al. (1993)	244	78
Alu	AluYa5a2	*FIX*	Hemophilia B	X	Wulff et al. (2000)	237	39
Alu	AluY	*FIX*	Hemophilia B	X	Li et al. (2001)	279	40
Alu	AluYc1	*GK*	Glycerol kinase deficiency (GKD)	X	Zhang et al. (2000)	241	74
Alu	AluYa5	*IL2RG*	X-linked (XSCID)	X	Hancks and Kazazian (2012)	N/A	N/A
Alu	AluY	*CRB1*	Retinitis pigmentosa (RP)	1	den Hollander et al. (1999)	244	70
Alu	Alu	*SERPINC1*	Type 1 autoimmune thyroid disease (ATD)	1	Beauchamp et al. (2000)	6	40
Alu	AluYa5	*ALMS1*	Alström syndrome	2	Hancks and Kazazian (2012)	257	76
Alu	AluJ	*MSH2*	Lynch syndrome or hereditary nonpolyposis colorectal cancer (HNPCC)	2	Kloor et al. (2004)	85	40
Alu	N/A	*MSH2*	Hereditary cancer	2	Qian et al. (2015)	N/A	N/A
Alu	AluYa5	*ZFHX1B*	Mowat-Wilson syndrome	2	Hancks and Kazazian (2012)	281	93
Alu	AluYb9	*BCHE*	Cholinesterase deficiency	3	Muratani et al. (1991)	289	38

(continued)

Table 1 (continued)

Element	Subfamily	Gene	Disease	Chr	Reference	Size (bp)	PolyA tail length (bp)
Alu	AluYa5	*CASR*	Familial hypocalciuric hypercalcemia and neonatal severe hyperparathyroidism (FHH and NSHPT)	3	Janicic et al. (1995)	280	93
Alu	AluYb8	*HESX1*	Anterior pituitary aplasia	3	Sobrier et al. (2005)	288	30
Alu	AluYb8	*OPA1*	Autosomal dominant optic atrophy (ADOA)	3	Gallus et al. (2010)	289	25
Alu	AluYa5	*MLVI2*	Associated with leukemia*	5	Anagnou et al. (1989)	280	26
Alu	AluYb8	*APC*	Hereditary desmoid disease (HDD)*	5	Halling et al. (1999)	278	40
Alu	N/A	*APC*	Hereditary cancer	5	Qian et al. (2015)	N/A	N/A
Alu	AluYb9	*APC*	Familial adenomatous polyposis (FAP)	5	Su et al. (2000)	93	60
Alu	AluY	*MCC*	hepatocellular carcinoma	5	Tighe et al. (2002)	N/A	N/A
Alu	AluYb8	*MAK*	Retinitis pigmentosa (RP)	6	Tucker et al. (2011)	281	57
Alu	AluYa5	*NT5C3*	Chronic hemolytic leukemia (CHL)	7	Manco et al. (2006)	281	36
Alu	AluY	*CFTR*	Cystic fibrosis	7	Chen et al. (2008)	46	57
Alu	AluYa5	*CFTR*	Cystic fibrosis	7	Chen et al. (2008)	281	56
Alu	AluYa5	*EYA1*	Brancio-oto-renal (BOR) syndrome	8	Abdelhak et al. (1997)	N/A	97
Alu	AluYb9	*LPL*	Lipoprotein disease (LPL) deficiency	8	Hancks and Kazazian (2012)	150	60
Alu	AluYb5/8	*CHD7*	CHARGE syndrome	8	Udaka et al. (2007)	75	100
Alu	AluYa5	*POMT1*	Walker Walburg syndrome	9	Bouchet et al. (2007)	290	53
Alu	AluYa5	*FGFR2*	Apert syndrome	10	Oldridge et al. (1999)	283	69
Alu	AluYb8	*FGFR2*	Apert syndrome	10	Oldridge et al. (1999)	288	47
Alu	AluYk13	*FGFR2*	Apert syndrome	10	Bochukova et al. (2009)	214	12
Alu	AluYa5	*FAS*	Autoimmune lymphoproliferative syndrome (ALPS)	10	Tighe et al. (2002)	281	33
Alu	AluYc1	*SERPING1*	Hereditary form of angioedma (HAE)	11	Stoppa-Lyonnet et al. (1990)	285	42
Alu	AluYa5	*HMBS*	Acute intermittant prophyria (AIP)	11	Mustajoki et al. (1999)	279	39
Alu	N/A	*ATM*	Hereditay cancer	11	Qian et al. (2015)	N/A	N/A

Alu	AluYa5	*GNPTAB*	Mucopolydosis type II (ML II)	12	Tappino et al. (2008)	279	17
Alu	AluYc1	*BRCA2*	Breast cancer	13	Miki et al. (1996)	281	62
Alu	N/A	*BRCA2*	Breast cancer	13	Qian et al. (2015)	N/A	N/A
Alu	N/A	*BRCA2*	Breast cancer	13	Qian et al. (2015)	N/A	N/A
Alu	N/A	*BRCA2*	Breast cancer	13	Qian et al. (2015)	N/A	N/A
Alu	N/A	*BRCA2*	Breast cancer	13	Qian et al. (2015)	N/A	N/A
Alu	N/A	*BRCA2*	Breast cancer	13	Qian et al. (2015)	N/A	N/A
Alu	N/A	*BRCA2*	Breast cancer	13	Qian et al. (2015)	N/A	N/A
Alu	N/A	*BRCA2*	Breast cancer	13	Qian et al. (2015)	N/A	N/A
Alu	N/A	*BRCA2*	Breast cancer	13	Qian et al. (2015)	N/A	N/A
Alu	AluYa5	*BRCA2*	Breast cancer	13	Teugels et al. (2005)	285	N/A
Alu	N/A	*PALB2*	Hereditary cancer	16	Qian et al. (2015)	N/A	N/A
Alu	AluYb8	*PMM2*	Congenital disorders of glycosylation type Ia (CDG-Ia)	16	Schollen et al. (2007)	263	10
Alu	AluYc1	*BRCA1*	Breast and ovarian cancer, familial	17	Peixoto et al. (2013)	191	60
Alu	N/A	*BRCA1*	Hereditary cancer	17	Qian et al. (2015)	N/A	N/A
Alu	AluS	*BRCA1*	Breast cancer	17	Teugels et al. (2005)	286	N/A
Alu	AluYa5	*NF1*	Neurofibromatosis type 1 (NF1, cancer)*	17	Wallace et al. (1991)	282	40
Alu	AluY	*NF1*	Neurofibromatosis type 1 (NF1, cancer)	17	Wimmer et al. (2011)	280	N/A
Alu	AluY	*NF1*	Neurofibromatosis type 1 (NF1, cancer)	17	Wimmer et al. (2011)	281	N/A
Alu	AluYa5	*NF1*	Neurofibromatosis type 1 (NF1, cancer)	17	Wimmer et al. (2011)	282	60
Alu	AluYa5	*NF1*	Neurofibromatosis type 1 (NF1, cancer)	17	Wimmer et al. (2011)	284	120
Alu	AluYa5	*NF1*	Neurofibromatosis type 1 (NF1, cancer)	17	Wimmer et al. (2011)	281	N/A
Alu	AluYa5	*NF1*	Neurofibromatosis type 1 (NF1, cancer)	17	Wimmer et al. (2011)	284	110
Alu	AluYa5	*NF1*	Neurofibromatosis type 1 (NF1, cancer)	17	Wimmer et al. (2011)	279	N/A
Alu	AluYa5	*NF1*	Neurofibromatosis type 1 (NF1, cancer)	17	Wimmer et al. (2011)	264	60-85
Alu	AluYb8	*NF1*	Neurofibromatosis type 1 (NF1, cancer)	17	Wimmer et al. (2011)	249	121

(continued)

Table 1 (continued)

Element	Subfamily	Gene	Disease	Chr	Reference	Size (bp)	PolyA tail length (bp)
Alu	AluYb8	*NF1*	Neurofibromatosis type 1 (NF1, cancer)	17	Wimmer et al. (2011)	288	N/A
Alu	AluYb8	*NF1*	Neurofibromatosis type 1 (NF1, cancer)	17	Wimmer et al. (2011)	289	120
Alu	AluYb8	*NF1*	Neurofibromatosis type 1 (NF1, cancer)	17	Wimmer et al. (2011)	288	78-178
Alu	AluYb8	*NF1*	Neurofibromatosis type 1 (NF1, cancer)	17	Wimmer et al. (2011)	288	118
Alu	AluYb8	*NF1*	Neurofibromatosis type 1 (NF1, cancer)	17	Wimmer et al. (2011)	268	121
LINE-1	L1 Ta	*CYBB*	Chronic granulomatous disease (CGD)	X	Meischl et al. (2000), Brouha et al. (2002)	1722	101
LINE-1	L1 Ta	*CYBB*	Chronic granulomatous disease (CGD)	X	Brouha et al. (2002)	836	69
LINE-1	L1 Ta	*CHM*	Choroideremia	X	Van Den Hurk et al. (2003)	6017	71
LINE-1	L1 Ta	*DMD*	Duchenne muscular dystrophy (DMD)	X	Musova et al. (2006)	452	41
LINE-1	L1 Ta	*DMD*	Duchenne muscular dystrophy (DMD)	X	Narita et al. (1993)	608	16
LINE-1	L1 Ta	*DMD*	Duchenne muscular dystrophy (DMD)	X	Hancks and Kazazian (2012)	1400	38
LINE-1	L1 Ta	*DMD*	Duchenne muscular dystrophy (DMD)	X	Holmes et al. (1994)	530	73
LINE-1	N/A	*DMD*	Duchenne muscular dystrophy (DMD)	X	E Bakker and G van Omenn (pers. comm.)	878	N/A
LINE-1	L1 Ta	*DMD*	Duchenne muscular dystrophy (DMD)	X	Qian et al. (2015), Callinan et al. (2006)	212	118
LINE-1	L1 Ta	*FVIII*	Hemophilia A	X	Mukherjee et al. (2004)	3800	54
LINE-1	L1 preTa	*FVIII*	Hemophilia A	X	Mukherjee et al. (2004)	2300	77
LINE-1	L1 Ta	*FIX*	Hemophilia B	X	Li et al. (2001)	463	68
LINE-1	L1 Ta	*FIX*	Hemophilia B	X	Mukherjee et al. (2004)	163	125
LINE-1	L1 Ta	*RP2*	X linked retinitis pigmentosa (XLRP)	X	Hancks and Kazazian (2012)	6000	64
LINE-1	L1 HS	*RPS6KA3*	Coffin-Lowry syndrome	X	Hancks and Kazazian (2012)	2800	N/A
LINE-1	N/A	*ABDH5*	Chanaric-Dorfman syndrome (CDS)	3	Sprecher (pers. comm.)	FL	N/A
LINE-1	N/A	*MLH1*	Hereditary cancer	3	Qian et al. (2015)	N/A	N/A

LINE-1	N/A	*MLH1*	Hereditary cancer	3	Qian et al. (2015)	N/A	N/A
LINE-1	L1 Ta	*APC*	Colon cancer	5	Callinan et al. (2006)	520	222
LINE-1	L1 Hs	*EYA1*	Branchio-oto-renal syndrome (BOR)	8	Morisada et al. (2010)	3756	None
LINE-1	L1 Ta	*ST18*	Hepatocellular carcinoma*	8	Shukla et al. (2013)	410	N/A
LINE-1	L1 Ta	*FKTN*	Fukuyama muscular dystrophy (FCMD)	9	Kondo-Iida et al. (1999)	1200	59
LINE-1	L1 Ta	*FKTN*	Fukuyama muscular dystrophy (FCMD)	9	Kondo-Iida et al. (1999)	3000	N/A
LINE-1	L1 Hs	*SETX*	Ataxia with oculomotor apraxia type 2 (AOA2)	9	Bernard et al. (2009)	1300	42
LINE-1	L1 Ta	*HBB*	β thalassemia	11	Lanikova et al. (2013)	6000	107
LINE-1	L1 Hs	*PDHX*	PHHc deficiency	11	Miné et al. (2007)	6086	67
LINE-1	L1 Ta	*SLCO1B3*	Rotor syndrome	12	Kagawa et al. (2015)	6100	N/A
LINE-1	L1 preTa	*NF1*	Neurofibromatosis type 1 (NF1, cancer)	17	Shukla et al. (2013)	1800	N/A
LINE-1	L1 Ta	*NF1*	Neurofibromatosis type 1 (NF1, cancer)	17	Shukla et al. (2013)	6000	N/A
LINE-1	N/A	*NF1*	Neurofibromatosis type 1 (NF1, cancer)	17	Shukla et al. (2013)	2200	N/A
LINE-1	L1 Ta	*PTEN*	Endometrial carcioma	10	Teugels et al. (2005)	112	N/A
SVA	F	*FVIX*	Hemophilia B	X	Nakamura et al. (2015)	2524	28
SVA	F_1	*SUZ1P*	Neurofibromatosis type 1 (NF1, cancer)*	17	Arca et al. (2002)	1700	23
SVA	F	*SUZ1P*	Neurofibromatosis type 1 (NF1, cancer)*	17	Arca et al. (2002)	1300	40
SVA	F	*BTK*	X linked agammaglobulinemia (XLA)	X	Arca et al. (2002)	251	92
SVA	F	*TAF1*	X linked dystonia-parkinonism syndrome (XDP)	X	Makino et al. (2007)	2627	62
SVA	E	*LDRAP1*	Autosomal recessive hypercholesterolaemia (ARH)	1	Arca et al. (2002)	2600	57
SVA	E	*SPTA1*	Hereditary elliptoytosis and hereditary pyropoikilocytosis (HE and HPP)	1	Hancks and Kazazian (2012)	632	50
SVA	F	*HLA-A*	Leukemia	6	Takasu et al. (2007)	2000	45
SVA	F	*PMS2*	Lynch syndrome	7	Tubio et al. (2014)	2200	N/A
SVA	E	*FKTN*	Fukuyama muscular dystrophy (FCMD)	9	Kobayashi et al. (1998), Taniguchi-Ikeda et al. (2011)	3023	32

(continued)

Table 1 (continued)

SVA	E	*PNPLA2*	Neuroal lipid storage disease with myopathy (NLSDM)	11	Akman et al. (2010)	1800	44
pA	N/A	*COL4A6*	Alport syndrome	X	Segal et al. (1999)	N/A	70
pA	N/A	*AGA*	Aspartylglucosaminuria (AGU)	4	Hancks and Kazazian (2012)	N/A	37
pA	N/A	*BRCA2*	breast cancer	13	Wang et al. (2001)	N/A	35
pA	N/A	*NF1*	Neurofibromatosis type 1 (NF1, cancer)	17	Wimmer et al. (2011)	N/A	120
PP**	*TMF1*	*CYBB*	Chronic granulomatous disease (CGD)	X	Gu et al. (2007)	5800	100

This table details an up-to-date list of all known disease-causing retrotransposon-mediated mutations. The diseases with a * following their name indicate insertions which are known to be somatic. In this table, the type of element and the location into which it inserted, the disease caused by the insertion, the size of the insertion, and the length of the poly (A) tail are listed. The ** indicates a processed pseudogene genome (via 5′and 3′ transduction) (Kaer and Speek 2013; Raiz et al. 2012). Transductions can occur from both non-reference and reference L1 elements and are a result of the weak poly(A) signal in the L1 element (Holmes et al. 1994; Goodier et al. 2000; Pickeral et al. 2000; Szak et al. 2002). Because of the weak poly (A) signal, RNA polymerase II reads through the L1 to the adjacent DNA following the 3′ end of the element. This process is estimated to occur in 15–23 % of all L1 mobilization events (Holmes et al. 1994; Goodier et al. 2000; Pickeral et al. 2000; Szak et al. 2002). An L1 mediated 3′ transduction of a novel noncoding gene into exon 67 of the dystrophin gene was observed; however, due to severe 5′ truncation of the element there was no recognizable L1 sequence present (Solyom et al. 2012a; Awano et al. 2010). When L1s carry regulatory sequences in the transduction, "exon shuffling" can also occur which can affect gene expression (Moran et al. 1999). Yet another mechanism by which L1 insertions can cause aberrant gene expression is "gene breaking" (Wheelan et al. 2005). For gene breaking to occur and L1 must insert into an intron in the antisense orientation and then split the associated transcript into two parts through the combined effects of the L1 polyadenylation signal and the L1 antisense promoter (Wheelan et al. 2005). TEs can also provide an alternative promoter for a gene following their insertion into a new location. A TE which is fixed in the genome or even polymorphic in the population can acquire mutations which enable the sequence to create a cryptic splice site (Varon et al. 2003) or it can undergo deletions which facilitate branch site recognition and result in *Alu* exonization (Meili et al. 2009).

In addition to insertional mutagenesis, retrotransposons can mediate ectopic recombination through non-allelic homologous recombination (NAHR) and non-homologous end joining (NHEJ) in the genome (Callinan et al. 2005; Kazazian 2004; Hedges and Deininger 2007; Belancio et al. 2008; Goodier and Kazazian 2008; Lee et al. 2012a). In fact, a frequently observed example of this process is also the most frequently observed translocation in the human genome where there is a recombination of two *Alu* sequences on chromosomes 11 and 22, respectively (Hill et al. 2000). Additionally, it has been demonstrated that *Alu* repeats are enriched in segmental duplication breakpoints (Bailey et al. 2003) and countless examples of NAHR mediated by *Alu* elements have been found (Belancio et al. 2008). Recently a broad analysis of pathogenic variants in Fanconi anemia genes found that up to 75 % of FANCA deletions are *Alu-Alu* mediated, predominantly mediated by NAHR due to *Alu*Y elements (Flynn et al. 2014). Occasionally, the homologous sequences of L1 elements cause mis-alignment during meiosis and result in NAHR, especially when elements are proximal to each other and in the same orientation (Kazazian 2004; Hedges and Deininger 2007; Belancio et al. 2008; Goodier and Kazazian 2008). A 520 kb deletion containing four genes occurred due to an L1 associated non-allelic recombination and caused Ellis von Creveld syndrome in a family (Temtamy et al. 2008). Other previous reports also noted a recombination between L1 elements flanking the PHKB gene (Burwinkel and Kilimann 1998) and a similar event occurred due to the same mechanism causing Alport syndrome diffuse leiomyomatosis (Segal et al. 1999). More recently, a deletion in the factor IX gene between two highly homologous L1 sequences seems to have occurred due to non-allelic homologous recombination between the two tandem L1s (Wu et al. 2014). SVA elements, through NAHR, are also responsible for disease-causing mutations occurring due to copy number changes with nonrecurrent breakpoints (Vogt et al. 2014). In a recent study by Vogt and colleagues, large NF1 deletions were studied and two of 17 deletions with nonrecurrent breakpoints occurred with the concomitant insertion of SVA elements at the deletion breakpoints (Vogt et al. 2014).

Many of the previously described disease-causing events associated with the presence of retrotransposons in the human genome occur during a post-zygotic stage (Vogt et al. 2014) or occur in the germline. Although diseases caused by insertion or aberrant recombination events have mostly been due to insertions prior to or during development, insertions occurring in somatic cells of diseased organisms have also been clearly exhibited. Although the somatic insertions appear to be occurring in various regions of the brain, a subset of normal tissues, and most epithelial cancers, it has yet to be determined to what extent the insertions are changing gene expression and potentially contributing to human disease. Somatic insertions in cancer will comprise the bulk of this book chapter and will be further discussed later in this text (Lee et al. 2012b).

2.1 *Conclusions*

Due to the various mechanisms through which repetitive elements can cause disease, the host has evolved many pathways to thwart the amplification of retrotransposons and thereby the potential mutations which come along with them. The defense mechanisms employed by the cell are diverse as they affect various aspects of the L1 life cycle. The previously discussed L1 control mechanisms display how the host uses multiple, if not redundant, mechanisms to control retrotransposon mobility and suggest when any of these mechanisms is not operating optimally L1 may be more active. In other words, a cell subject to aberrant expression of its protective mechanisms may be particularly susceptible to L1 somatic insertions that are inherently mutagenic.

3 L1 and Cancer

It follows logically that genetic instability caused by retrotransposition activity would be elevated in diseases where normal cellular check points during proliferation and DNA replication are absent, such as cancer. Indeed, many cancers have shown high L1 expression and a high occurrence of somatic insertions in patients evaluated thus far (Tubio et al. 2014; Solyom et al. 2012b; Ewing et al. 2015; Shukla et al. 2013; Doucet-O'Hare et al. 2015; Lee et al. 2012c; Rodić et al. 2015; Rodić et al. 2014; Helman et al. 2014). Although L1 activity in cancer, especially epithelial cancers, is prevalent, it is still unclear how much somatic retrotransposon insertions are contributing to oncogenesis. Furthermore, it is still unclear if the relationship between cancer and retrotransposition in epithelial cancers is due to cancer activating the process of retrotransposition or due to retrotransposition causing somatic mutations which contribute to tumor formation. Cancer is by no means a simple disease; in fact, it encompasses a wide-ranging group of more than 200 diseases that involve uninhibited proliferation of cells leading to tumor formation, in addition to several additional common features (Hanahan and Weinberg 2011, 2000). Epidemiological studies on twins suggest that environment plays a much clearer role in the process of tumorigenesis than genetics (Lichtenstein et al. 2000; Sorenson et al. 1988). For example, Sorenson and colleagues found that when an adoptive parent died as a result of cancer before the age of 50, the rate of mortality due to cancer for the adoptees increased (Sorenson et al. 1988). These findings suggest that a strong enough environmental mutagen will have a potent effect on individuals who live in the same environment despite differences in genetic background.

Furthermore, another study found that the overwhelming contributor to cancer development in twins was the environment (Lichtenstein et al. 2000). The authors found that even when they considered cancer for which there was statistically significant evidence of cancer heritability, most twin pairs were discordant for presence of the cancer (Lichtenstein et al. 2000). Environmental factors contribute to sporadic cancer occurrence as much as 58–82 %, as compared to the highest known genetic contribution to cancers, colorectal, breast, and prostate cancers which is 27–42 % (Lichtenstein et al. 2000). Mutations which contribute to cancer development are referred to as drivers, and mutations which accumulate due to the dysregulation of DNA replication and repair pathways are referred to as passengers. It is a continuous challenge in the study of cancer genetics to differentiate between driver and passenger mutations. The apparent dysregulation of L1 elements in cancer is only one of many sources of genetic aberrations that frequently contribute to cancer development. However, L1 elements and other retrotransposons have a large size effect upon insertion and due to their structure have multiple ways in which their newly acquired presence can disrupt gene expression and regulation. It is also telling that L1 mobilization has been observed in many different tumors (Lee et al. 2012b; Tubio et al. 2014; Solyom et al. 2012b; Ewing et al. 2015; Shukla et al. 2013; Doucet-O'Hare et al. 2015; Rodić et al. 2015; Helman et al. 2014; Miki et al. 1992; Iskow et al. 2010), cancer cell lines (Moran et al. 1996; Garcia-Perez et al. 2010; Ostertag et al. 2000), and during development (Garcia-Perez et al. 2007; Kano et al. 2009; Coufal et al. 2009). Due to the potentially substantial effect of an L1 insertion and the predominantly deleterious effects on host gene expression observed thus far (Shukla et al. 2013; Baillie et al. 2011; Evrony et al. 2012), L1 insertions may be more prone than other types of mutations to have an impact on tumorigenesis.

There are many carcinogenic environmental factors (Boffetta and Nyberg 2003) which have an impact on retrotransposon activity in cultured cells (Fornace and Mitchell 1986). Benzopyrenes, for example, are a risk factor for lung cancer, colorectal cancer, and breast cancer (Denissenko et al. 1996; Tabatabaei et al. 2010;

Rathore and Wang 2013) and have been shown to increase L1 retrotransposition in HeLa cells (Stribinskis and Ramos 2006). Exposure to certain metals such as cadmium, chromium VI, and nickel are risk factors for lung and breast cancer (Beveridge et al. 2010; Al-Qubaisi et al. 2013), and interestingly nickel has been shown to induce a higher rate of L1 retrotransposition (Kale et al. 2005). Another feature of many tumors is a higher level of free radicals involved in oxidative stress (Toyokuni et al. 1995), and oxidative stress has also been demonstrated to affect L1 activity (Giorgi et al. 2011). Furthermore, oxidative stress and DNA damage frequently occur as a result of cellular senescence and can also increase both retrotransposition rates and chromosomal instability, thereby potentially contributing to somatic mosaicism and cancer development (Maxwell et al. 2011; Jacobs et al. 2012; Laurie et al. 2012). It is certainly plausible that many more environmental factors which contribute to cancer development also activate L1 retrotransposition and thereby increase the probability of L1 generating an insertion which affects an oncogenic locus and contributes to tumorigenesis (Carreira et al. 2014). The role of environmental stress in L1 mobilization is developed in chapter "Retrotransposon-Driven Transcription and Cancer" of this book.

When a cancer genome is sequenced, tens or hundreds of thousands of single nucleotide variants, insertions, deletions, translocations, rearrangements, and other mutations may be found. In order to understand the role L1 mobilization plays in tumorigenesis, it is necessary to separate the winnow from the chafe, determine whether any somatic L1 insertions are present in the tumor, and absent from the normal tissues. To determine whether or not somatic insertions are contributing to tumor development, the insertions must be mapped in individuals with relevant disease. In 1992, Miki et al. mapped a somatic L1 insertion in a colorectal tumor in an exon of the APC gene (Miki et al. 1992). The somatic insertion was confirmed with Southern blot and because *APC* is the primary tumor suppressor gene in colorectal cancer and causes familial adenomatous polyposis (Kinzler et al. 1991a, b) it is reasonable to conclude that the somatic L1 insertion, which was found to be absent from normal colon, was sufficient to drive oncogenesis (Carreira et al. 2014). Although the preliminary discovery of an L1 insertion contributing to cancer occurred in the early 1990s, it was two decades before researchers returned to the topic to investigate the role of L1 in carcinogenesis. To date, only one other definitive somatic insertion has been found in the exon of a tumor suppressor gene. An insertion into an exon of the PTEN gene was discovered with whole genome and whole exome sequencing by Helman and colleagues in 2014 (Helman et al. 2014). High-throughput next-generation sequencing enabled researchers to examine the genomes of more individuals at one time and compare those genomes between the cancer and normal samples in addition to comparing individuals' genome differences. Due to the new technology available, many methods were subsequently specifically developed for assessing L1 activity in the genome, for detailed reviews see (Ray and Batzer 2011; Faulkner 2011). Prior to a paper from Iskow and colleagues, several groups were able to successfully identify novel L1 insertions; however, they used assays which were inherently low-throughput and which had high false positive rates (Sheen et al. 2000; Ovchinnikov et al. 2001; Badge et al. 2003). The initial high-throughput method utilized for the discovery of somatic insertions, termed "Transposon-seq" utilized digested genomic DNA using restriction enzymes which recognize the 3′ end of the L1 and *Alu* elements (Iskow et al. 2010). The authors linked adapters to the resulting fragments and amplified them with PCR to create retrotransposon-specific libraries (Iskow et al. 2010). In the initial efforts of the study, 38 ethnically diverse humans and 8 ATCC cell lines derived from human tumors were utilized to create libraries (Iskow et al. 2010). Approximately 4600 library fragments were cloned and sequenced with ABI capillary sequencing yielding 152 putative novel L1 insertion polymorphisms (Iskow et al. 2010). In order to ensure a low false positive rate, the authors applied specialized informatics to filter the datasets and identified high probability L1Ta insertion candidates (Iskow et al. 2010). The PCR validation rate for the insertions was 97 % with approximately a third of the insertions possessing a minor allele frequency (MAF) equal to or below 5 % (Iskow et al. 2010). Additionally, 47 "rare" insertions were found in very few individuals and 9 of them were only found in one cell line evaluated (Iskow et al. 2010). One of the nine rare insertions in only one cell line was deemed as a somatic insertion due to its presence in the tumor cell line and its absence from the normal cell line (Iskow et al. 2010). After finding the somatic insertion, the authors implemented their technique in a high-throughput fashion by acquiring 20 non-small cell lung cancers with matched normal tissues. Previous work in the mouse brain (Muotri et al. 2005) and in human neural stem cells (Coufal et al. 2009) suggested that L1 activity in the brain was highly active. Two types of brain tumors were also evaluated in the study including glioblastoma and medulloblastoma (Iskow et al. 2010) with five cases of each condition along with matched blood leukocyte controls (Iskow et al. 2010). The high-throughput version of "Transposon-seq" utilized barcoding sequences to assign a given sequence to specific samples within the sample pool sequenced with 454 pyrosequencing (Iskow et al. 2010). Following sequencing analysis, 1389 distinct L1 insertions were detected in the 30 samples assessed. After all the novel insertion candidates were compared to the human reference genome and to dbRIP (Wang et al. 2006), 650 putative novel L1 insertions remained, and 45 % of them had MAFs less than or equal to 5 %. Of all the individuals evaluated, 93 % of the genomes had at least one rare L1 insertion present in only a single human in the study. After screening the low frequency alleles with PCR assays, the authors found there were nine tumor-specific somatic L1 insertions present in their lung cancer cohort. Surprisingly, no somatic insertions were confirmed in the brain tumors evaluated. In 6 of the 20 lung tumors studied, somatic tumor-only insertions were confirmed. Lastly, the authors confirmed hypomethylation of many potentially active polymorphic L1 elements in the genomes of affected patients (Iskow et al. 2010). The hypomethylation present in the affected individuals suggests that one mechanism of L1 escape from host control in cancer is due to changes in methylation due to mutations in tumor suppressor genes.

Several years later, Lee et al. used a computational method, "Tea" for transposable element analyzer, to analyze whole genome paired end sequencing data from tumors and matching blood samples (Lee et al. 2012b). In this study, the authors performed a single nucleotide resolution analysis of retrotransposons in 43 high coverage whole-genome sequencing data sets from five cancer types (Lee et al. 2012b). The study samples consisted of colorectal tumors, ovarian tumors, prostate tumors, blood cancer, and brain cancer (Lee et al. 2012b). The authors identified 194 high-confidence putative somatic retrotransposon insertions in the samples of epithelial origin only, e.g., ovarian, prostate, and colorectal tumors (Lee et al. 2012b). Of the 194 high-confidence putative insertions, 183 of them were purported to be L1s, 10 Alu elements, and 1 endogenous retrovirus (ERV) (Lee et al. 2012b). It was later determined that the putative ERV insertion was likely caused by a microhomology-mediated break-induced repair mechanism (Hastings et al. 2009). With regard to the PCR and capillary sequencing validation of the predicted somatic insertions, 25/26 insertions were validated in colorectal cancer and 13/13 insertions validated in ovarian cancer with an overall rate of 97 % validation (Lee et al. 2012b). Finally, the authors noted that somatic and germline L1 insertion sites differed in genomic distribution as well as epigenetic characteristics. When comparing germline insertions to somatic insertions, germline events are depleted from genes significantly, likely due to strong negative selection on the events (Graham and Boissinot 2006). The authors assert that the retrotransposon insertions seem to provide a selective advantage in certain individuals and that the insertions occurred in genes commonly mutated in cancers and substantially disrupted their expression (Lee et al. 2012b).

In a publication from our own group in the same year, two high-throughput sequencing techniques which enrich for retrotransposons in different ways were utilized. L1-seq, developed by Adam Ewing (Ewing and Kazazian 2010), utilizes a hemi-specific PCR-based library construction method to enrich for the young, active subfamily of L1s in the genome. RC-seq (version 1), developed by the Faulkner lab (Baillie et al. 2011), uses probes designed to bind the 5′ and 3′ ends of L1 and SVA elements and probes tiled across the full length of an *Alu*. The probes are tiled on an array and the sheared genomic DNA is applied to the array as the relevant sequences bind. This DNA later has adapters ligated to it and is minimally amplified with PCR using only eight cycles (Baillie et al. 2011). Using L1-seq on two cohorts of 16 total colorectal cancer patients with matched tumor and normal tissues, 26/40 and 37/51 high stringency somatic insertions were identified and validated, respectively. An additional 9 out of 16 lower stringency insertions with lower read-counts and map scores were identified and validated between both cohorts as well. In total, L1-seq resulted in the 3′ validation of 69/107 putative tumor-specific somatic insertions and both 5′ and 3′ validation of 35 of said insertions. As is typical of both previous and follow-up studies, one tumor had 17 insertions present while three others had no insertions. Most of the insertions identified had severe 5′ truncation and averaged about 1 kb in size. Five of the 16 colorectal cancer patient samples were barcoded, pooled, and analyzed by shallow, multiplexed RC-seq. Using RC-seq, 8L1, 83 *Alu*, and 5 SVA somatic insertions different from those identified with L1-seq were predicted. Only one of the L1 insertions predicted was confirmed to be truly tumor-specific, and 11 high-confidence predicted L1 insertions identified by L1-seq were also identified with RC-seq. Of the remaining putative insertions, 6/8 L1s, 30/57 *Alu* elements, and 6/11 SVA elements were validated in both tumor and paired normal tissue.

A year later, Faulkner and colleagues published an updated version of RC-seq which was utilized to analyze retrotransposon activity in 19 hepatocellular carcinomas (HCC) (Shukla et al. 2013). The HCC cases consisted of fresh frozen tissue from patients positive for HBV or HCV and matched normal tissue. In the new version of RC-seq, a liquid phase capture was utilized to increase the number of probes available for binding to increase efficiency; furthermore the sequences of the probes used were also refined and edited to be a more effective pool for binding active elements. The optimized version of RC-seq produced a fourfold increase in reads which aligned to non-reference genome L1s per library sequenced. Twelve out of 17 potential somatic insertions were validated in tumor only with PCR and sequencing confirmed the L1 is active in HCC. No SVA or Alu element somatic insertions were confirmed in any of the patients; however, a single L1 insertion was confirmed in normal liver and was found to be absent from the corresponding tumor. The insertion into normal liver is surprising because it had been previously assumed that retrotransposition was not an active process in somatic tissues with the exception of the brain (Coufal et al. 2009; Baillie et al. 2011; Evrony et al. 2012; Cai et al. 2014). If somatic retrotransposition happens in the normal tissues of some individuals, it is possible that in those individuals it could cause mutations which lead to disease like cancer development. Interestingly, the authors noted three different germline insertions into the *MCC* gene, *m*utated in *c*olorectal *c*ancers, in three individuals with HCC. The germline insertions coincided with a strong reduction of *MCC* expression as confirmed with immunoblot and qRT PCR. Although this study did not definitively address whether or not somatic insertions contribute to tumorigenesis, it did present evidence that in some individuals inherited polymorphic L1 insertions may play a role. It seems plausible to assume that if a germline insertion can cause such a reaction, then so too can somatic insertions.

In 2014, yet another pipeline emerged for analyzing whole genome sequencing data from 200 tumor samples and their matched normal counterparts (Helman et al. 2014). The following cancers were analyzed in the study: lung adenocarcinoma, lung squamous cell carcinoma, ovarian carcinoma, rectal adenocarcinoma, colon adenocarcinoma, kidney clear-cell carcinoma, uterine corpus endometrioid carcinoma, head and neck squamous cell carcinoma, breast carcinoma, acute myeloid leukemia, and glioblastoma multiforme (Helman et al. 2014). The study identified 7724 unique, non-reference germline insertion sites and approximately 65 % of them are known retrotransposon insertion polymorphisms (RIPs) previously identified in other studies (Beck et al. 2010; Lee et al. 2012c; Iskow et al. 2010; Ewing and Kazazian 2010; Xing et al. 2009; Hormozdiari et al. 2013; Huang et al. 2010). In total, 810 putative retrotransposon insertions were predicted in the cancer samples and absent from normal samples. The candidate insertions exhibited the hallmarks of TPRT including target site duplications averaging 15 nucleotides in addition to a canonical endonuclease motif (Feng et al. 1996; Morrish et al. 2002). However, 47 putative somatic retrotransposition events were selected for experimental validation. The 47 tested insertions were predicted across 21 individuals and 4 tumor types. Thirty-nine of the insertions (83 %) were validated as tumor specific by PCR and sequencing of either the 5′ or the 3′ end. For 32 of the 47 insertions, evidence was present for both the 3′ and 5′ ends. Two of the putative somatic insertions were found to be germline, present in both tumor and normal, after the validation attempt. Six of the 47 putative somatic insertions were not amplified in either the normal or the tumor samples. Not unlike previous similar studies, it was noted that 97 % of the L1 somatic insertions are in the L1Hs subfamily. After considering which cancers exhibited L1 activity among their samples the authors noted that cancers of epithelial origin were the only ones which had detectable somatic retrotransposition events. Historically, nearly all cancers found to possess retrotransposon activity, in the form of newly acquired somatic insertions unique to the tumor, have been epithelial cancers. Interestingly, the authors also observed several 3′ transduction events from different regions of the genome in a single patient. The 3′ transductions are evidence that at least three different source L1 elements contributed to the somatic insertions in the cancer. In contrast to this finding, the authors also noted a patient in which a single L1 element caused at least four events, detected due to their 3′ transductions, into different areas of the genome. These findings seem to suggest two models for somatic retrotransposition activity in cancer. In some patients, a single hyperactive source element may insert itself into multiple genomic locations in the same tumor. In others, there may be several active source elements which contribute the somatic insertions present in the sample. It is also possible that both of these situations happen simultaneously in the same individual as well.

Later in 2014, a paper analyzed whole-genome sequencing data on 290 tumor and matched normal pairs consisting of 210 primary tumors, 52 metastatic tumors, and 28 cancer cell lines with matched normal cell lines (Tubio et al. 2014). The samples were obtained from 244 patients across 12 cancer types, including bladder, bone, breast, colon, head and neck, lung, pancreatic, prostate, and renal cancer as well as mesothelioma, melanoma, and glioma (Tubio et al. 2014). The algorithm used to analyze the sequencing data, "TraFiC," identified 2756 putative L1 retrotransposition events including both "solo" L1 events and 3′ transductions. PCR validation was attempted on 308 putative insertions and 259 insertions were confirmed with PCR and capillary sequencing (Tubio et al. 2014). The authors also observed a single patient with 22 somatic 3′ transduction events from a hyperactive L1 which mobilized many times in the same cancer (Tubio et al. 2014). The average insertion length was approximately 1 kb for insertions lacking a 5′ inversion, the TSDs averaged between 10 and 20 base-pairs, and 3′ transductions occurred in one-fourth of the cancer genomes evaluated (Tubio et al. 2014). Due to the abundance of 3′ transductions in many of the samples, the authors were able to conclude that few loci were driving the 3′ transductions in cancer (Tubio et al. 2014).

Recently, another paper analyzing whole-genome paired-end sequencing was published in which the authors studied 43 cases of esophageal adenocarcinoma (Paterson et al. 2015). The authors predicted an average of 16 insertions per tumor and a range of 0–153 insertions among the patients studied (Paterson et al. 2015). One-fifth of the L1 insertions found was predicted to have 5′ inversions and there were nine insertions identified with 3′ transductions. The authors also attempted to correlate p53 loss with L1 activity by evaluating p53 mutations in all of the patients. The authors observed a p53 mutation present in over 88 % of patients with esophageal adenocarcinoma patients studied; furthermore, two of the five cases where no p53 mutation were present were cases with no insertion (Paterson et al. 2015). In addition, it is noted that it is not possible to conclude that the L1s are only active in the tumor as the technique is not sensitive enough to detect potential events in the non-cancer cells due to their potentially highly polyclonal nature (Paterson et al. 2015). A serious deficit in this study was the lack of matched normal tissues for nearly all patients. Without matched normal tissue, it is impossible to know whether putative somatic insertions detected in the cancer are truly somatic or if they occurred in early development and are present throughout the tissue of interest. The continuous development of databases, such as euL1db (Mir et al. 2015), and the cataloguing common and rare germline insertions obtained from large-scale sequencing projects, such as the 1000 Genome Project (Stewart et al. 2011), will help to discriminate natural variants from somatic retrotransposition events.

Using a technique dubbed "Tip-seq," Rodic and colleagues studied 20 cases of pancreatic ductal adenocarcinoma (PDAC) to detect somatic L1 insertions present in the cancer and absent from normal pancreatic tissue (Rodić et al. 2015). The authors had previously described L1 protein ORF1 expression in up to 89 % of PDAC patients (Rodić et al. 2014). Tip-seq is a PCR-based L1 enrichment library preparation technique and it detected 268 somatic L1 insertions in the tumors of 18 patients evaluated which were absent from matched normal tissue (Rodić et al. 2015). A range of 0–65 insertions was detected in the patients and an average of 15 insertions per case was calculated (Rodić et al. 2015). There were 15 metastases which were evaluated with Tip-seq as well from 15 different patients and 242 insertions were detected in these samples (Rodić et al. 2015). In 13 of the cases where both a metastasis and a primary tumor from a patient were shared, 45 insertions were confirmed by PCR and capillary sequencing to be present in both tissues and absent from the normal tissue (Rodić et al. 2015). The expression of ORF1p in the samples subjected to Tip-seq correlated with the number of somatically acquired insertions per sample (Rodić et al. 2015). The authors reported 81 % of tested insertions validated with both PCR and capillary sequencing with all insertions being 5′ truncated and an average size of approximately 1 kb (Rodić et al. 2015). Finally, the authors noted two 3′ transductions among the validated insertions in the study (Rodić et al. 2015).

Ewing et al. published a study looking at multiple types of cancer including four colorectal cancer patients with matched colonic polyps and normal colon, seven patients with pancreatic ductal adenocarcinoma with matched normal, seven patients with gastric cancer and matched normal tissue, and eight testicular germ cell tumors with matched blood (Ewing et al. 2015). For eight of the aforementioned cases, metastatic tissues were available and evaluated as well (Ewing et al. 2015). Following L1-seq (Ewing and Kazazian 2010) and subsequent computational analysis 104 somatic heterozygous L1Hs insertions were validated by PCR and Sanger sequencing in the 18 gastrointestinal cancers and 1 insertion was validated in a single patient with a testicular germ cell tumor (Ewing et al. 2015). However, the most interesting finding in this article was insertions occurring in the polyps which precede the cancer development (Ewing et al. 2015). This pattern suggests that L1 is active in tissue before the cancer develops and certainly makes it seem more likely that L1 could contribute to the process of tumorigenesis.

All of the cancer studies performed to date have strongly established the hyperactivity of L1 in epithelial cancer; furthermore, many of the studies have established similar patterns with regard to retrotransposition. Several of the studies noted an average insertion size of approximately 1 kb likely due to the dramatic 5′ truncations which most of the validated insertions possess (Tubio et al. 2014; Doucet-O'Hare et al. 2015; Rodić et al. 2015). Thus far, nearly all the papers have reported target-site duplications in the same size range, approximately 10–20 nucleotides on average, and approximately 20 % of the insertions detected in cancer have 5′ inversions.

3.1 Conclusions

Although the activity of L1 elements in cancer has been firmly established and the events seem to adhere to most of the hallmarks of the process, it is still uncertain to what degree these elements play a role in carcinogenesis. Anything short of a glaringly obvious insertion disrupting a known tumor suppressor or activating an oncogene is a difficult sell to the scientific community as a cause or contributor to cancer. Furthermore, there is the possibility that the dysregulation of normal cellular processes in cancer may simply be enabling L1 activity due to differences in methylation or the under expression of host genes which normally suppress retrotransposons activity. The evidence contrasting the simple activation of elements due to cancer development is the confirmed somatic insertions in not only the precursor conditions to cancer, but also in normal tissues. Observations of validated somatic insertions in tissues which are the precursor to cancer were made in the recent publication by Ewing and colleagues. Although there is mounting evidence of somatic retrotransposition occurring in normal tissues, it has not been definitively shown that this activity leads to cancer. Like any other potential mutagen, retrotransposition likely leads to disease a certain percentage of the time regardless of the disease type. However, when retrotransposons are hyperactive in a tissue, like in cancer or potentially precancerous conditions, it may be more likely to be the cause of a mutation which leads down the path to cancer development. In the future, to distinguish between retrotransposons as passengers versus drivers, single-cell sequencing, database development and the acquisition of larger cohorts of patients will likely lead to an answer.

References

Abdelhak S, Kalatzis V, Heilig R, Compain S, Samson D, Vincent C et al (1997) Clustering of mutations responsible for branchio-oto-renal (BOR) syndrome in the eyes absent homologous region (eyaHR) of EYA1. Hum Mol Genet 6(13):2247–2255

Akman HO, Davidzon G, Tanji K, Macdermott EJ, Larsen L, Davidson MM et al (2010) Neutral lipid storage disease with subclinical myopathy due to a retrotransposal insertion in the PNPLA2 gene. Neuromuscul Disord 20(6):397–402

Al-Qubaisi MS, Rasedee A, Flaifel MH, Ahmad SHJ, Hussein-Al-Ali S, Hussein MZ et al (2013) Cytotoxicity of nickel zinc ferrite nanoparticles on cancer cells of epithelial origin. Int J Nanomedicine 8:2497–2508

Anagnou NP, Economou-Pachnis A, O'Brien SJ, Modi WS, Nienhuis AW, Tsichlis PN (1989) The human homolog of the Moloney leukemia virus integration 2 locus (MLV12) maps to band p14 of chromosome 5. Genomics 5(2):354–358

Apoil PA, Kuhlein E, Robert A, Rubie H, Blancher A (2007) HIGM syndrome caused by insertion of an AluYb8 element in exon 1 of the CD40LG gene. Immunogenetics 59(1):17–23

Aravin A, Gaidatzis D, Pfeffer S, Lagos-Quintana M, Landgraf P, Iovino N et al (2006) A novel class of small RNAs bind to MILI protein in mouse testes. Nature 442(7099):203–207

Aravin AA, Sachidanandam R, Bourc'his D, Schaefer C, Pezic D, Toth KF et al (2008) A piRNA pathway primed by individual transposons is linked to de novo DNA methylation in mice. Mol Cell 31(6):785–799

Arca M, Zuliani G, Wilund KR, Campagna F, Fellin R, Bertolini S et al (2002) Autosomal recessive hypercholesterolaemia in Sardinia, Italy, and mutations in ARH: a clinical and molecular genetic analysis. Lancet 359(9309):841–847

Arjan-Odedra S, Swanson CM, Sherer NM, Wolinsky SM, Malim MH (2012) Endogenous MOV10 inhibits the retrotransposition of endogenous retroelements but not the replication of exogenous retroviruses. Retrovirology 9:53

Awano H, Malueka RG, Yagi M, Okizuka Y, Takeshima Y, Matsuo M (2010) Contemporary retrotransposition of a novel non-coding gene induces exon-skipping in dystrophin mRNA. J Hum Genet 55(12):785–790

Baba Y, Huttenhower C, Nosho K, Tanaka N, Shima K, Hazra A et al (2010) Epigenomic diversity of colorectal cancer indicated by LINE-1 methylation in a database of 869 tumors. Mol Cancer 9:125

Badge RM, Alisch RS, Moran JV (2003) ATLAS: a system to selectively identify human-specific L1 insertions. Am J Hum Genet 72(4):823–838

Bailey JA, Liu G, Eichler EE (2003) An Alu transposition model for the origin and expansion of human segmental duplications. Am J Hum Genet 73(4):823–834

Baillie JK, Barnett MW, Upton KR, Gerhardt DJ, Richmond TA, De Sapio F et al (2011) Somatic retrotransposition alters the genetic landscape of the human brain. Nature 479:534–537

Batzer MA, Deininger PL (2002) Alu repeats and human genomic diversity. Nat Rev Genet 3(5):370–379

Beauchamp NJ, Makris M, Preston FE, Peake IR, Daly ME (2000) Major structural defects in the antithrombin gene in four families with type I antithrombin deficiency—partial/complete deletions and rearrangement of the antithrombin gene. Thromb Haemost 83(5):715–721

Beck CR, Collier P, Macfarlane C, Malig M, Kidd JM, Eichler EE et al (2010) LINE-1 retrotransposition activity in human genomes. Cell 141(7):1159–1170
Beck CR, Garcia-Perez JL, Badge RM, Moran JV (2011) LINE-1 elements in structural variation and disease. Annu Rev Genomics Hum Genet 12:187–215
Belancio VP, Hedges DJ, Deininger P (2008) Mammalian non-LTR retrotransposons: for better or worse, in sickness and in health. Genome Res 18(3):343–358
Bennett EA, Keller H, Mills RE, Schmidt S, Moran JV, Weichenrieder O et al (2008) Active Alu retrotransposons in the human genome. Genome Res 18(12):1875–1883
Bernard V, Minnerop M, Bürk K, Kreuz F, Gillessen-Kaesbach G, Zühlke C (2009) Exon deletions and intragenic insertions are not rare in ataxia with oculomotor apraxia 2. BMC Med Genet 10:87
Beveridge R, Pintos J, Parent M, Asselin J, Siemiatycki J (2010) Lung cancer risk associated with occupational exposure to nickel, chromium VI, and cadmium in two population-based case—control studies in Montreal. Am J Ind Med 485(53):476–485
Bochukova EG, Roscioli T, Hedges DJ, Taylor IB, Johnson D, David DJ et al (2009) Rare mutations of FGFR2 causing apert syndrome: identification of the first partial gene deletion, and an Alu element insertion from a new subfamily. Hum Mutat 30(2):204–211
Boffetta P, Nyberg F (2003) Contribution of environmental factors to cancer risk. Br Med Bull 68:71–94
Bogerd HP, Wiegand HL, Hulme AE, Garcia-Perez JL, O'Shea KS, Moran JV et al (2006a) Cellular inhibitors of long interspersed element 1 and Alu retrotransposition. Proc Natl Acad Sci U S A 103(23):8780–8785
Bogerd HP, Wiegand HL, Doehle BP, Lueders KK, Cullen BR (2006b) APOBEC3A and APOBEC3B are potent inhibitors of LTR-retrotransposon function in human cells. Nucleic Acids Res 34(1):89–95
Boissinot S, Chevret P, Furano AV (2000) L1 (LINE-1) retrotransposon evolution and amplification in recent human history. Mol Biol Evol 17(6):915–928
Boissinot S, Entezam A, Young L, Munson PJ, Furano AV (2004) The insertional history of an active family of L1 retrotransposons in humans. Genome Res 14(7):1221–1231
Bouchet C, Vuillaumier-Barrot S, Gonzales M, Boukari S, Le Bizec C, Fallet C et al (2007) Detection of an Alu insertion in the POMT1 gene from three French Walker Warburg syndrome families. Mol Genet Metab 90(1):93–96
Bourc'his D, Bestor TH (2004) Meiotic catastrophe and retrotransposon reactivation in male germ cells lacking Dnmt3L. Nature 431(7004):96–99
Brouha B, Meischl C, Ostertag E, de Boer M, Zhang Y, Neijens H et al (2002) Evidence consistent with human L1 retrotransposition in maternal meiosis I. Am J Hum Genet 71(2):327–336
Brouha B, Schustak J, Badge RM, Lutz-Prigge S, Farley AH, Moran JV et al (2003) Hot L1s account for the bulk of retrotransposition in the human population. Proc Natl Acad Sci U S A 100(9):5280–5285
Burwinkel B, Kilimann MW (1998) Unequal homologous recombination between LINE-1 elements as a mutational mechanism in human genetic disease. J Mol Biol 277(3):513–517
Cabot EL, Angeletti B, Usdin K, Furano AV (1997) Rapid evolution of a young L1 (LINE-1) clade in recently speciated Rattus taxa. J Mol Evol 45(4):412–423
Cai X, Evrony GD, Lehmann HS, Elhosary PC, Mehta BK, Poduri A et al (2014) Single-cell, genome-wide sequencing identifies clonal somatic copy-number variation in the human brain. Cell Rep 8(5):1280–1289
Callinan PA, Wang J, Herke SW, Garber RK, Liang P, Batzer MA (2005) Alu retrotransposition-mediated deletion. J Mol Biol 348(4):791–800
Callinan A, Batzer MA, Callinan PA (2006) Retrotransposable elements and human disease. Genome Dynam 1:104–115
Carreira PE, Richardson SR, Faulkner GJ (2014) L1 retrotransposons, cancer stem cells and oncogenesis. FEBS J 281(1):63–73

Casavant NC, Hardies SC (1994) The dynamics of murine LINE-1 subfamily amplification. J Mol Biol 241(3):390–397

Chen H, Lilley CE, Yu Q, Lee DV, Chou J, Narvaiza I et al (2006) APOBEC3A is a potent inhibitor of adeno-associated virus and retrotransposons. Curr Biol 16(5):480–485

Chen J-M, Masson E, Macek M, Raguénès O, Piskackova T, Fercot B et al (2008) Detection of two Alu insertions in the CFTR gene. J Cyst Fibros 7(1):37–43

Cheung VG, Cheung VG, Spielman RS, Spielman RS, Ewens KG, Ewens KG et al (2005) Mapping determinants of human gene expression by regional and genome-wide association. Nature 437(7063):1365–1369

Chow JC, Ciaudo C, Fazzari MJ, Mise N, Servant N, Glass JL et al (2010) LINE-1 activity in facultative heterochromatin formation during X chromosome inactivation. Cell 141(6):956–969

Claverie-Martin F, González-Acosta H, Flores C, Antón-Gamero M, García-Nieto V (2003) De novo insertion of an Alu sequence in the coding region of the CLCN5 gene results in Dent's disease. Hum Genet 113(6):480–485

Conley ME, Partain JD, Norland SM, Shurtleff SA, Kazazian HH (2005) Two independent retrotransposon insertions at the same site within the coding region of BTK. Hum Mutat 25(3): 324–325

Coufal NG, Garcia-Perez JL, Peng GE, Yeo GW, Mu Y, Lovci MT et al (2009) L1 retrotransposition in human neural progenitor cells. Nature 460(7259):1127–1131

Czech B, Malone CD, Zhou R, Stark A, Schlingeheyde C, Dus M et al (2008) An endogenous small interfering RNA pathway in Drosophila. Nature 453(7196):798–802

de Boer M, van Leeuwen K, Geissler J, Weemaes CM, van den Berg TK, Kuijpers TW et al (2014) Primary immunodeficiency caused by an exonized retroposed gene copy inserted in the CYBB gene. Hum Mutat 35(4):486–496

den Hollander AI, ten Brink JB, de Kok YJ, van Soest S, van den Born LI, van Driel MA et al (1999) Mutations in a human homologue of Drosophila crumbs cause retinitis pigmentosa (RP12). Nat Genet 23(2):217–221

Denissenko MF, Pao A, Tang M, Pfeifer GP (1996) Preferential formation of benzo[a]pyrene adducts at lung cancer mutational hotspots in p53. Science 274(5286):430–432

Doucet-O'Hare TT, Rodić N, Sharma R, Darbari I, Abril G, Choi JA et al (2015) LINE-1 expression and retrotransposition in Barrett's esophagus and esophageal carcinoma. Proc Natl Acad Sci U S A 112(35):201502474

Evrony GD, Cai X, Lee E, Hills LB, Elhosary PC, Lehmann HS et al (2012) Single-neuron sequencing analysis of l1 retrotransposition and somatic mutation in the human brain. Cell 151(3):483–496

Ewing AD, Kazazian HH (2010) High-throughput sequencing reveals extensive variation in human-specific L1 content in individual human genomes. Genome Res 20(9):1262–1270

Ewing AD, Gacita A, Wood LD, Ma F, Xing D, Manda SS et al (2015) Widespread somatic L1 retrotransposition occurs early during gastrointestinal cancer evolution. Genome Res 25(10):1536–1545

Faulkner GJ (2011) Retrotransposons: mobile and mutagenic from conception to death. FEBS Lett 585(11):1589–1594

Feng Q, Moran JV, Kazazian HH, Boeke JD (1996) Human L1 retrotransposon encodes a conserved endonuclease required for retrotransposition. Cell 87(5):905–916

Flynn EK, Kamat A, Lach FP, Donovan FX, Kimble DC, Narisu N et al (2014) Comprehensive analysis of pathogenic deletion variants in fanconi anemia genes. Hum Mutat 35(11):1342–1353

Fornace AJ, Mitchell JB (1986) Induction of B2 RNA polymerase III transcription by heat shock: enrichment for heat shock induced sequences in rodent cells by hybridization subtraction. Nucleic Acids Res 14(14):5793–5811

Gallus GN, Cardaioli E, Rufa A, Da Pozzo P, Bianchi S, D'Eramo C et al (2010) Alu-element insertion in an OPA1 intron sequence associated with autosomal dominant optic atrophy. Mol Vis 16:178–183
Ganguly A, Dunbar T, Chen P, Godmilow L, Ganguly T (2003) Exon skipping caused by an intronic insertion of a young Alu Yb9 element leads to severe hemophilia A. Hum Genet 113(4):348–352
Garcia-Perez JL, Marchetto MCN, Muotri AR, Coufal NG, Gage FH, O'Shea KS et al (2007) LINE-1 retrotransposition in human embryonic stem cells. Hum Mol Genet 16(13):1569–1577
Garcia-Perez JL, Morell M, Scheys JO, Kulpa DA, Morell S, Carter CC et al (2010) Epigenetic silencing of engineered L1 retrotransposition events in human embryonic carcinoma cells. Nature 466(7307):769–773
Giorgi G, Marcantonio P, Del Re B (2011) LINE-1 retrotransposition in human neuroblastoma cells is affected by oxidative stress. Cell Tissue Res 346(3):383–391
Goodier JL, Kazazian HH (2008) Retrotransposons revisited. The restraint and rehabilitation of parasites. Cell 135(1):23–35
Goodier JL, Ostertag EM, Kazazian HH (2000) Transduction of 3′-flanking sequences is common in L1 retrotransposition. Hum Mol Genet 9(4):653–657
Goodier JL, Ostertag EM, Du K, Kazazian HH Jr (2001) A novel active L1 retrotransposon subfamily in the mouse. Genome Res 11(10):1677–1685
Goodier JL, Cheung LE, Kazazian HH (2012) MOV10 RNA helicase is a potent inhibitor of retrotransposition in cells. PLoS Genet 8(10):e1002941
Goodier JL, Cheung LE, Kazazian HH (2013) Mapping the LINE1 ORF1 protein interactome reveals associated inhibitors of human retrotransposition. Nucleic Acids Res 41(15):7401–7419
Graham T, Boissinot S (2006) The genomic distribution of L1 elements: the role of insertion bias and natural selection. J Biomed Biotechnol 2006(1):75327
Green PM, Bagnall RD, Waseem NH, Giannelli F (2008) Haemophilia A mutations in the UK: results of screening one-third of the population. Br J Haematol 143(1):115–128
Grimaldi G, Skowronski J, Singer MF (1984) Defining the beginning and end of KpnI family segments. EMBO J 3(8):1753–1759
Gu Y, Kodama H, Watanabe S, Kikuchi N, Ishitsuka I, Ozawa H et al (2007) The first reported case of Menkes disease caused by an Alu insertion mutation. Brain Dev 29(2):105–108
Halling KC, Lazzaro CR, Honchel R, Bufill JA, Powell SM, Arndt CA et al (1999) Hereditary desmoid disease in a family with a germline Alu I repeat mutation of the APC gene. Hum Hered 49(2):97–102
Hanahan D, Weinberg RA (2000) The Hallmarks of Cancer Review evolve progressively from normalcy via a series of pre. Cell 100(1):57–70
Hanahan D, Weinberg RA (2011) Hallmarks of cancer: the next generation. Cell 144(5):646–674
Hancks DC, Kazazian HH (2012) Active human retrotransposons: variation and disease. Curr Opin Genet Dev 22(3):191–203
Hastings PJ, Ira G, Lupski JR (2009) A microhomology-mediated break-induced replication model for the origin of human copy number variation. PLoS Genet 5(1):e1000327
Hedges DJ, Deininger PL (2007) Inviting instability: transposable elements, double-strand breaks, and the maintenance of genome integrity. Mutat Res 616(1–2):46–59
Helman E, Lawrence MS, Stewart C, Sougnez C, Getz G, Meyerson M (2014) Somatic retrotransposition in human cancer revealed by whole-genome and exome sequencing. Genome Res 24(7):1053–1063
Heras SR, Macias S, Plass M, Fernandez N, Cano D, Eyras E et al (2013) The microprocessor controls the activity of mammalian retrotransposons. Nat Struct Mol Biol 20(10):1173–1181
Hill AS, Foot NJ, Chaplin TL, Young BD (2000) The most frequent constitutional translocation in humans, the t(11;22)(q23;q11) is due to a highly specific alu-mediated recombination. Hum Mol Genet 9(10):1525–1532

Holmes SE, Dombroski BA, Krebs CM, Boehm CD, Kazazian HH (1994) A new retrotransposable human L1 element from the LRE2 locus on chromosome 1q produces a chimaeric insertion. Nat Genet 7(2):143–148

Hormozdiari F, Konkel MK, Prado-Martinez J, Chiatante G, Herraez IH, Walker JA et al (2013) Rates and patterns of great ape retrotransposition. Proc Natl Acad Sci U S A 110(33):13457–13462

Huang CRL, Schneider AM, Lu Y, Niranjan T, Shen P, Robinson MA et al (2010) Mobile interspersed repeats are major structural variants in the human genome. Cell 141(7):1171–1182

Iskow RC, McCabe MT, Mills RE, Torene S, Pittard WS, Neuwald AF et al (2010) Natural mutagenesis of human genomes by endogenous retrotransposons. Cell 141(7):1253–1261

Jacobs KB, Yeager M, Zhou W, Wacholder S, Wang Z, Rodriguez-Santiago B et al (2012) Detectable clonal mosaicism and its relationship to aging and cancer. Nat Genet 44(6):651–658

Janicic N, Pausova Z, Cole DE, Hendy GN (1995) Insertion of an Alu sequence in the Ca(2+)-sensing receptor gene in familial hypocalciuric hypercalcemia and neonatal severe hyperparathyroidism. Am J Hum Genet 56(4):880–886

Kaer K, Speek M (2013) Retroelements in human disease. Gene 518(2):231–241

Kagawa T, Oka A, Kobayashi Y, Hiasa Y, Kitamura T, Sakugawa H et al (2015) Recessive inheritance of population-specific intronic LINE-1 insertion causes a rotor syndrome phenotype. Hum Mutat 36(3):327–332

Kale SP, Moore L, Deininger PL, Roy-Engel AM (2005) Heavy metals stimulate human LINE-1 retrotransposition. Int J Environ Res Public Health 2(1):14–23

Kano H, Godoy I, Courtney C, Vetter MR, Gerton GL, Ostertag EM et al (2009) L1 retrotransposition occurs mainly in embryogenesis and creates somatic mosaicism. Genes Dev 23(11):1303–1312

Kato Y, Kaneda M, Hata K, Kumaki K, Hisano M, Kohara Y et al (2007) Role of the Dnmt3 family in de novo methylation of imprinted and repetitive sequences during male germ cell development in the mouse. Hum Mol Genet 16(19):2272–2280

Kazazian HH (2004) Mobile elements: drivers of genome evolution. Science 303(5664):1626–1632

Kazazian HH, Wong C, Youssoufian H, Scott AF, Phillips DG, Antonarakis SE (1988) Haemophilia A resulting from de novo insertion of L1 sequences represents a novel mechanism for mutation in man. Nature 332(6160):164–166

Khan H, Smit A, Boissinot S (2006) Molecular evolution and tempo of amplification of human LINE-1 retrotransposons since the origin of primates. Genome Res 16(1):78–87

Kinzler KW, Nilbert MC, Su LK, Vogelstein B, Bryan TM, Levy DB et al (1991a) Identification of FAP locus genes from chromosome 5q21. Science 253(5020):661–665

Kinzler KW, Nilbert MC, Vogelstein B, Bryan TM, Levy DB, Smith KJ et al (1991b) Identification of a gene located at chromosome 5q21 that is mutated in colorectal cancers. Science 251(4999):1366–1370

Kloor M, Sutter C, Wentzensen N, Cremer FW, Buckowitz A, Keller M et al (2004) A large MSH2 Alu insertion mutation causes HNPCC in a German kindred. Hum Genet [Internet] 115(5):432–438

Kobayashi K, Nakahori Y, Miyake M, Matsumura K, Kondo-Iida E, Nomura Y et al (1998) An ancient retrotransposal insertion causes Fukuyama-type congenital muscular dystrophy. Nature 394(6691):388–392

Kondo-Iida E, Kobayashi K, Watanabe M, Sasaki J, Kumagai T, Koide H et al (1999) Novel mutations and genotype-phenotype relationships in 107 families with Fukuyama-type congenital muscular dystrophy (FCMD). Hum Mol Genet 8(12):2303–2309

Kutsche K, Ressler B, Katzera HG, Orth U, Gillessen-Kaesbach G, Morlot S et al (2002) Characterization of breakpoint sequences of five rearrangements in L1CAM and ABCD1 (ALD) genes. Hum Mutat 19(5):526–535

Lander E, Linton L, Birren B, Nusbaum C, Zody M, Baldwin J et al (2001) Initial sequencing and analysis of the human genome. Nature 409(6822):860–921

Lanikova L, Kucerova J, Indrak K, Divoka M, Issa J-P, Papayannopoulou T et al (2013) β-Thalassemia due to intronic LINE-1 insertion in the β-globin gene (HBB): molecular mechanisms underlying reduced transcript levels of the β-globin(L1) allele. Hum Mutat 34(10):1361–1365

Laurie CC, Laurie CA, Rice K, Doheny KF, Zelnick LR, McHugh CP et al (2012) Detectable clonal mosaicism from birth to old age and its relationship to cancer. Nat Genet Internet 44(6):642–650

Lee J, Ha J, Son SY, Han K (2012a) Human genomic deletions generated by SVA-associated events. Comp Funct Genomics 2012:807270

Lee E, Iskow R, Yang L, Gokcumen O, Haseley P, Luquette LJ et al (2012b) Landscape of somatic retrotransposition in human cancers. Science 337(6097):967–971

Lee E, Iskow R, Yang L, Gokcumen O, Haseley P, Luquette LJ et al (2012c) Analysis of somatic retrotransposition in human cancers. BMC Proc 6:O23

Levin HL, Moran JV (2011) Dynamic interactions between transposable elements and their hosts. Nat Rev Genet 12(9):615–627

Li X, Scaringe WA, Hill KA, Roberts S, Mengos A, Careri D et al (2001) Frequency of recent retrotransposition events in the human factor IX gene. Hum Mutat 17(6):511–519

Lichtenstein P, Holm NV, Verkasalo PK, Iliadou A, Kaprio J, Koskenvuo M, Pukkala E, Skytthe A, Hemminki K (2000) Environmental and heritable factors in the causation of cancer-analyses of Cohorts of Twins from Sweden, Denmark, and Finland. N Engl J Med 343(2):78–85

Liu C, Zhang X, Huang F, Yang B, Li J, Liu B et al (2012) APOBEC3G inhibits microRNA-mediated repression of translation by interfering with the interaction between Argonaute-2 and MOV10. J Biol Chem 287(35):29373–29383

Lutz SM, Vincent BJ, Kazazian HH, Batzer MA, Moran JV (2003) Allelic heterogeneity in LINE-1 retrotransposition activity. Am J Hum Genet 73(6):1431–1437

Macias S, Plass M, Stajuda A, Michlewski G, Eyras E, Cáceres JF (2012) DGCR8 HITS-CLIP reveals novel functions for the microprocessor. Nat Struct Mol Biol 19(8):760–766

Makino S, Kaji R, Ando S, Tomizawa M, Yasuno K, Goto S et al (2007) Reduced neuron-specific expression of the TAF1 gene is associated with X-linked dystonia-parkinsonism. Am J Hum Genet 80(3):393–406

Malone CD, Hannon GJ (2009) Small RNAs as guardians of the genome. Cell 136(4):656–668

Manco L, Relvas L, Pinto C (2006) Molecular characterization of five Portuguese patients with pyrimidine 5'-nucleotidase deficient hemolytic anemia showing three new P5'NI mutations. Haematologica 91(2):2–3

Masson E, Hammel P, Garceau C, Bénech C, Quéméner-Redon S, Chen J-M et al (2013) Characterization of two deletions of the CTRC locus. Mol Genet Metab 109(3):296–300

Maxwell PH, Burhans WC, Curcio MJ (2011) Retrotransposition is associated with genome instability during chronological aging. Proc Natl Acad Sci U S A 108(51):20376–20381

Meili D, Kralovicova J, Zagalak J, Bonafe L, Fiori L, Blau N et al (2009) Disease-causing mutations improving the branch site and polypyrimidine tract: Pseudoexon activation of LINE-2 and antisense alu lacking the poly(T)-Tail. Hum Mutat 30(5):823–831

Meischl C, Boer M, Ahlin A, Roos D (2000) A new exon created by intronic insertion of a rearranged LINE-1 element as the cause of chronic granulomatous disease. Eur J Hum Genet 8(9):697–703

Miki Y, Nishisho I, Horii A, Miyoshi Y, Utsunomiya J, Kinzler KW et al (1992) Disruption of the APC gene by a retrotransposal insertion of L1 sequence in a colon cancer. Cancer Res 52(3):643–645

Miki Y, Katagiri T, Kasumi F, Yoshimoto T, Nakamura Y (1996) Mutation analysis in the BRCA2 gene in primary breast cancers. Nat Genet 13(2):245–247

Mills RE, Bennett EA, Iskow RC, Devine SE (2007) Which transposable elements are active in the human genome? Trends Genet 23(4):183–191

Miné M, Chen J-M, Brivet M, Desguerre I, Marchant D, de Lonlay P et al (2007) A large genomic deletion in the PDHX gene caused by the retrotranspositional insertion of a full-length LINE-1 element. Hum Mutat 28(2):137–142

Mir AA, Philippe C, Cristofari G (2015) euL1db: the European database of L1HS retrotransposition insertions in humans. Nucleic Acids Res 43(Database issue):D43–D47

Moran JV, Holmes SE, Naas TP, DeBerardinis RJ, Boeke JD, Kazazian HH (1996) High frequency retrotransposition in cultured mammalian cells. Cell 87(5):917–927

Moran JV, DeBerardinis RJ, Kazazian HH (1999) Exon shuffling by L1 retrotransposition. Science 283(5407):1530–1534

Morisada N, Rendtorff ND, Nozu K, Morishita T, Miyakawa T, Matsumoto T et al (2010) Branchio-oto-renal syndrome caused by partial EYA1 deletion due to LINE-1 insertion. Pediatr Nephrol 25(7):1343–1348

Morrish TA, Gilbert N, Myers JS, Vincent BJ, Stamato TD, Taccioli GE et al (2002) DNA repair mediated by endonuclease-independent LINE-1 retrotransposition. Nat Genet 31(2):159–165

Muckenfuss H, Hamdorf M, Held U, Perkovic M, Löwer J, Cichutek K et al (2006) APOBEC3 proteins inhibit human LINE-1 retrotransposition. J Biol Chem 281(31):22161–22172

Mukherjee S, Mukhopadhyay A, Banerjee D, Chandak GR, Ray K (2004) Molecular pathology of haemophilia B: identification of five novel mutations including a LINE 1 insertion in Indian patients. Haemophilia 10(3):259–263

Muotri AR, Chu VT, Marchetto MCN, Deng W, Moran JV, Gage FH (2005) Somatic mosaicism in neuronal precursor cells mediated by L1 retrotransposition. Nature 435(7044):903–910

Muotri AR, Marchetto MCN, Coufal NG, Oefner R, Yeo G, Nakashima K et al (2010) L1 retrotransposition in neurons is modulated by MeCP2. Nature 468(7322):443–446

Muratani K, Hada T, Yamamoto Y, Kaneko T, Shigeto Y, Ohue T et al (1991) Inactivation of the cholinesterase gene by Alu insertion: possible mechanism for human gene transposition. Proc Natl Acad Sci U S A 88(24):11315–11319

Musova Z, Hedvicakova P, Mohrmann M, Tesarova M, Krepelova A, Zeman J et al (2006) A novel insertion of a rearranged L1 element in exon 44 of the dystrophin gene: further evidence for possible bias in retroposon integration. Biochem Biophys Res Commun 347(1):145–149

Mustajoki S, Ahola H, Mustajoki P, Kauppinen R (1999) Insertion of Alu element responsible for acute intermittent porphyria. Hum Mutat 13(6):431–438

Myers JS, Vincent BJ, Udall H, Watkins WS, Morrish TA, Kilroy GE et al (2002) A comprehensive analysis of recently integrated human Ta L1 elements. Am J Hum Genet 71(2):312–326

Nakamura Y, Murata M, Takagi Y, Kozuka T, Nakata Y, Hasebe R et al (2015) SVA retrotransposition in exon 6 of the coagulation factor IX gene causing severe hemophilia B. Int J Hematol 102(1):134–139

Narita N, Nishio H, Kitoh Y, Ishikawa Y, Ishikawa Y, Minami R et al (1993) Insertion of a 5′ truncated L1 element into the 3′ end of exon 44 of the dystrophin gene resulted in skipping of the exon during splicing in a case of Duchenne muscular dystrophy. J Clin Invest 91(5):1862–1867

Oldridge M, Zackai EH, McDonald-McGinn DM, Iseki S, Morriss-Kay GM, Twigg SR et al (1999) De novo alu-element insertions in FGFR2 identify a distinct pathological basis for Apert syndrome. Am J Hum Genet 64(2):446–461

Ostertag EM, Prak ET, DeBerardinis RJ, Moran JV, Kazazian HH (2000) Determination of L1 retrotransposition kinetics in cultured cells. Nucleic Acids Res 28(6):1418–1423

Ovchinnikov I, Troxel AB, Swergold GD (2001) Genomic characterization of recent human LINE-1 insertions: evidence supporting random insertion. Genome Res 11(12):2050–2058

Paterson AL, Weaver JMJ, Eldridge MD, Tavaré S, Fitzgerald RC, Edwards P, OCCAMs Consortium (2015) Mobile element insertions are frequent in oesophageal adenocarcinomas and can mislead paired-end sequencing analysis. BMC Genomics 16(1):473

Peixoto A, Pinheiro M, Massena L, Santos C, Pinto P, Rocha P et al (2013) Genomic characterization of two large Alu-mediated rearrangements of the BRCA1 gene. J Hum Genet 58(2):78–83

Perng W, Mora-Plazas M, Marín C, Rozek LS, Baylin A, Villamor E (2013) A prospective study of LINE-1DNA methylation and development of adiposity in school-age children. PLoS One 8(4):1–7
Pickeral OK, Makałowski W, Boguski MS, Boeke JD (2000) Frequent human genomic DNA transduction driven by line-1 retrotransposition. Genome Res 10(4):411–415
Pizarro JG, Cristofari G (2016) Post-transcriptional control of LINE-1 retrotransposition by cellular host factors in somatic cells. Front Cell Dev Biol 4:14. doi:10.3389/fcell.2016.00014
Qian Y, Mancini-DiNardo D, Judkins T, Cox HC, Daniels C, Holladay J, Ryder M, Coffee B, Bowles KR, Roa BB (2015) Identification of retrotransposons insertion mutations in hereditary cancer. Myriad Genetics, Inc., Salt Lake City, UT
Raiz J, Damert A, Chira S, Held U, Klawitter S, Hamdorf M et al (2012) The non-autonomous retrotransposon SVA is trans-mobilized by the human LINE-1 protein machinery. Nucleic Acids Res 40(4):1666–1683
Rathore K, Wang HCR (2013) Mesenchymal and stem-like cell properties targeted in suppression of chronically-induced breast cell carcinogenesis. Cancer Lett 333(1):113–123
Ray DA, Batzer MA (2011) Reading TE leaves: new approaches to the identification of transposable element insertions. Genome Res 21(6):813–820
Rodić N, Sharma R, Sharma R, Zampella J, Dai L, Taylor MS et al (2014) Long interspersed element-1 protein expression is a hallmark of many human cancers. Am J Pathol 184(5):1280–1286
Rodić N, Steranka JP, Makohon-Moore A, Moyer A, Shen P, Sharma R et al (2015) Retrotransposon insertions in the clonal evolution of pancreatic ductal adenocarcinoma. Nat Med 21(9):1060–1064
Schollen E, Keldermans L, Foulquier F, Briones P, Chabas A, Sánchez-Valverde F et al (2007) Characterization of two unusual truncating PMM2 mutations in two CDG-Ia patients. Mol Genet Metab 90(4):408–413
Segal Y, Peissel B, Renieri A, de Marchi M, Ballabio A, Pei Y et al (1999) LINE-1 elements at the sites of molecular rearrangements in Alport syndrome-diffuse leiomyomatosis. Am J Hum Genet 64(1):62–69
Seleme M d C, Vetter MR, Cordaux R, Bastone L, Batzer MA, Kazazian HH (2006) Extensive individual variation in L1 retrotransposition capability contributes to human genetic diversity. Proc Natl Acad Sci U S A 103(17):6611–6616
Sheen F, Sherry ST, Risch GM, Robichaux M, Nasidze I, Stoneking M et al (2000) Reading between the LINEs: human genomic variation induced by LINE-1 retrotransposition. Genome Res 10(10):1496–1508
Shukla R, Upton KR, Muñoz-Lopez M, Gerhardt DJ, Fisher ME, Nguyen T et al (2013) Endogenous retrotransposition activates oncogenic pathways in hepatocellular carcinoma. Cell 153(1):101–111
Sijen T, Plasterk RHA (2003) Transposon silencing in the Caenorhabditis elegans germ line by natural RNAi. Nature 426(6964):310–314
Skowronski J, Fanning TG, Singer MF (1988) Unit-length line-1 transcripts in human teratocarcinoma cells. Mol Cell Biol 8(4):1385–1397
Slotkin RK, Vaughn M, Borges F, Tanurdzić M, Becker JD, Feijó JA et al (2009) Epigenetic reprogramming and small RNA silencing of transposable elements in pollen. Cell 136(3):461–472
Smit AF, Tóth G, Riggs AD, Jurka J (1995) Ancestral, mammalian-wide subfamilies of LINE-1 repetitive sequences. J Mol Biol 246(3):401–417
Sobrier M-L, Netchine I, Heinrichs C, Thibaud N, Vié-Luton M-P, Van Vliet G et al (2005) Alu-element insertion in the homeodomain of HESX1 and aplasia of the anterior pituitary. Hum Mutat 25(5):503
Solyom S, Ewing AD, Hancks DC, Takeshima Y, Awano H, Matsuo M et al (2012a) Pathogenic orphan transduction created by a nonreference LINE-1 retrotransposon. Hum Mutat 33(2):369–371

Solyom S, Ewing AD, Rahrmann EP, Doucet T, Nelson HH, Burns MB et al (2012b) Extensive somatic L1 retrotransposition in colorectal tumors. Genome Res 22(12):2328–2338
Sorenson TIA, Nielsen GG, Andersen PKTT (1988) Genetic and environmental influences on premature death in adult adoptees. N Engl J Med 318(12):727–732
Stenglein MD, Harris RS (2006) APOBEC3B and APOBEC3F inhibit L1 retrotransposition by a DNA deamination-independent mechanism. J Biol Chem 281(25):16837–16841
Stetson DB, Ko JS, Heidmann T, Medzhitov R (2008) Trex1 prevents cell-intrinsic initiation of autoimmunity. Cell 134(4):587–598
Stewart C, Jural D, Stromberg MP, Walker JA, Konkel ML, Stutz AM, Urban AE, Grubert F, Lam HY, Lee WP, Busby M, Indap AR, Garrison E, Huff C, Xing J, Snyder MP, Jorde LB, Batzer MA, Korbel JO, Marth GT, 1000 Genoms Project (2011) A comprehensive map of mobile element insertion polymorphisms in humans. PLoS Genet 7(8):e1002236
Stoppa-Lyonnet D, Carter PE, Meo T, Tosi M (1990) Clusters of intragenic Alu repeats predispose the human C1 inhibitor locus to deleterious rearrangements. Proc Natl Acad Sci U S A 87(4):1551–1555
Stribinskis V, Ramos KS (2006) Activation of human long interspersed nuclear element 1 retrotransposition by benzo(a)pyrene, an ubiquitous environmental carcinogen. Cancer Res 66(5):2616–2620
Su LK, Steinbach G, Sawyer JC, Hindi M, Ward PA, Lynch PM (2000) Genomic rearrangements of the APC tumor-suppressor gene in familial adenomatous polyposis. Hum Genet 106(1):101–107
Sukarova E, Dimovski AJ, Tchacarova P, Petkov GH, Efremov GD (2001) An Alu insert as the cause of a severe form of hemophilia A. Acta Haematol 106:126–129
Szak ST, Pickeral OK, Makalowski W, Boguski MS, Landsman D, Boeke JD (2002) Molecular archeology of L1 insertions in the human genome. Genome Biol 3(10):research0052
Tabatabaei SM, Heyworth JS, Knuiman MW, Fritschi L (2010) Dietary benzo[a]pyrene intake from meat and the risk of colorectal cancer. Cancer Epidemiol Biomarkers Prev 19(12):3182–3184
Takasu M, Hayashi R, Maruya E, Ota M, Imura K, Kougo K et al (2007) Deletion of entire HLA-A gene accompanied by an insertion of a retrotransposon. Tissue Antigens 70(2):144–150
Taniguchi-Ikeda M, Kobayashi K, Kanagawa M, Yu C, Mori K, Oda T et al (2011) Pathogenic exon-trapping by SVA retrotransposon and rescue in Fukuyama muscular dystrophy. Nature 478:127–131
Tappino B, Regis S, Corsolini F, Filocamo M (2008) An Alu insertion in compound heterozygosity with a microduplication in GNPTAB gene underlies Mucolipidosis II. Mol Genet Metab 93(2):129–133
Temtamy SA, Aglan MS, Valencia M, Cocchi G, Pacheco M, Ashour AM et al (2008) Long interspersed nuclear element-1 (LINE1)-mediated deletion of EVC, EVC2, C4orf6, and STK32B in ellis-van Creveld syndrome with borderline intelligence. Hum Mutat 29(7):931–938
Teugels E, De Brakeleer S, Goelen G, Lissens W, Sermijn E, De Grève J (2005) De novo Alu element insertions targeted to a sequence common to the BRCA1 and BRCA2 genes. Hum Mutat 26(3):284
Tighe PJ, Stevens SE, Dempsey S, Le Deist F, Rieux-Laucat F, Edgar JDM (2002) Inactivation of the Fas gene by Alu insertion: retrotransposition in an intron causing splicing variation and autoimmune lymphoproliferative syndrome. Genes Immun 3(Suppl 1):S66–S70
Toyokuni S, Okamoto K, Yodoi J, Hiai H (1995) Persistent oxidative stress in cancer. FEBS Lett 358(1):1–3
Tubio JMC, Li Y, Ju YS, Martincorena I, Cooke SL, Tojo M et al (2014) Mobile DNA in cancer. Extensive transduction of nonrepetitive DNA mediated by L1 retrotransposition in cancer genomes. Science 345(6196):1251343

Tucker BA, Scheetz TE, Mullins RF, DeLuca AP, Hoffmann JM, Johnston RM et al (2011) Exome sequencing and analysis of induced pluripotent stem cells identify the cilia-related gene male germ cell-associated kinase (MAK) as a cause of retinitis pigmentosa. Proc Natl Acad Sci U S A 108(34):E569–E576

Udaka T, Okamoto N, Aramaki M, Torii C, Kosaki R, Hosokai N et al (2007) An Alu retrotransposition-mediated deletion of CHD7 in a patient with CHARGE syndrome. Am J Med Genet A 143(7):721–726

Van Den Hurk JAJM, Van De Pol DJR, Wissinger B, Van Driel MA, Hoefsloot LH, De Wijs IJ et al (2003) Novel types of mutation in the choroideremia (CHM) gene: a full-length L1 insertion and an intronic mutation activating a cryptic exon. Hum Genet 113(3):268–275

Van Valen L (1973) A new evolutionary theory. Evol Theory 1(1):1–30

Varon R, Gooding R, Steglich C, Marns L, Tang H, Angelicheva D et al (2003) Partial deficiency of the C-terminal-domain phosphatase of RNA polymerase II is associated with congenital cataracts facial dysmorphism neuropathy syndrome. Nat Genet 35(2):185–189

Vidaud D, Tartary M, Costa J (1993) Nucleotide substitutions at the-6 position in the promoter region of the factor IX gene result in different severity of hemophilia B Leyden: consequences for genetic counseling. Hum Genet 91(3):241–244

Vogt J, Bengesser K, Claes KB, Wimmer K, Mautner V-F, van Minkelen R et al (2014) SVA retrotransposon insertion-associated deletion represents a novel mutational mechanism underlying large genomic copy number changes with non-recurrent breakpoints. Genome Biol 15(6):R80

Wallace MR, Andersen LB, Saulino AM, Gregory PE, Glover TW, Collins FS (1991) A de novo Alu insertion results in neurofibromatosis type 1. Nature 353(6347):864–866

Walsh CP, Chaillet JR, Bestor TH (1998) Transcription of IAP endogenous retroviruses is constrained by cytosine methylation. Nat Genet 20(2):116–117

Wang T, Lerer I, Gueta Z, Sagi M, Kadouri L, Peretz T et al (2001) A deletion/insertion mutation in the BRCA2 gene in a breast cancer family: a possible role of the Alu-polyA tail in the evolution of the deletion. Genes Chromosom Cancer 31(1):91–95

Wang J, Song L, Grover D, Azrak S, Batzer MA, Liang P (2006) dbRIP: a highly integrated database of retrotransposon insertion polymorphisms in humans. Hum Mutat 27:323–329

Watanabe T, Takeda A, Tsukiyama T, Mise K, Okuno T, Sasaki H et al (2006) Identification and characterization of two novel classes of small RNAs in the mouse germline: retrotransposon-derived siRNAs in oocytes and germline small RNAs in testes. Genes Dev 20(13):1732–1743

Watanabe T, Totoki Y, Toyoda A, Kaneda M, Kuramochi-Miyagawa S, Obata Y et al (2008) Endogenous siRNAs from naturally formed dsRNAs regulate transcripts in mouse oocytes. Nature 453(7194):539–543

Wheelan SJ, Aizawa Y, Han JS, Boeke JD (2005) Gene-breaking: A new paradigm for human retrotransposon-mediated gene evolution. Genome Res 15(8):1073–1078

Wimmer K, Callens T, Wernstedt A, Messiaen L (2011) The NF1 gene contains hotspots for L1 endonuclease-dependent de novo insertion. PLoS Genet 7(11):e1002371

Woodcock DM, Lawler CB, Linsenmeyer ME, Doherty JP, Warren WD (1997) Asymmetric methylation in the hypermethylated CpG promoter region of the human L1 retrotransposon. J Biol Chem 272(12):7810–7816

Wu X, Lu Y, Ding Q, You G, Dai J, Xi X et al (2014) Characterisation of large F9 deletions in seven unrelated patients with severe haemophilia B. Thromb Haemost 112(3):459–465

Wulff K, Gazda H, Schröder W, Robicka-Milewska R, Herrmann FH (2000) Identification of a novel large F9 gene mutation-an insertion of an Alu repeated DNA element in exon e of the factor 9 gene. Hum Mutat 15(3):299

Xing J, Zhang Y, Han K, Salem AH, Sen SK, Huff CD et al (2009) Mobile elements create structural variation: analysis of a complete human genome. Genome Res 19(9):1516–1526

Yang N, Kazazian HH (2006) L1 retrotransposition is suppressed by endogenously encoded small interfering RNAs in human cultured cells. Nat Struct Mol Biol 13(9):763–771

Yoder JA, Walsh CP, Bestor TH (1997) Cytosine methylation and the ecology of intragenomic parasites. Trends Genet 13(8):335–340

Zhang Y, Dipple KM, Vilain E, Huang BL, Finlayson G, Therrell BL et al (2000) AluY insertion (IVS4-52ins316alu) in the glycerol kinase gene from an individual with benign glycerol kinase deficiency. Hum Mutat 15(4):316–323

The Mobilisation of Processed Transcripts in Germline and Somatic Tissues

Adam D. Ewing

1 The Annotation of Retrogenes in Reference Genome Assemblies

The term 'processed pseudogene' was first used to describe a copy of a 5S RNA gene in *Xenopus laevis* which was very similar to the known 5S RNA sequence, but with a number of changes leading the authors to label the pseudogene as a 'relic of evolution' (Jacq et al. 1977). A number of publications followed, describing processed pseudogenes in mice, rats, and humans, reviewed in 1985 by Vanin (Vanin 1985). In that review, Vanin proposes that the term *pseudogene* only be used to refer to copies that are non-functional but related in sequence to their progenitor or 'parent' gene. It is important when considering the potential impact of re-integrated processed transcripts to have a standardised nomenclature with reference to their potential for functionality (Mighell et al. 2000): therefore we adopt the convention where the term *retrogene* refers to a functional copy, *retropseudogene* refers to a non-functional processed pseudogene, and *retrocopy* refers to the sequence reverse-transcribed from the processed transcript without specifying whether the result is functional or not (Vinckenbosch et al. 2006; Kaessmann et al. 2009).

The repetitive nature of many plant and animal genomes was suggested from early hybridisation and dissociation experiments (Britten and Kohne 1968), and confirmed by the sequencing and assembly of mammalian genomes, beginning with human in 2001 (Lander et al. 2001) and mouse in 2002 (Mouse Genome Sequencing Consortium et al. 2002). There are many sequences that exist in more than one copy, which is one of many reasons why genome assembly following shotgun sequencing is not straightforward (Pevzner et al. 2001). This problem is further exacerbated today when current genome sequencing technology yields billions of reads even shorter than those used

A.D. Ewing (✉)
Mater Research Institute—University of Queensland, Brisbane, QLD, Australia
e-mail: adam.ewing@mater.uq.edu.au

G. Cristofari (ed.), *Human Retrotransposons in Health and Disease*,
DOI 10.1007/978-3-319-48344-3_4

to assemble the first draft of the human genome reference (Alkan et al. 2011). The existence of multiple copies of many exon sequences is not only problematic for accurate genome assembly of the gene sequences, but it is also an important consideration when designing probes for hybridisation or PCR experiments (Menon et al. 1991; Hurteau and Spivack 2002), and when mapping RNA-seq reads to the reference genome (Schrider et al. 2011b). Therefore, it is practically important to have a catalogue or annotation of which genes have retrocopies and where in the reference genome the retrocopies are located. A number of approaches have yielded retrocopy annotations in assembled reference genomes (Zhang et al. 2002, 2003; Torrents et al. 2003; Emerson et al. 2004; Marques et al. 2005; Baertsch et al. 2008; Carelli et al. 2016). The list maintained at pseudogene.org (Karro et al. 2007) is a frequently used example, and is regularly updated with new reference genome assemblies. In its simplest form, the process for finding retrocopies in an assembled genome reference is reasonably straightforward, provided that some gene annotation exists in the form of protein or mRNA sequences. The sequences corresponding to processed messenger RNAs (i.e. without introns) are aligned to the reference assembly via a sequence aligner such as BLAST (Altschul et al. 1990), and post-processed into gene copy and retrocopy predictions through a series of filters and heuristics to further categorise them and remove false positives that might be due to transposable elements or low-complexity matches. The results for the human genome vary between about 4000 (Marques et al. 2005; Baertsch et al. 2008) and about 8000 (Zhang et al. 2003) retrocopies or more (Torrents et al. 2003), with the variability likely due to a combination of dataset choice, alignment parameters, and filtering methods. When choosing a pseudogene annotation to use, the nature of the application should be considered: studies of pseudogene functionalization might choose a more stringent filtering, whereas studies that require a pseudogene list as a filter for false positives might consider using a more extensive list of candidates.

Surveys of retrocopies in the human genome allow analysis of global patterns of retrocopy acquisition. From these surveys, it is immediately obvious that some genes produce more retrocopies than others. This appears to be related to genes that are highly and broadly expressed, and genes that are expressed in the germline (Vinckenbosch et al. 2006). The latter criteria should be expected, as for retrocopies to accumulate they must be heritable. This includes genes such as GAPDH (Liu et al. 2009), ribosomal genes (Zhang et al. 2002) including cytochrome c (Zhang and Gerstein 2003), and olfactory receptors (Glusman et al. 2001; Gilad et al. 2005). Another observation is the high level of retrocopy 'traffic' from the X chromosome onto autosomes relative to random expectation (Emerson et al. 2004; Vinckenbosch et al. 2006; Marques et al. 2005). There are a number of examples of 'out-of-X' retrogenes critical for fertility (Rohozinski and Bishop 2004; Bradley et al. 2004; Rohozinski et al. 2006), providing evidence that this increase in autosomal retrogenes originating from the X chromosome is related to meiotic male sex chromosome inactivation (MSCI), specifically to compensate for the silencing of critical genes during MSCI. This phenomenon has also been explored in the other logical direction: the divergence between out-of-X genes and their parents can be used to date the onset of MSCI in therian evolution to ~180 MYA (Potrzebowski et al. 2008).

2 Processed Transcripts as Non-autonomous Mobile DNA

The genomes of humans, mice, and most other mammalian species are colonised by a family of non-LTR retrotransposons known as Long INterspersed Elements, or LINEs. LINEs are autonomous retroelements, meaning that they encode all the machinery necessary for their own mobilisation through target-primed reverse transcription (TPRT) (Luan et al. 1993). While there seems to be a preference for LINE-1 elements mobilising their own transcripts (*cis*-preference) (Kulpa and Moran 2006), they have also been shown to mobilise other RNA species *in trans*. This includes non-autonomous retroelements, which in humans include *Alu* elements (Dewannieux et al. 2003), SVA elements (Hancks et al. 2011; Raiz et al. 2012), and most topically processed mRNAs (Esnault et al. 2000). The insertion mechanism is why retrogenes lack introns: TPRT operates on processed transcripts. This is clear from the cultured cell retrotransposition assay, which uses a globin intron in an antisense-driven reporter gene which when processed by the spliceosome yields an intact reporter transcript which is subsequently reverse-transcribed and integrated at the endonuclease cut site (Moran et al. 1996). Additionally, recent evidence indicates a requirement for polyadenlyation at the 3′ end of sequences retroposed by LINE-1 (Doucet et al. 2015), which is generally satisfied by the poly-A signals found at the 3′ end of most genes (Colgan and Manley 1997).

In total, the activity of LINE elements in *cis* and in *trans* is responsible for ~25 % of the human genome and some 4000–8000 processed pseudogenes depending on the annotation method, respectively. The number of processed pseudogene annotations differs considerably across species (Navarro and Galante 2015; Carelli et al. 2016). Interestingly, but perhaps not unexpectedly, species which lack substantial non-LTR retrotransposition activity, or whose retrotransposable elements operate through a mechanism non-conducive to mobilising mRNAs *in trans*, accumulate relatively few retrocopies. Examples include platypus (Warren et al. 2008; Carelli et al. 2016) and chicken (International Chicken Genome Sequencing Consortium 2004; Suh 2015; Carelli et al. 2016): the chicken genome is populated by CR1 elements which may require a specific sequence on the 3′ end for mobilisation to occur (Suh 2015), and the platypus genome contains active LINE-2 elements (Warren et al. 2008) which also share 3′ sequence with the non-autonomous MIR elements they mobilise (Jurka et al. 2005), perhaps suggesting that in these species, specific sequences are necessary for recognition and *trans* mobilisation, which would generally preclude mobilisation of processed gene transcripts.

3 Pseudogenes as a Source for Evolutionary Innovation

Gene duplication was first proposed as a key mechanism feeding evolutionary innovation by Susumu Ohno in his 1970 book “Evolution by Gene Duplication” (Ohno 1970). A number of features of gene retrocopies suggest that they are important for driving the evolution of the coding and non-coding repertoire of the genome.

Firstly, many retrocopies have coding potential immediately upon retrotransposition as they are derived directly from processed mRNAs—of course, many of these new copies will not be functional due to mutations introduced by the transcription or reverse transcription, or due to the propensity for TPRT to generate truncated transcripts (Pavlícek et al. 2002). Most new retrocopies are probably transcriptionally inactive upon retrotransposition as the promoters that drove their parent genes are generally not included in the duplicated mRNAs. This is an important consideration for evolutionary innovation: the lack of the parental promoter means new retrocopies will, in some instances, be driven by promoters and regulated by enhancers which lead to expression in a different tissue context than the parent gene. An example of this is the *jingwei* gene in *Drosophila teissieri* and *D. yakuba* which appeared about 2.5 Ma ago when a processed transcript originating from the alcohol dehydrogenase (*Adh*) gene was retrotransposed into a new location where it attracted transcription originating from upstream exons (Long and Langley 1993). This ability for retrocopies to create functional fusion products is the basis for another evolutionary innovation, TRIM5-CypA gene fusions which restrict infection by retroviruses such as HIV and MLV (Sayah et al. 2004; Nisole et al. 2004; Yap et al. 2004). Insertions of a cyclophilin A (CypA) retrocopy into the TRIM5α gene have occurred at least twice, independently, in owl monkeys (Sayah et al. 2004; Nisole et al. 2004; Virgen et al. 2008) and in rhesus species where it has been detected in both rhesus macaque (Wilson et al. 2008) and pigtail macaque (Virgen et al. 2008).

As previously mentioned, some retrogenes compensate for genes silenced during MSCI, and these provide interesting examples of how retrocopies might take on functional roles similar to the parent gene but in a different temporal context. The first documented example is phosphoglycerate kinase (PGK), a critical enzyme for glycolysis (McCarrey and Thomas 1987): PGK-1 is expressed in all somatic tissues and in the premeiotic germline, but after MSCI the autosomal testis-specific copy of PGK, PGK-2 on chromosome 6, takes over during the later stages of spermatogenesis. Additional more recently discovered examples include UTP14c and UTP14b, which are autosomal retrogenes originating from a parent gene on the X chromosome and are critical for spermatogenesis in mice (Rohozinski and Bishop 2004; Bradley et al. 2004) and in humans (Rohozinski et al. 2006).

Transcriptional activity has been detected corresponding to a myriad number of retrocopies (Vinckenbosch et al. 2006; Baertsch et al. 2008; Kalyana-Sundaram et al. 2012; Carelli et al. 2016). Detecting transcription of duplicated genes requires special precautions, given that the sequences of the retrocopy-derived transcripts will have strong similarity to transcripts originating from the parent gene locus. In general, this problem is approached through the use of unique regions within the retrocopy that differ from the parent gene through the accumulation of mutations. Examination of transcribed retrocopies suggests multiple mechanisms for the establishment of new promoters. These include 'piggybacking' on the transcriptional activity of other genes or transcribed elements nearby the insertion site, inclusion of the promoter region from the parent transcript, and *de novo* origination of new promoters. While it has been suggested that pre-existing promoters nearby the retrocopy insertion sites account for many of the transcribed retrocopies (Vinckenbosch et al. 2006), a recent study suggests that 'enhancer-like' regions enriched for CpG islands and active histone marks account for the origins of a large majority of retrocopy-associated transcripts (Carelli et al. 2016). Transcribed pseudogenes have the potential to affect the transcript levels of their parent genes through a number of documented mechanisms. Most of these involve antisense RNAs produced by retrocopies which are able to duplex with the transcript originating from the parent gene. Examples of this duplex formation include nitric oxide synthase (NOS) from snail neurons (Korneev et al. 1999). Further studies provide a mechanism for the duplex decreasing parent transcript levels: a pair of papers identified many genes regulated in mouse oocytes via a mechanism where dicer-mediated generation of regulatory siRNAs from dsRNAs derived from the RNA-RNA duplex between transcribed retrocopies and their parent genes (Watanabe et al. 2008; Tam et al. 2008). Additional evidence for this retrocopy-mediated means of affecting parent gene transcript level includes a study in *Trypanosoma brucei*, again showing dicer-mediated formation of siRNAs from the retrocopy:parent dsRNA duplex (Wen et al. 2011). This effect can also point in the opposite direction: endogenous miRNAs which would normally regulate a canonical protein-coding gene can be 'absorbed' by retrocopy-derived transcripts present in the same cell. A key example of this involves PTEN, an important tumour suppressor disrupted in many tumours: loss of an expressed PTEN pseudogene (PTENP1) leads to miRNAs which were normally binding to the PTENP1 transcript to bind and downregulate PTEN, thus suppressing the tumour suppressor and contributing to tumour growth (Poliseno et al. 2010). In addition to these small RNA-mediated mechanisms, an effect of the retrocopy-derived transcript on parent gene transcript stability has also been reported (Hirotsune et al. 2003). In an interesting functional example of a retrocopy insertion interfering with its parent gene, Ostrander and colleagues (Parker et al. 2009) identified a novel expressed retrocopy of *Fgf4* which reduced the expression of the parent gene, leading to the chondrodysplastic (i.e. 'short legged') phenotype associated with several popular dog breeds.

4 Gene Retrocopy Insertion Polymorphisms

Recent LINE insertions and insertions of the other non-autonomous retrotransposons mobilised by LINEs often exist as insertional polymorphisms: copies present in one or more individuals but not fixed across all human populations. Current estimates for the rate at which new insertions accumulate in the human germline range from 1 new insertion for every 100–200 live births for LINE-1 elements (Xing et al. 2009; Ewing and Kazazian 2010; Huang et al. 2010; Stewart et al. 2011), to 1 in 20–50 births for Alu elements (Cordaux et al. 2006; Xing et al. 2009; Stewart et al. 2011) and 1 in 500–800 births for the seemingly more quiescent SVA elements (Xing et al. 2009; Stewart et al. 2011). As these mutations accumulate, they are passed along to subsequent generations as segregating variation, often linked to other nearby variants by virtue of becoming part of a common haplotype (Kuhn et al. 2014; Sudmant et al. 2015). As methods for detecting insertions absent from reference genomes have advanced, many thousands of these retrotransposon insertion polymorphisms (RIPs) have been catalogued through dozens of studies (Ewing 2015). It stands to reason that since gene retrocopies arise through non-autonomous LINE-mediated retrotransposition, and since retrotransposons are actively accumulating in the genomes of virtually all plant and animal species, gene-derived retrocopies should also be accumulating in genomes that have active non-LTR retrotransposons, depending on the sequence requirements for non-autonomous retrotransposition.

Initial observations of non-reference spliced introns present in DNA from flies (Schrider et al. 2011a) and observations of non-reference retrocopies in humans accompanying larger studies of structural variation (Conrad et al. 2009; Lam et al. 2010; Karakoc et al. 2012) provided evidence that many retrocopies may exist that are not fixed in populations. These studies were followed by a series of papers specifically focused on identifying non-reference retrocopy insertions from various WGS data sets (Ewing et al. 2013; Schrider et al. 2013; Abyzov et al. 2013; Cooke et al. 2014; Kabza et al. 2015), although the exact terminology for differentially present processed transcript insertions varied. Terminology across these studies includes gene retrocopy insertion polymorphism (GRIP) (Ewing et al. 2013), retroduplication variant (RDV) (Abyzov et al. 2013; Kabza et al. 2015), and retroCNV (Schrider et al. 2011a, 2013). In this chapter, we use the term 'GRIP' to refer to polymorphisms where an insertion is shared between multiple people at some allele frequency, and gene retrocopy insertion (GRI) to refer to instances where the allele frequency is not known or not relevant, as is the case for somatically acquired GRIs. Each of these studies used whole-genome sequence (WGS) data to identify GRIPs across multiple individuals, and two (Ewing et al. 2013; Cooke et al. 2014) also examine patient-matched pairs of tumour and normal genomes to identify instances where processed transcripts were retrotransposed somatically. The findings in terms of number of insertions, species, and approach are summarised in Table 1.

The bioinformatic approach to GRIP discovery closely mirrors the approach for detecting transposable elements in WGS data, reviewed in Ewing (2015) and shown in Fig. 1b, c. Detection of transposable elements mostly uses two signals to identify insertion locations from short paired-end sequence data: discordant read mapping and split read (or junction read) mapping. Discordant read mappings are those where the two ends of a read pair to not map in a way that is consistent with expectation set by the library preparation protocol. These might be cases where the ends best map to different chromosomes, distal locations on the same chromosome, or in the wrong orientation. Split reads are based on the direct detection of non-reference junctions in reads or contigs assembled from reads: one region of the split read maps to the insertion site while the other maps to the sequence that was inserted but not reflected in the reference genome assembly. There are some important differences between detecting gene retrocopy insertions and detecting transposable element insertions. Processed transcripts lack introns, as do the resulting insertions, so the presence of non-reference junctions is an indicator of a gene retrocopy insertion (Fig. 1b, c). In general, the hallmarks of retrotransposition apply to detection of novel retrocopy insertions and should be considered in approaches to their detection: most have target-site duplications (TSDs), the presence of a poly-A tail, and ideally

Table 1 Differential retrocopy insertions discovered among individual genomes

Study	Species	Samples	In reference	Genome junction	Exon junction only	Somatic
Cooke et al. (2014)	Human	660	n/a	31	11	42
Abyzov et al. (2013)	Human	968	27	37[a]	111	n/a
Schrider et al. (2013)	Human	17 (Discovery)[b]	18	21	52	n/a
Ewing et al. (2013)	Human	1024	10	51	n/a	3
Ewing et al. 2013	Chimpanzee	10	n/a	19	n/a	n/a
Ewing et al. (2013)	Mouse	17 Lines	n/a	755	n/a	n/a
Schrider et al. (2011a)	Drosophila	37 Lines	n/a	n/a	34	n/a

'In reference' denotes insertions present in the human reference genome assembly but absent from one or more individual genomes

'Genome junction' refers to insertion with a known genomic location

'Exon junction only' refers to reported retrocopy insertions only evident from detection of non-reference exon-exon junctions

Somatic insertions, if present, are included in the total number of detected insertions

[a]Counted TMEM126B with 2 insertions as an additional insertion in addition to the 36 presented in this study

[b]Larger set of samples was used for genotyping

detection at both the 5′ and 3′ junctions between the insertion and the reference genome (Fig. 1b). Many of the GRI-focused studies (Schrider et al. 2011a, 2013; Abyzov et al. 2013; Cooke et al. 2014; Kabza et al. 2015) also report events using exon-exon junctions as evidence for the existence of a retrocopy, although the genomic location of the GRI is unknown—this method of detection may be associated with a higher false-positive rate (Richardson et al. 2014).

In addition to cataloging this source of genomic variation, these studies have contributed various pieces of information concerning the biology of retrocopies and insights into retrotransposon biology in general. Schrider et al. found that an appreciable amount of gene copy number variation in *Drosophila melanogaster* can be explained by GRIPs (Schrider et al. 2011a), and in another study showed that GRIPs in humans could create chimeric transcripts (Schrider et al. 2013). Retrocopy-associated chimeric transcripts were also observed for some somatically acquired insertions (Cooke et al. 2014). Applying some basic assumptions about effective population size and sharing of GRIP alleles between individuals, we were able to conservatively estimate the rate of new GRIP occurrence at 1 for about every 5000 live births (Ewing et al. 2013). In another study, the parent gene's relationship with cell cycle was used to infer that retrotransposition may occur more often during the transition from M to G1 phase (Abyzov et al. 2013). Given that retrotransposon insertions are known to cause disease through insertional mutagenesis (Kazazian et al. 1988; Hancks and Kazazian 2012), it stands to reason that

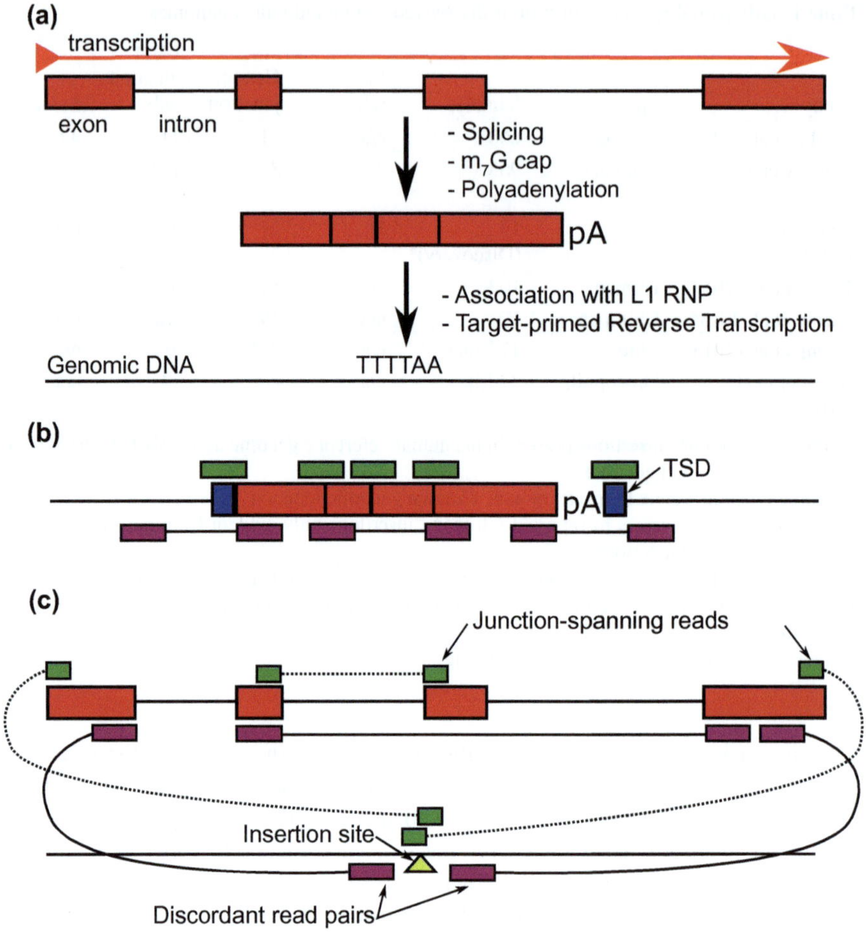

Fig. 1 Detection of non-reference retrocopy insertions from whole-genome sequencing. (**a**) Mechanism of retrocopy insertion: retrocopies are derived from L1-mediated retrotransposition of processed transcripts. Because mobilisation is L1 mediated, insertions occur preferentially at canonical 'TTTTAA' endonuclease motifs. (**b**) Read mapping signatures of retrocopies, shown with retrocopy present. (**c**) Read mapping signatures indicating the presence of a non-reference retrocopy, shown relative to the empty insertion site

retrocopy insertions might also play a similar role in a limited number of instances. Indeed, a retrotransposed copy of TMP1 mRNA was found in the first intron of the CYBB gene in a patient affected by X-linked chronic granulomatous disease (CGD) (de Boer et al. 2014). This led to the inclusion through exonisation of an additional sequence between CYBB exons 1 and 2, which introduced a premature stop codon. In summary, gene retrocopy insertions are a mechanistically interesting, and in a few cases medically relevant, form of insertional structural variation that should be considered along with retrotransposon insertions and DNA repair-mediated events.

References

Abyzov A et al (2013) Analysis of variable retroduplications in human populations suggests coupling of retrotransposition to cell division. Genome Res 23(12):2042–2052

Alkan C, Sajjadian S, Eichler EE (2011) Limitations of next-generation genome sequence assembly. Nat Methods 8(1):61–65

Altschul SF et al (1990) Basic local alignment search tool. J Mol Biol 215(3):403–410

Baertsch R et al (2008) Retrocopy contributions to the evolution of the human genome. BMC Genomics 9:466

Bradley J et al (2004) An X-to-autosome retrogene is required for spermatogenesis in mice. Nat Genet 36(8):872–876

Britten RJ, Kohne DE (1968) Repeated sequences in DNA. Hundreds of thousands of copies of DNA sequences have been incorporated into the genomes of higher organisms. Science 161(3841):529–540

Carelli FN et al (2016) The life history of retrocopies illuminates the evolution of new mammalian genes. Genome Res 26(3):301–314. doi:10.1101/gr.198473.115

Colgan DF, Manley JL (1997) Mechanism and regulation of mRNA polyadenylation. Genes Dev 11(21):2755–2766

Conrad DF et al (2009) Origins and functional impact of copy number variation in the human genome. Nature 464(7289):704–712

Cooke SL et al (2014) Processed pseudogenes acquired somatically during cancer development. Nat Commun 5:3644

Cordaux R et al (2006) Estimating the retrotransposition rate of human Alu elements. Gene 373:134–137

de Boer M et al (2014) Primary immunodeficiency caused by an exonized retroposed gene copy inserted in the CYBB gene. Hum Mutat 35(4):486–496

Dewannieux M, Esnault C, Heidmann T (2003) LINE-mediated retrotransposition of marked Alu sequences. Nat Genet 35(1):41–48

Doucet AJ et al (2015) A 3′ Poly(A) tract is required for LINE-1 retrotransposition. Mol Cell 60(5):728–741

Emerson JJ, Kaessmann H, Betrán E et al (2004) Extensive gene traffic on the mammalian X chromosome. Science 303(5657):537–540

Esnault C, Maestre J, Heidmann T (2000) Human LINE retrotransposons generate processed pseudogenes. Nat Genet 24(4):363–367

Ewing AD (2015) Transposable element detection from whole genome sequence data. Mob DNA 6:24

Ewing AD, Kazazian HH (2010) High-throughput sequencing reveals extensive variation in human-specific L1 content in individual human genomes. Genome Res 20(9):1262–1270

Ewing AD et al (2013) Retrotransposition of gene transcripts leads to structural variation in mammalian genomes. Genome Biol 14(3):R22

Gilad Y, Man O, Glusman G (2005) A comparison of the human and chimpanzee olfactory receptor gene repertoires. Genome Res 15(2):224–230

Glusman G et al (2001) The complete human olfactory subgenome. Genome Res 11(5):685–702

Hancks DC, Kazazian HH Jr (2012) Active human retrotransposons: variation and disease. Curr Opin Genet Dev 22(3):191–203

Hancks DC et al (2011) Retrotransposition of marked SVA elements by human L1s in cultured cells. Hum Mol Genet 20(17):3386–3400

Hirotsune S et al (2003) An expressed pseudogene regulates the messenger-RNA stability of its homologous coding gene. Nature 423(6935):91–96

Huang CRL et al (2010) Mobile interspersed repeats are major structural variants in the human genome. Cell 141(7):1171–1182

Hurteau GJ, Spivack SD (2002) mRNA-specific reverse transcription-polymerase chain reaction from human tissue extracts. Anal Biochem 307(2):304–315

International Chicken Genome Sequencing Consortium (2004) Sequence and comparative analysis of the chicken genome provide unique perspectives on vertebrate evolution. Nature 432(7018):695–716

Jacq C, Miller JR, Brownlee GG (1977) A pseudogene structure in 5S DNA of Xenopus laevis. Cell 12(1):109–120

Jurka J et al (2005) Repbase Update, a database of eukaryotic repetitive elements. Cytogenet Genome Res 110(1–4):462–467

Kabza M et al (2015) Inter-population differences in retrogene loss and expression in humans. PLoS Genet 11(10):e1005579

Kaessmann H, Vinckenbosch N, Long M (2009) RNA-based gene duplication: mechanistic and evolutionary insights. Nat Rev Genet 10(1):19–31

Kalyana-Sundaram S et al (2012) Expressed pseudogenes in the transcriptional landscape of human cancers. Cell 149(7):1622–1634

Karakoc E et al (2012) Detection of structural variants and indels within exome data. Nat Methods 9(2):176–178

Karro JE et al (2007) Pseudogene.org: a comprehensive database and comparison platform for pseudogene annotation. Nucleic Acids Res 35(Database issue):D55–D60

Kazazian HH et al (1988) Haemophilia A resulting from de novo insertion of L1 sequences represents a novel mechanism for mutation in man. Nature 332(6160):164–166

Korneev SA, Park JH, O'Shea M (1999) Neuronal expression of neural nitric oxide synthase (nNOS) protein is suppressed by an antisense RNA transcribed from an NOS pseudogene. J Neurosci 19(18):7711–7720

Kuhn A et al (2014) Linkage disequilibrium and signatures of positive selection around LINE-1 retrotransposons in the human genome. Proc Natl Acad Sci U S A 111(22):8131–8136

Kulpa DA, Moran JV (2006) Cis-preferential LINE-1 reverse transcriptase activity in ribonucleoprotein particles. Nat Struct Mol Biol 13(7):655–660

Lam HYK et al (2010) Nucleotide-resolution analysis of structural variants using BreakSeq and a breakpoint library. Nat Biotechnol 28(1):47–55

Lander ES et al (2001) Initial sequencing and analysis of the human genome. Nature 409(6822):860–921

Liu Y-J et al (2009) Comprehensive analysis of the pseudogenes of glycolytic enzymes in vertebrates: the anomalously high number of GAPDH pseudogenes highlights a recent burst of retrotrans-positional activity. BMC Genomics 10:480

Long M, Langley CH (1993) Natural selection and the origin of jingwei, a chimeric processed functional gene in Drosophila. Science 260(5104):91–95

Luan DD et al (1993) Reverse transcription of R2Bm RNA is primed by a nick at the chromosomal target site: a mechanism for non-LTR retrotransposition. Cell 72(4):595–605

Marques AC et al (2005) Emergence of young human genes after a burst of retroposition in primates. PLoS Biol 3(11):e357

McCarrey JR, Thomas K (1987) Human testis-specific PGK gene lacks introns and possesses characteristics of a processed gene. Nature 326(6112):501–505

Menon RS et al (1991) RT-PCR artifacts from processed pseudogenes. PCR Methods Appl 1(1):70–71

Mighell AJ et al (2000) Vertebrate pseudogenes. FEBS Lett 468(2–3):109–114

Moran JV et al (1996) High frequency retrotransposition in cultured mammalian cells. Cell 87(5):917–927

Mouse Genome Sequencing Consortium et al (2002) Initial sequencing and comparative analysis of the mouse genome. Nature 420(6915):520–562

Navarro FCP, Galante PAF (2015) A genome-wide landscape of retrocopies in primate genomes. Genome Biol Evol 7(8):2265–2275

Nisole S et al (2004) A Trim5-cyclophilin A fusion protein found in owl monkey kidney cells can restrict HIV-1. Proc Natl Acad Sci U S A 101(36):13324–13328

Ohno S (1970) Evolution by gene duplication. Springer, Berlin

Parker HG et al (2009) An expressed fgf4 retrogene is associated with breed-defining chondrodysplasia in domestic dogs. Science 325(5943):995–998

Pavlícek A et al (2002) Length distribution of long interspersed nucleotide elements (LINEs) and processed pseudogenes of human endogenous retroviruses: implications for retrotransposition and pseudogene detection. Gene 300(1–2):189–194

Pevzner PA, Tang H, Waterman MS (2001) An Eulerian path approach to DNA fragment assembly. Proc Natl Acad Sci U S A 98(17):9748–9753

Poliseno L et al (2010) A coding-independent function of gene and pseudogene mRNAs regulates tumour biology. Nature 465(7301):1033–1038

Potrzebowski L et al (2008) Chromosomal gene movements reflect the recent origin and biology of therian sex chromosomes. PLoS Biol 6(4):e80

Raiz J et al (2012) The non-autonomous retrotransposon SVA is trans-mobilized by the human LINE-1 protein machinery. Nucleic Acids Res 40(4):1666–1683

Richardson SR, Salvador-Palomeque C, Faulkner GJ (2014) Diversity through duplication: whole-genome sequencing reveals novel gene retrocopies in the human population. Bioessays 36(5):475–481

Rohozinski J, Bishop CE (2004) The mouse juvenile spermatogonial depletion (jsd) phenotype is due to a mutation in the X-derived retrogene, mUtp14b. Proc Natl Acad Sci U S A 101(32):11695–11700

Rohozinski J, Lamb DJ, Bishop CE (2006) UTP14c is a recently acquired retrogene associated with spermatogenesis and fertility in man. Biol Reprod 74(4):644–651

Sayah DM et al (2004) Cyclophilin A retrotransposition into TRIM5 explains owl monkey resistance to HIV-1. Nature 430(6999):569–573

Schrider DR, Stevens K et al (2011a) Genome-wide analysis of retrogene polymorphisms in Drosophila melanogaster. Genome Res 21(12):2087–2095

Schrider DR, Gout J-F, Hahn MW (2011b) Very few RNA and DNA sequence differences in the human transcriptome. PLoS One 6(10):e25842

Schrider DR et al (2013) Gene copy-number polymorphism caused by retrotransposition in humans. PLoS Genet 9(1):e1003242

Stewart C et al (2011) A comprehensive map of mobile element insertion polymorphisms in humans. PLoS Genet 7(8):e1002236

Sudmant PH et al (2015) An integrated map of structural variation in 2,504 human genomes. Nature 526(7571):75–81

Suh A (2015) The specific requirements for CR1 retrotransposition explain the scarcity of retrogenes in birds. J Mol Evol 81(1–2):18–20

Tam OH et al (2008) Pseudogene-derived small interfering RNAs regulate gene expression in mouse oocytes. Nature 453(7194):534–538

Torrents D et al (2003) A genome-wide survey of human pseudogenes. Genome Res 13(12):2559–2567

Vanin EF (1985) Processed pseudogenes: characteristics and evolution. Annu Rev Genet 19:253–272

Vinckenbosch N, Dupanloup I, Kaessmann H (2006) Evolutionary fate of retroposed gene copies in the human genome. Proc Natl Acad Sci U S A 103(9):3220–3225

Virgen CA et al (2008) Independent genesis of chimeric TRIM5-cyclophilin proteins in two primate species. Proc Natl Acad Sci U S A 105(9):3563–3568

Warren WC et al (2008) Genome analysis of the platypus reveals unique signatures of evolution. Nature 453(7192):175–183
Watanabe T et al (2008) Endogenous siRNAs from naturally formed dsRNAs regulate transcripts in mouse oocytes. Nature 453(7194):539–543
Wen Y-Z et al (2011) Pseudogene-derived small interference RNAs regulate gene expression in African Trypanosoma brucei. Proc Natl Acad Sci U S A 108(20):8345–8350
Wilson SJ et al (2008) Independent evolution of an antiviral TRIMCyp in rhesus macaques. Proc Natl Acad Sci U S A 105(9):3557–3562
Xing J et al (2009) Mobile elements create structural variation: analysis of a complete human genome. Genome Res 19(9):1516–1526
Yap MW et al (2004) Trim5alpha protein restricts both HIV-1 and murine leukemia virus. Proc Natl Acad Sci U S A 101(29):10786–10791
Zhang Z, Gerstein M (2003) The human genome has 49 cytochrome c pseudogenes, including a relic of a primordial gene that still functions in mouse. Gene 312:61–72
Zhang Z, Harrison P, Gerstein M (2002) Identification and analysis of over 2000 ribosomal protein pseudogenes in the human genome. Genome Res 12(10):1466–1482
Zhang Z et al (2003) Millions of years of evolution preserved: a comprehensive catalog of the processed pseudogenes in the human genome. Genome Res 13(12):2541–2558

Neuronal Genome Plasticity: Retrotransposons, Environment and Disease

Marie-Jeanne H.C. Kempen, Gabriela O. Bodea, and Geoffrey J. Faulkner

1 Introduction

Transposable elements (TEs) play a central role in genome evolution and genetic innovation, as first proposed by Barbara McClintock's seminal work describing somatic transposition in maize, and many subsequent studies (Johnson and Guigo 2014; Feschotte 2008; Oliver and Greene 2011; Bourque 2009; Sasaki et al. 2008; Böhne et al. 2008; Hutchins and Pei 2015; Casacuberta and González 2013; McClintock 1950). The remnants of now inactive TEs pervade most eukaryotic genomes and, in some cases, carry out biological functions that favour the host cell, a phenomenon called 'exaptation' (Bejerano et al. 2006; Jordan et al. 2003; Jacques et al. 2013; Gifford et al. 2013; Kelley et al. 2014; Fort et al. 2014; Faulkner et al. 2009). In humans, the only class of TE still able to mobilise autonomously is the retrotransposon LINE-1 (L1). A full-length L1 is a transcribed 6 kb genetic unit (Grimaldi et al. 1984) that encodes two proteins essential for L1 mobility (called ORF1p and ORF2p) (Moran et al. 1996; Scott et al. 1987; Singer et al. 1993), as well as an unusual antisense open reading frame (ORF0) of unclear relevance to L1 retrotransposition (Denli et al. 2015). Although ~500,000 L1 copies comprise 17 % of human genomic DNA, nearly all of these copies are now immobile due to 5′ truncations, internal rearrangements and mutations (Lander et al. 2001). As a result, ~100 L1 copies remain retrotransposition competent (Sassaman et al. 1997) and, of these, only a small number, dubbed 'hot' L1s, account for the vast majority of new

M.-J. H.C. Kempen • G.O. Bodea
Mater Research Institute, University of Queensland, Brisbane, QLD, Australia
e-mail: marie.kempen@mater.uq.edu.au; gabriela.bodea@mater.uq.edu.au

G.J. Faulkner (✉)
Queensland Brain Institute, University of Queensland, Brisbane, QLD, Australia
e-mail: faulknergj@gmail.com

G. Cristofari (ed.), *Human Retrotransposons in Health and Disease*,
DOI 10.1007/978-3-319-48344-3_5

L1 retrotransposition events observed in human populations (Brouha et al. 2003; Beck et al. 2010). The L1 proteins can also recognise and mobilise non-autonomous retrotransposons, such as *Alu* and SINE-VNTR-ALU (SVA) elements, in *trans* (Dewannieux et al. 2003; Belancio et al. 2010; Garcia-Perez et al. 2007a; Wei et al. 2001; Doucet et al. 2015). Until recently, it was considered that mammalian cells only allowed somatic L1 retrotransposition during early embryonic development and under pathological circumstances, such as cancer (Kazazian et al. 1988; Garcia-Perez et al. 2007b; Kano et al. 2009; Trelogan and Martin 1995; Iskow et al. 2010; Miki et al. 1992). However, Muotri et al., analysing the transcriptional profiles of multipotent neural progenitor cells (NPC), were the first to discover that L1 transcripts were also expressed in the brain under normal conditions (Muotri et al. 2005). Here we consider the last decade of discoveries relating to L1 activity in the mammalian brain that followed on from the key findings of Muotri et al. We particularly emphasise the role of the environment and neurological disease in modulating neuronal L1 retrotransposition, as this is arguably the clearest route available to understand the functional significance of L1 mobilisation in the brain.

2 Detecting Retrotransposition in the Neuronal Lineage

The developmental timing of neuronal L1 retrotransposition is decisive in determining how many somatic L1 insertions are found per neuron, and how many neurons each L1 insertion is found in. It is now well established that L1 mobilisation occurs during neuronal differentiation, when neural stem cells (NSCs) commit to neuronal progenitor cells (NPCs), and potentially in mature, postmitotic neurons. The key findings supporting this conclusion are primarily based on in vitro and in vivo measurements of L1 activity using transgenic L1 elements, and in vivo studies of endogenous L1 behaviour. Cultured adult rat NPCs, as well as human NPCs derived from foetal brain stem cells, each support retrotransposition of a human L1 element bearing an enhanced green fluorescence protein (EGFP) reporter cassette during the early stages of neuronal differentiation (Muotri et al. 2005; Coufal et al. 2009). The L1-EGFP cassette contains the gene encoding EGFP in reverse orientation to the L1 transcript. Due to an interruption of the EGFP gene by an intron in the same transcriptional orientation as the L1, EGFP-positive cells only arise when L1 retrotransposition is completed and the EGFP intron is removed from the RNA intermediate before reverse transcription (Ostertag 2000). Additionally, endogenous L1 mRNAs are detectable in human NPCs (Coufal et al. 2009). The cells that support retrotransposition events and contain endogenous L1 transcripts present a multipotent NSC phenotype with bias towards neuronal differentiation (Muotri et al. 2005; Coufal et al. 2009). L1 insertions can occur within neuronal genes and thereby have the potential to cause gene expression changes (Muotri et al. 2005; Klawitter et al. 2016; Han et al. 2004; Upton et al. 2015). As well as during adult neurogenesis, L1 retrotransposition occurs during early embryonic development, as found in human embryonic stem cells (hESCs) (Garcia-Perez et al. 2007b) and transgenic L1-EGFP mice where an engineered human L1 is under the control of a native L1 promoter ($L1_{RP}$) (Muotri et al. 2005).

Coufal et al. subsequently developed an L1 copy number variation (CNV) assay based on qPCR which, when applied to human central nervous system (CNS) and other somatic tissues, displayed an overall elevation of L1 copy number in the CNS (Coufal et al. 2009), consistent with substantial full-length and processed L1 mRNA expression occurring in the brain (Faulkner et al. 2009; Belancio et al. 2010; Tyekucheva et al. 2011). This higher L1 copy number is particularly observed in the hippocampal dentate gyrus (DG) (Coufal et al. 2009; Baillie et al. 2011). It is notable that although the engineered L1-EGFP and L1 CNV assays provide a window into endogenous L1 activity in the brain, they also carry considerable drawbacks. For example, the L1-EGFP assay requires reverse transcription of the sizeable EGFP cassette, at a minimum, to observe EGFP-positive cells, and the EGFP promoter is subject to host genome silencing (Garcia-Perez et al. 2010). The L1 CNV assay, by contrast, measures endogenous L1 genome content, but is primarily useful as an indicator of relative L1 copy number, and does not provide the genomic locations of L1 integration sites.

High-throughput DNA sequencing can overcome these issues by allowing the detection and genomic localisation of endogenous L1 variants. Briefly, this usually involves sequencing genomic DNA in order to identify L1 integration sites in brain tissue that are not present in matched non-brain tissue (e.g. liver or heart). Subsequently these data is cross referenced to databases containing known polymorphic insertions (Baillie et al. 2011; Kurnosov et al. 2015; Mir et al. 2015) to gain further confidence in predicted somatic L1 variants. To facilitate higher sequencing depth at L1 insertion sites, DNA can be enriched prior to sequencing. Retrotransposon capture sequencing (RC-seq), for instance, is a hybridisation-based method developed to enrich sequencing libraries for fragments containing L1 junctions (Baillie et al. 2011; Upton et al. 2015; Shukla et al. 2013). Using RC-seq, Baillie et al. again identified the hippocampus as a region prone to somatic L1 retrotransposition (Baillie et al. 2011), corroborating the earlier Coufal et al. study (Coufal et al. 2009). Interestingly, the hippocampus is one of the primary brain regions where neurogenesis is maintained in adulthood (Eriksson et al. 1998), which is consistent with the finding that L1 activity becomes more prominent during neurogenesis and neuronal differentiation (Coufal et al. 2009; Muotri et al. 2005). Investigating the genome-wide integration site pattern of detected somatic L1 insertions, Baillie et al. found an overrepresentation of insertions in some protein-coding loci, specifically the introns of neurobiological genes, corroborating a preliminary observation made by Muotri et al. based on genomic mapping of L1-EGFP insertions (Baillie et al. 2011; Muotri et al. 2005).

That the hippocampus is a major source of adult neurogenesis, and provides a substantial contribution to behavioural phenotypes (Kim et al. 2015; McDonald and Hong 2013), combined with Baillie et al.'s finding that somatic L1 insertions primarily occur in gene-rich regions, is stunning because in this setting the chances of an L1 insertion leading to phenotypic change are greatly increased (Richardson et al. 2014). However, the rate at which L1 mobilisation takes place in neurons is still unclear. Single-cell genomic analyses, where DNA is obtained from individual cells and then massively amplified, estimate that 1 L1 insertion is found per 300 neurons (Evrony et al. 2012), through to multiple insertions per cell (Upton et al. 2015). The last study aiming to resolve this issue reported 13.7 somatic L1 insertions per hippocampal neuron (Upton et al. 2015), leaving the chances of functional consequences relatively high.

3 Insertional Impact and Regulation of Retrotransposons

The integration of new L1 insertions, or other TEs, into genes can significantly impact gene expression by constraining or differentially regulating transcription or altering the encoded protein. The consequences of an L1 insertion depend on the characteristics of the insertion (full-length or 5′ truncated, sense or antisense to the gene) and the cellular environment, including the response of the host cell to the insertion. For example, L1 insertions in the sense orientation to a gene are expected to be more detrimental to that gene than an antisense insertion because RNA polymerase II struggles to process the L1 sequence in sense (Chen et al. 2006; Han and Boeke 2004; Han et al. 2004). This is the primary explanation for a strong depletion of sense-oriented L1 insertions in protein-coding genes in the human reference genome (Ewing and Kazazian 2011). New L1 insertions can, however, impact host gene expression via many routes, and generate phenotypes (Beck et al. 2011). This is nicely illustrated in two distinct mouse models: the spastic mouse and the Orleans reeler. The spastic mouse contains a homozygous mutation in the brain-expressed *glycine receptor β subunit-encoding* (*Glyrb*) gene. This mutation results in defects of the glycine signalling pathway and subsequent motor deficiency and is the consequence of a full-length L1 insertion in intron 5 of the *Glyrb* gene, leading to aberrant splicing of the pre-mRNA by skipping of exon 5 (Mülhardt et al. 1994; Kingsmore et al. 1994). As the L1 insertion solely affects splicing of the adult isoform of the receptor subunit ($GlyR_A$), the spastic phenotype only becomes apparent around 2 weeks of age, when a developmental switch from the neonatal isoform ($GlyR_N$) to $GlyR_A$ takes place (Becker 1990). By comparison, the Orleans reeler mouse has a full-length L1 insertion into an exon of the *Reelin* (*Reln*) gene, inducing exon skipping (D'Arcangelo et al. 1995). Exon skipping leads to a frame shift that causes a 220 bp deletion of the *Reln* mRNA, which encodes a truncated protein that is secreted inefficiently (de Bergeyck et al. 1997; Takahara et al. 1996). As Reln is an extracellular signalling protein required for the regulation of neuronal migration, deficiency in its secretion leads to a severe impairment of neuronal migration and, as a consequence, cortical and cerebellar delamination and subsequent typical neurological symptoms. These archetypal examples of germline L1 retrotransposition leading to neuronal phenotypes, in the Orleans reeler and spastic mouse, point to the possible consequences of somatic L1 retrotransposition occurring during neurogenesis. Unsurprisingly, the host genome has evolved several mechanisms to limit L1 mobilisation in germ cells, and the neuronal lineage (Fig. 1).

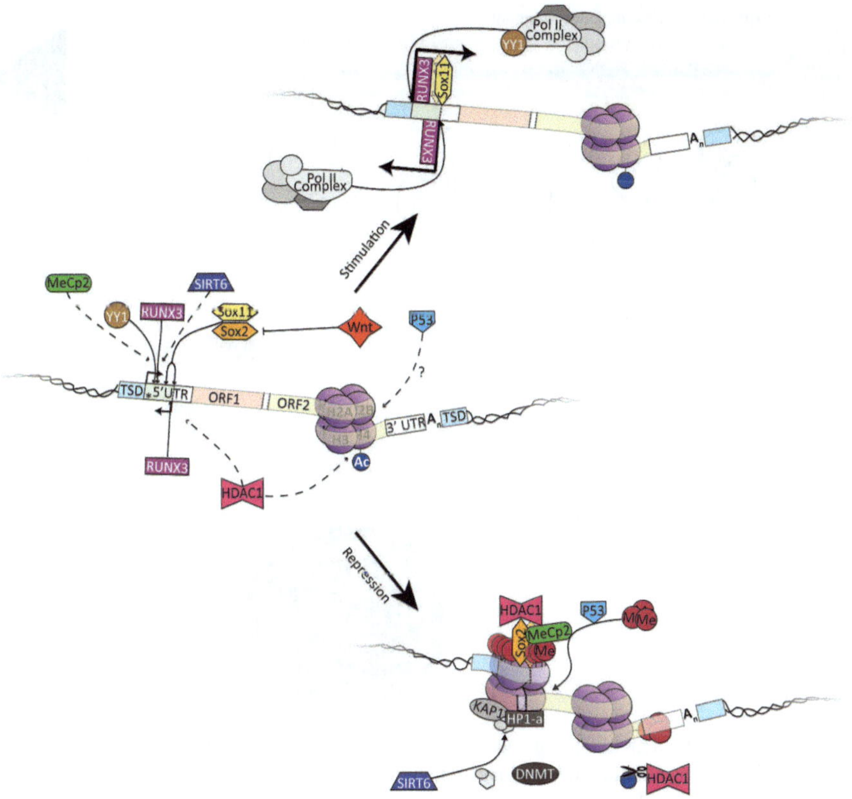

Fig. 1 L1 regulation is complex and dynamic. Numerous proteins, including YY1, RUNX3, SRY (Sox2 and 11), HDAC1, MeCP2, SIRT6 and P53, regulate L1 activity via epigenetic modifications, and through transcriptional stimulation/repression

Methylation of the L1 promoter region is the first line of defence for cells to guard against potentially deleterious L1 mobilisation (Hata and Sakaki 1997). Methyl CpG-binding protein 2 (MeCP2), a protein required for DNA methylation-mediated gene repression and mainly expressed in mature neurons (Fig. 2), is closely involved in inhibiting L1 activity. MeCP2 knockdown correlates with an increase in L1 promoter activity (Muotri et al. 2010). Under normal circumstances, MeCP2 binds methylated CpG dinucleotides and interacts with histone deacetylase protein (HDAC) and SIN3A corepressor complex resulting in blockage of transcription factors, histone deacetylation and methylation (Fig. 1) (Fuks et al. 2003; Nan et al. 1998). Inhibition of an MeCP2-interacting protein, HDAC1, by valproic acid enhances the transcriptional activity of L1 (Lennartsson et al. 2015). This indicates that HDAC1 is also involved in L1 repression (Fig. 1). HDAC1 dysfunction is known to play a role in psychiatric disorders, specifically schizophrenia, suggesting a potential mechanism underlying the symptoms experienced by these patients (Weïwer et al. 2013). The mono-ADP ribosyltransferase enzyme, Sirtuin 6 (SIRT6), another deacetylase,

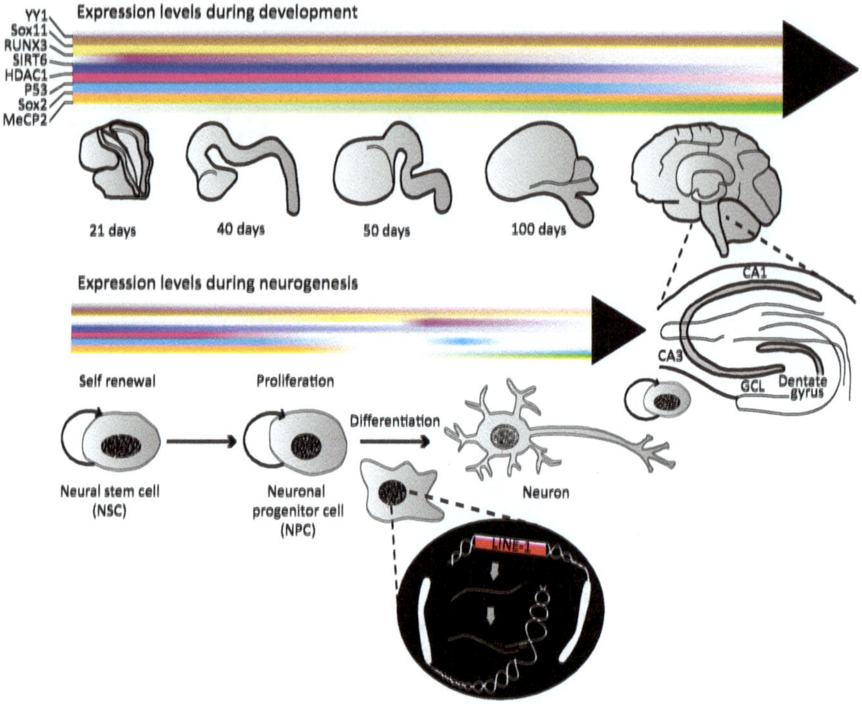

Fig. 2 Dynamic L1 activity during neurogenesis. The factors illustrated in Fig. 1 are involved in proliferation, differentiation and neuronal function. L1 expression is, as a result, differentially regulated during brain development as well as early and adult neurogenesis, resulting in potentially dynamic L1 activity, and mobilisation, during these stages (*CA 1, 3* cornu ammonis 1 and 3, *GCL* granule cell layer)

is suggested to inhibit L1 transcription by promoting heterochromatin formation (Van Meter et al. 2014). SIRT6 localises to the L1 promoter and, interestingly, appears to be displaced during aging as well as in oxidative stress conditions, circumstances known to enhance TE activity (Li et al. 2013). Although L1 is silenced by the MeCP2 complex and other mechanisms in most tissues, the brain exhibits significantly lower L1 methylation than matched skin samples (Coufal et al. 2009). Furthermore, during cell differentiation the L1 promoter tends to be demethylated (Muotri et al. 2010) potentially creating a brief window for retrotransposition to take place (Kano et al. 2009; Muotri et al. 2005).

Beyond epigenetic suppression, L1 can be regulated by transcription factors (TFs) expressed in neural cells. For instance, Ying Yang 1 (YY1), a zinc finger protein TF, strongly and predominantly expressed in neurons (Rylski et al. 2008), is involved in neuronal differentiation (Fig. 2) (reviewed in He and Casaccia-Bonnefil 2008) and facilitates L1 transcription, potentially by directing the RNA polymerase II (pol II) complex to its proper binding site (Fig. 1) (Becker et al. 1993; Athanikar et al. 2004). Members of the sex-determining region Y (SRY) protein family can also impact L1 activity. SRY-box 2 (Sox2) can inhibit L1 transcription (Kuwabara et al. 2009; Coufal et al. 2009; Muotri et al. 2005), while Sox11 is suggested to stimulate L1 activity (Tchénio et al. 2000). During embryonic and adult neurogenesis Sox2 is involved in maintenance of the multipotent state of NSCs and NPCs (Graham et al. 2003; Heinrich et al. 2014; Ring et al. 2012). By contrast, Sox11 is mainly expressed in non-proliferative, committed neuronal cells in the neurogenic niches of the adult brain, where it acts as a transcriptional activator of several neuronal genes (Haslinger et al. 2009; Mu et al. 2012; Bergsland et al. 2006). Another TF, runt-related transcription factor 3 (RUNX3), which is involved in neurogenesis, development and survival of proprioceptive neurons, stimulates the L1 promoter region (Yang et al. 2003; Inoue et al. 2008; Lallemend et al. 2012). Finally, p53 supresses L1 retrotransposition through its involvement in H3K9 trimethylation (H3K9me3), a silencing marker, which has been found to occur at the L1 enhancer region (Wylie et al. 2015; Harris et al. 2009). P53 expression is found in proliferating and newly formed neurons where it helps regulate proliferation and differentiation (reviewed in Tedeschi and Di Giovanni 2009). Thus, L1 activity in the brain is regulated by TFs essential to neurogenesis. It remains unclear as to whether this is by coincidence or because L1, a molecular parasite, has found a niche where it is derepressed as part of the greater cascade of gene regulation governing neurogenesis.

As new L1 insertions attract epigenetic suppression and carry TF-binding sites, the integration of an L1 into introns or intergenic regions upstream of protein-coding genes can alter the expression pattern of those genes. For example, 79 protein-coding genes were shown by Kuwabara et al. to present SRY-binding sites from L1 insertions occurring proximal to their transcription start sites in the human genome (Kuwabara et al. 2009). In these cases, transcriptional activation or suppression of L1 by one of the members of the SRY family may lead to the activation or suppression of the downstream protein-coding gene. That the regulatory factors described above play a role in neurogenesis and differentiation may suggest that L1 can influence these processes by, for example, genetically reprogramming differentiating cells (Spadafora 2015; Peaston et al. 2004; Muotri et al. 2005). It follows that L1 mobilisation in the brain is proposed as a source of neuron functional diversity (Muotri et al. 2005; Baillie et al. 2011; Upton et al. 2015; Singer et al. 2010; Coufal et al. 2009; Richardson et al. 2014). Hypothetically, if L1 causes genome plasticity in neurons, it may provide itself, and the host organism, extra capacity to adapt to its environment (Casacuberta and González 2013; Oliver and Greene 2011) at the cost of, perhaps, occasional catastrophic consequences for the individual, including neurological disorders (reviewed in Reilly et al. 2013).

4 Environmental Influences upon L1 Activity

Barbara McClintock was the first to propose the "genomic shock" hypothesis, speculating that environmental factors have the ability to stimulate the activity of TEs (McClintock 1984). Since then, numerous studies have aimed to address this hypothesis for environmental/cellular changes ranging from stress and toxic agents to voluntary physical activity. Although these studies have often reported enhanced L1 activity, we must emphasise that many of these observations require replication.

Preliminary experiments suggest that heavy metals may, for instance, modulate L1 mobilisation. Mercury (Hg), nickel (Ni) and cadmium (Cd) exposure appear to increase L1 retrotransposition (El-Sawy et al. 2005; Kale et al. 2005, 2006). The particulate, water-insoluble forms of these heavy metals (mercury sulfide (HgS), nickel oxide (NiO) and cadmium sulfide (CdS)) increase L1 mobilisation in HeLa cells (Kale et al. 2005). Exploring the effect of the soluble forms of these substances produces slightly different results for mercury ($HgCl_2$) (Habibi et al. 2014). No difference in L1 promoter activity, transcription or putative genomic L1 integration is detected for non-neuronal cells, including HeLa cells, after $HgCl_2$ exposure. By contrast, a neuroblastoma cell line (NB) does potentially show an increase in all of these measurements. The soluble form of nickel ($NiCl_2$) and cadmium ($CdCl_2$) however generates similar results to those of their particulates (Kale et al. 2006; El-Sawy et al. 2005). Examination of these phenomena reveals that L1 endonuclease activity associated with the increase in L1 mobilisation does not contribute to the toxicity observed for CdS or $CdCl_2$ (Kale et al. 2006). Furthermore, the increased L1 retrotransposition resulting from $NiCl_2$ exposure is not mediated by enhancement of L1 promoter activity (El-Sawy et al. 2005); also the direct genotoxicity of CdS and $NiCl_2$, which could potentially facilitate L1 insertion into DNA double-stranded breaks (DSBs), is not causative (El-Sawy et al. 2005; Kale et al. 2006). Instead, it appears that the influence of Ni and Cd on the displacement of magnesium (Mg) and zinc (Zn) cofactors induces L1 activity, as is demonstrated by the abolishment of this effect after Mg and Zn supplementation (El-Sawy et al. 2005; Kale et al. 2006).

L1 activity induced by other genotoxic agents, such as benzo[a]pyrene (BaP), an aromatic hydrocarbon produced by wood burning and found in coal tar and automobile exhaust fumes, is plausibly dependent on their ability to induce DNA damage (Stribinskis and Ramos 2006). This potentially reflects cellular attempts to recruit L1 as a compensatory mechanism, either to induce apoptosis via genome instability triggered via ORF2p activity or to use the ability of L1 to repair DNA damage through EN-independent L1 integration (Stribinskis and Ramos 2006; Morrish et al. 2002; Teng et al. 1996). Morrish et al. described enhanced levels of retrotransposition of an EN-incompetent L1 in cell lines lacking DNA repair mechanisms (Morrish et al. 2002). However, induced DSBs in cell lines with intact DNA repair mechanisms were not found to increase retrotransposition of an EN-incompetent L1 (Farkash et al. 2006). Coufal et al. further reported that mutations inactivating the function of both non-homologous end joining (NHEJ) and p53 are required for efficient EN-incompetent L1 retrotransposition (Coufal et al. 2011). Therefore, L1 could be used by the cell to mediate the repair of DNA damage, but exclusively in cells suffering from NHEJ and p53 dysfunction. Finally, oxidative stress, which can result from a number of natural stimuli as well as toxic agents, appears to increase retrotransposition of an L1 reporter in cultured neuroblastoma cells (Giorgi et al. 2011), an interesting finding considering that the brain is a metabolic hotspot. Despite the studies described above, it remains unclear whether L1 can function as a cellular buffer against the environmental impact of toxic agents. Additionally, the observed influence of environmental factors may be dependent on the exact characteristics of the chosen stimulus, as well as the cell type investigated. As a result, more extensive investigation is required in this area, particularly for primary neuronal cells, as most data obtained thus far has been from immortalised cancer cell lines.

Although environmental factors impact neurogenesis (Koehl 2015) and, as the above-mentioned literature suggests, may also alter L1 activity, it remains to be proven whether environmental perturbation during neuronal differentiation leaves L1 more prone to mobilise. The only substantive data in this area is from a 2009 study by Muotri et al.: using transgenic mice carrying the human L1-EGFP reporter construct, they found that voluntary exercise resulted in an increase in EGFP-positive cells in the brain (Muotri et al. 2009). However, these EGFP-positive cells were not only found in the hippocampus where exercise was shown to lead to a significant increase in NPC proliferation and newborn neurons, providing the opportunity for L1 to mobilise, but also in the cerebellum, a non-neurogenic area. L1 retrotransposition in the cerebellum was an intriguing observation because it either indicated that L1 could jump in postmitotic neurons or that the detected EGFP was found in cells born elsewhere that migrated to the cerebellum and then underwent derepression of the EGFP cassette in mature neurons due to chromatin remodelling. Hence, it is difficult to conclude whether exercise led to an increased detection of L1 insertions due to increased L1 mobilisation, neurogenic rate, chromatin accessibility or a combination of these factors. Muotri et al.'s experiments therefore highlight difficulties in attributing phenotypic effects to L1 mobilisation in vivo, but do at least favour speculation that L1 can mediate neuronal genome plasticity in response to environmental changes.

5 Retrotransposon Involvement in Neurological Disorders

Traumatic early life events and chronic stress are major risk factors for the development of a range of neurological disorders (Bagot et al. 2014). If L1 is reactive to environmental stressors, it could play a potentially important role in the development or exacerbation of neurological diseases. Here we highlight the intriguing findings in this area while noting that there are no certain causative links at this stage established between any brain disorder and somatic L1 retrotransposition.

5.1 Retrotransposons in Neurodevelopmental and Neurodegenerative Disorders

Neurological disorders resulting from inherited or spontaneous genetic mutations can reproducibly present upregulation of L1 retrotransposition in the brain. In particularly, recent works have revealed that L1 copy number is elevated in Rett syndrome (RTT) and ataxia telangiectasia (AT) patient brains (Coufal et al. 2011; Muotri et al. 2010).

RTT is a progressive and devastating disease predominantly associated with mutation of the MeCP2 gene, characterised by a range of neurological problems from ataxia to autism and usually developing before 2 years of age (Amir et al. 1999). As noted above, MeCP2 is involved in transcriptional repression by binding methylated DNA and inducing histone methylation and deacetylation. MeCP2 is highly expressed in mature neuronal nuclei (Fig. 2) and, when mutated, is associated with aberrant epigenetic profiles, potentially explaining the severe CNS defects seen in RTT (Shahbazian 2002; Gabel et al. 2015). Although Yu et al. established that MeCP2 influences L1 promoter activity and L1 retrotransposition in transformed cell lines (Yu et al. 2001), Muotri et al. brought this work forward by showing that MeCP2 knockout in mouse neuroepithelial cells increases L1 promoter activity fourfold (Muotri et al. 2010). This result was specific for the reduction of MeCP2 and was not found for methyl CpG-binding domain protein 1 (MBD1), a protein from the same family but with a different DNA specificity. L1-EGFP transgenic mice deficient for MeCP2 also showed increased L1 retrotransposition compared to wild-type animals, with the strongest effects found in the cerebellum, striatum and hippocampus. Muotri et al. also found, using the L1 qPCR assay, a marked increase in L1 ORF2 copy number but not the L1 5′UTR, perhaps indicating that new L1 retrotransposition events were characterised by substantial 5′ truncations. NPCs produced from induced pluripotent stem cells (iPSCs) derived from RTT patient fibroblasts supported a higher (twofold) retrotransposition rate of the L1-EGFP reporter compared to unaffected controls. Altogether, this seminal work from Muotri et al. showed conclusively that L1 activity was higher in RTT patients than in controls. L1 insertion site mapping with single-cell genomics (Upton et al. 2015) would be a valuable future strategy to demonstrate differential L1 activity in RTT. It also should be considered that a wild-type phenotype can be rescued in a conditional mutant mouse RTT model (Guy et al. 2007), raising an important question as to whether elevated L1 activity impacts RTT phenotype.

AT patients suffer from a loss-of-function mutation in the ATM gene, a 350 kDa serine/threonine kinase (Taylor et al. 2015). The most severe and typical form of AT cases start to show symptoms between 1 and 2 years of age. ATM dysfunction leads to neuronal degeneration, immunodeficiency, chromosomal instability and a predisposition to cancer (Shiloh 2001). Under normal circumstances, ATM phosphorylates downstream factors as CHK2, p53, BRCA1 and the MRN complex (MRE11, Rad50 and NBS1) in response to the presence of double-stranded DNA breaks, which activates DNA damage checkpoint and cell cycle arrest leading to the repair of damaged DNA or p53-mediated apoptosis. NPCs produced from hESCs and carrying ATM mutations present a two- to fourfold increase of L1-EGFP retrotransposition but do not show a significant difference in promoter activity or endogenous ORF1p levels (Coufal et al. 2011). Even though L1 retrotransposition may be toxic for cells (Wallace et al. 2008), no difference in survival rates, nor cell cycle or cell division pattern, is observed in ATM-mutated versus wild-type cells (Symer et al. 2002; Coufal et al. 2011; Haoudi et al. 2004). These findings led to speculation that ATM-deficient cells might have a survival advantage due to higher tolerance for L1-induced toxicity (Coufal et al. 2011). Further experiments are required to address whether this is the case, and why ATM mutations result in higher L1 retrotransposition. Notably, ATM mutation appears to lead to longer L1 insertions, possibly due to the role of ATM in cellular DNA repair, which may interfere with L1 retrotransposition in wild-type cells. Although the L1 CNV assay revealed an increase in L1 content in post-mortem hippocampal neurons from AT patients compared to age/gender-matched healthy individuals, single-cell genomic analyses are again required to corroborate this result, and identify if the endogenous L1 insertions generated are longer, or follow a different genome-wide integration pattern compared to wild-type cells.

Although RTT and AT present the clearest evidence of unusual retrotransposon activity in the brain, TEs have also been observed to undergo derepression in neurodegenerative disorders commonly associated with aging (Bollati et al. 2011). For example, TAR DNA-binding protein 43 (TDP-43) dysfunction, a hallmark for a number of neurodegenerative disorders such as amyotrophic lateral sclerosis (ALS), frontotemporal lobar degeneration (FTLD) and Alzheimer's disease, is found to correspond to higher transcription levels of LINEs, SINEs and LTRs, the three major classes of TEs, in mice (Li et al. 2012). Brain samples with TDP-43 dysfunction, from transgenic mouse models as well as human FTLD patients, show a reduced association between this protein and a range of TE-derived transcripts. These particular TE transcripts are the same transcripts identified as being upregulated in response to TDP-43 dysfunction, indicating that this multifunctional RNA-binding protein might play a role in the regulation of TEs in somatic tissue. Furthermore, aging itself has been found to lead to activation of transposable elements (Li et al. 2013; Van Meter et al. 2014), suggesting that the increase of TE transcripts and potential copy number in the genome may lay at the base of the development of these neurodegenerative disorders.

5.2 *Do Retrotransposons Link Environmental and Genetic Risk Factors in Psychiatric Disorders?*

In RTT and AT, where the mutated gene responsible for pathophysiology also influences L1 activity, the detected increase in L1 expression and, potentially, L1 copy number is most likely a direct effect of the main driver mutation in these diseases. However, abnormal retrotransposon activity in the brain has also been observed for several psychiatric disorders where the interaction between genetic and environmental risk factors, or solely environmental factors, is considered central to disease aetiology. For example, use of methamphetamines and cocaine, a major risk factor for the development of psychiatric disorders (Akindipe et al. 2014; Zhornitsky et al. 2015), with the potential to turn into substance-use disorder (SUD), is suggested to lead to L1 activation (Okudaira et al. 2014; Maze et al. 2011). This effect is observed using an engineered L1 reporter in vitro for neuronal, but not non-neuronal, cell lines (Okudaira et al. 2014). Furthermore, the abolishment of unusual L1 mobilisation after knockdown of the cAMP response element-binding protein (CREB) suggests that this neuronal response to methamphetamine and cocaine is CREB dependent. Investigation of the mechanism underlying this phenomenon identified an enhanced recruitment of L1 ORF1p to chromatin-rich fractions, without increasing the total expression of L1 mRNA or ORF1p. This startling result suggests that methamphetamine and cocaine use may elevate L1 mobilisation by recruiting L1 ORF1p to the chromatin in a CREB-dependent manner, facilitating L1 integration into the genome. This in turn could induce changes in chromatic structures and gene expression, with the potential to lead to psychiatric disorders.

Post-traumatic stress disorder (PTSD), a disorder closely related to SUD (Jacobsen et al. 2001), has been found to lead to differential epigenetic regulation of L1 as well as *Alu* copies in the genome (Rusiecki et al. 2012). PTSD and SUD share numerous cellular circuits and signalling pathways in their pathophysiology, due to similar involvement of the learning and memory system (reviewed in Tipps et al. 2014). PTSD is an anxiety disorder characterised by persistent re-experiences of a past traumatic event or events, often accompanied by memory and concentration problems, anxiety, panic attacks, insomnia, substance abuse and/or depressive symptoms (American Psychiatric Association 2013). PTSD patients present gene expression signatures not found in controls (Segman et al. 2005). Multiple studies have shown that epigenetic alterations play an important role in facilitating changes in gene expression associated with the formation and persistence of memory (Kwapis and Wood 2014; Zovkic and Sweatt 2013). Changes in methylation levels of L1 and *Alu* in soldiers pre- and post-deployment, of which a subset developed PTSD after their return, have been detected, potentially reflecting resilience or vulnerability factors to PTSD development (Rusiecki et al. 2012). Increased methylation of L1 was detected in the control group post-deployment compared to pre-deployment which, to speculate, might be a result of the body's response to stress-mediated L1 activation (Li and Schmid 2001). By contrast, a pre-existing abundance of *Alu* methylation in cases compared to controls might reflect a potential vulnerability to stress or a protective effect of hypomethylation. Specific patterns of *Alu* expression have been previously linked to physiological stress responses, with perhaps functional consequences (Berger et al. 2014; Pandey et al. 2011; Li and Schmid 2001). Hypermethylation may prevent *Alu* from fulfilling a protective function, although the mechanism involved is unknown at this stage.

More recently, Bundo et al. investigated L1 CNV in schizophrenia (SCZ), major depression (MD) and bipolar disorder (BD), detecting increased L1 copy number in the prefrontal cortex (PFC) of patients suffering from SCZ compared to healthy controls (Bundo et al. 2014). SCZ is a multifactorial disorder with a typical onset between late puberty and early adulthood and characterised by a chronic and dynamic progression, with genes and environment playing important aetiological roles (Brown 2011). Diagnosis of SCZ is based on a collection of positive, negative and cognitive symptoms, persisting over a period of time (American Psychiatric Association 2013). The PFC is considered to be involved in SCZ symptomology and show differential gene expression when patients are compared to controls, making the finding of Bundo et al. particularly interesting (Kimoto et al. 2014; Joshi et al. 2014; Farzan et al. 2010; Guillozet-Bongaarts et al. 2014). Repeating the L1 CNV analysis using solely neuronal cells yielded a more prominent difference, suggesting that the phenomenon is neuron specific (Bundo et al. 2014). In order to investigate the contribution of genetic factors, Bundo et al. assessed L1 CNV in neurons derived from iPSCs of patients suffering from a rare variant of schizophrenia caused by a 22q11 deletion, one of the highest genetic risk factors. This resulted in the detection of a consistent increase in L1 copy number in the neuronal cells of patients. Furthermore, the influence of environmental risk factors was explored by determining L1 CNV in the PFC of two established SZ animal models and, consistently, higher L1 copy number was detected in both models. Several environmental risk factors for the development of schizophrenia, such as metal exposure and drug use (Modabbernia et al. 2016; Akindipe et al. 2014), were discussed above to also influence L1 activity, making it plausible that L1 would be involved in the development of this disorder.

Although these studies of L1 activity in psychiatric disorders are correlative, they do suggest that L1 mobilisation may be more than a secondary effect of abnormal neurobiology. Particularly impressive were the experiments by Bundo et al. showing that L1 content is increased in SCZ patient samples, iPSC-derived neurons and SCZ animal models. Consistent L1 upregulation in SCZ across very diverse experimental systems indicates a close association between disease phenotype and ectopic L1 activity, though it remains unknown whether L1 plays an active role in the manifestation of SCZ symptoms or is merely a passenger. Given that L1 can influence genome stability, as well as gene transcription, and is responsive to environmental cues, it is plausible that subtle genetic differences arise in genes related to SCZ symptomology. Alternatively, inability to control L1 activity is at the least emblematic of neuronal genome vulnerability and instability. A great deal of more future research is required in this area to make any substantive conclusions regarding the functional role of L1 mobilisation in SCZ and other psychiatric disorders.

6 Conclusion and Future Directions

Somatic L1 retrotransposition is now well established to occur in the neuronal lineage. The field also has a reasonable idea of how this process is regulated, by MeCP2 and other factors. However, we lack even basic understanding of how L1 mobilisation in the brain impacts normal neurobiology, let alone neuronal phenotype in psychiatric, neurodevelopmental or neurodegenerative disorders. As a result, the significance of L1 retrotransposition to brain function is still largely unclear. To move forward in this area, we require improved resolution of the precise timing and cell specificity of retrotransposition during embryonic and adult neurogenesis as well as, potentially, in mature neurons. These parameters are prerequisites to define the contribution of L1 mobilisation to neuronal genome diversity. Moreover, despite advances in single-cell genomics, it is currently not possible to assay the genome and transcriptome of the same individual neuron, precluding detection of gene expression changes associated with somatic L1 insertions. One alternative approach in this area would be to use newly developed genome editing tools (e.g. CRISPR-Cas9) (Wright et al. 2016) to artificially introduce L1 insertions found in patient samples into homogenous neuronal cultures in vitro, or into transgenic animal models. This could facilitate a more comprehensive analysis of how individual L1 insertions alter normal neuronal physiology and, potentially, behaviour. Moreover, although L1 deregulation has been found in several neurological disorders, the mechanisms through which L1 retrotransposition could impact disease symptomology remain largely unexplored. Therefore, the role of L1-derived genomic mosaicism in neurobiology remains unclear, despite its obvious appeal as a foundation for complex brain functions (e.g. memory formation), and as an aetiological factor in the dysregulation of those functions.

References

Akindipe T, Wilson D, Stein DJ (2014) Psychiatric disorders in individuals with methamphetamine dependence: prevalence and risk factors. Metab Brain Dis 29(2):351–357

American Psychiatric Association (2013) Diagnostic and statistical manual of mental disorders, 5th edn. American Psychiatric Association, Arlington

Amir RE et al (1999) Rett syndrome is caused by mutations in X-linked MECP2, encoding methyl-CpG-binding protein 2. Nat Genet 23(2):185–188

Athanikar JN, Badge RM, Moran JV (2004) A YY1-binding site is required for accurate human LINE-1 transcription initiation. Nucleic Acids Res 32(13):3846–3855

Bagot RC et al (2014) Epigenetic signaling in psychiatric disorders: stress and depression. Dialogues Clin Neurosci 16(3):281–295

Baillie JK et al (2011) Somatic retrotransposition alters the genetic landscape of the human brain. Nature 479(7374):534–537

Beck CR et al (2010) LINE-1 retrotransposition activity in human genomes. Cell 141(7): 1159–1170

Beck CR et al (2011) LINE-1 elements in structural variation and disease. Annu Rev Genomics Hum Genet 12:187–215

Becker CM (1990) Disorders of the inhibitory glycine receptor: the spastic mouse. FASEB J 4(10):2767–2774

Becker KG et al (1993) Binding of the ubiquitous nuclear transcription factor YY1 to a cis regulatory sequence in the human LINE-1 transposable element. Hum Mol Genet 2(10):1697–1702

Bejerano G et al (2006) A distal enhancer and an ultraconserved exon are derived from a novel retroposon. Nature 441(7089):87–90

Belancio VP et al (2010) Somatic expression of LINE-1 elements in human tissues. Nucleic Acids Res 38(12):3909–3922

Berger A et al (2014) Direct binding of the Alu binding protein dimer SRP9/14 to 40S ribosomal subunits promotes stress granule formation and is regulated by Alu RNA. Nucleic Acids Res 42(17):11203–11217

Bergsland M et al (2006) The establishment of neuronal properties is controlled by Sox4 and Sox11. Genes Dev 20(24):3475–3486

Böhne A et al (2008) Transposable elements as drivers of genomic and biological diversity in vertebrates. Chromosome Res 16(1):203–215

Bollati V et al (2011) DNA methylation in repetitive elements and Alzheimer disease. Brain Behav Immun 25(6):1078–1083

Bourque G (2009) Transposable elements in gene regulation and in the evolution of vertebrate genomes. Curr Opin Genet Dev 19(6):607–612

Brouha B et al (2003) Hot L1s account for the bulk of retrotransposition in the human population. Proc Natl Acad Sci U S A 100(9):5280–5285

Brown AS (2011) The environment and susceptibility to schizophrenia. Prog Neurobiol 93(1):23–58

Bundo M et al (2014) Increased L1 retrotransposition in the neuronal genome in schizophrenia. Neuron 81(2):306–313

Casacuberta E, González J (2013) The impact of transposable elements in environmental adaptation. Mol Ecol 22(6):1503–1517

Chen J, Rattner A, Nathans J (2006) Effects of L1 retrotransposon insertion on transcript processing, localization and accumulation: lessons from the retinal degeneration 7 mouse and implications for the genomic ecology of L1 elements. Hum Mol Genet 15(13):2146–2156

Coufal NG et al (2009) L1 retrotransposition in human neural progenitor cells. Nature 460(7259): 1127–1131

Coufal NG et al (2011) Ataxia telangiectasia mutated (ATM) modulates long interspersed element-1 (L1) retrotransposition in human neural stem cells. Proc Natl Acad Sci U S A 108(51): 20382–20387

D'Arcangelo G et al (1995) A protein related to extracellular matrix proteins deleted in the mouse mutant reeler. Nature 374(6524):719–723

de Bergeyck V et al (1997) A truncated Reelin protein is produced but not secreted in the "Orleans" reeler mutation (Relnrl-Orl). Mol Brain Res 50(1–2):85–90

Denli AM et al (2015) Primate-specific ORF0 contributes to retrotransposon-mediated diversity. Cell 163(3):583–593

Dewannieux M, Esnault C, Heidmann T (2003) LINE-mediated retrotransposition of marked Alu sequences. Nat Genet 35(1):41–48

Doucet AJ et al (2015) A 3′ Poly(A) tract is required for LINE-1 retrotransposition. Mol Cell 60(5):728–741

El-Sawy M et al (2005) Nickel stimulates L1 retrotransposition by a post-transcriptional mechanism. J Mol Biol 354(2):246–257

Eriksson PS et al (1998) Neurogenesis in the adult human hippocampus. Nat Med 4(11): 1313–1317

Evrony GD et al (2012) Single-neuron sequencing analysis of L1 retrotransposition and somatic mutation in the human brain. Cell 151(3):483–496

Ewing AD, Kazazian HH (2011) Whole-genome resequencing allows detection of many rare LINE-1 insertion alleles in humans. Genome Res 21(6):985–990

Farkash EA et al (2006) Gamma radiation increases endonuclease-dependent L1 retrotransposition in a cultured cell assay. Nucleic Acids Res 34(4):1196–1204

Farzan F et al (2010) Evidence for gamma inhibition deficits in the dorsolateral prefrontal cortex of patients with schizophrenia. Brain 133(Pt 5):1505–1514

Faulkner GJ et al (2009) The regulated retrotransposon transcriptome of mammalian cells. Nat Genet 41(5):563–571

Feschotte C (2008) Transposable elements and the evolution of regulatory networks. Nat Rev Genet 9(5):397–405

Fort A et al (2014) Deep transcriptome profiling of mammalian stem cells supports a regulatory role for retrotransposons in pluripotency maintenance. Nat Genet 46(6):558–566

Fuks F et al (2003) The methyl-CpG-binding protein MeCP2 links DNA methylation to histone methylation. J Biol Chem 278(6):4035–4040

Gabel HW et al (2015) Disruption of DNA-methylation-dependent long gene repression in Rett syndrome. Nature 522(7554):89–93
Garcia-Perez JL et al (2007a) Distinct mechanisms for trans-mediated mobilization of cellular RNAs by the LINE-1 reverse transcriptase. Genome Res 17(5):602–611
Garcia-Perez JL et al (2007b) LINE-1 retrotransposition in human embryonic stem cells. Hum Mol Genet 16(13):1569–1577
Garcia-Perez JL et al (2010) Epigenetic silencing of engineered L1 retrotransposition events in human embryonic carcinoma cells. Nature 466(7307):769–773
Gifford WD, Pfaff SL, Macfarlan TS (2013) Transposable elements as genetic regulatory substrates in early development. Trends Cell Biol 23(5):218–226
Giorgi G, Marcantonio P, Del Re B (2011) LINE-1 retrotransposition in human neuroblastoma cells is affected by oxidative stress. Cell Tissue Res 346(3):383–391
Graham V et al (2003) SOX2 functions to maintain neural progenitor identity. Neuron 39(5):749–765
Grimaldi G, Skowronski J, Singer M (1984) Defining the beginning and end of KpnI family segments. EMBO J 3(8):1753–1759
Guillozet-Bongaarts AL et al (2014) Altered gene expression in the dorsolateral prefrontal cortex of individuals with schizophrenia. Mol Psychiatry 19(4):478–485
Guy J et al (2007) Reversal of neurological defects in a mouse model of Rett syndrome. Science 315(5815):1143–1147
Habibi L et al (2014) Mercury specifically induces LINE-1 activity in a human neuroblastoma cell line. Mutat Res 759:9–20
Han JS, Boeke JD (2004) A highly active synthetic mammalian retrotransposon. Nature 429(6989):314–318
Han JS, Szak ST, Boeke JD (2004) Transcriptional disruption by the L1 retrotransposon and implications for mammalian transcriptomes. Nature 429(6989):268–274
Haoudi A et al (2004) Retrotransposition-competent human LINE-1 induces apoptosis in cancer cells with intact p53. J Biomed Biotechnol 2004(4):185–194
Harris CR et al (2009) p53 responsive elements in human retrotransposons. Oncogene 28(44):3857–3865
Haslinger A et al (2009) Expression of Sox11 in adult neurogenic niches suggests a stage-specific role in adult neurogenesis. Eur J Neurosci 29(11):2103–2114
Hata K, Sakaki Y (1997) Identification of critical CpG sites for repression of L1 transcription by DNA methylation. Gene 189(2):227–234
He Y, Casaccia-Bonnefil P (2008) The Yin and Yang of YY1 in the nervous system. J Neurochem 106(4):1493–1502
Heinrich C et al (2014) Sox2-mediated conversion of NG2 glia into induced neurons in the injured adult cerebral cortex. Stem Cell Reports 3(6):1000–1014
Hutchins AP, Pei D (2015) Transposable elements at the center of the crossroads between embryogenesis, embryonic stem cells, reprogramming, and long non-coding RNAs. Sci Bull (Beijing) 60(20):1722–1733
Inoue K, Shiga T, Ito Y (2008) Runx transcription factors in neuronal development. Neural Dev 3(1):20
Iskow RC et al (2010) Natural mutagenesis of human genomes by endogenous retrotransposons. Cell 141(7):1253–1261
Jacobsen LK, Southwick SM, Kosten TR (2001) Substance use disorders in patients with posttraumatic stress disorder: a review of the literature. Am J Psychiatry 158(8):1184–1190
Jacques P-É, Jeyakani J, Bourque G (2013) The majority of primate-specific regulatory sequences are derived from transposable elements. PLoS Genet 9(5):e1003504
Johnson R, Guigo R (2014) The RIDL hypothesis: transposable elements as functional domains of long noncoding RNAs. RNA 20(7):959–976
Jordan IK et al (2003) Origin of a substantial fraction of human regulatory sequences from transposable elements. Trends Genet 19(2):68–72

Joshi D, Fullerton JM, Weickert CS (2014) Elevated ErbB4 mRNA is related to interneuron deficit in prefrontal cortex in schizophrenia. J Psychiatr Res 53:125–132

Kale SP et al (2005) Heavy metals stimulate human LINE-1 retrotransposition. Int J Environ Res Public Health 2(1):14–23

Kale SP et al (2006) The L1 retrotranspositional stimulation by particulate and soluble cadmium exposure is independent of the generation of DNA breaks. Int J Environ Res Public Health 3(2):121–128

Kano H et al (2009) L1 retrotransposition occurs mainly in embryogenesis and creates somatic mosaicism. Genes Dev 23(11):1303–1312

Kazazian HH et al (1988) Haemophilia A resulting from de novo insertion of L1 sequences represents a novel mechanism for mutation in man. Nature 332(6160):164–166

Kelley DR et al (2014) Transposable elements modulate human RNA abundance and splicing via specific RNA-protein interactions. Genome Biol 15(12):537

Kim S et al (2015) Memory, scene construction, and the human hippocampus. Proc Natl Acad Sci U S A 112(15):4767–4772

Kimoto S, Bazmi HH, Lewis DA (2014) Lower expression of glutamic acid decarboxylase 67 in the prefrontal cortex in schizophrenia: contribution of altered regulation by Zif268. Am J Psychiatry 171(9):969–978

Kingsmore SF et al (1994) Glycine receptor beta-subunit gene mutation in spastic mouse associated with LINE-1 element insertion. Nat Genet 7(2):136–141

Klawitter S et al (2016) Reprogramming triggers endogenous L1 and Alu retrotransposition in human induced pluripotent stem cells. Nat Commun 7:10286

Koehl M (2015) Gene-environment interaction in programming hippocampal plasticity: focus on adult neurogenesis. Front Mol Neurosci 8:41

Kurnosov AA et al (2015) The evidence for increased L1 activity in the site of human adult brain neurogenesis. PLoS One 10(2):e0117854

Kuwabara T et al (2009) Wnt-mediated activation of NeuroD1 and retro-elements during adult neurogenesis. Nat Neurosci 12(9):1097–1105

Kwapis JL, Wood MA (2014) Epigenetic mechanisms in fear conditioning: implications for treating post-traumatic stress disorder. Trends Neurosci 37(12):706–720

Lallemend F et al (2012) Positional differences of axon growth rates between sensory neurons encoded by Runx3. EMBO J 31(18):3718–3729

Lander E et al (2001) Initial sequencing and analysis of the human genome. Nature 409(6822):860–921

Lennartsson A et al (2015) Remodeling of retrotransposon elements during epigenetic induction of adult visual cortical plasticity by HDAC inhibitors. Epigenetics Chromatin 8(1):55

Li T-H, Schmid CW (2001) Differential stress induction of individual Alu loci: implications for transcription and retrotransposition. Gene 276(1–2):135–141

Li W et al (2012) Transposable elements in TDP-43-mediated neurodegenerative disorders. PLoS One 7(9):e44099

Li W et al (2013) Activation of transposable elements during aging and neuronal decline in Drosophila. Nat Neurosci 16(5):529–531

Maze I et al (2011) Cocaine dynamically regulates heterochromatin and repetitive element unsilencing in nucleus accumbens. Proc Natl Acad Sci U S A 108(7):3035–3040

McClintock B (1950) The origin and behavior of mutable loci in maize. Proc Natl Acad Sci U S A 36(6):344–355

McClintock B (1984) The significance of responses of the genome to challenge. Science 226(4676):792–801

McDonald RJ, Hong NS (2013) How does a specific learning and memory system in the mammalian brain gain control of behavior? Hippocampus 23(11):1084–1102

Miki Y et al (1992) Disruption of the APC gene by a retrotransposal insertion of L1 sequence in a colon cancer. Cancer Res 52(3):643–645

Mir AA, Philippe C, Cristofari G (2015) euL1db: the European database of L1HS retrotransposon insertions in humans. Nucleic Acids Res 43(Database issue):D43–D47
Modabbernia A, Arora M, Reichenberg A (2016) Environmental exposure to metals, neurodevelopment, and psychosis. Curr Opin Pediatr 28(2):243–249
Moran JV et al (1996) High frequency retrotransposition in cultured mammalian cells. Cell 87(5):917–927
Morrish TA et al (2002) DNA repair mediated by endonuclease-independent LINE-1 retrotransposition. Nat Genet 31(2):159–165
Mu L et al (2012) SoxC transcription factors Are required for neuronal differentiation in adult hippocampal neurogenesis. J Neurosci 32(9):3067–3080
Mülhardt C et al (1994) The spastic mouse: aberrant splicing of glycine receptor beta subunit mRNA caused by intronic insertion of L1 element. Neuron 13(4):1003–1015
Muotri AR et al (2005) Somatic mosaicism in neuronal precursor cells mediated by L1 retrotransposition. Nature 435(7044):903–910
Muotri AR et al (2009) Environmental influence on L1 retrotransposons in the adult hippocampus. Hippocampus 19(10):1002–1007
Muotri AR et al (2010) L1 retrotransposition in neurons is modulated by MeCP2. Nature 468(7322):443–446
Nan X et al (1998) Transcriptional repression by the methyl-CpG-binding protein MeCP2 involves a histone deacetylase complex. Nature 393(6683):386–389
Okudaira N, Ishizaka Y, Nishio H (2014) Retrotransposition of long interspersed element 1 induced by methamphetamine or cocaine. J Biol Chem 289(37):25476–25485
Oliver KR, Greene WK (2011) Mobile DNA and the TE-Thrust hypothesis: supporting evidence from the primates. Mob DNA 2(1):8
Ostertag EM (2000) Determination of L1 retrotransposition kinetics in cultured cells. Nucleic Acids Res 28(6):1418–1423
Pandey R et al (2011) Heat shock factor binding in Alu repeats expands its involvement in stress through an antisense mechanism. Genome Biol 12(11):R117
Peaston AE et al (2004) Retrotransposons regulate host genes in mouse oocytes and preimplantation embryos. Dev Cell 7(4):597–606
Reilly MT et al (2013) The role of transposable elements in health and diseases of the central nervous system. J Neurosci 33(45):17577–17586
Richardson SR, Morell S, Faulkner GJ (2014) L1 retrotransposons and somatic mosaicism in the brain. Annu Rev Genet 48:1–27
Ring KL et al (2012) Direct reprogramming of mouse and human fibroblasts into multipotent neural stem cells with a single factor. Cell Stem Cell 11(1):100–109
Rusiecki JA et al (2012) DNA methylation in repetitive elements and post-traumatic stress disorder: a case-control study of US military service members. Epigenomics 4(1):29–40
Rylski M et al (2008) Yin Yang 1 expression in the adult rodent brain. Neurochem Res 33(12):2556–2564
Sasaki T et al (2008) Possible involvement of SINEs in mammalian-specific brain formation. Proc Natl Acad Sci U S A 105(11):4220–4225
Sassaman DM et al (1997) Many human L1 elements are capable of retrotransposition. Nat Genet 16(1):37–43
Scott AF et al (1987) Origin of the human L1 elements: proposed progenitor genes deduced from a consensus DNA sequence. Genomics 1(2):113–125
Segman RH et al (2005) Peripheral blood mononuclear cell gene expression profiles identify emergent post-traumatic stress disorder among trauma survivors. Mol Psychiatry 10(5):500–513, 425
Shahbazian MD (2002) Insight into Rett syndrome: MeCP2 levels display tissue- and cell-specific differences and correlate with neuronal maturation. Hum Mol Genet 11(2):115–124
Shiloh Y (2001) ATM (ataxia telangiectasia mutated): expanding roles in the DNA damage response and cellular homeostasis. Biochem Soc Trans 29(6):661–666
Shukla R et al (2013) Endogenous retrotransposition activates oncogenic pathways in hepatocellular carcinoma. Cell 153(1):101–111

Singer MF et al (1993) LINE-1: a human transposable element. Gene 135(1–2):183–188
Singer T et al (2010) LINE-1 retrotransposons: mediators of somatic variation in neuronal genomes? Trends Neurosci 33(8):345–354
Spadafora C (2015) A LINE-1-encoded reverse transcriptase-dependent regulatory mechanism is active in embryogenesis and tumorigenesis. Ann N Y Acad Sci 1341:164–171
Stribinskis V, Ramos KS (2006) Activation of human long interspersed nuclear element 1 retrotransposition by benzo(a)pyrene, an ubiquitous environmental carcinogen. Cancer Res 66(5):2616–2620
Symer DE et al (2002) Human L1 retrotransposition is associated with genetic instability in vivo. Cell 110:327–338
Takahara T et al (1996) Dysfunction of the Orleans reeler gene arising from exon skipping due to transposition of a full-length copy of an active L1 sequence into the skipped exon. Hum Mol Genet 5(7):989–993
Taylor AMR et al (2015) Ataxia telangiectasia: more variation at clinical and cellular levels. Clin Genet 87(3):199–208
Tchénio T, Casella JF, Heidmann T (2000) Members of the SRY family regulate the human LINE retrotransposons. Nucleic Acids Res 28(2):411–415
Tedeschi A, Di Giovanni S (2009) The non-apoptotic role of p53 in neuronal biology: enlightening the dark side of the moon. EMBO Rep 10(6):576–583
Teng SC, Kim B, Gabriel A (1996) Retrotransposon reverse-transcriptase-mediated repair of chromosomal breaks. Nature 383(6601):641–644
Tipps ME, Raybuck JD, Lattal KM (2014) Substance abuse, memory, and post-traumatic stress disorder. Neurobiol Learn Mem 112:87–100
Trelogan SA, Martin SL (1995) Tightly regulated, developmentally specific expression of the first open reading frame from LINE-1 during mouse embryogenesis. Proc Natl Acad Sci 92(5):1520–1524
Tyekucheva S et al (2011) Establishing the baseline level of repetitive element expression in the human cortex. BMC Genomics 12:495
Upton KR et al (2015) Ubiquitous L1 mosaicism in hippocampal neurons. Cell 161(2):228–239
Van Meter M et al (2014) SIRT6 represses LINE1 retrotransposons by ribosylating KAP1 but this repression fails with stress and age. Nat Commun 5:5011
Wallace NA, Belancio VP, Deininger PL (2008) L1 mobile element expression causes multiple types of toxicity. Gene 419(1–2):75–81
Wei W et al (2001) Human L1 retrotransposition: cis preference versus trans complementation. Mol Cell Biol 21(4):1429–1439
Weïwer M et al (2013) Therapeutic potential of isoform selective HDAC inhibitors for the treatment of schizophrenia. Future Med Chem 5(13):1491–1508
Wright AV, Nuñez JK, Doudna JA (2016) Biology and applications of CRISPR systems: harnessing nature's toolbox for genome engineering. Cell 164(1–2):29–44
Wylie A et al (2015) p53 genes function to restrain mobile elements. Genes Dev 30(1):64–77
Yang N et al (2003) An important role for RUNX3 in human L1 transcription and retrotransposition. Nucleic Acids Res 31(16):4929–4940
Yu F et al (2001) Methyl-CpG-binding protein 2 represses LINE-1 expression and retrotransposition but not Alu transcription. Nucleic Acids Res 29(21):4493–4501
Zhornitsky S et al (2015) Psychopathology in substance use disorder patients with and without substance-induced psychosis. J Addict 2015:843762
Zovkic IB, Sweatt JD (2013) Epigenetic mechanisms in learned fear: implications for PTSD. Neuropsychopharmacology 38(1):77–93

Activity of Retrotransposons in Stem Cells and Differentiated Cells

Angela Macia, Leon Tejwani, Pinar Mesci, Alysson Muotri, and Jose L. Garcia-Perez

1 Introduction

Mobile genetic elements are present in virtually all-eukaryotic genomes examined to date, from plants to mammals, and their activity has generated diverse types of effects in their hosts over evolution. While a typical cellular gene resides at a discrete chromosomal locus, TEs are present in multiple copies across the genome and thus influencing its integrity and organization. In general, the majority of the mobile DNA load in a mammalian genome consists of evolutionary old or "fixed" TE insertions unable to further mobilize due to the accumulation of mutations and rearrangements over genome evolution (Lander et al. 2001; Richardson et al. 2015). However, a minor fraction typically retains the ability to mobilize in genomes (Mills et al. 2007). The percentage of genome occupied by TEs and their ongoing activity varies widely between organisms. In humans, TE-derived sequences may account for up to two-thirds of our genome (de Koning et al. 2011), and the human genome contains TEs with very different structures and modes of amplification, including DNA transposons and retrotransposons (Goodier and Kazazian 2008). DNA transposons, which

A. Macia (✉) • L. Tejwani • P. Mesci • A. Muotri
Department of Pediatrics/Rady Children's Hospital San Diego, University of California San Diego, La Jolla, CA 92093, USA
e-mail: amaciaortega@ucsd.edu; angela.macia1@gmail.com; leon.tejwani@yahoo.com; pmesci@ucsd.edu; muotri@ucsd.edu

J.L. Garcia-Perez (✉)
Department of Genomic Medicine, GENYO, Centre for Genomics & Oncology (Pfizer—University of Granada & Andalusian Regional Government), PTS Granada, Avda. de la Ilustración 114, PTS Granada, Granada 18016, Spain

Medical Research Council Human Genetics Unit, Institute of Genetics and Molecular Medicine, University of Edinburgh, Western General Hospital, Edinburgh EH4 2XU, UK
e-mail: Jose.Garcia-Perez@igmm.ed.ac.uk; jlgp@genyo.es

G. Cristofari (ed.), *Human Retrotransposons in Health and Disease*,
DOI 10.1007/978-3-319-48344-3_6

replicate by a cut-and-paste mechanism, are excised from one location and reintegrated into a new genomic location using a transposase activity (reviewed in (Munoz-Lopez and Garcia-Perez 2010; Tellier et al. 2015)). These elements are widely distributed in nature and are usually found in plants and lower-order organisms. Although often no longer active in mammals (Richardson et al. 2015), bats contain active DNA transposons in their genomes suggesting that their activity continues to generate genetic diversity in some mammals (Ray et al. 2008). By contrast, DNA transposons account for only 3 % of the human genome and are no longer active (Lander et al. 2001). On the other hand, retrotransposons are DNA sequences that move through an intermediate RNA by a copy-and-paste process that results in their amplification in genomes over evolution (Moran and Gilbert 2002). Indeed, the majority of TEs found in both humans and mice are retrotransposons. Retrotransposons are classified into long terminal repeat (LTR) or non-LTR retrotransposons depending on the presence/absence of LTR sequences. LTR retrotransposons (also known as endogenous retroviruses or ERVs) comprise approximately 8 % of the human genome and their structure resembles simple retroviruses (recently reviewed in (Mager and Stoye 2015)). Despite their prevalence in the human genome, de novo LTR-retrotransposon mobilization events in humans remain uncharacterized, suggesting they are no longer active in our genome (Mager and Stoye 2015). Why LTR-retroelements are no longer active in humans is a constant source of research and debate. In this article, we will focus on the non-LTR class of retrotransposons present in the human genome, as this is the class that is currently active in our genome (Mills et al. 2007; Richardson et al. 2015).

Within the non-LTR retrotransposons class, a minor fraction of LINE-1s and Short Interspersed Elements (SINEs) remains active in the human genome; non-LTR retrotransposons can be also classified in autonomous (LINE-1s) and non-autonomous (SINEs) retrotransposons respectively. LINE-1 or L1 is the most common superfamily of autonomous retrotransposons in mammals and constitute approximately 21 % of the human genome (Lander et al. 2001). Active LINE-1 elements are considered autonomous elements as they encode the minimal proteins required to mediate their mobilization by a process termed retrotransposition. On the other hand, *Alu* and SVA (SINE-VNTR-Alus) non-LTR retrotransposons are two types of SINEs that have found an effective way to amplify in the human genome without coding enzymatic machinery, as they both rely on L1-encoded proteins to mediate their own mobilization (Dewannieux et al. 2003; Hancks et al. 2011; Raiz et al. 2011). *Alu* is present in more than 1 million copies in the human genome and together with SVA elements represents more than 10 % of our genome. Thus, due to their ongoing activity and as a result of retrotransposon mobilization our genetic configuration is susceptible to changes over time, and new insertions can generate new genomic variants by a combination of DNA deletions, epigenetic alterations, or chromosomal rearrangements among other mechanisms. Indeed, recent retrotransposon insertions and TE-mediated recombination processes in humans have been linked to more than a 100 human genetic disorders (Hancks and Kazazian 2012). In sum, non-LTR retrotransposon activity is a constant source of genetic variation among humans.

2 Retrotransposition of Active LINE-1 Elements in the Human Genome

Human LINE-1s comprise more than 20 % of our genome and the 500,000 copies of L1s present in the human genome can be subclassified in several families. Since the emergence of apes, L1 elements have evolved and amplified rapidly during the last 25 Ma, generating five distinct L1 subfamilies (L1PA5 to L1PA1) (Furano 2000). The human-specific L1 family, L1Hs, is the only current family of active LINE-1s in humans, generating genomic variability among humans (Lander et al. 2001). Indeed, the completion of the Human Reference Genome (HRG) in combination with functional assays in cultured human cells has revealed that the average human genome contains approximately 80–100 L1 copies that are able to mobilize, termed retrotransposition-competent L1s (or RC-L1s (Brouha et al. 2003; Sassaman et al. 1997)). A full-length active RC-L1Hs is typically 6.0 kb in length (Scott et al. 1987) and contains a 5′ untranslated region (UTR), possesses up to three open reading frames (ORF0, ORF1, and ORF2) and ends in a short 3′ UTR followed by a poly(A) tail (Fig. 1). The 5′ UTR from human LINE-1s contains both sense and antisense RNA polymerase II promoter activity (Athanikar et al. 2004; Swergold 1990; Becker et al. 1993; Speek 2001; Macia et al. 2011). The sense promoter activity of the L1-5′ UTR is key to generate the sense L1 mRNA transcript that is used to translate LINE-1 encoded proteins and that later serves as a template to generate a new LINE-1 insertion during retrotransposition. Within this sense L1 mRNA, active LINE-1s contain two ORFs named ORF1 and ORF2. ORF1 codes for a 40 kDa protein (ORF1p) that has RNA binding and nucleic acid chaperone activities (Hohjoh and Singer 1996, 1997a, b; Martin and Bushman 2001; Khazina and Weichenrieder 2009). ORF2 encodes a 150 kDa protein (ORF2p) with DNA endonuclease (EN) and reverse transcriptase (RT) activities (Alisch et al. 2006; Feng et al. 1996; Mathias et al. 1991). ORF1p and ORF2p are strictly required for L1 retrotransposition (Moran et al. 1996). Intriguingly, and despite its conservation in mammalian LINE-1 elements (Furano 2000), the 3′ UTR of human LINE-1s is not strictly required for retrotransposition, at least using an engineered retrotransposition assay (Moran et al. 1996). However, the presence of a poly(A) tail is required for efficient LINE-1 retrotransposition (Doucet et al. 2015), both by recruiting ORF2p (Doucet et al. 2015) and facilitating the initiation of reverse transcription (Monot et al. 2013). More recently, and due to the presence of a conserved antisense promoter in full-length LINE-1s (Macia et al. 2011), a third ORF, named ORF0, has been discovered in selected human LINE-1s (Denli et al. 2015). Indeed, ORF0 is a primate-specific

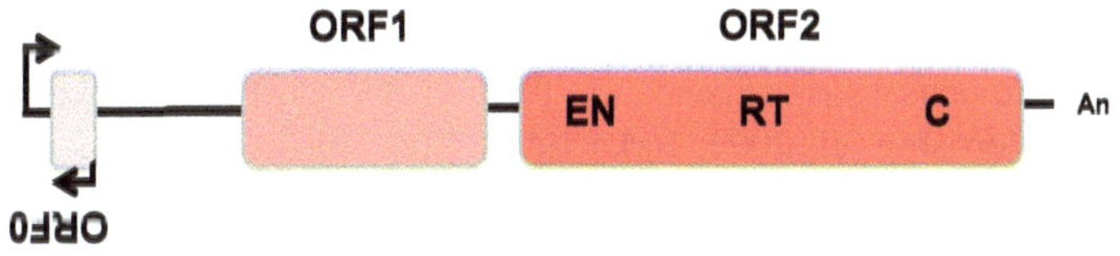

Fig. 1 Structure of active human LINE-1 elements. Details are provided in the main text

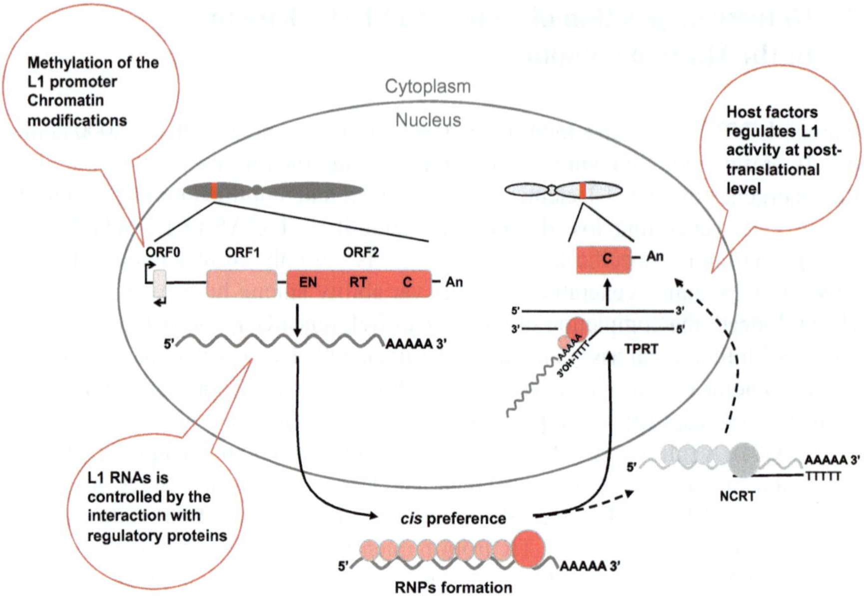

Fig. 2 Control of L1 Retrotransposition. Model of the retrotransposition cycle displaying a summary of the main cellular pathways controlling L1 replication

antisense ORF located in the RNA transcript generated by the antisense L1 promoter (Denli et al. 2015). ORF0 is present in approximately 800 loci in the human genome and it can increase retrotransposition of LINE-1s using an engineered based assay (Denli et al. 2015), although its mechanism of action remains to be determined.

A round of L1 retrotransposition starts with the production of a full-length L1 mRNA using the sense L1 promoter (Swergold 1990) (Fig. 2); the transcript is exported to the cytoplasm where ORF1p and ORF2p are translated by cap-dependent and unconventional termination-reinitiation mechanisms, respectively (Alisch et al. 2006; Dmitriev et al. 2007). Notably, both ORF1p and ORF2p preferentially associate with their own RNA in a *cis*-preference manner (Wei et al. 2001), leading to the formation of a ribonucleoprotein particle (RNP). The assembly of the L1 RNP is strictly necessary for L1 retrotransposition (Kulpa and Moran 2005, 2006; Doucet et al. 2010; Martin 1991; Hohjoh and Singer 1996; Wei et al. 2001). Next, the L1-RNP complex gains access to the nucleus by a process that does not require cell division (Kubo et al. 2006). Once in the nucleus, integration of a new L1 copy occurs by a mechanism known as target primed reverse transcription (TPRT), initially described for R2Bm retrotransposons (Luan et al. 1993; Cost et al. 2002). During TPRT, the EN activity of L1-ORF2p makes a single strand nick in the genomic DNA at a degenerate consensus sequence (5′-TTTT/A-3′) (Cost and Boeke 1998; Cost et al. 2002; Morrish et al. 2002; Jurka 1997). The exposed 3′-OH is then used to prime L1 cDNA synthesis using the RT activity of L1-ORF2p and using the L1 mRNA as a template during polymerization (Luan et al. 1993; Cost et al. 2002). Notably, the efficiency of the initiation of reverse transcription is thought to be influenced by the 10 last nucleotides of the target DNA, with 4 Ts being the most optimal sequence configuration (Monot et al. 2013). Second strand cleavage and subsequent cDNA synthesis will generate a new L1 insertion, although how these steps occur at the molecular level requires elucidation. Thus, TPRT will give rise to a new L1 inserted elsewhere in the genome, although the vast majority of de novo L1 integrations are 5′ truncated (Grimaldi and Singer 1983). Although the process of 5′ truncation is not fully understood, it has been demonstrated that it is influenced by DNA-repair processes (Coufal et al. 2011) rather than reflecting an inherent limitation of the L1 encoded RT activity; notably, it has been demonstrated that the RT activity encoded by a LINE element from the silkworm is very processive (Bibillo and Eickbush 2002) and new L1 insertions can be full-length. In addition to 5′ truncation, other alterations of the L1 sequence are often observed in new retrotransposition events: it has been described that new L1 insertions can produced inverted/deleted L1 structures (a process termed Twin Priming (Ostertag and Kazazian 2001)) and can also be accompanied by flanking genomic sequences (3′ or 5′ transductions) (Gilbert et al. 2002; Moran et al. 1999). Thus, there is variation in the structures generated during retrotransposition.

In general, L1 retrotransposition is initiated by cleaving genomic DNA through the action of the L1-encoded EN. However, it has been described that a preexisting free 3′-OH may also be used by the L1-RNP to initiate retrotransposition. This pathway of retrotransposition is termed endonuclease-independent (ENi) L1 retrotransposition and is thought to occur when L1 use genomic DNA lesions (i.e., free 3′ OH ends) to initiate TPRT. However, efficient ENi L1-retrotransposition has only been described in cells defective for both the non-homologous end-joining (NHEJ) pathway of DNA repair and p53 function (Morrish et al. 2002, 2007; Eickbush 2002; Coufal et al. 2011). In NHEJ-mutant cells, it has been proposed that L1s can parasitize unrepaired DNA sites present in these cells to initiate retrotransposition, leading to insertions with altered structures (Morrish et al. 2002, 2007; Eickbush 2002; Coufal et al. 2011). Finally, and although TPRT is the canonical mechanism of retrotransposition, it is tempting to speculate that in selected circumstances reverse transcription by the L1 encoded RT could also occur in the cytoplasm, in an analogy to retroviruses and LTR-retrotransposons (Telesnitsky and Goff 1997). However, how the L1 RT would prime cDNA synthesis (non-canonical reverse transcription or NCRT, Fig. 2) and how this cytoplasmic DNA may be integrated in the nucleus remain to be determined.

3 The Impact of LINE-1 Retrotransposition in the Human Genome

New LINE-1 retrotransposition events can impact the genome in multiple ways (reviewed in (Cordaux and Batzer 2009; Kaer and Speek 2013; Beck et al. 2011; Belancio et al. 2009; Belancio et al. 2008a; Richardson et al. 2015)). Briefly, from simply disrupting an exon (Kazazian et al. 1988) to inducing gross alterations of the insertion site (Mine et al. 2007), there is a myriad of ways in which a new L1 insertion can impact the genome. Additionally, a new retrotransposition event can disrupt and modify a gene structure by a process known as "gene breaking," but can also affect gene expression by adding promoter sequences or polyadenylation signals, by altering the chromatin status of nearby sequences, by altering splicing patterns, etc. (Ostertag and Kazazian 2001; Wheelan et al. 2005; Belancio et al. 2008b; Kazazian et al. 1988; Han et al. 2004; Garcia-Perez et al. 2010; Goodier and Kazazian 2008; Matlik et al. 2006; Nigumann et al. 2002; Speek 2001). In sum, the impact of each insertion will be influenced by where in the genome the insertion took place and what type of L1 structure is generated at the insertion site. Indeed, L1s are a recurrent source of genomic novelty, and can provide a unique configuration to mammalian genomes.

4 LINE-1 Regulation by Host Factors

Despite the presence of approximately 80–100 RC-L1s that are able to move in our genome, it is also clear that the human genome is in a constant battle with TEs to prevent their amplification and exacerbated activity, as this could be negative at the individual level. The process of 5′ truncation is a good example of how our genome prevents the accumulation of active L1s over evolution (Goodier and Kazazian 2008). Indeed, there are many layers of retrotransposition control, from transcription to the latest stages of TPRT (Heras et al. 2014). In general, DNA methylation, chromatin remodeling, and post-transcriptional regulation of L1 mRNAs are the main actors to control retrotransposition (Fig. 2).

The epigenetic regulation of L1 expression is perhaps the most efficient manner to reduce retrotransposition over evolution; thus, different mechanisms have evolved in mammalian cells to recruit the silencing machinery necessary to control and regulate L1 expression and its subsequent activity. Similarly, it is also likely that L1s have evolved to avoid the inhibitory mechanisms created by the host, as recently demonstrated (Jacobs et al. 2014). Cytosine methylation of the canonical CpG island located in the promoter region of L1 elements regulate its transcription (Thayer et al. 1993), and this process is governed by the action of the DNA methyltransferase-3 like (DNMT3L) protein in germ cells (Bourc'his and Bestor 2004). However, when genomes are hypomethylated during normal biological processes, alternative mechanisms to regulate L1 retrotransposition must be present in these cells (Munoz-Lopez et al. 2011; Castro-Diaz et al. 2014). It is well known that during embryogenesis a wave of DNA hypomethylation results in TE expression (Munoz-Lopez et al. 2011; Castro-Diaz et al. 2014). Consistently, both human embryonic stem cells (hESCs) and induced pluripotent stem cells (hiPSCs) show significant hypomethylation of L1 promoters, resulting in the accumulation of L1 mRNAs in these pluripotent cell types (Shen et al. 2006; Garcia-Perez et al. 2010; Munoz-Lopez et al. 2011; Wissing et al. 2012). Thus, if L1s are regulated during embryogenesis, other epigenetic mechanisms may exist in these cells. Indeed, KRAB-containing zinc finger protein (KRAB-ZFPs), a mediator of heterochromatin formation, binds the 5′ UTR of old L1 subfamilies regulating L1 expression in pluripotent cells (Castro-Diaz et al. 2014; Jacobs et al. 2014). Additionally, other modifications like 5-hydroxymethylcytosine (5hmC) induced by the TET family of proteins are also located within L1 sequences (Branco et al. 2012), although their functional significance in regulating L1 expression is less clear.

However, the epigenetic silencing of TE expression is not the only strategy used by the host to control L1 retrotransposition. The microprocessor complex (Drosha-DGCR8), which processes pre-miRNAs to generate miRNAs, has been recently described to regulate LINE-1 mRNA levels (Heras et al. 2013). Indeed, the microprocessor complex has been shown to mediate the degradation of L1 mRNAs and thus could control the rate of L1 retrotransposition in human cells (Heras et al. 2013). Other proteins involved in L1 RNA regulation are Piwi proteins. Piwi proteins are highly conserved across evolution and specifically interact with piwi-interacting small RNAs (piRNAs), inducing TE RNA degradation by a complex mechanism (Heras et al. 2014; Aravin et al. 2007, 2008). Additional host factors such as MOV10 (RNA helicase) or RNAse L are also known for being potent restriction factors of L1 expression and retrotransposition (Goodier et al. 2012; Zhang et al. 2014). More recently, the use of epitope-tagged L1 constructs has allowed to generate a list of host factors that interact with L1-RNPs; however, their role on L1 regulation remains to be discovered for most identified factors (Goodier et al. 2013; Moldovan and Moran 2015; Taylor et al. 2013). One of these factors, ZAP, is a zinc-finger protein that targets positive and negative strand RNA viruses, as well as some DNA viruses. Remarkably, ZAP can also interact with human retrotransposons. ZAP associates with L1-RNPs and strongly restricts retrotransposition in cell culture assays by a mechanism that affects L1-RNP integrity (Goodier et al. 2015). Notably, many of the identified host factors interacting with L1-RNPs may act at later stages, specifically during TPRT in the nucleus, and may regulate retrotransposition by still unknown mechanisms. Indeed, APOBEC3 (A3) is a family of cytidine deaminase enzymes that are known to limit HIV infection and can also regulate L1 retrotransposition, likely by deamination and editing L1 sequences (Schumann 2007; Wissing et al. 2011; Marchetto et al. 2013; Richardson et al. 2014b). Other factors that may regulate TPRT include *Proliferating cell nuclear antigen* (PCNA); PCNA functions as a scaffold protein during DNA replication and during DNA repair and it has been proposed to promote ORF2p loading on the genomic DNA, to act as an ORF2p processivity factor, and/or to help resolving the integration process after reverse transcription (Taylor et al. 2013). In sum, we are just starting to uncover new host factors that control and regulate L1 activity during the retrotransposition cycle.

Finally, it is becoming more evident that there might be a connection of L1 retrotransposition with selected genetic disorders, including Ataxia Telangiectasia (Coufal et al. 2011), Rett Syndrome (Muotri et al. 2010), and Aicardi-Goutieres Syndrome (AGS), among others (see below). Intriguingly, the AGS-related enzymes SAMHD1 and TREX1, dNTP phosphohydrolase, and 3′ repair exonuclease, respectively, are known to control the mobilization of L1 retrotransposons (Zhao et al. 2013; Stetson et al. 2008). Interestingly, mutations in either enzyme result in the accumulation of DNA fragments, which may trigger autoimmunity and therefore cause AGS (Volkman and Stetson 2014). This topic of research is in its infancy, and more research will be required to definitively learn how TE and TE-derived products could contribute to the molecular basis of these human genetic disorders.

5 L1 Activity in Pluripotent and Somatic Cells

Historically, TEs were catalogued as "junk DNA" (Orgel and Crick 1980); however when the Kazazian laboratory identified the first mutagenic L1 insertion in 1988 (Kazazian et al. 1988) this view started to change, and it became undeniable that active TEs can impact the function of the human genome. Since 1988, and as L1 insertions presumably occur randomly in the genome, mutagenic L1 retrotransposition events have been implicated in a variety of human diseases such as Hemophilia A, X-linked retinitis pigmentosa, or Duchenne muscular dystrophy (Hancks and Kazazian 2012). The generation of a human genetic disorder by retrotransposition is likely a by-product of their inherent capability to generate new transmissible L1 insertions in man. To ensure its evolutionary success, L1 elements may generate new copies in a cell type that will guarantee its transmission to the next generation; thus, early embryogenesis and germ cells are cell types where new L1 insertions could accumulate. Indeed, the work from different laboratories has demonstrated that most heritable L1 insertions accumulate during early embryonic development in both mouse models and using in vitro pluripotent human cells (An et al. 2006; Babushok et al. 2006; Garcia-Perez et al. 2007; van den Hurk et al. 2007; Kano et al. 2009; Levin and Moran 2011). From an evolutionary point of view, these L1 insertions can perpetuate LINE-1 presence within the human genome. However, recent research has demonstrated that LINE-1s are also active in selected somatic cell types; thus, the load of retrotransposition in humans is not restricted to early embryogenesis. Indeed, several studies have shown that L1 expression and retrotransposition can occur in selected tumors and in the human brain (Fig. 3)

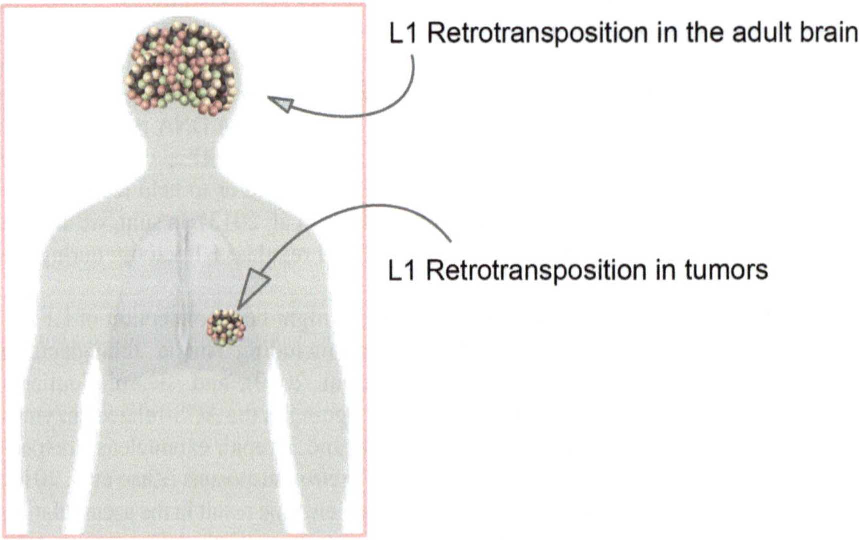

Fig. 3 Schematic representation of somatic L1 retrotransposition. The adult brain and tumors have the highest L1 retrotransposition efficiency in humans

(Muotri et al. 2005; Coufal et al. 2009; Baillie et al. 2011; Evrony et al. 2012; Goodier 2014; Miki et al. 1992) (Carreira et al. 2014; Iskow et al. 2010). From an evolutionary point of view, the somatic accumulation of new L1 copies does not represent a straightforward selective advantage. However, it is tempting to speculate that this variability in somatic genomes may increase our plasticity over the life of an individual. In the following sections, we will discuss known facts about somatic retrotransposition of TE in humans but will also discuss the rate of accumulation and genomic impact in the human genome in a healthy or diseased background.

5.1 L1 Retrotransposition During Embryogenesis: L1s Can Be Transmitted to the Next Generation

As discussed above, and as a TE, L1s can be considered as the prototype of "selfish DNA" whose sole function is to accumulate more copies of themselves in newborns. To do that, L1s generate new insertions in cells that can transmit genetic information to the next generation. Notably, pluripotent genomes are characterized by their low content in DNA methylation especially after fertilization, which correlates with elevated levels of L1 mRNA expression. For a better understanding of L1 biology during embryogenesis, hESC and iPSCs, which mimic early stages of human development, have been used to model this process in vitro, due to the difficulty of conducting research with human embryos. hESCs are pluripotent cells derived from the inner cell mass of human blastocysts that can self-renew indefinitely and undergo differentiation to generate the three embryonic germ layers (Thomson et al. 1998). Notably, it has been described that hESCs overexpress a constellation of L1 RNA derived sequences, including both potentially active and inactive copies of L1 and *Alu* elements (Garcia-Perez et al. 2010; Macia et al. 2011; Wissing et al. 2012). Importantly, the expression level of L1 in these cells inversely correlates with the level of methylation of the L1 promoter (Garcia-Perez et al. 2007; Munoz-Lopez et al. 2011, 2012). Additionally, the use of an engineered L1 retrotransposition assay (Moran et al. 1996) has demonstrated that the minimal set of host factors required to accumulate new L1 insertions is present in hESCs and hiPSCs (Garcia-Perez et al. 2007, 2010; Klawitter et al. 2016; Wissing et al. 2011, 2012). The retrotransposition assay relies on the use of engineered L1s tagged with a reporter cassette that can only be activated after a round of retrotransposition (Moran et al. 1996). Thus, these data support the notion that new heritable L1 insertions can accumulate during early embryogenesis (van den Hurk et al. 2007). The rate of endogenous L1 retrotransposition at this developmental stage remains to be determined. However, estimates have suggested that the global rate of inherited L1 retrotransposition events in humans may be as high as 1/50 (Kazazian 1999; Xing et al. 2009; Ewing and Kazazian 2010; Huang et al. 2010).

More recently, the development of hiPSCs from somatic differentiated cells allowed researchers to study the overall rate of engineered L1 retrotransposition in a variety of genetic backgrounds (healthy or diseased). Others and we have demonstrated that hiPSCs, as hESCs, are characterized by expressing elevated levels of endogenous L1 RNPs due to the severe hypomethylation of the L1 promoter (Wissing et al. 2012; Friedli et al. 2014). In addition, the retrotransposition of engineered L1s in hiPSCs is 10–15-fold more efficient than in fibroblasts, indicating that hiPSCs are also a good model to study L1 biology (Wissing et al. 2012). Additionally, the characterization of engineered L1 insertions in hiPSCs and hESCs revealed canonical hallmarks associated with L1 insertions (Garcia-Perez et al. 2007; Wissing et al. 2012). Notably, the presence of restriction factors such as APOBEC and Piwi proteins in hESCs and iPSCs indicates that not only the molecular machinery required to retrotranspose is present in these cells, but also several major mechanisms acting to regulate retrotransposition (Marchetto et al. 2013; Wissing et al. 2011). More recently, to elucidate if the genome-wide remodeling of epigenetics marks, which occurs upon cellular reprogramming, activates endogenous L1 retrotransposition, researchers applied retrotransposon capture sequencing (RC-seq) to identify and map potential new L1 insertions that might accumulate in the genome of hiPSCs and hESCs, either during reprogramming or long-term cultivation (Klawitter et al. 2016). Notably, this demonstrated that endogenous non-LTR retrotransposons are active in both hESCs and hiPSCs (Klawitter et al. 2016). Intriguingly, an unexpected fraction of characterized L1 insertions in hiPSCs are full-length insertions, which suggests that endogenous L1 retrotransposition events have the potential to impact hiPSC genomes and perhaps compromise their future application in biomedicine (Klawitter et al. 2016; Gore et al. 2011). These observations suggest that L1 expression may be a marker of pluripotent cells. However, a number of open questions remain and will require additional studies in the future. For example: Can endogenous retrotransposition lead to altered hiPSCs phenotypes? What is the rate of endogenous L1 retrotransposition in humans? When does retrotransposition start upon fertilization?

5.2 L1 Retrotransposition in Human Tumors: The Chicken or the Egg Argument

Genomic instability, driver point mutations and karyotype abnormalities are major driving forces of tumorigenesis (Ellsworth et al. 2004). The role that TEs may play in the initiation of tumorigenesis might be merely speculative; despite this, there are good examples of this possible relationship. For example, the disruption of a tumor suppressor gene caused by a somatic or early embryonic TE insertion could facilitate additional genomic insults that may lead to cancer. Indeed, in 1992, it was described that a new L1 retrotransposition event disrupted the *APC* gene in a patient affected with colon cancer, suggesting that this could be a tumorigenic L1 insertion (Miki et al. 1992). Thus, in this particular case, the L1 insertion in the *APC* gene could be causative of the colorectal tumor, independently of its somatic or early embryogenesis origin. This study was likely the tip of the iceberg; indeed, the genomics revolution has revealed ongoing retrotransposition in most epithelial cancers examined to date (reviewed in (Carreira et al. 2014)). However, whether somatic retrotransposition of L1s in healthy tissues is a common process that may normally participate in tumorigenesis is far from clear, as only selected cells within the human brain have been shown to support endogenous retrotransposition in healthy conditions (reviewed in (Richardson et al. 2014a); see below). Indeed, it is tempting to speculate that somatic retrotransposition of L1s in the brain is the exception more than the generic rule, although further studies are required to firmly determine if retrotransposition can occur in other cell types or tissues present in a healthy human body. Related to this, recent studies suggest that only a small number of L1 copies (some being polymorphic between individual) escape cellular silencing mechanisms in a cell-type specific manner, and might be associated with increased cancer risk (Philippe et al. 2016; Scott et al. 2016).

Furthermore, it is possible that ongoing L1 retrotransposition can also impact tumor progression. Several studies have shown an inverse correlation between the DNA methylation level of the L1 promoter and L1 expression, and intriguingly also with a poor clinical prognosis in cancer patients (Rhee et al. 2015; Inamura et al. 2014; Suter et al. 2004; Cho et al. 2007). Similarly, a correlation between low levels of L1 promoter methylation and high levels of nuclear L1 ORF1p expression with poor clinical outcomes has been observed (Harris et al. 2010). Intriguingly, L1 ORF1p is expressed at low or undetectable levels in most adult tissues and at early tumorigenesis stages; however L1 ORF1p expression is exacerbated in several neoplasms including lung, ovarian, or prostate carcinomas (Belancio et al. 2010a; Rodic et al. 2014). The use of L1 expression or CpG methylation as a cancer biomarker is discussed in Chap. 12 of this book. In sum, it is still unclear whether reactivation of L1 expression by CpG hypomethylation is a cause or consequence of tumor progression. Indeed, it is likely that DNA hypomethylation on tumors reflects rapid cell division and thus results in loss of DNA methylation patterns on L1 promoters, which lead to L1 mRNA overexpression. As DNA hypomethylation in tumors typically occurs in a genome-wide manner, it is likely that a fraction of overexpressed L1s might correspond to mRNAs from active RC-L1s; next, and if the tumor cells express the host factors required for L1 retrotransposition to occur, it is likely that these RC-L1s may generate new insertions. As in any mutagenic loop, the more insertions produced, the higher the probability to disrupt a gene that may affect tumor progression. This model rather implicates L1 retrotransposition in the late stages of cancer development. However, this model was very recently challenged by the detection of de novo L1 somatic insertions in precancerous colorectal and esophageal lesions or even in normal tissue (Doucet-O'Hare et al. 2015; Ewing et al. 2015; Scott et al. 2016). Thus, studying the extent of this phenomenon will require the development of more sensitive techniques to detect L1 insertions in a small number of cells or even in single cells.

Aside from the putative role that L1 retrotransposition may exert during tumor initiation, in general most tumors are characterized by their inherent genomic instability. Notably, studies in cultured tumor cells have demonstrated that a fraction of new L1 retrotransposition events are associated with genomic instability processes (Symer et al. 2002; Gilbert et al. 2002). On the other hand, it was also described that

L1 ORF2p expression, a key protein in the retrotransposition process, is associated with DNA damage formation, genotoxicity, and a marked increase of both intra- and inter-chromosomal translocations (Gasior et al. 2006; Wallace et al. 2008; Belancio et al. 2010b; Lin et al. 2009). Consistently, altered expression patterns of ORF1p and ORF2p have been found within invasive cancers, which is again associated with poorer patient survival (Chen et al. 2012).

Over the past years, several methods have appeared to precisely map new L1 retrotransposition events in human DNAs (Doucet and Kazazian 2016; Sanchez-Luque et al. 2016). Briefly, these methods were either based on previous established methods to identify young and likely RC-L1s (Badge et al. 2003) or have exploited hybridization/capture protocols to enrich for L1-fragments (Baillie et al. 2011). These protocols were coupled to next-generation DNA sequencing (NGS) platforms, greatly facilitating the identification of new L1 retrotransposition events in human DNA, although astringent controls are required at all steps (Evrony et al. 2016; Goodier 2014). Using NGS, and for the first time, Iskow and colleagues reported nine tumor-specific somatic L1 insertions in 6/20 primary lung carcinomas (Iskow et al. 2010). Since then, other studies have reached similar conclusions using other tumor types and other sequencing approaches (reviewed in (Carreira et al. 2014; Goodier 2014)). More recently, other studies have exploited Whole Genome Sequencing data (WGS) from tumor–normal tissue pairs to identify somatic L1 insertions in tumors. Using this approach, Lee et al. discovered 183 insertions in colorectal, ovarian, and prostatic carcinomas (Lee et al. 2012). Interestingly, the analysis of multiple cancer types has revealed that somatic retrotransposition occurs preferentially in cancers with an epithelial origin. Notably, these cells have the potential to switch from an epithelial to mesenchymal phenotype, in which migrating cells would trigger the metastasis. Epithelial cells can also be transformed to yield cancer stem cells, where an oncogenic L1 insertion in this cell type would increase the probability to spread the mutation (Carreira et al. 2014). Thus, it is reasonable to propose that the plasticity of epithelial tumors could explain their permissiveness for L1 activity. In sum, it is well established that L1s are actively retrotransposing in tumor cells but it is less clear if new L1 insertions represent merely passenger mutations or if they have the capability to affect tumor progression.

Finally, and from a therapeutic point of view, the use of RT inhibitors (RTis) has been proposed to treat cancer patients, using commercially available drugs against HIV infection that also inhibit L1 retrotransposition (Jones et al. 2008; Dai et al. 2011). However, a proper clinical trial must be established to definitively conclude if L1 retrotransposition could be a valuable pharmacological target to treat cancer.

5.3 L1 Retrotransposition in Adult Healthy Tissues

Over the past years, several laboratories have used complementary approaches and experimental models to conclude that there is ongoing somatic retrotransposition of L1s in selected brain cells. However, many questions remain unsolved regarding the landscape of somatic retrotransposition in other human tissues. Given their potential relationship with tumorigenesis, understanding the impact of this process on the cellular physiology should be a high research priority. Although the ongoing activity of L1s is well supported in tumors and in the healthy human brain, it is currently unknown whether the somatic activity of L1 is an active process regulated by the developmental potency of the host cell. In a recent study, researchers investigated L1 expression in a broad spectrum of normal human tissues including adult stem cells (Belancio et al. 2010b). Belancio et al. found that tissues from esophagus, prostate, stomach, and heart muscle expression levels were similar to those detected in HeLa cells. In contrast, L1 expression levels in adrenal gland, spleen, kidney, and cervix were below the sensitivity of the assay (Belancio et al. 2010b). However, we simply do not know which fraction of these L1 RNAs correspond to mRNAs from RC-L1s or if these cells support the mobilization of RC-L1s. In sum, whether L1 somatic retrotransposition is an active process in human tissues or whether it is restricted to the brain remains unknown.

5.4 L1 Retrotransposition in the Adult Brain

The first evidence demonstrating that L1s are able to mobilize in mammalian brains used engineered human L1-EGFP reporter assays in both cultured rodent cells and animal models (Muotri et al. 2005) (see also Chap. 5). Remarkably, Muotri and colleagues demonstrated that Neuronal Progenitor Cells (NPCs) isolated from the adult rat hippocampus were able to accommodate the retrotransposition of human L1 retrotransposons with elevated frequency and that a fraction of these engineered L1 insertions occurred into neuronal expressed genes (Muotri et al. 2005). These initial observations have been further reinforced by additional experiments done in vivo with L1-EGFP transgenic mice, in which cells containing new insertions could be directly visualized in the rodent brain. This animal model identified that some de novo human L1 retrotransposition events co-localized with a postmitotic neuronal marker (NeuN) in striatum, cortex, hypothalamus, hilus, cerebellum, ventricles, amygdala, and hippocampus (Muotri et al. 2005). Interestingly, it was later reported a higher level of L1-EGFP expressing cells after voluntary exercise in transgenic mice versus controls (Muotri et al. 2009). This increase in L1-EGFP expression rate could be explained by an increase of neurogenesis where new L1 events will accumulate, or by the de-repression of previously silenced L1 insertions (Muotri et al. 2009). However, this seminal study demonstrated for the first time that the mammalian brain is a mosaic of genomes.

Next, similar studies were extended to humans (Coufal et al. 2009, 2011). Coufal and colleagues isolated NPCs from human fetal brain or use hESCs to derive NPCs and conducted retrotransposition assays; surprisingly, this study revealed that human NPCs could accumulate new L1 insertions with a very high frequency, even higher than previously reported in tumor cell lines (Coufal et al. 2009). Additionally, an increase in the copy number of endogenous L1s was found in several regions of the adult human brain compared to other somatic tissues (liver and heart) isolated from healthy donors, further suggesting ongoing retrotransposition of L1s in the human brain (Coufal et al. 2009). However, these qPCR estimates were not valid to establish a rate of brain retrotransposition, as the assay compared a plasmid to genomic DNA among other caveats. However, and with these caveats, it was observed that the adult hippocampus, a major neurogenic niche in the brain, contained elevated L1 copies compared with other brain regions. Thus, it is tempting to speculate that L1 retrotransposition may occur frequently during the formation of the central nervous system (CNS) and later during neurogenesis in the adult brain. Intriguingly, this mobilization in individual cells could lead to neuron-to-neuron variation and might be partially responsible for the generation of somatic mosaicism; however the functional consequences of these findings remain unknown (Richardson et al. 2014a). What is certainly clear is that NPCs are very permissive for L1 retrotransposition, even higher than cancer cell lines (Coufal et al. 2009).

Next, and due to the development of NGS, several independent laboratories have subsequently demonstrated that the human brain is indeed made of a mosaic of genomes, although there is a serious debate about what the real frequency of retrotransposition in the human brain might be (Baillie et al. 2011; Evrony et al. 2012, 2015, 2016; Upton et al. 2015). In 2011, a high-throughput methodology was used to identify de novo L1 insertions in three individual's hippocampus (Baillie et al. 2011). Briefly, Baillie and colleagues developed a new capture method named RC-seq that relies on a low number of PCR cycles and thus is less prone to artifacts (Sanchez-Luque et al. 2016). In RC-seq, the DNA junctions between retrotransposons and adjacent genomic regions are enriched by hybridization/capturing, followed by paired-end sequencing and alignment to the human genome to identify L1 insertions absent from the HRG. Using RC-seq, these researchers were able to identify numerous somatic L1 insertions inserted in protein-coding genes differentially expressed in the brain; however this could be explained by the fact that neuron-related genes have a larger average gene length in humans. However, it is also possible that the chromatin status of brain cells might be influencing where retrotransposition events in the brain may occur (Thomas et al. 2012). In this study, thousands of somatic L1 insertions were identified, although only a small fraction of insertions was thereof validated (Baillie et al. 2011).

Next, single-cell technology was exploited to determine the timing and load of L1 retrotransposition events in the brain. In 2012, Evrony and colleagues exploited single cell genomics and a modified L1-seq (Ewing and Kazazian 2010) approach to identify and validate the first somatic full-length L1 insertion reported in the human brain (Evrony et al. 2012). However, and in stark difference with the previous study, Evrony et al. reported an average of 1.1 somatic L1 insertions per neuron, and 0.6 unique somatic insertions per neuron from cerebral cortex and caudate nucleus of three normal individuals (Evrony et al. 2012). However, in this study only 5/81 insertions were validated. More recently, Upton and colleagues adapted RC-seq to single cells, and concluded that an estimated 13.7 and 6.5 somatic L1 insertions occurred per hippocampal neuron and glial cell, respectively (Upton et al. 2015). Notably, hippocampal L1 insertions were specifically enriched in transcribed neuronal stem cell enhancers and hippocampus genes (Upton et al. 2015). Thus, these data suggest that somatic retrotransposition is ubiquitous in the human brain, and that perhaps every neuron of the human brain may contain a unique genome (Upton et al. 2015). However, and very recently, a controversial debate has appeared regarding what the real rate of retrotransposition in the brain might be (Evrony et al. 2016). Simply stated these studies used very different approaches and bioinformatics analyses that in part may explain part of the discrepancy between both studies. However, none of these studies validated the full spectrum of identified L1 insertions. That said, a recent study used a very different approach (reprogramming of mature neuronal cells) to biologically amplify genomes and used WGS to map the full repertoire of mutations accumulated in neuronal cells over time (Hazen et al. 2016). In this study, authors identify a small number of putative de novo L1 insertions in MT neurons, and some insertions were PCR validated. Notably, this alternative study concluded that up to 1.3 de novo somatic L1 insertions could be accumulated in MT neuronal cells from the rodent brain (Hazen et al. 2016). However, further thereof studies are required to resolve what the real ratio of L1 retrotransposition in the human brain might be and whether different brain areas have different capabilities to support retrotransposition of human L1s.

No matter what the real number of somatic L1 insertions per neuron might be, the biological impact of neuronal retrotransposition remains an open question. It is undeniable that the adult brain contains approximately 86 billions of neurons (Azevedo et al. 2009), with likely more than 10^{15} connections. Thus, and although the role of L1 retrotransposition in generating ubiquitous neuronal diversity has been recently challenged (Evrony et al. 2016), additional thereof studies must be conducted in order to infer the impact of somatic retrotransposition in the brain. This is especially important as a subtle change in a genome can have a dramatic impact in any biological process (i.e., a point mutation can be fully detrimental to a cell) but also because we simply lack real information about what the impact of retrotransposition in the brain might be. Additionally, speculating that the somatic retrotransposition of LINE-1s might not be significant enough to allow retrotransposition to manifest a role on brain biology (Evrony et al. 2016) is not based on functional analyses and it is likely detrimental for this field of research. Notably, long ago TEs were classified as merely "junk DNA" but it is now undeniable that TEs have dramatically impacted the structure of the human genome, as well as its evolution and genome-wide regulation. Thus, reusing the concept "junk DNA" for the putative role that the somatic retrotransposition of L1s may have on brain biology is inaccurate, at least at present. In sum, more research on this topic must be conducted before concluding that L1 somatic retrotransposition is involved (or not) in any biological process operating on the human brain, no matter what the real rate of retrotransposition in the brain might be (Evrony et al. 2016).

5.5 *L1 Retrotransposition in the Adult Brain of Model Organisms*

More recently, studies suggested that L1s are more active in somatic tissues during the course of aging (see also Chap. 13). It has been shown that Sirtuin 6 (SIRT6) is a powerful repressor of L1 activity (Van Meter et al. 2014). SIRT6 is a protein deacetylase and mono-ADP ribosyltransferase that promotes chromatin silencing through KAP1 (KRAB-associated protein 1), and facilitates DNA repair. Notably, SIRT6 KO mice develop a severe premature aging syndrome, characterized by genomic instability, while mice overexpressing SIRT6 exhibit extended lifespans. Intriguingly, it has been demonstrated that SIRT6 silences L1 by binding to its promoter and recruiting repressive heterochromatin factors, regulating the expression of these retroelements (Van Meter et al. 2014). In this study, authors found that SIRT6 is depleted from L1 loci in aged cells, allowing the reactivation of these previously silenced retroelements. Thus, upon aging and DNA damage, SIRT6 leaves L1 promoters and localizes to sites presumably containing DNA breaks (Van Meter et al. 2014). Similarly, in another study conducted in *Saccharomyces cerevisiae*, the authors described that the mobility of the Ty1 LTR-retrotransposon is elevated in mother cells versus their daughter cells (Patterson et al. 2015). Overall, these data suggest that retrotransposons can become more active during the course of aging; however, many questions remain unanswered regarding the relationship between retrotransposons and age-related neurodegenerative disorders.

Notably, somatic retrotransposition in the brain might not be exclusive to mammals and recent studies strongly suggest that retrotransposition also occurs in the Drosophila brain (Perrat et al. 2013). Specifically, retrotransposon expression and retrotransposition of LTR-retrotransposons have been described in αβ neurons from the mushroom body, a brain structure critical for olfactory memory in Drosophila. The authors of this study observed that Piwi-interacting RNA (piRNA) proteins were less abundant in this type of neuron, and was inversely correlated with elevated retrotransposon expression in the brain. In order to identify somatic mobilization events, whole genome sequencing (WGS) and paired-end deep sequencing were performed and more than 200 somatic insertions were identified. Consistently, Li and colleagues reported retrotransposition in Drosophila brains using a LTR-retrotransposon gypsy-TRAP engineered system where neural retrotransposition activation may contribute to neuronal decline with age, also in mushroom body neurons (Li et al. 2013).

In sum, it is tempting to speculate that on one hand retrotransposition may provide a source of somatic diversity in the brain as a normal part of brain physiology. However, it is unknown whether mutations caused by new TE insertions could alter the brain circuitries or even contribute to neuronal decline in humans and other organisms. On the other hand, in mammals, L1 elements are de-repressed in a variety of neurodegenerative disorders or models, suggesting that deregulation of TEs could exacerbate some aspects of these diseases or even have a major causative role. Thus, the impact of TEs in the brain may be much higher than previously anticipated but functional testing of these hypotheses requires further experimentation.

5.6 *L1 Retrotransposition and Brain Disorders*

Consistent with a potential role of L1 activity in the pathophysiology of brain disorders, several of these diseases are associated with mutations in factors also known to regulate retrotransposition. Sox2 is a negative regulator of neuronal differentiation but is known to interact with the L1 promoter limiting its retrotransposition (Kuwabara et al. 2009). Additionally, methyl CpG binding protein 2 (MeCP2) also regulates L1 expression and retrotransposition in NPCs and neurons of mouse and human brains (Muotri et al. 2010). Notably, de novo *MECP2* mutations in humans cause Rett Syndrome (RTT), a progressive neurological disorder being considered part of the autism spectrum disorders (ASD) (Marchetto et al. 2010). Both Mecp2 KO mouse and *MECP2* deficient human cells seem to accommodate increased endogenous L1 retrotransposition (using a copy number qPCR assay). Thus, it is possible that L1 retrotransposition rates in Rett Syndrome patients might be higher, although the biological significance of the increased retrotransposition rate remains to be determined.

Similarly, other host factors involved in DNA repair pathways that act to regulate L1 retrotransposition may have a differential impact on brain biology when mutated. Ataxia telangiectasia mutated (ATM) is a serine/threonine protein kinase that is activated by DNA double-strand breaks (Coufal et al. 2011). Mutations in the *ATM* gene cause Ataxia telangiectasia disorder, characterized by progressive neuronal degeneration, immunodeficiency, and predisposition to cancer. Notably, ATM-deficient hESCs, NPCs, and human fetal neural progenitor cells accumulate more L1 insertions per cell and/or longer L1 insertions (Coufal et al. 2011). Notably, these differences were also observed in an in vivo mouse model of ATM. Thus, ATM may be implicated in the recognition, resolution, and/ or repair of DNA breaks generated normally during L1 integration. However, whether altered levels of L1 retrotransposition as observed in ATM cells is involved in some of the symptoms observed in ATM patients remains to be determined.

Other candidate L1 regulator in the brain is the TAR DNA binding protein-43 (TDP-43). Mutations in *TDP-43* have been associated with amyotrophic lateral sclerosis (ALS) and frontotemporal lobar degeneration (FTLD). Notably, TDP-43 can naturally bind to L1-derived RNA transcripts (Li et al. 2012). Interestingly, binding to L1 RNAs is reduced in patients with FTLD, resulting in a dramatic up-regulation of L1s (Li et al. 2012). This phenomenon might not only involve L1 elements but other TEs. Indeed, a recent study demonstrated that human endogenous retroviruses (HERVs) could play a role in ALS. Specifically, activation of HERV-K was reported in cortical and spinal neurons of sporadic ALS patients (sALS), reducing number of cells and causing neurite retraction. Notably, expression of HERV-K is also regulated by TDP-43, which binds to its LTR sequences (Li et al. 2015).

Finally, a recent study reported a higher content of L1 copy number in patients with schizophrenia (Bundo et al. 2014). The increase in L1 copy number observed in schizophrenic patients was observed in neurons from the prefrontal cortex of patients and in iPSCs-derived neurons containing 22q11 deletions (22q11-del), one of the highest risk factors for schizophrenia patients (Karayiorgou et al. 2010).

Notably, when the site of new insertions was mapped, L1 insertions were preferentially enriched in or near genes related to synaptic function and neuropsychiatric diseases (Bundo et al. 2014). However, none of the L1 insertions identified in this study were validated, and further studies are required to conclude that retrotransposition might be differential in schizophrenia. Finally, this study also demonstrated an increase in L1 copy number in an animal model of schizophrenia, induced by maternal immune activation with poly-I:C in mice or epidermal growth factor in macaques, suggesting that L1 might be involved in the pathogenesis of schizophrenia.

In sum, a growing list of research suggests that TEs might be involved, directly or indirectly, in the molecular basis of several neurodegenerative disorders and future studies will surely help to determine whether LINE-1s might be a new pharmacological target in these and others neurodegenerative disorders.

5.7 L1 Retrotransposition and Brain Homeostasis

The expression of L1s has also been linked to several psychopathological conditions. One example is drug consumption. Indeed, it has been shown that when pregnant rats were treated with cocaine (starting on the second day of pregnancy), the born pups had increased L1 mRNA levels in selected organs such as heart, spinal cord, and the brain (Voskresenskiy and Sun 2008). Additionally, researchers have described changes in the heterochromatin status of brain cells of cocaine-treated mouse. Indeed, cocaine use can cause a decrease of the H3K9me3 enrichment (a mark of transcriptional repression) at specific genomic repeats such as L1s, which could result in increased L1 expression in nucleus accumbens (Maze et al. 2010). Additionally, further studies using another drug of abuse, morphine, detected global DNA methylation changes in the CpG island of L1 promoters; additionally, morphine treatment have shown to increase L1 mRNA expression in cultured neuronal SH-SY5Y cells (Trivedi et al. 2014). Thus, and although very preliminary, these studies suggest that the consumption of selected addictive drugs may affect the normal homeostasis of L1 expression in the mammalian brain.

Similarly, stress represents another environmental factor related with changes in epigenetic marks associated with L1 expression. Indeed, Hunter and colleagues described an increase of the repressive histone H3 lysine 9 trimethylation (H3K9me3) in the hippocampus after the implementation of acute stress in rats (Hunter et al. 2012). Surprisingly, L1 expression was increased in the cerebellum of stressed rats compared to controls. Interestingly, studies in human have also shown variations in L1 DNA methylation levels between veterans with post-traumatic stress disorder (PTSD) and combat-deployed controls in the US military service members (Rusiecki et al. 2012). Specifically, L1 was found to be hypomethylated in PTSD cases versus controls. Altogether, these studies suggest that stress could be influencing L1 expression and perhaps retrotransposition in the human brain. If stress is a risk factor for future mental illnesses that could be developed by people who had experienced PTSD remains to be determined. Interestingly, using a computational analysis that can predict phenotypic changes associated with genetic modifications, a recent study investigated the influence of somatic retrotransposition on brain metabolites, including neurotransmitters (Abrusan 2012). This study revealed that somatic retrotransposition could influence the biosynthesis of more than 250 metabolites including dopamine, serotonin, and glutamate which could indeed contribute to the development of many neurodegenerative and neurodevelopmental diseases such as Parkinson's in this case, schizophrenia or even autism (Abrusan 2012).

In conclusion, although L1 expression, methylation status, and retrotransposition have been associated with several pathological conditions in the brain such as drug addiction, schizophrenia, or PTSD, further investigation is required in order to establish if L1 can indeed participate in the disease onset and/or progression. Thus, future studies should uncover the role of L1 in diseased brains and whether L1 activity can be the cause or the consequence of a given pathological condition.

5.8 L1 Retrotransposition and Neuroinflammation

In addition to its contribution to somatic mosaicism and disease, recent research has discovered that L1s may also play a role in the induction of inflammation in various conditions. More specifically, intracellular intermediates of the L1 retrotransposition cycle might act as key effectors of inflammation. In patients with an autoimmune condition such as rheumatoid arthritis (RA), AGS, and systemic lupus erythematosus (SLE), altered levels of L1 intermediates in affected tissues have been documented (Ali et al. 2003; Stetson et al. 2008; Pokatayev et al. 2016). While the exact contribution of these intermediates to inflammation has yet to be fully determined, it is becoming more evident that TEs may interact with key immune pathways. In some type I interferonopathies such as AGS, it is thought that levels of L1-derived nucleic acid species are elevated as a result of the compromised function of key host factors that are involved in cytosolic nucleic acid processing and degradation (Burdette and Vance 2013; Woo et al. 2014; Stetson and Medzhitov 2006). Thus, the inability to clear these anomalous cytosolic nucleic acid species may lead to the activation of endogenous antiviral nucleic acid sensing mechanisms that ultimately could initiate an immune response. Indeed, and by acting through various innate DNA-sensing mechanisms that are stimulated in response to deregulated retroelement production, certain cells have the ability to produce immune molecules, even in the absence of an intact immune system. For example, in AGS mice it has been shown that abnormal cytosolic DNA (presumably L1 sequences) triggers a STING-dependent type I interferon and interferon stimulated gene (ISG) activation through the DNA sensor cGAS in various peripheral tissues including the heart and kidney, but not in the brain (Pokatayev et al. 2016). By either blocking the formation of L1 intermediates using reverse transcriptase inhibitors or disrupting the DNA-sensing mechanism through STING knockout, inflammation is reduced and overall survival is enhanced. Interestingly, while human AGS patients demonstrate a severe psychomotor retardation and neuroinflammation, the AGS mice do not display any neurological deficit. Perhaps a similar type I interferon and ISG signature can be seen in AGS human brain tissue but not in the mouse brain, explaining the differences in presentation of neurological phenotypes between the different species. However, part of these data based on the use of nucleoside-analogs reverse transcriptase inhibitors should be reevaluated, as it was recently demonstrated that some reverse transcriptase inhibitors have intrinsic anti-inflammatory properties (Fowler et al. 2014).

However, we speculate that in order to attenuate an undesired immune response and to prevent initiation of disease through retrotransposition, our genome has likely evolved various mechanisms to suppress harmful endogenous mobile elements but many of these mechanisms may also be involved in antiviral functions. Interestingly, many of these dual-functioning antiviral/antiretroelement host factors are ISGs, some of which are produced as a result of immune activation by L1 retroelements and contribute to inflammation. The zinc-finger antiviral protein (ZAP) is an ISG that targets specific viruses by functioning as a cofactor with the RNA exosome to degrade viral RNAs (Guo et al. 2007). However, it has also been shown that ZAP can function to regulate human L1 activity through stress granule association with the L1 RNA. Overexpression of ZAP results in strong restriction of L1 retrotransposition while depletion allows enhanced retrotransposition (Goodier et al. 2015; Moldovan and Moran 2015). In sum, the interactions between molecules like ZAP and L1 display a glimpse of the complex interplay between regulatory immune molecules involved in inflammation such as ISGs and IFNs and mobile elements. Thus, we propose that normally beneficial negative feedback mechanisms that restore proper L1 levels through ISGs may actually serve to cause adverse effects, including severe inflammation, which could further exacerbate disease pathology.

Acknowledgements J.L.G.P's lab is supported by CICE-FEDER-P12-CTS-2256, Plan Nacional de I+D+I 2008–2011 and 2013–2016 (FIS-FEDER-PI11/01489 and FIS-FEDER-PI14/02152), PCIN-2014-115-ERA-NET NEURON II, the European Research Council (ERC-Consolidator ERC-STG-2012-233764), by an International Early Career Scientist grant from the Howard Hughes Medical Institute (IECS-55007420) and by The Wellcome Trust-University of Edinburgh Institutional Strategic Support Fund (ISFF2).

References

Abrusan G (2012) Somatic transposition in the brain has the potential to influence the biosynthesis of metabolites involved in Parkinson's disease and schizophrenia. Biol Direct 7:41. doi:10.1186/1745-6150-7-41

Ali M, Veale DJ, Reece RJ, Quinn M, Henshaw K, Zanders ED, Markham AF, Emery P, Isaacs JD (2003) Overexpression of transcripts containing LINE-1 in the synovia of patients with rheumatoid arthritis. Ann Rheum Dis 62(7):663–666

Alisch RS, Garcia-Perez JL, Muotri AR, Gage FH, Moran JV (2006) Unconventional translation of mammalian LINE-1 retrotransposons. Genes Dev 20(2):210–224

An W, Han JS, Wheelan SJ, Davis ES, Coombes CE, Ye P, Triplett C, Boeke JD (2006) Active retrotransposition by a synthetic L1 element in mice. Proc Natl Acad Sci U S A 103(49):18662–18667. doi:10.1073/pnas.0605300103

Aravin AA, Hannon GJ, Brennecke J (2007) The Piwi-piRNA pathway provides an adaptive defense in the transposon arms race. Science 318(5851):761–764. doi:10.1126/science.1146484

Aravin AA, Sachidanandam R, Bourc'his D, Schaefer C, Pezic D, Toth KF, Bestor T, Hannon GJ (2008) A piRNA pathway primed by individual transposons is linked to de novo DNA methylation in mice. Mol Cell 31(6):785–799. doi:10.1016/j.molcel.2008.09.003

Athanikar JN, Badge RM, Moran JV (2004) A YY1-binding site is required for accurate human LINE-1 transcription initiation. Nucleic Acids Res 32(13):3846–3855

Azevedo FA, Carvalho LR, Grinberg LT, Farfel JM, Ferretti RE, Leite RE, Jacob Filho W, Lent R, Herculano-Houzel S (2009) Equal numbers of neuronal and nonneuronal cells make the human brain an isometrically scaled-up primate brain. J Comp Neurol 513(5):532–541. doi:10.1002/cne.21974

Babushok DV, Ostertag EM, Courtney CE, Choi JM, Kazazian HH Jr (2006) L1 integration in a transgenic mouse model. Genome Res 16(2):240–250

Badge RM, Alisch RS, Moran JV (2003) ATLAS: a system to selectively identify human-specific L1 insertions. Am J Hum Genet 72(4):823–838

Baillie JK, Barnett MW, Upton KR, Gerhardt DJ, Richmond TA, De Sapio F, Brennan PM, Rizzu P, Smith S, Fell M, Talbot RT, Gustincich S, Freeman TC, Mattick JS, Hume DA, Heutink P, Carninci P, Jeddeloh JA, Faulkner GJ (2011) Somatic retrotransposition alters the genetic landscape of the human brain. Nature 479(7374):534–537. doi:10.1038/nature10531

Beck CR, Garcia-Perez JL, Badge RM, Moran JV (2011) LINE-1 elements in structural variation and disease. Annu Rev Genomics Hum Genet 12:187–215. doi:10.1146/annurev-genom-082509-141802

Becker KG, Swergold GD, Ozato K, Thayer RE (1993) Binding of the ubiquitous nuclear transcription factor YY1 to a cis regulatory sequence in the human LINE-1 transposable element. Hum Mol Genet 2(10):1697–1702

Belancio VP, Hedges DJ, Deininger P (2008a) Mammalian non-LTR retrotransposons: for better or worse, in sickness and in health. Genome Res 18(3):343–358

Belancio VP, Roy-Engel AM, Deininger P (2008b) The impact of multiple splice sites in human L1 elements. Gene 411(1–2):38–45. doi:10.1016/j.gene.2007.12.022

Belancio VP, Deininger P, Roy-Engel AM (2009) LINE dancing in the human genome: transposable elements and disease. Genome Med 1(10):97

Belancio VP, Roy-Engel AM, Deininger PL (2010a) All y'all need to know 'bout retroelements in cancer. Semin Cancer Biol 20(4):200–210. doi:10.1016/j.semcancer.2010.06.001

Belancio VP, Roy-Engel AM, Pochampally RR, Deininger P (2010b) Somatic expression of LINE-1 elements in human tissues. Nucleic Acids Res 38(12):3909–3922. doi:10.1093/nar/gkq132

Bibillo A, Eickbush TH (2002) High processivity of the reverse transcriptase from a non-long terminal repeat retrotransposon. J Biol Chem 277(38):34836–34845

Bourc'his D, Bestor TH (2004) Meiotic catastrophe and retrotransposon reactivation in male germ cells lacking Dnmt3L. Nature 431(7004):96–99

Branco MR, Ficz G, Reik W (2012) Uncovering the role of 5-hydroxymethylcytosine in the epigenome. Nat Rev Genet 13(1):7–13. doi:10.1038/nrg3080

Brouha B, Schustak J, Badge RM, Lutz-Prigge S, Farley AH, Moran JV, Kazazian HH Jr (2003) Hot L1s account for the bulk of retrotransposition in the human population. Proc Natl Acad Sci U S A 100(9):5280–5285

Bundo M, Toyoshima M, Okada Y, Akamatsu W, Ueda J, Nemoto-Miyauchi T, Sunaga F, Toritsuka M, Ikawa D, Kakita A, Kato M, Kasai K, Kishimoto T, Nawa H, Okano H, Yoshikawa T, Kato T, Iwamoto K (2014) Increased l1 retrotransposition in the neuronal genome in schizophrenia. Neuron 81(2):306–313. doi:10.1016/j.neuron.2013.10.053

Burdette DL, Vance RE (2013) STING and the innate immune response to nucleic acids in the cytosol. Nat Immunol 14(1):19–26. doi:10.1038/ni.2491

Carreira PE, Richardson SR, Faulkner GJ (2014) L1 retrotransposons, cancer stem cells and oncogenesis. FEBS J 281(1):63–73. doi:10.1111/febs.12601

Castro-Diaz N, Ecco G, Coluccio A, Kapopoulou A, Yazdanpanah B, Friedli M, Duc J, Jang SM, Turelli P, Trono D (2014) Evolutionally dynamic L1 regulation in embryonic stem cells. Genes Dev 28(13):1397–1409. doi:10.1101/gad.241661.114

Chen L, Dahlstrom JE, Chandra A, Board P, Rangasamy D (2012) Prognostic value of LINE-1 retrotransposon expression and its subcellular localization in breast cancer. Breast Cancer Res Treat 136(1):129–142. doi:10.1007/s10549-012-2246-7

Cho NY, Kim BH, Choi M, Yoo EJ, Moon KC, Cho YM, Kim D, Kang GH (2007) Hypermethylation of CpG island loci and hypomethylation of LINE-1 and Alu repeats in prostate adenocarcinoma and their relationship to clinicopathological features. J Pathol 211(3):269–277. doi:10.1002/path.2106

Cordaux R, Batzer MA (2009) The impact of retrotransposons on human genome evolution. Nat Rev Genet 10(10):691–703. doi:10.1038/nrg2640

Cost GJ, Boeke JD (1998) Targeting of human retrotransposon integration is directed by the specificity of the L1 endonuclease for regions of unusual DNA structure. Biochemistry 37(51):18081–18093

Cost GJ, Feng Q, Jacquier A, Boeke JD (2002) Human L1 element target-primed reverse transcription in vitro. Embo J 21(21):5899–5910

Coufal NG, Garcia-Perez JL, Peng GE, Yeo GW, Mu Y, Lovci MT, Morell M, O'Shea KS, Moran JV, Gage FH (2009) L1 retrotransposition in human neural progenitor cells. Nature 460(7259):1127–1131

Coufal NG, Garcia-Perez JL, Peng GE, Marchetto MC, Muotri AR, Mu Y, Carson CT, Macia A, Moran JV, Gage FH (2011) Ataxia telangiectasia mutated (ATM) modulates long interspersed element-1 (L1) retrotransposition in human neural stem cells. Proc Natl Acad Sci U S A 108(51):20382–20387. doi:10.1073/pnas.1100273108

Dai L, Huang Q, Boeke JD (2011) Effect of reverse transcriptase inhibitors on LINE-1 and Ty1 reverse transcriptase activities and on LINE-1 retrotransposition. BMC Biochem 12:18. doi:10.1186/1471-2091-12-18

de Koning AP, Gu W, Castoe TA, Batzer MA, Pollock DD (2011) Repetitive elements may comprise over two-thirds of the human genome. PLoS Genet 7(12):e1002384. doi:10.1371/journal.pgen.1002384

Denli AM, Narvaiza I, Kerman BE, Pena M, Benner C, Marchetto MC, Diedrich JK, Aslanian A, Ma J, Moresco JJ, Moore L, Hunter T, Saghatelian A, Gage FH (2015) Primate-specific ORF0 contributes to retrotransposon-mediated diversity. Cell 163(3):583–593. doi:10.1016/j.cell.2015.09.025

Dewannieux M, Esnault C, Heidmann T (2003) LINE-mediated retrotransposition of marked Alu sequences. Nat Genet 35(1):41–48

Dmitriev SE, Andreev DE, Terenin IM, Olovnikov IA, Prassolov VS, Merrick WC, Shatsky IN (2007) Efficient translation initiation directed by the 900-nucleotide-long and GC-rich 5′ untranslated region of the human retrotransposon LINE-1 mRNA is strictly cap dependent rather than internal ribosome entry site mediated. Mol Cell Biol 27(13):4685–4697

Doucet TT, Kazazian HH Jr (2016) Long interspersed element sequencing (L1-Seq): a method to identify somatic LINE-1 insertions in the human genome. Methods Mol Biol 1400:79–93. doi:10.1007/978-1-4939-3372-3_5

Doucet AJ, Hulme AE, Sahinovic E, Kulpa DA, Moldovan JB, Kopera HC, Athanikar JN, Hasnaoui M, Bucheton A, Moran JV, Gilbert N (2010) Characterization of LINE-1 ribonucleoprotein particles. PLoS Genet 6(10). doi:10.1371/journal.pgen.1001150

Doucet AJ, Wilusz JE, Miyoshi T, Liu Y, Moran JV (2015) A 3′ Poly(A) tract is required for LINE-1 retrotransposition. Mol Cell 60(5):728–741. doi:10.1016/j.molcel.2015.10.012

Doucet-O'Hare TT, Rodic N, Sharma R, Darbari I, Abril G, Choi JA, Young Ahn J, Cheng Y, Anders RA, Burns KH, Meltzer SJ, Kazazian HH Jr (2015) LINE-1 expression and retrotransposition in Barrett's esophagus and esophageal carcinoma. Proc Natl Acad Sci U S A 112(35):E4894–E4900. doi:10.1073/pnas.1502474112

Eickbush TH (2002) Repair by retrotransposition. Nat Genet 31(2):126–127

Ellsworth DL, Ellsworth RE, Liebman MN, Hooke JA, Shriver CD (2004) Genomic instability in histologically normal breast tissues: implications for carcinogenesis. Lancet Oncol 5(12):753–758. doi:10.1016/S1470-2045(04)01653-5

Evrony GD, Cai X, Lee E, Hills LB, Elhosary PC, Lehmann HS, Parker JJ, Atabay KD, Gilmore EC, Poduri A, Park PJ, Walsh CA (2012) Single-neuron sequencing analysis of L1

retrotransposition and somatic mutation in the human brain. Cell 151(3):483–496. doi:10.1016/j.cell.2012.09.035
Evrony GD, Lee E, Mehta BK, Benjamini Y, Johnson RM, Cai X, Yang L, Haseley P, Lehmann HS, Park PJ, Walsh CA (2015) Cell lineage analysis in human brain using endogenous retroelements. Neuron 85(1):49–59. doi:10.1016/j.neuron.2014.12.028
Evrony GD, Lee E, Park PJ, Walsh CA (2016) Resolving rates of mutation in the brain using single-neuron genomics. Elife 5. doi:10.7554/eLife.12966
Ewing AD, Kazazian HH Jr (2010) High-throughput sequencing reveals extensive variation in human-specific L1 content in individual human genomes. Genome Res 20(9):1262–1270. doi:10.1101/gr.106419.110
Ewing AD, Gacita A, Wood LD, Ma F, Xing D, Kim MS, Manda SS, Abril G, Pereira G, Makohon-Moore A, Looijenga LH, Gillis AJ, Hruban RH, Anders RA, Romans KE, Pandey A, Iacobuzio-Donahue CA, Vogelstein B, Kinzler KW, Kazazian HH Jr, Solyom S (2015) Widespread somatic L1 retrotransposition occurs early during gastrointestinal cancer evolution. Genome Res 25(10):1536–1545. doi:10.1101/gr.196238.115
Feng Q, Moran JV, Kazazian HH Jr, Boeke JD (1996) Human L1 retrotransposon encodes a conserved endonuclease required for retrotransposition. Cell 87(5):905–916
Fowler BJ, Gelfand BD, Kim Y, Kerur N, Tarallo V, Hirano Y, Amarnath S, Fowler DH, Radwan M, Young MT, Pittman K, Kubes P, Agarwal HK, Parang K, Hinton DR, Bastos-Carvalho A, Li S, Yasuma T, Mizutani T, Yasuma R, Wright C, Ambati J (2014) Nucleoside reverse transcriptase inhibitors possess intrinsic anti-inflammatory activity. Science 346(6212):1000–1003. doi:10.1126/science.1261754
Friedli M, Turelli P, Kapopoulou A, Rauwel B, Castro-Diaz N, Rowe HM, Ecco G, Unzu C, Planet E, Lombardo A, Mangeat B, Wildhaber BE, Naldini L, Trono D (2014) Loss of transcriptional control over endogenous retroelements during reprogramming to pluripotency. Genome Res 24(8):1251–1259. doi:10.1101/gr.172809.114
Furano AV (2000) The biological properties and evolutionary dynamics of mammalian LINE-1 retrotransposons. Prog Nucleic Acid Res Mol Biol 64:255–294
Garcia-Perez JL, Marchetto MC, Muotri AR, Coufal NG, Gage FH, O'Shea KS, Moran JV (2007) LINE-1 retrotransposition in human embryonic stem cells. Hum Mol Genet 16(13):1569–1577. doi:10.1093/hmg/ddm105
Garcia-Perez JL, Morell M, Scheys JO, Kulpa DA, Morell S, Carter CC, Hammer GD, Collins KL, O'Shea KS, Menendez P, Moran JV (2010) Epigenetic silencing of engineered L1 retrotransposition events in human embryonic carcinoma cells. Nature 466(7307):769–773. doi:10.1038/nature09209
Gasior SL, Wakeman TP, Xu B, Deininger PL (2006) The human LINE-1 retrotransposon creates DNA double-strand breaks. J Mol Biol 357(5):1383–1393. doi:10.1016/j.jmb.2006.01.089
Gilbert N, Lutz-Prigge S, Moran JV (2002) Genomic deletions created upon LINE-1 retrotransposition. Cell 110(3):315–325
Goodier JL (2014) Retrotransposition in tumors and brains. Mob DNA 5(1):11. doi:10.1186/1759-8753-5-11
Goodier JL, Kazazian HH (2008) Retrotransposons revisited: the restraint and rehabilitation of parasites. Cell 135(1):23–35
Goodier JL, Cheung LE, Kazazian HH Jr (2012) MOV10 RNA helicase is a potent inhibitor of retrotransposition in cells. PLoS Genet 8(10):e1002941. doi:10.1371/journal.pgen.1002941
Goodier JL, Cheung LE, Kazazian HH Jr (2013) Mapping the LINE1 ORF1 protein interactome reveals associated inhibitors of human retrotransposition. Nucleic Acids Res 41(15):7401–7419. doi:10.1093/nar/gkt512
Goodier JL, Pereira GC, Cheung LE, Rose RJ, Kazazian HH Jr (2015) The broad-spectrum antiviral protein ZAP restricts human retrotransposition. PLoS Genet 11(5):e1005252. doi:10.1371/journal.pgen.1005252
Gore A, Li Z, Fung HL, Young JE, Agarwal S, Antosiewicz-Bourget J, Canto I, Giorgetti A, Israel MA, Kiskinis E, Lee JH, Loh YH, Manos PD, Montserrat N, Panopoulos AD, Ruiz S, Wilbert

ML, Yu J, Kirkness EF, Izpisua Belmonte JC, Rossi DJ, Thomson JA, Eggan K, Daley GQ, Goldstein LS, Zhang K (2011) Somatic coding mutations in human induced pluripotent stem cells. Nature 471(7336):63–67. doi:10.1038/nature09805

Grimaldi G, Singer MF (1983) Members of the KpnI family of long interspersed repeated sequences join and interrupt alpha-satellite in the monkey genome. Nucleic Acids Res 11(2):321–338

Guo X, Ma J, Sun J, Gao G (2007) The zinc-finger antiviral protein recruits the RNA processing exosome to degrade the target mRNA. Proc Natl Acad Sci U S A 104(1):151–156. doi:10.1073/pnas.0607063104

Han JS, Szak ST, Boeke JD (2004) Transcriptional disruption by the L1 retrotransposon and implications for mammalian transcriptomes. Nature 429(6989):268–274

Hancks DC, Kazazian HH Jr (2012) Active human retrotransposons: variation and disease. Curr Opin Genet Dev 22(3):191–203. doi:10.1016/j.gde.2012.02.006

Hancks DC, Goodier JL, Mandal PK, Cheung LE, Kazazian HH (2011) Retrotransposition of marked SVA elements by human L1s in cultured cells. Hum Mol Genet 20(17):3386–3400

Harris CR, Normart R, Yang Q, Stevenson E, Haffty BG, Ganesan S, Cordon-Cardo C, Levine AJ, Tang LH (2010) Association of nuclear localization of a long interspersed nuclear element-1 protein in breast tumors with poor prognostic outcomes. Genes Cancer 1(2):115–124. doi:10.1177/1947601909360812

Hazen JL, Faust GG, Rodriguez AR, Ferguson WC, Shumilina S, Clark RA, Boland MJ, Martin G, Chubukov P, Tsunemoto RK, Torkamani A, Kupriyanov S, Hall IM, Baldwin KK (2016) The complete genome sequences, unique mutational spectra, and developmental potency of adult neurons revealed by cloning. Neuron 89(6):1223–1236. doi:10.1016/j.neuron.2016.02.004

Heras SR, Macias S, Plass M, Fernandez N, Cano D, Eyras E, Garcia-Perez JL, Caceres JF (2013) The microprocessor controls the activity of mammalian retrotransposons. Nat Struct Mol Biol 20(10):1173–1181. doi:10.1038/nsmb.2658

Heras SR, Macias S, Caceres JF, Garcia-Perez JL (2014) Control of mammalian retrotransposons by cellular RNA processing activities. Mob Genet Elements 4:e28439

Hohjoh H, Singer MF (1996) Cytoplasmic ribonucleoprotein complexes containing human LINE-1 protein and RNA. Embo J 15(3):630–639

Hohjoh H, Singer MF (1997a) Ribonuclease and high salt sensitivity of the ribonucleoprotein complex formed by the human LINE-1 retrotransposon. J Mol Biol 271(1):7–12

Hohjoh H, Singer MF (1997b) Sequence-specific single-strand RNA binding protein encoded by the human LINE-1 retrotransposon. Embo J 16(19):6034–6043

Huang CR, Schneider AM, Lu Y, Niranjan T, Shen P, Robinson MA, Steranka JP, Valle D, Civin CI, Wang T, Wheelan SJ, Ji H, Boeke JD, Burns KH (2010) Mobile interspersed repeats are major structural variants in the human genome. Cell 141(7):1171–1182. doi:10.1016/j.cell.2010.05.026

Hunter RG, Murakami G, Dewell S, Seligsohn M, Baker ME, Datson NA, McEwen BS, Pfaff DW (2012) Acute stress and hippocampal histone H3 lysine 9 trimethylation, a retrotransposon silencing response. Proc Natl Acad Sci U S A 109(43):17657–17662. doi:10.1073/pnas.1215810109

Inamura K, Yamauchi M, Nishihara R, Lochhead P, Qian ZR, Kuchiba A, Kim SA, Mima K, Sukawa Y, Jung S, Zhang X, Wu K, Cho E, Chan AT, Meyerhardt JA, Harris CC, Fuchs CS, Ogino S (2014) Tumor LINE-1 methylation level and microsatellite instability in relation to colorectal cancer prognosis. J Natl Cancer Inst 106 (9). doi:10.1093/jnci/dju195

Iskow RC, McCabe MT, Mills RE, Torene S, Pittard WS, Neuwald AF, Van Meir EG, Vertino PM, Devine SE (2010) Natural mutagenesis of human genomes by endogenous retrotransposons. Cell 141(7):1253–1261

Jacobs FM, Greenberg D, Nguyen N, Haeussler M, Ewing AD, Katzman S, Paten B, Salama SR, Haussler D (2014) An evolutionary arms race between KRAB zinc-finger genes ZNF91/93 and SVA/L1 retrotransposons. Nature 516:242–245. doi:10.1038/nature13760

Jones RB, Garrison KE, Wong JC, Duan EH, Nixon DF, Ostrowski MA (2008) Nucleoside analogue reverse transcriptase inhibitors differentially inhibit human LINE-1 retrotransposition. PLoS One 3(2):e1547

Jurka J (1997) Sequence patterns indicate an enzymatic involvement in integration of mammalian retroposons. Proc Natl Acad Sci U S A 94(5):1872–1877

Kaer K, Speek M (2013) Retroelements in human disease. Gene 518(2):231–241. doi:10.1016/j.gene.2013.01.008

Kano H, Godoy I, Courtney C, Vetter MR, Gerton GL, Ostertag EM, Kazazian HH Jr (2009) L1 retrotransposition occurs mainly in embryogenesis and creates somatic mosaicism. Genes Dev 23(11):1303–1312. doi:10.1101/gad.1803909

Karayiorgou M, Simon TJ, Gogos JA (2010) 22q11.2 microdeletions: linking DNA structural variation to brain dysfunction and schizophrenia. Nat Rev Neurosci 11(6):402–416. doi:10.1038/nrn2841

Kazazian HH Jr (1999) An estimated frequency of endogenous insertional mutations in humans. Nat Genet 22(2):130

Kazazian HH Jr, Wong C, Youssoufian H, Scott AF, Phillips DG, Antonarakis SE (1988) Haemophilia A resulting from de novo insertion of L1 sequences represents a novel mechanism for mutation in man. Nature 332(6160):164–166. doi:10.1038/332164a0

Khazina E, Weichenrieder O (2009) Non-LTR retrotransposons encode noncanonical RRM domains in their first open reading frame. Proc Natl Acad Sci U S A 106(3):731–736. doi:10.1073/pnas.0809964106

Klawitter S, Fuchs NV, Upton KR, Munoz-Lopez M, Shukla R, Wang J, Garcia-Canadas M, Lopez-Ruiz C, Gerhardt DJ, Sebe A, Grabundzija I, Merkert S, Gerdes P, Pulgarin JA, Bock A, Held U, Witthuhn A, Haase A, Sarkadi B, Lower J, Wolvetang EJ, Martin U, Ivics Z, Izsvak Z, Garcia-Perez JL, Faulkner GJ, Schumann GG (2016) Reprogramming triggers endogenous L1 and Alu retrotransposition in human induced pluripotent stem cells. Nat Commun 7:10286. doi:10.1038/ncomms10286

Kubo S, Seleme Mdel C, Soifer HS, Perez JL, Moran JV, Kazazian HH Jr, Kasahara N (2006) L1 retrotransposition in nondividing and primary human somatic cells. Proc Natl Acad Sci U S A 103(21):8036–8041

Kulpa DA, Moran JV (2005) Ribonucleoprotein particle formation is necessary but not sufficient for LINE-1 retrotransposition. Hum Mol Genet 14(21):3237–3248. doi:10.1093/hmg/ddi354

Kulpa DA, Moran JV (2006) Cis-preferential LINE-1 reverse transcriptase activity in ribonucleoprotein particles. Nat Struct Mol Biol 13(7):655–660. doi:10.1038/nsmb1107

Kuwabara T, Hsieh J, Muotri A, Yeo G, Warashina M, Lie DC, Moore L, Nakashima K, Asashima M, Gage FH (2009) Wnt-mediated activation of NeuroD1 and retro-elements during adult neurogenesis. Nat Neurosci 12(9):1097–1105. doi:10.1038/nn.2360

Lander ES, Linton LM, Birren B, Nusbaum C, Zody MC, Baldwin J, Devon K, Dewar K, Doyle M, FitzHugh W, Funke R, Gage D, Harris K, Heaford A, Howland J, Kann L, Lehoczky J, LeVine R, McEwan P, McKernan K, Meldrim J, Mesirov JP, Miranda C, Morris W, Naylor J, Raymond C, Rosetti M, Santos R, Sheridan A, Sougnez C, Stange-Thomann N, Stojanovic N, Subramanian A, Wyman D, Rogers J, Sulston J, Ainscough R, Beck S, Bentley D, Burton J, Clee C, Carter N, Coulson A, Deadman R, Deloukas P, Dunham A, Dunham I, Durbin R, French L, Grafham D, Gregory S, Hubbard T, Humphray S, Hunt A, Jones M, Lloyd C, McMurray A, Matthews L, Mercer S, Milne S, Mullikin JC, Mungall A, Plumb R, Ross M, Shownkeen R, Sims S, Waterston RH, Wilson RK, Hillier LW, McPherson JD, Marra MA, Mardis ER, Fulton LA, Chinwalla AT, Pepin KH, Gish WR, Chissoe SL, Wendl MC, Delehaunty KD, Miner TL, Delehaunty A, Kramer JB, Cook LL, Fulton RS, Johnson DL, Minx PJ, Clifton SW, Hawkins T, Branscomb E, Predki P, Richardson P, Wenning S, Slezak T, Doggett N, Cheng JF, Olsen A, Lucas S, Elkin C, Uberbacher E, Frazier M, Gibbs RA, Muzny DM, Scherer SE, Bouck JB, Sodergren EJ, Worley KC, Rives CM, Gorrell JH, Metzker ML, Naylor SL, Kucherlapati RS, Nelson DL, Weinstock GM, Sakaki Y, Fujiyama A, Hattori M, Yada T, Toyoda A, Itoh T, Kawagoe C, Watanabe H, Totoki Y, Taylor T, Weissenbach J, Heilig R, Saurin W, Artiguenave F, Brottier P, Bruls T, Pelletier E, Robert C, Wincker P, Smith DR, Doucette-Stamm L, Rubenfield M, Weinstock K, Lee HM, Dubois J, Rosenthal A, Platzer M, Nyakatura G, Taudien S, Rump A, Yang H, Yu J, Wang J, Huang G, Gu J, Hood L, Rowen L, Madan A, Qin S, Davis RW, Federspiel NA, Abola AP, Proctor MJ, Myers RM, Schmutz J, Dickson M, Grimwood J, Cox DR, Olson MV, Kaul R, Shimizu N, Kawasaki K, Minoshima S, Evans GA,

Athanasiou M, Schultz R, Roe BA, Chen F, Pan H, Ramser J, Lehrach H, Reinhardt R, McCombie WR, de la Bastide M, Dedhia N, Blocker H, Hornischer K, Nordsiek G, Agarwala R, Aravind L, Bailey JA, Bateman A, Batzoglou S, Birney E, Bork P, Brown DG, Burge CB, Cerutti L, Chen HC, Church D, Clamp M, Copley RR, Doerks T, Eddy SR, Eichler EE, Furey TS, Galagan J, Gilbert JG, Harmon C, Hayashizaki Y, Haussler D, Hermjakob H, Hokamp K, Jang W, Johnson LS, Jones TA, Kasif S, Kaspryzk A, Kennedy S, Kent WJ, Kitts P, Koonin EV, Korf I, Kulp D, Lancet D, Lowe TM, McLysaght A, Mikkelsen T, Moran JV, Mulder N, Pollara VJ, Ponting CP, Schuler G, Schultz J, Slater G, Smit AF, Stupka E, Szustakowski J, Thierry-Mieg D, Thierry-Mieg J, Wagner L, Wallis J, Wheeler R, Williams A, Wolf YI, Wolfe KH, Yang SP, Yeh RF, Collins F, Guyer MS, Peterson J, Felsenfeld A, Wetterstrand KA, Patrinos A, Morgan MJ, de Jong P, Catanese JJ, Osoegawa K, Shizuya H, Choi S, Chen YJ (2001) Initial sequencing and analysis of the human genome. Nature 409(6822):860–921. doi:10.1038/35057062

Lee E, Iskow R, Yang L, Gokcumen O, Haseley P, Luquette LJ 3rd, Lohr JG, Harris CC, Ding L, Wilson RK, Wheeler DA, Gibbs RA, Kucherlapati R, Lee C, Kharchenko PV, Park PJ (2012) Landscape of somatic retrotransposition in human cancers. Science 337(6097):967–971. doi:10.1126/science.1222077

Levin HL, Moran JV (2011) Dynamic interactions between transposable elements and their hosts. Nat Rev Genet 12(9):615–627. doi:10.1038/nrg3030

Li W, Jin Y, Prazak L, Hammell M, Dubnau J (2012) Transposable elements in TDP-43-mediated neurodegenerative disorders. PLoS One 7(9):e44099. doi:10.1371/journal.pone.0044099

Li W, Prazak L, Chatterjee N, Gruninger S, Krug L, Theodorou D, Dubnau J (2013) Activation of transposable elements during aging and neuronal decline in Drosophila. Nat Neurosci 16(5):529–531. doi:10.1038/nn.3368

Li W, Lee MH, Henderson L, Tyagi R, Bachani M, Steiner J, Campanac E, Hoffman DA, von Geldern G, Johnson K, Maric D, Morris HD, Lentz M, Pak K, Mammen A, Ostrow L, Rothstein J, Nath A (2015) Human endogenous retrovirus-K contributes to motor neuron disease. Sci Transl Med 7(307):307ra153. doi:10.1126/scitranslmed.aac8201

Lin C, Yang L, Tanasa B, Hutt K, Ju BG, Ohgi K, Zhang J, Rose DW, Fu XD, Glass CK, Rosenfeld MG (2009) Nuclear receptor-induced chromosomal proximity and DNA breaks underlie specific translocations in cancer. Cell 139(6):1069–1083. doi:10.1016/j.cell.2009.11.030

Luan DD, Korman MH, Jakubczak JL, Eickbush TH (1993) Reverse transcription of R2Bm RNA is primed by a nick at the chromosomal target site: a mechanism for non-LTR retrotransposition. Cell 72(4):595–605

Macia A, Munoz-Lopez M, Cortes JL, Hastings RK, Morell S, Lucena-Aguilar G, Marchal JA, Badge RM, Garcia-Perez JL (2011) Epigenetic control of retrotransposon expression in human embryonic stem cells. Mol Cell Biol 31(2):300–316. doi:10.1128/MCB.00561-10

Mager DL, Stoye JP (2015) Mammalian endogenous retroviruses. Microbiol Spectr 3(1):MDNA3-0009-2014. doi:10.1128/microbiolspec.MDNA3-0009-2014

Marchetto MC, Carromeu C, Acab A, Yu D, Yeo GW, Mu Y, Chen G, Gage FH, Muotri AR (2010) A model for neural development and treatment of Rett syndrome using human induced pluripotent stem cells. Cell 143(4):527–539. doi:10.1016/j.cell.2010.10.016

Marchetto MC, Narvaiza I, Denli AM, Benner C, Lazzarini TA, Nathanson JL, Paquola AC, Desai KN, Herai RH, Weitzman MD, Yeo GW, Muotri AR, Gage FH (2013) Differential L1 regulation in pluripotent stem cells of humans and apes. Nature 503(7477):525–529. doi:10.1038/nature12686

Martin SL (1991) Ribonucleoprotein particles with LINE-1 RNA in mouse embryonal carcinoma cells. Mol Cell Biol 11(9):4804–4807

Martin SL, Bushman FD (2001) Nucleic acid chaperone activity of the ORF1 protein from the mouse LINE-1 retrotransposon. Mol Cell Biol 21(2):467–475

Mathias SL, Scott AF, Kazazian HH Jr, Boeke JD, Gabriel A (1991) Reverse transcriptase encoded by a human transposable element. Science 254(5039):1808–1810

Matlik K, Redik K, Speek M (2006) L1 antisense promoter drives tissue-specific transcription of human genes. J Biomed Biotechnol 2006(1):71753. doi:10.1155/JBB/2006/71753

Maze I, Covington HE 3rd, Dietz DM, LaPlant Q, Renthal W, Russo SJ, Mechanic M, Mouzon E, Neve RL, Haggarty SJ, Ren Y, Sampath SC, Hurd YL, Greengard P, Tarakhovsky A, Schaefer

A, Nestler EJ (2010) Essential role of the histone methyltransferase G9a in cocaine-induced plasticity. Science 327(5962):213–216. doi:10.1126/science.1179438

Miki Y, Nishisho I, Horii A, Miyoshi Y, Utsunomiya J, Kinzler KW, Vogelstein B, Nakamura Y (1992) Disruption of the APC gene by a retrotransposal insertion of L1 sequence in a colon cancer. Cancer Res 52(3):643–645

Mills RE, Bennett EA, Iskow RC, Devine SE (2007) Which transposable elements are active in the human genome? Trends Genet 23(4):183–191

Mine M, Chen JM, Brivet M, Desguerre I, Marchant D, de Lonlay P, Bernard A, Ferec C, Abitbol M, Ricquier D, Marsac C (2007) A large genomic deletion in the PDHX gene caused by the retrotranspositional insertion of a full-length LINE-1 element. Hum Mutat 28(2):137–142

Moldovan JB, Moran JV (2015) The zinc-finger antiviral protein ZAP inhibits LINE and Alu retrotransposition. PLoS Genet 11(5):e1005121. doi:10.1371/journal.pgen.1005121

Monot C, Kuciak M, Viollet S, Mir AA, Gabus C, Darlix JL, Cristofari G (2013) The specificity and flexibility of l1 reverse transcription priming at imperfect T-tracts. PLoS Genet 9(5):e1003499. doi:10.1371/journal.pgen.1003499

Moran JV, Gilbert N (2002) Mammalian LINE-1 retrotransposons and related elements. In: Craig N, Craggie R, Gellert M, Lambowitz A (eds) Mobile DNA II. ASM Press, Washington, DC

Moran JV, Holmes SE, Naas TP, DeBerardinis RJ, Boeke JD, Kazazian HH Jr (1996) High frequency retrotransposition in cultured mammalian cells. Cell 87(5):917–927

Moran JV, DeBerardinis RJ, Kazazian HH Jr (1999) Exon shuffling by L1 retrotransposition. Science 283(5407):1530–1534

Morrish TA, Gilbert N, Myers JS, Vincent BJ, Stamato TD, Taccioli GE, Batzer MA, Moran JV (2002) DNA repair mediated by endonuclease-independent LINE-1 retrotransposition. Nat Genet 31(2):159–165. doi:10.1038/ng898

Morrish TA, Garcia-Perez JL, Stamato TD, Taccioli GE, Sekiguchi J, Moran JV (2007) Endonuclease-independent LINE-1 retrotransposition at mammalian telomeres. Nature 446(7132):208–212. doi:10.1038/nature05560

Munoz-Lopez M, Garcia-Perez JL (2010) DNA transposons: nature and applications in genomics. Curr Genomics 11(2):115–128. doi:10.2174/138920210790886871

Munoz-Lopez M, Macia A, Garcia-Canadas M, Badge RM, Garcia-Perez JL (2011) An epi [c] genetic battle: LINE-1 retrotransposons and intragenomic conflict in humans. Mob Genet Elements 1(2):122–127. doi:10.4161/mge.1.2.16730

Munoz-Lopez M, Garcia-Canadas M, Macia A, Morell S, Garcia-Perez JL (2012) Analysis of LINE-1 expression in human pluripotent cells. Methods Mol Biol 873:113–125. doi:10.1007/978-1-61779-794-1_7

Muotri AR, Chu VT, Marchetto MC, Deng W, Moran JV, Gage FH (2005) Somatic mosaicism in neuronal precursor cells mediated by L1 retrotransposition. Nature 435(7044):903–910

Muotri AR, Zhao C, Marchetto MC, Gage FH (2009) Environmental influence on L1 retrotransposons in the adult hippocampus. Hippocampus 19(10):1002–1007. doi:10.1002/hipo.20564

Muotri AR, Marchetto MC, Coufal NG, Oefner R, Yeo G, Nakashima K, Gage FH (2010) L1 retrotransposition in neurons is modulated by MeCP2. Nature 468(7322):443–446. doi:10.1038/nature09544

Nigumann P, Redik K, Matlik K, Speek M (2002) Many human genes are transcribed from the antisense promoter of L1 retrotransposon. Genomics 79(5):628–634. doi:10.1006/geno.2002.6758

Orgel LE, Crick FH (1980) Selfish DNA: the ultimate parasite. Nature 284(5757):604–607

Ostertag EM, Kazazian HH Jr (2001) Twin priming: a proposed mechanism for the creation of inversions in L1 retrotransposition. Genome Res 11(12):2059–2065

Patterson MN, Scannapieco AE, Au PH, Dorsey S, Royer CA, Maxwell PH (2015) Preferential retrotransposition in aging yeast mother cells is correlated with increased genome instability. DNA Repair 34:18–27. doi:10.1016/j.dnarep.2015.07.004

Perrat PN, DasGupta S, Wang J, Theurkauf W, Weng Z, Rosbash M, Waddell S (2013) Transposition-driven genomic heterogeneity in the Drosophila brain. Science 340(6128):91–95. doi:10.1126/science.1231965

Philippe C, Vargas-Landin DB, Doucet AJ, van Essen D, Vera-Otarola J, Kuciak M, Corbin A, Nigumann P, Cristofari G (2016) Activation of individual L1 retrotransposon instances is restricted to cell-type dependent permissive loci. Elife 5:e13926. doi:10.7554/eLife.13926

Pokatayev V, Hasin N, Chon H, Cerritelli SM, Sakhuja K, Ward JM, Morris HD, Yan N, Crouch RJ (2016) RNase H2 catalytic core Aicardi-Goutieres syndrome-related mutant invokes cGAS-STING innate immune-sensing pathway in mice. J Exp Med 213:329–336. doi:10.1084/jem.20151464

Raiz J, Damert A, Chira S, Held U, Klawitter S, Hamdorf M, Lower J, Stratling WH, Lower R, Schumann GG (2011) The non-autonomous retrotransposon SVA is trans-mobilized by the human LINE-1 protein machinery. Nucleic Acids Res 40(4):1666–1683

Ray DA, Feschotte C, Pagan HJ, Smith JD, Pritham EJ, Arensburger P, Atkinson PW, Craig NL (2008) Multiple waves of recent DNA transposon activity in the bat, Myotis lucifugus. Genome Res 18(5):717–728. doi:10.1101/gr.071886.107

Rhee YY, Lee TH, Song YS, Wen X, Kim H, Jheon S, Lee CT, Kim J, Cho NY, Chung JH, Kang GH (2015) Prognostic significance of promoter CpG island hypermethylation and repetitive DNA hypomethylation in stage I lung adenocarcinoma. Virchows Arch 466(6):675–683. doi:10.1007/s00428-015-1749-0

Richardson SR, Morell S, Faulkner GJ (2014a) L1 retrotransposons and somatic mosaicism in the brain. Annu Rev Genet 48:1–27. doi:10.1146/annurev-genet-120213-092412

Richardson SR, Narvaiza I, Planegger RA, Weitzman MD, Moran JV (2014b) APOBEC3A deaminates transiently exposed single-strand DNA during LINE-1 retrotransposition. Elife 3:e02008

Richardson SR, Doucet AJ, Kopera HC, Moldovan JB, Garcia-Perez JL, Moran JV (2015) The influence of LINE-1 and SINE retrotransposons on mammalian genomes. Microbiol Spectr 3(2):MDNA3-0061-2014. doi:10.1128/microbiolspec.MDNA3-0061-2014

Rodic N, Sharma R, Sharma R, Zampella J, Dai L, Taylor MS, Hruban RH, Iacobuzio-Donahue CA, Maitra A, Torbenson MS, Goggins M, Shih Ie M, Duffield AS, Montgomery EA, Gabrielson E, Netto GJ, Lotan TL, De Marzo AM, Westra W, Binder ZA, Orr BA, Gallia GL, Eberhart CG, Boeke JD, Harris CR, Burns KH (2014) Long interspersed element-1 protein expression is a hallmark of many human cancers. Am J Pathol 184(5):1280–1286. doi:10.1016/j.ajpath.2014.01.007

Rusiecki JA, Chen L, Srikantan V, Zhang L, Yan L, Polin ML, Baccarelli A (2012) DNA methylation in repetitive elements and post-traumatic stress disorder: a case-control study of US military service members. Epigenomics 4(1):29–40. doi:10.2217/epi.11.116

Sanchez-Luque FJ, Richardson SR, Faulkner GJ (2016) Retrotransposon capture sequencing (RC-Seq): a targeted, high-throughput approach to resolve somatic L1 retrotransposition in humans. Methods Mol Biol 1400:47–77. doi:10.1007/978-1-4939-3372-3_4

Sassaman DM, Dombroski BA, Moran JV, Kimberland ML, Naas TP, DeBerardinis RJ, Gabriel A, Swergold GD, Kazazian HH Jr (1997) Many human L1 elements are capable of retrotransposition. Nat Genet 16(1):37–43

Schumann GG (2007) APOBEC3 proteins: major players in intracellular defence against LINE-1-mediated retrotransposition. Biochem Soc Trans 35(Pt 3):637–642. doi:10.1042/BST0350637

Scott AF, Schmeckpeper BJ, Abdelrazik M, Comey CT, O'Hara B, Rossiter JP, Cooley T, Heath P, Smith KD, Margolet L (1987) Origin of the human L1 elements: proposed progenitor genes deduced from a consensus DNA sequence. Genomics 1(2):113–125

Scott EC, Gardner EJ, Masood A, Chuang NT, Vertino PM, Devine SE (2016) A hot L1 retrotransposon evades somatic repression and initiates human colorectal cancer. Genome Res 26(6):745–755. doi:10.1101/gr.201814.115

Shen Y, Chow J, Wang Z, Fan G (2006) Abnormal CpG island methylation occurs during in vitro differentiation of human embryonic stem cells. Hum Mol Genet 15(17):2623–2635. doi:10.1093/hmg/ddl188

Speek M (2001) Antisense promoter of human L1 retrotransposon drives transcription of adjacent cellular genes. Mol Cell Biol 21(6):1973–1985

Stetson DB, Medzhitov R (2006) Recognition of cytosolic DNA activates an IRF3-dependent innate immune response. Immunity 24(1):93–103. doi:10.1016/j.immuni.2005.12.003

Stetson DB, Ko JS, Heidmann T, Medzhitov R (2008) Trex1 prevents cell-intrinsic initiation of autoimmunity. Cell 134(4):587–598

Suter CM, Martin DI, Ward RL (2004) Hypomethylation of L1 retrotransposons in colorectal cancer and adjacent normal tissue. Int J Colorectal Dis 19(2):95–101. doi:10.1007/s00384-003-0539-3

Swergold GD (1990) Identification, characterization, and cell specificity of a human LINE-1 promoter. Mol Cell Biol 10(12):6718–6729

Symer DE, Connelly C, Szak ST, Caputo EM, Cost GJ, Parmigiani G, Boeke JD (2002) Human l1 retrotransposition is associated with genetic instability in vivo. Cell 110(3):327–338

Taylor MS, Lacava J, Mita P, Molloy KR, Huang CR, Li D, Adney EM, Jiang H, Burns KH, Chait BT, Rout MP, Boeke JD, Dai L (2013) Affinity proteomics reveals human host factors implicated in discrete stages of LINE-1 retrotransposition. Cell 155(5):1034–1048. doi:10.1016/j.cell.2013.10.021

Telesnitsky A, Goff SP (1997) Reverse transcriptase and the generation of retroviral DNA. In: Coffin JM, Hughes SH, Varmus HE (eds) Retroviruses. Cold Spring Harbor, New York

Tellier M, Bouuaert CC, Chalmers R (2015) Mariner and the ITm superfamily of transposons. Microbiol Spectr 3(2):MDNA3-0033-2014. doi:10.1128/microbiolspec.MDNA3-0033-2014

Thayer RE, Singer MF, Fanning TG (1993) Undermethylation of specific LINE-1 sequences in human cells producing a LINE-1-encoded protein. Gene 133(2):273–277

Thomas CA, Paquola AC, Muotri AR (2012) LINE-1 retrotransposition in the nervous system. Annu Rev Cell Dev Biol 28:555–573. doi:10.1146/annurev-cellbio-101011-155822

Thomson JA, Itskovitz-Eldor J, Shapiro SS, Waknitz MA, Swiergiel JJ, Marshall VS, Jones JM (1998) Embryonic stem cell lines derived from human blastocysts. Science 282(5391):1145–1147

Trivedi M, Shah J, Hodgson N, Byun HM, Deth R (2014) Morphine induces redox-based changes in global DNA methylation and retrotransposon transcription by inhibition of excitatory amino acid transporter type 3-mediated cysteine uptake. Mol Pharmacol 85(5):747–757. doi:10.1124/mol.114.091728

Upton KR, Gerhardt DJ, Jesuadian JS, Richardson SR, Sanchez-Luque FJ, Bodea GO, Ewing AD, Salvador-Palomeque C, van der Knaap MS, Brennan PM, Vanderver A, Faulkner GJ (2015) Ubiquitous L1 mosaicism in hippocampal neurons. Cell 161(2):228–239. doi:10.1016/j.cell.2015.03.026

van den Hurk JA, Meij IC, Seleme MC, Kano H, Nikopoulos K, Hoefsloot LH, Sistermans EA, de Wijs IJ, Mukhopadhyay A, Plomp AS, de Jong PT, Kazazian HH, Cremers FP (2007) L1 retrotransposition can occur early in human embryonic development. Hum Mol Genet 16(13):1587–1592. doi:10.1093/hmg/ddm108

Van Meter M, Kashyap M, Rezazadeh S, Geneva AJ, Morello TD, Seluanov A, Gorbunova V (2014) SIRT6 represses LINE1 retrotransposons by ribosylating KAP1 but this repression fails with stress and age. Nat Commun 5:5011. doi:10.1038/ncomms6011

Volkman HE, Stetson DB (2014) The enemy within: endogenous retroelements and autoimmune disease. Nat Immunol 15(5):415–422. doi:10.1038/ni.2872

Voskresenskiy AM, Sun LS (2008) The housekeeping gene (GA3PDH) and the long interspersed nuclear element (LINE) in the blood and organs of rats treated with cocaine. Ann N Y Acad Sci 1137:309–315. doi:10.1196/annals.1448.045

Wallace NA, Belancio VP, Deininger PL (2008) L1 mobile element expression causes multiple types of toxicity. Gene 419(1–2):75–81. doi:10.1016/j.gene.2008.04.013

Wei W, Gilbert N, Ooi SL, Lawler JF, Ostertag EM, Kazazian HH, Boeke JD, Moran JV (2001) Human L1 retrotransposition: cis preference versus trans complementation. Mol Cell Biol 21(4):1429–1439

Wheelan SJ, Aizawa Y, Han JS, Boeke JD (2005) Gene-breaking: a new paradigm for human retrotransposon-mediated gene evolution. Genome Res 15(8):1073–1078. doi:10.1101/gr.3688905

Wissing S, Montano M, Garcia-Perez JL, Moran JV, Greene WC (2011) Endogenous APOBEC3B restricts LINE-1 retrotransposition in transformed cells and human embryonic stem cells. J Biol Chem 286(42):36427–36437. doi:10.1074/jbc.M111.251058

Wissing S, Munoz-Lopez M, Macia A, Yang Z, Montano M, Collins W, Garcia-Perez JL, Moran JV, Greene WC (2012) Reprogramming somatic cells into iPS cells activates LINE-1 retroelement mobility. Hum Mol Genet 21(1):208–218. doi:10.1093/hmg/ddr455

Woo SR, Fuertes MB, Corrales L, Spranger S, Furdyna MJ, Leung MY, Duggan R, Wang Y, Barber GN, Fitzgerald KA, Alegre ML, Gajewski TF (2014) STING-dependent cytosolic DNA sensing mediates innate immune recognition of immunogenic tumors. Immunity 41(5):830–842. doi:10.1016/j.immuni.2014.10.017

Xing J, Zhang Y, Han K, Salem AH, Sen SK, Huff CD, Zhou Q, Kirkness EF, Levy S, Batzer MA, Jorde LB (2009) Mobile elements create structural variation: analysis of a complete human genome. Genome Res 19(9):1516–1526

Zhang A, Dong B, Doucet AJ, Moldovan JB, Moran JV, Silverman RH (2014) RNase L restricts the mobility of engineered retrotransposons in cultured human cells. Nucleic Acids Res 42(6):3803–3820. doi:10.1093/nar/gkt1308

Zhao K, Du J, Han X, Goodier JL, Li P, Zhou X, Wei W, Evans SL, Li L, Zhang W, Cheung LE, Wang G, Kazazian HH Jr, Yu XF (2013) Modulation of LINE-1 and Alu/SVA retrotransposition by Aicardi-Goutieres syndrome-related SAMHD1. Cell Rep 4(6):1108–1115. doi:10.1016/j.celrep.2013.08.019

Environment, Cellular Signaling, and L1 Activity

Catherine M. Ade, Geraldine Servant, Maria E. Morales, and Astrid M. Roy-Engel

Preface

Transposable elements (TEs) or transposons are discrete pieces of DNA capable of either mobilizing themselves or creating a copy of themselves to insert into a new genomic location. Although misleading, TEs are often referred to as "junk DNA." However, TEs are far from innocuous, having a profound impact on genomes and functioning as drivers of mutation and evolution. Interestingly, TEs respond to environmental stimuli. Since the initial discovery of "jumping genes" in maize by Barbara McClintock (McCLINTOCK, Proc Natl Acad Sci U S A 36:344–355, 1950), transposons have been shown to be widespread across species. McClintock observed and characterized chromosomal rearrangements in maize created by transposition events stimulated by ionizing radiation treatment. Thus, she was the first to postulate that transposition activation can be induced in response to stress in order for the cells to survive to DNA damage (Fedoroff, Proc Natl Acad Sci U S A 109: 20200–20203, 2012). More recent research supports McClintock's hypothesis that transposons play an important role in the adaptation to environmental stress (Capy et al., Heredity 85(Pt 2):101–106, 2000; Grandbastien, J Soc Biol 198:425–432, 2004; Foster, Crit Rev Biochem Mol Biol 42:373–397, 2007; Taruscio and Mantovani, Cytogenet Genome Res 105:351–362, 2004; Cho et al., Shock 30:105–116, 2008; Chadha and Sharma, PLoS ONE 9: e94415, 2014; Stapley et al., Mol Ecol 24:2241–2252, 2015). This chapter focuses on the currently active human TEs,

C.M. Ade
Department of Cellular and Molecular Biology, Tulane University, New Orleans, LA 70118, USA

G. Servant • M.E. Morales • A.M. Roy-Engel (✉)
Department of Epidemiology and Tulane Cancer, Tulane University, 1700 Tulane Ave. LCRC 815, New Orleans, LA 70112, USA
e-mail: aengel@tulane.edu

G. Cristofari (ed.), *Human Retrotransposons in Health and Disease*,
DOI 10.1007/978-3-319-48344-3_7

the retroelements, specifically on the Long INterspersed Element-1 (LINE-1 or L1) and its response to environmental stimuli. The chapter is divided into four basic sections: an introduction to the human retroelements, the impact of L1 on human health, regulatory mechanisms of L1 activity, (i.e., host defense mechanisms), and the impact of environmental factors on L1 activity. To conclude the chapter, three basic mechanisms of how L1 is affected by the environment are proposed: (1) by altering epigenetic silencing mechanisms; (2) by increasing functional L1 transcripts; and (3) by deregulating host defense strategies that prevent L1 insertion.

1 Introduction

1.1 L1 is the Main Driver of Retrotransposition in the Human Genome

Conservative estimates indicate that 45 % of the human genome is composed of transposable element sequences, making these genetic elements highly successful in mammalian genomes (Lander et al. 2001). There are two classes of transposable elements: the class I (retrotransposons) and the class II (DNA transposons). The class II DNA transposons insert into new genomic locations through a "cut and paste" mechanism and they represent the true "jumping genes." Conversely, the currently active (in the human genome) class I retrotransposons insert into new genomic locations through a copy and paste mechanism using an RNA intermediate. Thus, the original retroelement does not "jump" per se, but it generates a new copy of itself that inserts elsewhere in the genome. Interestingly, many examples of TEs that respond to stress belong to class I, e.g., the plant LTR retrotransposons BARE-1 and Ty1 in yeast (Mourier et al. 2014; Casacuberta and Gonzalez 2013).

One of the most successful mobile elements in the human genome is the non-LTR retroelement Long INterspersed Element 1 (LINE-1 or L1). There are over 500,000 L1 copies which comprises about approximately 17 % of the human genome (Lander et al. 2001; Smit 1999). However, most of the L1 copies are short 5′ truncated sequences with only less than 5000 being a full-length L1 element. Furthermore, once inserted, the new copies tend to accrue inactivating mutations. Therefore, it is estimated that on average the human genome will only have between 80 and 100 retrotranspositionally competent full-length L1 elements (Beck et al. 2010). Further analyses of the human genome revealed that active L1 copies significantly vary in their retrotransposition efficiency with as few as 10 L1 elements being highly active. These highly active L1 elements are responsible for the bulk of de novo polymorphic insertions, earning the moniker "Hot L1s" (Brouha et al. 2003).

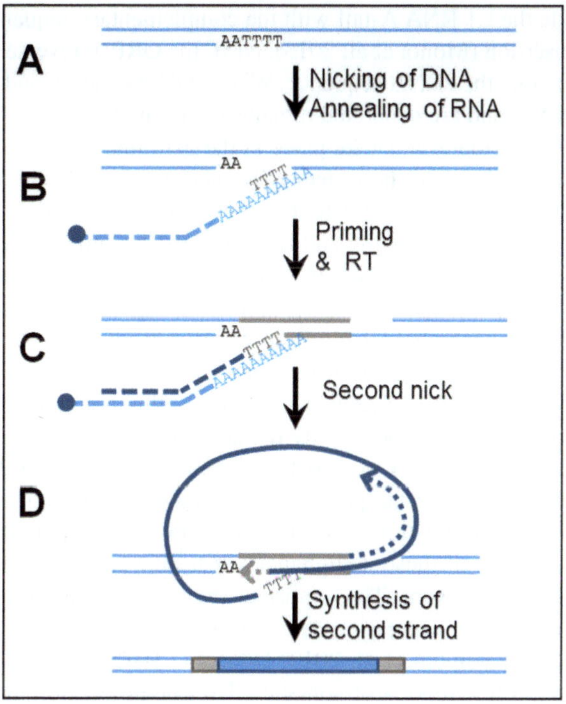

Fig. 1 Target Primed Reverse Transcription (TPRT) process. A schematic representation of the TPRT step of the retrotransposition process is shown. (**a**) The genomic DNA contains an AT-rich sequence that is cleaved by the ORF2p and shown on the *bottom* strand (consensus 5′-TTTTAA-3′). (**b**) A genomic nick is thought to occur between the T and the A, creating a single strand with available Ts that base pair with the poly-A tail of the RNA (*dashed light blue line*) and provide the priming site for (**c**) reverse transcription by ORF2p. The cDNA (depicted as a *dashed dark blue line*) is then synthetized. (**d**) The generation of the second DNA strand and final steps of insertion are currently undefined but a double strand break is formed in order to complete the process

1.2 L1 Structure

An active, full-length human L1 element is ~6000 base pair-long and contains two open reading frames coding for the proteins ORF1p and ORF2p. L1 transcripts are bicistronic where both proteins get translated from the same RNA molecule (Smit 1999). ORF1p interacts with the L1 transcript and is proposed to be a nucleic acid chaperone. ORF2p has endonuclease and reverse transcriptase functions, both required for L1 retrotransposition (Mathias et al. 1991; Feng et al. 1996). The generation of new L1 copies occurs through a process termed retrotransposition and uses a Target Primed Reverse Transcription (TPRT; Fig. 1) mechanism for insertion (Luan et al. 1993; Christensen et al. 2006). Briefly, the endonuclease function of the L1 ORF2p nicks the genome at an AT-rich sequence. This nick exposes thymine bases to provide the L1 RNA A-tail with the complementary sequence to base pair at the site of insertion (Monot et al. 2013). Next, the ORF2p reverse transcribes the L1 RNA to generate the cDNA sequence. When and how the second DNA nick and the second DNA strand are generated remains poorly understood.

L1 retrotransposition events take place in the germline and/or early in embryogenesis and passed down to the progeny (Branciforte and Martin 1994; Trelogan and Martin 1995). Estimates calculate that new L1 insertions can occur in between 1/100 and 1/200 live births (Ewing and Kazazian 2010; Cordaux and Batzer 2009). Most of these new L1 insertions are thought to arise from the few active polymorphic "Hot L1" elements harbored in an individual's genome (Streva et al. 2015). More recent data indicate that L1 retrotransposition events are not restricted to the germline, as L1 RNA, protein, and de novo insertions have been detected in somatic tissues (Belancio et al. 2010; Ergun et al. 2004; Baillie et al. 2011). However, up until now, reports of L1 somatic activity remain scarce. The paucity of reports may be either due to a genuine lack of somatic L1 retrotransposition or due to the difficulty of detecting single events present within hundreds of thousands of somatic cells. Somatic retrotransposition events in cancers may be easier to detect due to the clonal nature of some of these malignancies that may enrich the number of cells containing the somatic event. Thus, most of the current examples of somatic L1 mobilization and protein expression derive from detailed analyses of tumor tissues (Streva et al. 2015; Solyom et al. 2012; Doucet-O'Hare et al. 2015; Rodic et al. 2014). However, a few reports of somatic L1 retrotransposition events in normal tissues exist, specifically in the brain (Evrony et al. 2012). In addition, it may be possible that under certain conditions (stress or an environmental exposure) L1 elements get activated (i.e., de-repressed) in somatic tissues. Further evaluation is needed to confirm this hypothesis.

2 Impact of L1 on Human Health

L1 functions as an efficient genomic mutagen, as insertions into a gene can disrupt its function. Multiple examples of L1 insertional mutagenesis that contribute to a large diversity of genetic diseases have been reported (reviewed in (Kaer and Speek 2013), see also Chap. 3 of this book).

2.1 Insertional Mutagenesis

2.1.1 Germline Mutagenesis

The first description in humans of an endogenous L1 causing a disease by insertional mutagenesis was reported in 1988. In that case, an L1 had inserted into the coagulation Factor VIII gene, causing hemophilia A (Kazazian et al. 1988). Since then, many descriptions of genetic diseases caused by de novo L1 insertions have been reported

(Kaer and Speek 2013). The number of L1 insertions in the human genome continues to increase. This observation strongly supports the accepted notion that insertions occur in the germline, which are then passed on to the progeny. However, recent work indicates that some insertions may occur very early during embryogenesis. In one report, van den Hurk et al. (2007) reported an example of an early embryonic L1 insertion in a human who exhibited both germline and somatic mosaicism for an L1 insertion. Furthermore, Kano et al. (2009) showed that human and mouse L1 transgenes produce more retrotransposition events in early embryogenesis in mice and rats than in the germline. Interestingly, Garcia-Perez et al. (2007) showed that embryonic stem cells can support retrotransposition of an engineered L1. However, the proportion of these events contributing to L1 somatic mosaicism in humans remains unknown.

2.1.2 Somatic Mutagenesis

Laboratory studies provide some evidence for somatic activity of L1 elements in human and mouse cells. Although artificial, engineered L1 elements are able to retrotranspose in multiple cell lines derived from a large variety of somatic tissues. Although not all cell lines support L1 retrotransposition, this suggests that under the appropriate conditions, L1 elements can mobilize in somatic cells. In addition, there are a few in vivo reports from animal models that support somatic activity. Using engineered human L1 transgenes, Muotri et al. (2005) reported that L1 retrotransposition occurred in neuronal precursors in mouse brains, specifically the hippocampus (Muotri et al. 2005, 2010). In addition, An et al. (2006, 2008), using two different transgenic L1 mouse model systems, also observed L1 somatic activity in a variety of tissues in both models. These findings indicate that L1 elements are occasionally active in somatic tissue and could therefore contribute to somatic mutagenesis. Detection of de novo L1 inserts in the cancer genome provides indirect evidence of somatic mobilization (Solyom et al. 2012; Iskow et al. 2010; Lee et al. 2012a; Tubio et al. 2014; Ewing et al. 2015). Currently, little is known about the regulation, frequency, and relevance of such retrotransposition events. In some cancers, L1 insertions occurred in genes that are frequently mutated in tumors, possibly implicating a potential role of L1 insertional mutagenesis in cancer (Lee et al. 2012a). Furthermore, one report shows that a somatic L1 insertion in the APC tumor suppressor gene played a critical role in the initiation of a human colorectal cancer (Scott et al. 2016). However, in most of the cancer cases studied to date, the data are unclear if these cancer-specific L1 inserts are driver or passenger events.

2.2 NAHR

Copy-number variation (CNV) contributes significantly both to human genetic variation and disease (Sebat et al. 2004; Stankiewicz and Lupski 2010). Non-allelic homologous recombination (NAHR), occurring during meiosis, is the most common mechanism underlying the formation of recurrent CNVs in humans (Gu et al.

2008; Chen et al. 2010). NAHR events contribute to genomic deletions, reciprocal duplications, inversions, and inter- or intrachromosomal translocations (Stankiewicz and Lupski 2010; Dittwald et al. 2013). NAHR events between L1 elements have been previously reported (Han et al. 2008; Startek et al. 2015; Burwinkel and Kilimann 1998). Comparative analysis of human and chimpanzee genomes, verified by wet-lab analyses, identified 73 human-specific L1 recombination-associated deletion events, of which 55 (75 %) were classified as NAHR events (Han et al. 2007). Furthermore, several L1-L1 NAHR events have been identified as the causal factor in several germline diseases (Temtamy et al. 2008; Szafranski et al. 2013; Belancio et al. 2009).

In addition to insertional mutagenesis and NAHR, L1 induced double strand breaks (DSBs) have been implicated as an underlying factor promoting specific translocations in cancer (Lin et al. 2009). Furthermore, L1-induced DSBs near repetitive sequences, such as Alu elements, also contribute to genetic instability through NAHR between Alu elements (Morales et al. 2015a). This topic is covered in great details in Chap. 10. Interestingly, repetitive elements are reported to be enriched at deletion breakpoints (Vissers et al. 2009). However, the contribution of L1-induced DSBs to larger genomic rearrangements and other mutagenic events remains uncharacterized and current estimates likely represent an underestimation.

2.3 Expression of L1 Proteins Affects Cellular Homeostasis

One of the side consequences of the expression of L1 proteins is the mobilization of RNAs from other mobile elements (e.g., Alu and SVA), as well as mRNA from cellular genes (retropseudogenes). Thus, L1 has indirectly contributed to the insertion of over one million Alu elements (Lander et al. 2001), ~2700 SVA elements (Wang et al. 2005), and ~33,000 pseudogenes (Goncalves et al. 2000). Furthermore, the retrotransposition of these RNAs driven by L1 also contributes to insertional mutagenesis and disease (Kaer and Speek 2013).

Other effects have been reported in association with the expression of L1 proteins. The exogenous expression of L1 generates a greater number of DSBs than the predicted numbers of successful insertions, suggesting an intrinsic inefficiency in the integration process (Gasior et al. 2006; Belgnaoui et al. 2006). Because DSBs are deleterious, overexpression of ORF2p is toxic to cells (Kines et al. 2014, 2016). Furthermore, the impact of the chronic induction of DSBs appears to alter cellular response as an adaptation to the L1-related insults (Wallace et al. 2010).

3 Mechanisms of Host Repression of Mobile Elements/ Regulatory Mechanisms of L1 Activity

Host defense mechanisms mostly comprise preventive measures, as there are no specific mechanisms designed to remove TE inserts. A large variety of strategies to limit mobile element activity have been reported. Up until now, the majority of the

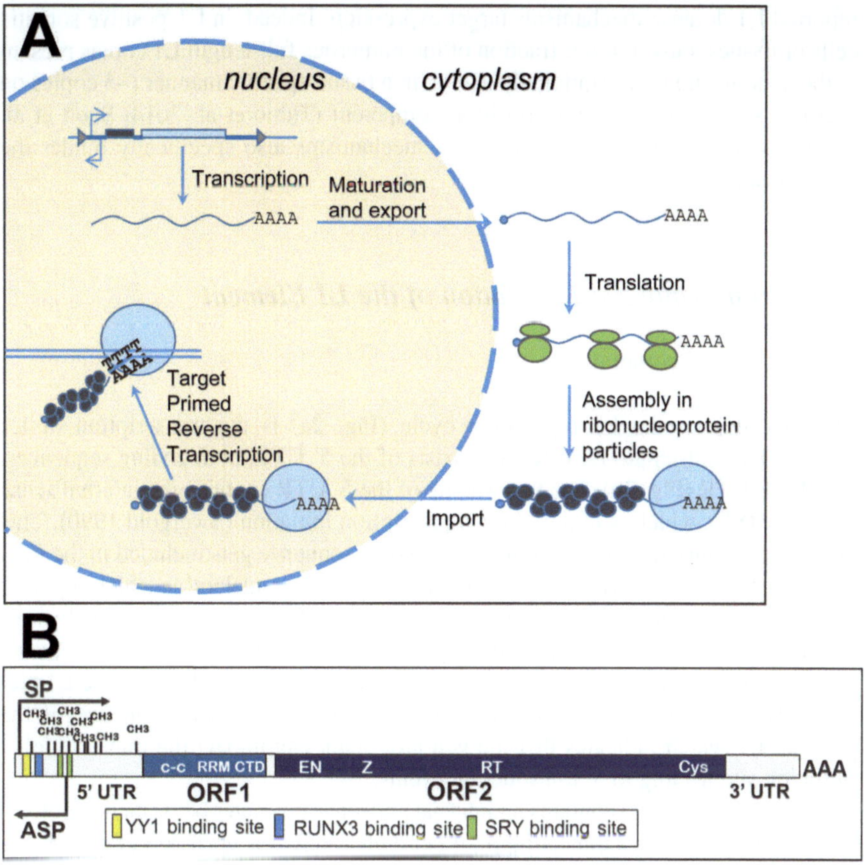

Fig. 2 (**a**) *L1 replication cycle*. Schematic representation of the main steps of the L1 retrotransposition cycle. Briefly, an L1 retrotransposon is transcribed from the L1 sense promoter and produces a full-length mRNA, which is capped, polyadenylated, and exported to the cytoplasm. The L1 proteins are then generated and assembled with the mRNA to form a ribonucleoprotein particle. The L1 ribonucleoprotein (RNP) somehow localizes to the nucleus where the insertion of a new copy of the retroelement into the genome occurs using the TPRT process (cf. Fig. 1). (**b**) *Schematic of the L1 element structure and promoter*. The full-length L1 element comprises five main components: the 5′ UTR, two open reading frames (ORF1 and ORF2) and inter-ORF region, the 3′ UTR, and ends with a poly-A tract (AAA). The 5′ UTR contains the self-transcribing promoter (SP) function, a CpG island that is usually methylated (represented as CH3), one RUNX3 (*blue*), one YY1 (*yellow*), and two SRY (*green*) binding sites. Antisense promoter (ASP) activity is also present in the 5′ UTR. ORF1 (*light blue box*) has a coiled-coil (c-c) domain, an RNA recognition motif (RRM), and a C-terminal domain (CTD). ORF2 (*dark blue box*) encodes the endonuclease (EN) and the reverse transcriptase (RT) activities, as well as a cysteine-rich (cys) motif, whose function is unknown, but is important for the L1 retrotransposition. The 3′ UTR contains the poly-A signal

reported L1 defense mechanisms target expression. Indeed, in L1-positive somatic cells or tissues, only a minor fraction of the numerous full-length L1 copies present in the genome are transcriptionally active, in a tissue-specific manner (~5 copies on average, some being retrotransposition-competent (Tubio et al. 2014; Scott et al. 2016; Philippe et al. 2016)). Additional mechanisms also specifically hinder the insertional process itself.

3.1 Transcriptional Regulation of the L1 Element

3.1.1 The L1 Promoter

The first step of the L1 replication cycle (Fig. 2a) is the transcription of L1 mRNA. The full-length L1 mRNA consists of the 5′ UTR, both coding sequences, and the 3′ UTR (Fig. 2b). The first 100 bp of the 5′ UTR contains the internal sense promoter (SP), which is required for transcription initiation (Swergold 1990). This promoter is unusual, as it self-transcribes and its sequence gets included in the final L1 transcript. L1 transcripts show variability in their first nucleotide which suggests that L1 can use different transcription initiation sites (Lavie et al. 2004). Analysis of L1 inserts has identified the occasional presence of additional nucleotides between the tandem site duplications (TSD) and the beginning of the 5′ UTR of the L1 element. These sequences correspond to the 5′ flanking genomic sequence of the parental L1 which indicates that the genomic flank can impact the transcriptional initiation site leading to 5′ transduction events.

The L1 5′ UTR also contains an internal antisense promoter between nucleotides 400 and 600 that drives transcription of genomic sequences upstream of the L1 element in a wide variety of normal tissues (Speek 2001; Yang and Kazazian 2006). Sequences located between nucleotide 600 and the end of the 5′ UTR are required to enhance this antisense promoter activity (Yang and Kazazian 2006). The discovery of chimeric mRNAs that contain part of the antisense L1 5′ UTR sequence plus additional exons suggests that the L1 antisense promoter can serve as an alternative promoter for expression of upstream adjacent genes (see Chap. 11). Furthermore, the antisense promoter activity can also generate antisense complementary RNAs for both the L1 and the adjacent genes. One study reports L1-generated antisense RNAs to two annotated genes, COL11A1 and BOLL (Matlik and Redik 2006), which could have regulatory effects on these genes. In addition, the synthesis of antisense L1 5′ UTR mRNA is associated with the decrease of L1 expression and retrotransposition, likely through the induction of RNA interference (see below) (Yang and Kazazian 2006; Soifer et al. 2005). However, it is unknown if the presence of either the antisense promoter or the enhancer sequences directly interferes with the L1 transcription.

3.1.2 Transcriptional Factors

Transcription of the L1 mRNA is critical for retrotransposition as the transcript serves as the template for reverse transcription in the insertion process, as well as the RNA used for translation of the ORF1 and ORF2 proteins (Wei et al. 2001). Therefore, the first cellular line of defense to protect genomic integrity is to limit L1 transcription. Several transcription factors have been shown to influence the transcriptional efficiency of L1 through binding to the 5′ UTR of L1 (Fig. 2b). The L1 5′ UTR interacts with the SRY family of transcription factors (Tchenio et al. 2000), YY1 (Athanikar et al. 2004), RUNX3 (Yang et al. 2003), and p53 (Harris et al. 2009). SRY factors bind to two central regions (nucleotides 472–477 and 572–577) (Tchenio et al. 2000), YY1 binds nucleotides 13–21 (Minakami et al. 1992; Becker et al. 1993), while RUNX3 binds to nucleotides 83–101 (Yang et al. 2003).

The tumor suppressor gene p53 is often referred to as "the guardian of the genome," as this gene is induced in response to stress in order to prevent genomic alterations (Yoshida and Miki 2010). In cancer cells, p53 is responsible for activating apoptotic signaling cascades in response to L1 retrotransposition events, resulting in cell death (Belgnaoui et al. 2006; Haoudi et al. 2004). Interestingly, a more recent study found p53 responsive elements in the 5′ UTR of younger genomic L1 elements (Harris et al. 2009). This study showed that p53 protein can bind to these sequences and that the overexpression of p53 is associated with an increase in L1 transcription. These findings suggest that the p53-mediated increase in L1 expression may be a strategy employed by cells in order to induce genomic instability in an attempt to reactivate p53, thereby stimulating apoptosis in affected cells.

3.1.3 L1 Silencing by the Promoter Methylation

The human L1 5′ UTR shows a relative conservation of transcription factor binding sites and CpG sites (Lee et al. 2012b), suggesting an important role in transcription. Although L1 elements are A-rich, the 5′ UTR contains 34 CpG sites spread over a 371 bp CpG island (Hata and Sakaki 1997). DNA methylation of the 5′ UTR is one of the strongest and best characterized regulatory mechanisms of L1 activity (Thayer et al. 1993). Growing evidence suggests that hypomethylation of L1 elements in somatic tissues may play a significant role in deregulating L1 (Chalitchagorn et al. 2004; Roman-Gomez et al. 2005; Suter et al. 2004), particularly in cancers (Roman-Gomez et al. 2005; Chalitchagorn et al. 2004; Perrin et al. 2007).

Several DNA methyltransferases are reported to restrict TEs (Zamudio and Bourc'his 2010). In particular, the DNA methyltransferase 3-like, Dnmt3L plays an important role repressing mobile elements during embryonic reprogramming (Kato et al. 2007). In Dnmt3L knock-out mice, loss of this gene prevents the de novo methylation of L1 elements after reprogramming, leading to their reactivation (Bourc'his and Bestor 2004). This repression mechanism of L1 expression involves members of the methyl-CpG binding domain proteins (MBD), specifically MeCP2 and MBD1 (Muotri et al. 2010; Yu et al. 2001). One study using cervical cancer cells showed that the MeCP2 protein binds to the methylated L1 5′ UTR and induces a decrease in promoter activity (Yu et al. 2001). In addition, a MeCP2 knockout transgenic mouse showed more L1 insertions from an engineered L1 transgene in the hippocampus when compared to control mice (Muotri et al. 2010). The study also reports that female human patients with the neurodevelopmental disorder Rett syndrome (who have a deficiency of MeCP2) have small but significant increases of L1 insertions in their hippocampi. However, the significance of these observations and correlation with this disorder is unclear at this time.

3.2 Posttranscriptional Regulation of the L1 Element

3.2.1 Destabilization of L1 mRNA by Small RNAs

Experimental data suggest that RNA interference downregulates L1 retrotransposition (Soifer et al. 2005; Horman et al. 2006; Yang and Kazazian 2006). The presence of sense and antisense promoters in the 5′ UTR of L1 elements (Fig. 2b) can generate complementary transcripts to form double stranded (ds) L1 RNAs. Yang et al. reported the detection of L1 dsRNAs, as well as small interfering RNAs (L1 siRNAs), in several cultured cell lines, which corresponded to both strands of the L1 5′ UTR (Yang and Kazazian 2006). Furthermore, these L1 siRNAs are shown to have a regulatory effect by posttranscriptionally degrading L1 RNA in cultured cells, which directly correlates with the antisense promoter activity (Yang and Kazazian 2006).

In addition to siRNA regulation, one microRNA, miRNA-128, is reported to be a regulator of L1 expression by directly binding to the L1 mRNA (Hamdorf et al. 2015). The authors propose that miRNA-128 binds to the L1 mRNA at a specific sequence located in the ORF2 coding region which correlates with a decrease in L1 ORF1p expression and in L1 retrotransposition rate. A previous study has already showed that the microprocessor, an enzymatic complex involved in the miRNA biogenesis, can directly cleave L1 mRNA in vitro, suggesting the inhibition of L1 retrotransposition by the degradation of L1 mRNA (Heras et al. 2013). However, the authors did not provide the evidence of a direct role of the microprocessor on L1 regulation and this inhibition can also result in the inhibition of a factor required for L1 expression.

Another category of small RNAs regulating retrotransposons at both transcriptional and posttranscriptional levels in germ line cells are the PIWI-interacting piRNAs. The role of PIWI proteins and the piRNA pathway is well characterized in the regulation of transposable element activity in *Drosophila* (for review, see (Thomson and Lin 2009; Luteijn and Ketting 2013)). However, in mammalian cells, the regulation of transposable elements by PIWI proteins is less characterized. In contrast to *Drosophila*, the proportion of piRNAs corresponding to transposable elements varies depending on the stage of gametogenesis with the highest proportion of piRNAs detected in the early developmental stages (Aravin et al. 2008). Studies show that the deficiency of one of the mouse PIWI proteins, *Miwi2*, causes defects in the formation of male gametes and an arrest during meiosis (Carmell et al. 2007). Interestingly, the *Miwi2* mutation also correlated with the demethylation of the promoters of transposons in mice and the activation of their expression in germline cells. Regulation of retrotransposons in the germline is detailed in Chaps. 1 and 2 of this book.

3.2.2 Processing of the L1 mRNA

The L1 transcript is capped, polyadenylated, and can be spliced (Moran et al. 1996; Faulkner et al. 2009; Perepelitsa-Belancio and Deininger 2003; Belancio et al. 2006). An effective mechanism to regulate the amount of L1 expression is through attenuation by posttranscriptional processes, such as premature polyadenylation and splicing of coding sequences (Perepelitsa-Belancio and Deininger 2003; Belancio et al. 2006). Not surprisingly, polyadenylation of the L1 transcript is needed for efficient L1 retrotransposition. Replacing the poly-A tail with a structured unrelated sequence causes a dramatic decrease in the retrotransposition rate (Doucet et al. 2015). Interestingly, addition of 16 adenosines or 16 cytosines to the structured end allows for retrotransposition to occur at very low rates. Notably, this study also reports that the poly-A tail is needed for an efficient interaction of the L1-ORF2p and the L1 mRNA (Doucet et al. 2015). In addition, the poly-A tail also provides the priming site for reverse transcription during TPRT (Monot et al. 2013) (see Fig. 1). The requirement of a poly-A tract at the 3′ end of the retrotransposon transcript is not unique to L1. All the sequences mobilized by the L1 protein contain a poly-A tract at their 3′ end (Boeke 1997; Roy-Engel 2012). Interestingly, the length of the poly-A-tract determines retrotransposition efficiency (Roy-Engel et al. 2002; Dewannieux and Heidmann 2005). In addition, poly-A binding proteins (PABPs) also play a role in L1 and, possibly, SINE retrotransposition (Roy-Engel et al. 2002; Dai et al. 2012; Muddashetty et al. 2002). These proteins specifically bind to the poly-A tracts of mRNA and play a dual role. In the nucleus, they stimulate the synthesis of the mRNA poly-A tail, and in the cytoplasm they increase the stability of mRNAs and the translation rate (reviewed in (Gray et al. 2015)). In addition, Dai et al. have suggested that the binding of PABPs to the poly-A tail may be needed for the import of L1 mRNA and proteins into the nucleus for retrotransposition to occur (Dai et al. 2012).

An L1 transcript contains multiple cryptic splicing sites and alternative polyadenylation signals that can modify the size of the RNA (Perepelitsa-Belancio and Deininger 2003; Belancio et al. 2006). The processing of the L1 transcript yields a variety of smaller L1 RNA species that are easily detected by Northern blot analyses. The amount of L1 processing differs between different human tissues, with each tissue type showing varying amounts of full-length L1 mRNA and smaller RNA species (Belancio et al. 2010). Splicing and premature polyadenylation of the L1 mRNA proves to be an efficient mechanism for regulating L1, as most of the processed transcripts produced would be unable to support L1 retrotransposition. However, a subset of the spliced products do contain coding sequences that could allow for the expression of ORF2p, which is capable of generating deleterious DSBs in addition to mediating Alu retrotransposition (Belancio et al. 2010).

3.3 *Translational and Posttranslational Regulation of L1 Elements*

3.3.1 Translation of L1 Proteins

L1 elements generate a bicistronic transcript that contains the 5′ UTR promoter and two open reading frames (ORF1 and ORF2); each contains their own stop codons and are separated by a noncoding sequence (inter-ORF region). L1 mRNA is exported to the cytoplasm where the translation of L1 proteins occurs (Fig. 2a). Translation of the L1 bicistronic transcript yield higher amounts of ORF1p than ORF2p (Dai et al. 2014). This effect is proposed to function as a built-in mechanism to control and regulate expression of ORF2p and limit its damaging effects (Alisch et al. 2006). The mechanism of translation of the second ORF (ORF2) is unclear, but some rules have been identified. The Moran laboratory has described that an unconventional termination/reinitiation mechanism is used in the synthesis of ORF2p (Alisch et al. 2006). The authors show that the ORF1 coding sequence and the inter-ORF region are not required for the expression of ORF2p, but the two open reading frames need to be separated by a stop codon in order for ORF2p to be translated. Additionally, any secondary structure in the L1 RNA, such as a hairpin, stalls the ribosome at the end of ORF1 and inhibits ORF2p expression, excluding an internal ribosome entry sequence (IRES)-mediated mechanism.

3.3.2 L1 Ribonucleoprotein Particle Formation

After synthesis, both L1 proteins are reported to associate with the L1 mRNA to form ribonucleoprotein particles (RNPs) in the cytoplasm (Doucet et al. 2010). ORF1p forms a trimeric structure that binds directly to RNAs (Khazina et al. 2011). Little is known about the interaction between ORF2p and the RNA, but mutations in ORF2p affect L1 RNP formation (Doucet et al. 2010). Additionally, ORF1p and ORF2p do not appear to interact directly, but both proteins are thought to remain associated through their interaction with the L1 RNA (Taylor et al. 2013).

ORF1p phosphorylation seems to be a necessary modification for L1 retrotransposition to occur, as mutations in several phosphorylation sites reduce L1 retrotransposition without modifying protein stability (Cook et al. 2015). However, phosphorylation is not required for the RNA binding and chaperone activities of the ORF1 protein. Thus, it is unclear how this posttranslational modification modulates ORF1p activity. ORF1p phosphorylation is hypothesized to be required for the formation of the L1 RNPs by altering the interaction either within the trimeric ORF1p complex or with other uncharacterized interacting proteins. In addition, phosphorylation of ORF1p seems to be required for the control of L1 retrotransposition by the circadian rhythm (see below) (deHaro et al. 2014).

The overexpression of L1 results in the sequestration of the L1 proteins and mRNA in the stress granules (Goodier et al. 2007). Stress granules are usually aggregations of proteins and mRNA molecules in the cytoplasm that form in order to store and protect mRNA in response to an environmental stress (Anderson and Kedersha 2006). Currently, it is unclear if localization to the stress granules is part of the L1 amplification cycle or is a dead end event. Because L1 RNPs need to reach the nucleus in order for a new copy to insert into the genome, localization to the stress granules may function as a limiting step in order to restrict L1 retrotransposition.

The analysis of proteins interacting with ORF1p or ORF2p has provided a valuable insight into the characterization of some of the cellular factors controlling L1 retrotransposition (Taylor et al. 2013; Goodier et al. 2015; Moldovan and Moran 2015). A large number of co-immunoprecipitated proteins identified are still under investigation in order to evaluate their potential role in L1 regulation. Interestingly, some of these proteins, such as the Zinc finger Antiviral Protein (ZAP) and the RNA helicases MOV10 and UPF1 also inhibit retroviral replication (Kurosaki and Maquat 2016; Meister et al. 2005; Xuan et al. 2013; Mao et al. 2013). ZAP, MOV10, and UPF1 proteins are involved in the degradation of viral RNAs in the stress granule (Kurosaki and Maquat 2016; Goodier et al. 2012; Gregersen et al. 2014). Overexpression of these three proteins disorganizes the L1 RNP structure and destabilizes L1 proteins and L1 mRNA. Intriguingly, overexpressed ZAP and MOV10 proteins co-localize with ORF1p and L1 mRNA in cytoplasmic granules (Goodier et al. 2012, 2015; Moldovan and Moran 2015; Li et al. 2013). A comprehensive list of the identified proteins associating with L1 components is available (Pizarro and Cristofari 2016).

3.4 Regulation of L1 Integration into the Genome

Although there are no specific mechanisms to remove L1 inserts, cells have developed a variety of preventive defense strategies. The following section will explore the identified defense mechanisms that have been reported to prevent the insertion of a new L1 copy into the genome.

3.4.1 APOBEC3: Sequestration of L1 RNP/Inhibition of the cDNA Synthesis

Apolipoprotein B mRNA-editing enzyme 3 (APOBEC3) proteins, particularly APOBEC3A (A3A) and APOBEC3C (A3C), prevent retrotransposition of endogenous L1 elements, although the exact mechanism is unclear (Chen et al. 2006; Muckenfuss et al. 2006; Bogerd et al. 2006; Kinomoto et al. 2007). The enzymes of the APOBEC3 family are known to restrict the replication of retroviruses and LTR retrotransposons by catalyzing the deamination of cytidine nucleotides to uridine during the synthesis of the first strand of cDNA. This nucleotide conversion leads to either the degradation of the cDNA or the integration of a mutated and inactive provirus or LTR retrotransposon. However, the replication of LTR retrotransposons differs from the non-LTR retrotransposons because the reverse transcription of the mRNA into cDNA occurs in the cytoplasm and not in the nucleus at the site of insertion. In other words, LTR retrotransposons do not use TPRT (Fig. 1) for integration. Thus, APOBEC3 enzymes appear to restrict the retrotransposition cycle of L1 elements in a different manner than how they restrict retroviruses. In addition, the deaminase activity does not seem to be required for the inhibition of L1 retrotransposition by the A3C protein. In fact, no evidence of editing has been found when evaluating the sequences of new L1 inserts (Kinomoto et al. 2007). Horn et al. (Horn et al. 2014) showed that overexpression of A3C restricts L1 retrotransposition in a deaminase-independent manner. However, the dimerization and RNA binding domains of A3C protein are both required. Additionally, the A3C proteins interact with L1 RNPs in cytoplasmic granules. Thus, it appears that A3C uses a mechanism of RNA sequestration similar to that of A3G (Chiu et al. 2006) in order to block reverse transcription in the nucleus (Horn et al. 2014). A study from the Moran lab reported that the overexpression of A3A protein does not inhibit cDNA synthesis, because the hybrid RNA-cDNA seems to protect the single stranded DNA from deaminase activity (Richardson et al. 2014). Instead, the enzyme edits the single stranded DNA that is exposed during the insertion process in order to limit L1 insertions (Richardson et al. 2014). Up to date, very little is known about the recruitment mechanism of the APOBEC3 enzymes to the L1 insertion site.

3.4.2 DNA Repair Proteins: Control of L1 Insertion Process

The L1 insertion process is not completely understood, particularly which proteins participate in the process. However during the insertion process, a double strand break (DSB) has to be created in order for the de novo L1 copy to insert into the genome. Gasior et al. (2006) have shown that the overexpression of ORF2p is associated with an induction of DSBs. Unrepaired DSBs are deleterious lesions that can lead to cell death. Cells have developed multiple mechanisms for repairing DSBs in an effort to preserve genomic integrity. Several of the proteins involved in DNA DSB repair pathways regulate L1 retrotransposition. Proteins from the non-homologous end joining (NHEJ) pathway, such as XRCC4 and DNA PKcs, appear to have a role in the integration of L1, generating longer L1 inserts in NHEJ-deficient DT40 cells (Suzuki et al. 2009). Additionally, mutations in NHEJ proteins increase the number of endonuclease-independent L1 retrotransposition events in a p53-deficient environment (Morrish et al. 2002). Endonuclease-independent insertions are unconventional, as they do not show the typical hallmarks of retrotransposed insertions, such as target site duplications and a poly-A tails. This suggests that the in the absence of a functional NHEJ pathway insertions occurred at preexisting DSBs.

The Ataxia Telangiectasia Mutated (ATM) protein kinase, which is involved in the recognition and repair signaling of DNA DSBs (reviewed in (Guleria and Chandna 2016)), is also involved in the regulation of L1 retrotransposition.

However, the exact regulatory mechanism remains to be elucidated as two separate reports using different approaches describe opposite effects of ATM on L1 retrotransposition. Using established cell lines derived from a patient carrying the ATM mutation, Gasior et al. reported that ATM is required for retrotransposition, as no de novo L1 insertion events were detected in ATM-deficient cells (Gasior et al. 2006). In contrast, Coufal et al. concluded that ATM restricts retrotransposition from data obtained from shRNA ATM knock-down experiments (Coufal et al. 2011). Variation between cell lines and approaches used may explain the discordant conclusions.

The heterodimer ERCC1/XPF has been shown to repress L1 retrotransposition (Gasior et al. 2008). The ERCC1/XPF complex is a specific endonuclease that recognizes 3′ flap DNA structures and cleaves DNA at the junction of the single stranded and double stranded DNA (de Laat et al. 1998). The inhibition of L1 retrotransposition by ERCC1/XPF suggests that the complex interacts with the elongating cDNA, an intermediate of the TPRT process, which forms a DNA flap structure (Fig. 1). Cleavage of the elongating cDNA by ERCC1/XPF effectively would prevent the integration of a new L1 copy. Furthermore, other proteins from the nucleotide excision repair (NER) pathway also limit L1 retrotransposition (Servant et al. 2016). XPD, XPA, and XPC are involved in limiting L1 retrotransposition. Interestingly, recovered L1 inserts from NER-deficient cells contained abnormally large duplications at the site of insertion, suggesting that NER proteins may also play a role in the normal L1 insertion process (Servant et al. 2016).

The PCNA protein, a cofactor of DNA polymerases, interacts with ORF2p (Taylor et al. 2013). Furthermore, ORF2p contains a highly conserved region between the endonuclease and reverse transcriptase domain known to be a PCNA Interacting Protein motif, or PIP box. The ORF2p-PCNA interaction depends on the endonuclease activity of ORF2p, and the loss of the interaction between the two proteins correlates with a fourfold decrease in the rate of L1 retrotransposition. Because PCNA is involved in both genome replication and DNA synthesis during DNA lesion repair (reviewed in (Wang 2014)), the specific interaction between ORF2p and PCNA suggests a potential role of PCNA in the TRPT process, either in the synthesis of the second strand of L1 cDNA or in the ligation step.

3.5 *Overview Host Regulation of L1*

Overall, there are multiple steps during the L1 life cycle that host factors can interact with L1 to regulate its activity. Knowing when or how cellular factors regulate L1 provides a foundation to understand how environmental factors may influence L1 activity. For example, exposure to compounds known to inhibit DNA repair pathways is likely to stimulate L1 activity. As an overview, Fig. 3 shows a simplified summary of some the known host regulatory factors and their interactions with different steps of the L1 cycle.

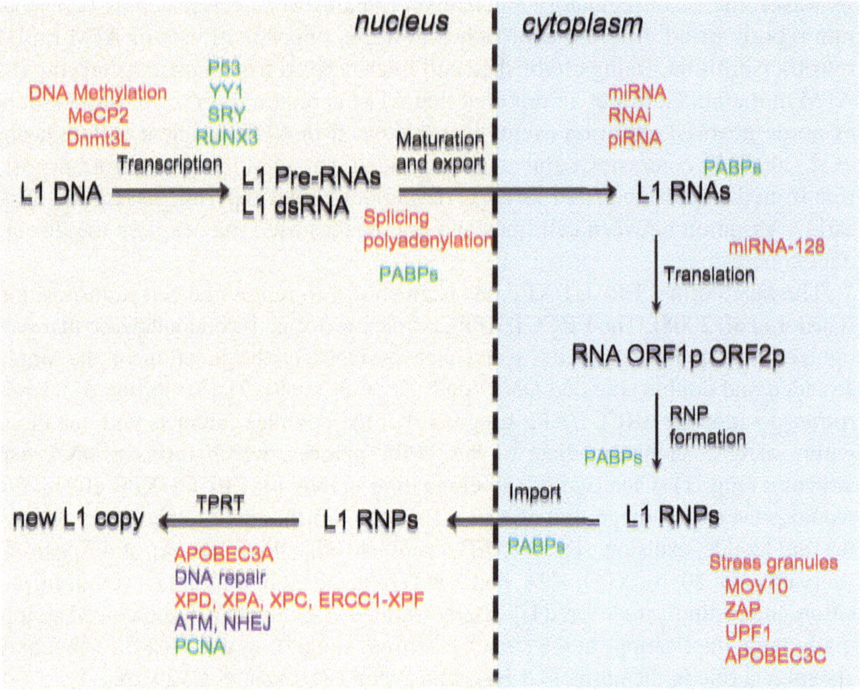

Fig. 3 Schematic overview of the cellular factors that interact with the different steps of L1 retrotransposition cycle. Known cellular factors with positive effects on L1 activity are shown in green, while the factors that reduce L1 activity are in red. Factors with both positive and negative effects on L1 retrotransposition are shown in purple

4 Environmental Influences on L1

4.1 Introduction

A large variety of environmental factors, including therapeutic drugs, have the potential to influence L1 activity by either directly interacting with the L1 components or indirectly by interfering with the cellular mechanisms that regulate L1. A direct effect is usually observed when the external factor changes expression of the L1 machinery L1 expression, leading to altered retrotransposition rates. In contrast, indirect effects do not necessarily increase L1 expression, but instead they deregulate cellular control mechanisms that affect L1 retrotransposition efficiency. Because of L1's genetic damage potential, it is of particular interest when exposure to environmental stimuli increases L1 retrotransposition rates.

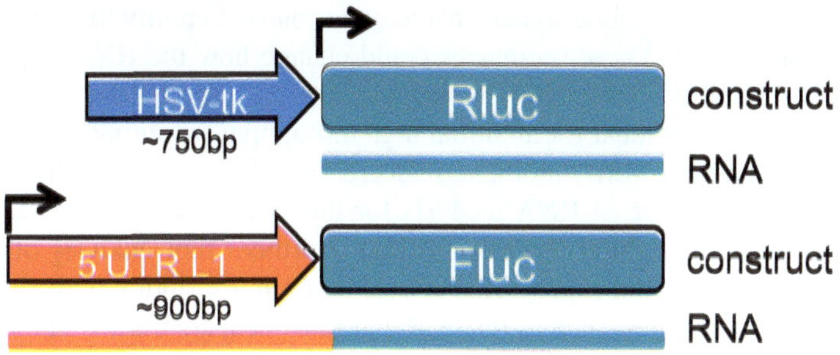

Fig. 4 Transcripts generated by the L1 promoter-reporter constructs differ from the transcripts generated by the internal control construct. For illustration, a schematic of the constructs used by Terasaki et al. (2013) are shown. The reference control plasmid driven by the HSV-tk promoter (*top*) and the L1 5′ UTR-driven plasmid (*bottom*) are shown with their corresponding transcription start sites (*arrows*). The RNA generated by each construct is shown underneath the construct. Note that the L1 driven construct will generate a larger transcript due to the inclusion of the L1 promoter (*orange* portion ~900 bp) in the RNA sequence

To better understand and interpret the existing data, it is first important to discuss the limitations of the assays used to evaluate exposure effects on L1 activity. The three most commonly used assays reported in the literature are: (1) bisulfite sequencing to evaluate L1 promoter methylation status, (2) L1 promoter-luciferase assay to evaluate L1 promoter activity (El Sawy et al. 2005; Terasaki et al. 2013), and (3) the L1 retrotransposition assay (Moran et al. 1996; Rangwala and Kazazian 2009).

Although bisulfite sequencing is a well-established approach to evaluate methylation status, several critical factors need to be considered when studying the promoters of the genomic L1 elements. Of the approximately 7000 full-length L1 elements, only about 300 belong to the currently active L1Hs subfamily (Khan et al. 2006). Most L1 promoters in the genome belong to inactive, older elements that, even if expressed, are far less likely to have an impact on the cell. Thus, distinguishing between the active and inactive loci may be important when evaluating methylation status of L1 promoters. In other words, demethylation of the promoter of a "hot" active L1 is significantly more biologically interesting than the demethylation of a dead L1 fossil. Additionally, promoter hypomethylation is not always synonymous with an increase in L1 expression. Therefore, promoter methylation status does not necessarily correlate with L1 retrotransposition.

The next approach, the L1 promoter-luciferase assay, is the application of a standard approach used to evaluate promoters. In this approach, the L1 5′ UTR (~900 bp) is cloned upstream of the firefly luciferase reporter gene and reporter output is measured in the presence/absence of the compound of interest and normalized to a control (e.g., HSV-tk-Rluc). A limitation of this approach is that the expressed L1 is not in the natural genomic context losing the ability to evaluate exposure effects on tissue-specific regulation of L1 expression (Tubio et al. 2014; Scott et al. 2016; Philippe et al. 2016). In addition, there is a difference between the transcript generated by the control and the L1 promoter driven plasmid. Specifically, the L1 promoter sequence will be included in the transcript (Fig. 4). The concern is that the 5′

UTR contains multiple splice signals that may respond differently to the presence of the compound tested. These sequences could change how the RNA is processed, affecting the stability of the RNA molecules, altering the final luciferase output. Although this assay is ideal for an initial high throughput screening of compounds, further confirmation to verify an increase/decrease of L1 expression should be performed. The importance of RNA analysis for this type of L1 reporter assays has been previously highlighted (Belancio 2011). Northern blot analysis of the full-length L1 element to confirm reported effects on L1 transcription is likely the best option. Other approaches, such as RT-PCR and RNA-seq, are limited in their ability to distinguish between *bona fide* L1 transcripts and processed L1 RNAs or other transcripts containing L1 sequences (discussed in (Deininger and Belancio 2016)).

Finally, the retrotransposition assay is designed to evaluate L1 activity in cultured cells, which relies on an engineered L1 that will generate an L1 insert tagged with a marker. Usually this marker confers resistance to an antibiotic (neomycin or blasticidin), but fluorescence (Ostertag et al. 2000) and luciferases (Terasaki et al. 2013) have also been used. A limitation of this assay is its dependence on cell viability and the cell's efficiency to express the marker. Compounds that affect these parameters will skew the results. Thus, controls to evaluate the effect of the compound on cell viability or marker expression are needed for proper interpretation of results. Because of these nuances, in this chapter we will cautiously recapitulate the published data, but the reader needs to keep in mind that in many instances further corroboration is probably needed.

4.2 Effectors of L1 Expression

As previously described in this chapter, the L1 promoter contains several conserved transcription factor binding sites and a CpG island (Lee et al. 2012b). Although any environmental stimuli could influence L1 transcription through any of these components, the most commonly studied effect is the impact of exposure on L1 methylation status.

4.2.1 Changes in the L1 Promoter Epigenetic Status

Compounds that directly alter methylation or affect DNA methylation maintenance enzymes could influence L1 expression levels (Bourc'his and Bestor 2004). Due to L1's abundance in the genome, many studies use L1 methylation as a proxy to evaluate the global methylation status of a cell or tissue. Because of this, there are numerous studies evaluating the methylation status of L1 in a variety of diseases, as well as studies in cultured cells that were exposed to selected compounds. In particular, L1 methylation has been analyzed in a large variety of cancers and blood samples from cancer patients (Miousse and Koturbash 2015; Kitkumthorn and Mutirangura 2011; Kitkumthorn et al. 2012). The list of evaluated compounds and

Table 1 A few selected examples of exposures reported to influence methylation status of the L1 promoter (direction of the arrow indicate increase or decrease)

Factor	Study	Effect on L1 promoter	References
phthalate	Test neonates and cord blood.	↓L1 methylation Weak inverse correlation with concentration	Huen et al. (2016)
	In utero exposure		
Arsenic	In vitro exposure of lymphoblastoid cells	↓L1 methylation	Intarasunanont et al. (2012)
Cadmium	Non-smoking women in Argentina	↓L1 methylation Weak inverse correlation with concentration	Hossain et al. (2012)
Cadmium	Blood from exposed individuals of Mae Sot Thailand	↑L1 methylation	Virani et al. (2016)
Airborne benzene	Individuals exposed (gas station attendants)	↓L1 methylation	Bollati et al. (2007a, b)
BPA: bisphenol A	Blood and sperm from factory workers.	↓L1 methylation inverse correlation with urine BPA	Miao et al. (2014)
Tobacco smoke	Oral mucosal cells from smokers	Altered methylation patterns of L1	Wangsri et al. (2012)
Trihalomethane	Granulocyte DNA from exposed individuals	↓L1 methylation was associated with bladder cancer risk	Salas et al. (2014)
Triclosan	In vitro exposure HepG2 cells	↓L1 methylation	Zeng et al. (2016)
Tobacco and alcohol	Neonates	Altered L1 methylation levels	Wilhelm-Benartzi et al. (2012)
	In utero exposure		
H_2O_2 reactive oxygen species	In vitro exposure bladder and kidney cells	↓L1 methylation	Kloypan et al. (2015)
Lead	Umbilical cord DNA	↓L1 methylation	Pilsner et al. (2009)
Xenoestrogens	Placenta DNA	Altered L1 methylation levels	Vilahur et al. (2014)
Pyrazinamide	Rat livers	↓L1 methylation	Kovalenko et al. (2007)
5-aza-2′-deoxicytidine	Treatment of patients with leukemia	↓L1 methylation	Yang and Kazazian (2006a)
etoposide	In vitro exposure of cultured cells	↓L1 methylation	Hagan et al. (2003)
BaP: benzo-alpha pyrene	Cervical cancer cells under prolonged exposure	↓L1 methylation	Teneng et al. (2011)
tamoxifen	Rat induced hepatocarcinogenesis	↓L1 methylation	Tryndyak et al. (2007)
Cocaine	Brain regin nucleus accumbens of mice	↓L1 H3K9me3	Maze et al. (2011)

exposures that promote hypomethylation of the L1 promoter includes a large variety of chemical agents from smoking to therapeutic agents. Table 1 highlights a few of the studies with reported effects on L1 promoter methylation. Most of these studies only show an association with L1 promoter hypomethylation and not causality. However, one study using cervical cancer cells demonstrated that prolonged treatment with the carcinogen Benzo-alpha-pyrene (BaP) changed the methylation status of the L1 promoter and increased L1 expression (Teneng et al. 2011; Stribinskis and Ramos 2006). The authors proposed that BaP inhibits the assembly of the methylation machinery (DNMT1 and DNMT3A) to induce hypomethylation (Weisenberger and Romano 1999). In addition, the authors also indicated that BaP exposure promoted the enrichment of histones with the hallmarks for open chromatin at the L1 promoter (Teneng et al. 2011). Unfortunately, the vast majority of L1 methylation studies to date lack the analysis of L1 RNA levels, limiting the interpretation of the results (Deininger and Belancio 2016). Until validation studies are performed, it is difficult to determine which of the compounds reported have an effect on L1 biology. In addition to CpG methylation, exposure to chemicals that alter histone methylation can also affect L1 expression. For example, cocaine has been reported to decrease the H3K9me3 at L1 loci and to increase L1 expression (Maze et al. 2011).

4.2.2 Effectors of L1 Transcription

The strength of the L1 promoter can vary in a tissue-specific manner depending on the availability of the required transcription factors in that tissue. Thus, the exposure to agents or compounds affecting L1 transcription factor function or expression can alter the transcriptional efficiency of the element. For example, cadmium and cobalt

Table 2 Compounds affecting reporter expression driven the L1 5′ UTR

Study	Compounds	References
Morales part 1: steroid hormone-like (lacZ)	Serum, testosterone, dihydrotestosterone, organochloride pesticides	Morales et al. (2002)
Morales et al. part 2 : stressors (lacZ)	Tetradecanoylphorbol 13-acetate (TPA), UV light, heat shock	Morales et al. (2003)
Terasaki et al. (firefly luciferase)	Mitomycin C, actinomycin D, cisplatin, 6-Thioguanine, campothecin, merbarone, BaP, vinblastine, cytochalasin D, Diethyl malate, phorone, Thenoyltrifluoroacetone, citrinin, cyclosporine A, etomoxir,exol, Bezafibrate, Fenofibrate, Fluvastatin, Pravastatin, Aminoglutethimide, Metyrapone, Acetylsalicylic acid, Diflunisal, Flufenamic acid, Mefenamic acid, Salicylamide, Sulindac, Sulindac, Guaiacol glyceryl ether, Nifedipine, Ticlopidine, BaP, merbarone	Terasaki et al. (2013)

Table 3 Compounds affecting L1 retrotransposition in culture

Factor	Effect on L1	References
UV	↑Retrotransposition	Teneng et al. (2007), Servomaa and Rytomaa (1988, 1990), Tanaka et al. (2012)
Heat shock	↑Retrotransposition	Puszyk et al. (2013)
Ionizing radiation	↑Retrotransposition	Servomaa and Rytomaa (1988, 1990), Tanaka et al. (2012), Luzhna et al. (2015), Farkash et al. (2006)
Gamma radiation	↑Retrotransposition	Farkash et al. (2006)
H_2O_2 (oxidative stress)	↑Retrotransposition	Giorgi et al. (2011)
Mitomycin C	↑Retrotransposition	Terasaki et al. (2013)
Etmoxir	↑Retrotransposition	Terasaki et al. (2013)
Salicylamide	↑Retrotransposition	Terasaki et al. (2013)
WY-14643	↑Retrotransposition	Terasaki et al. (2013)
Morphine	↑Retrotransposition	Okudaira et al. (2016)
Cocaine	↑Retrotransposition	Okudaira et al. (2014)
Heterocyclic amines (HCAs)	↑Retrotransposition	Okudaira et al. (2010)
Fentanyl citrate	↑Retrotransposition	Okudaira et al. (2016)
Methamphetamine	↑Retrotransposition in vitro, increased expression rat brain	Okudaira et al. (2014), Moszczynska et al. (2015)
FICZ (6-formylindolo[3,2-b] carbazole)	↑Retrotransposition	Okudaira et al. (2010)
X-ray	↑Retrotransposition	Banaz-Yasar et al. (2012)
Aluminum	↑Retrotransposition	Karimi et al. (2014a)
Iron	↑Retrotransposition	Habibi et al. (2013a)
Copper	↑Retrotransposition	Habibi et al. (2013a)
Arsenic trioxide	↑Retrotransposition	Karimi et al. (2014b)
Mercury	↑Retrotransposition	Kale et al. (2005)
Nickel	↑Retrotransposition	El Sawy et al. (2005)
Cadmium	↑Retrotransposition	Kale et al. (2005)
BaP	↑Retrotransposition	Stribinskis and Ramos (2006)
DMBA+TPA	↑Retrotransposition	Okudaira et al. (2011)
Cigarette smoke extract (CSE)	Detection of ORF2 in primary human lung fibroblast	Miglino et al. (2012)
Circadian disruption/ melatonin	↑Retrotransposition	deHaro et al. (2014)
Reprogramming	↑Retrotransposition	Klawitter et al. (2016), Wissing et al. (2012), Arokium et al. (2014)

have been shown to upregulate a variety of transcriptional factors, including SRY and YY1, in primary normal human bronchial epithelial cells (Glahn et al. 2008). Several genotoxic agents activate p53 (Nelson and Kastan 1994), which may in turn bind to the L1 promoter (Harris et al. 2009). Furthermore, some responses to different exposures have been reported to be "context specific" or cell specific (Teneng et al. 2007). For example, 2,3,7,8-tetrachlorodibenzo-p-dioxin (TCDD, also known as "dioxin," a contaminant in agent orange) and UV gamma irradiation induced L1 expression only in one cell type tested (Lu et al. 2000). Alternatively, environmental stress conditions can affect the ability of cellular proteins to alter their ability to process L1 RNA transcripts. For example, exposure to ultraviolet-C alters RNA splicing patterns by diminishing the presence of the nuclear splicing factor hSlu7 via changes in its nuclear/cytoplasmic localization (Shomron et al. 2005). In addition, UV-light is also reported to reduce the expression of several small nuclear RNAs involved in RNA splicing (Thirunavukkarasu et al. 1988). Multiple studies have evaluated the effect of UV-light on L1 mobility. UV exposure was shown to increase reverse transcriptase activities in cell lines that express L1 (Deragon et al. 1990). One study showed a small increase in human L1 promoter activity after exposure to UV light (Morales et al. 2003). This modest increase in promoter activity could potentially lead to an increase in L1 mRNA levels if the promoter was driving transcription of full-length L1 elements. Furthermore, L1 ORF2 mRNA was upregulated in cells exposed to UV light (Banerjee et al. 2005).

A variety of compounds (Table 2) have been shown to influence/alter the activity of the L1 promoter by using the previously described luciferase reporter assay (Fig. 4). Two reports from the same research group investigated the effects of exposures to hormone-like or hormone precursor agents (part 1) and stressors (part 2) on the L1 promoter (Morales et al. 2002, 2003). In part 1, the authors suggest that the L1 promoter may respond to the treatment with testosterone, dihydrotestosterone, and organochloride pesticides (Morales et al. 2002). In part 2, the authors reported that UV, heat shock, and TPA affected expression, but X-rays and hydrogen peroxide showed no increase in reporter gene activity (Morales et al. 2003). However, both Morales et al. studies (Morales et al. 2002, 2003) did not evaluate RNA levels, the effect of the exposure on a full-length L1 element and they lacked critical controls for toxicity and growth. Thus, further studies are needed to validate any of these results. The study by Terasaki et al. (Terasaki et al. 2013) also tested some of the compounds using the L1 retrotransposition assay (see Table 3) providing an additional support of their observations.

A few studies have used the full-length tagged L1 vector to evaluate the effect of different compounds on L1 expression. For example, cobalt treatments increased L1 mRNA in tissue culture experiments but retrotransposition activity remained unaffected (El Sawy et al. 2005; Habibi et al. 2014). In addition, studies on the environmental carcinogen BaP in both human and mouse tissue culture assays showed that exposure increased both L1 RNA levels and L1 activity (Stribinskis and Ramos 2006; Lu et al. 2000). However, studies on how any of these exposures affect endogenous L1 transcription are currently unavailable.

4.3 Effectors of L1 Retrotransposition

A large variety of compounds and environmental exposures (e.g., stress) have been reported to have some effect on L1 retrotransposition in cultured cells. Table 3 compiles the wide range of compounds tested that showed some increase in L1 activity relative to a control.

4.3.1 DNA Damaging Agents, Drugs, and Therapeutics

The effects of stress inducers on transposable element activation have been reported in many organisms (Capy et al. 2000). One report has shown increased L1 activity upon heat shock in vivo due to the deregulation of the PLZF-mediated epigenetic control of L1 (Puszyk et al. 2013). However, most studies have focused on DNA damaging agents. The DNA damage caused by UV light (thymidine dimerization) and ionizing radiation (DNA breaks) dramatically induce L1 activity in cultured rat chloroleukemia cells (Servomaa and Rytomaa 1988, 1990; Tanaka et al. 2012; Luzhna et al. 2015; Farkash et al. 2006). However, not all types of DNA damaging agents increase L1 retrotransposition. Studies in human cells demonstrate that diverse DNA damaging agents can exhibit very different effects on L1 mobilization (El Sawy et al. 2005; Kale et al. 2005; Kale et al. 2006; Farkash and Prak 2006) and reviewed in (Farkash and Prak 2006). For example, gamma irradiation stimulates L1 activity (Farkash et al. 2006), but the chemotherapeutic agents, cisplatin (a DNA crosslinking agent), calicheamicin γ (a DNA cleaving agent), and camptothecin (a topoisomerase inhibitor) do not increase L1 retrotransposition (Terasaki et al. 2013). Similarly, DNA damaging heavy metals like nickel and cadmium also stimulate L1 mobilization, which is not observed with exposures to paraquat (a pesticide that generates reactive oxygen species), and etoposide (a topoisomerase inhibitor and anticancer drug) (El Sawy et al. 2005; Kale et al. 2005, 2006). Because of the diversity in the response to DNA damaging agents, it is unlikely that the natural endonuclease-dependent L1 insertions benefit from having additional DNA breaks in the genome. Instead, these agents are likely to increase L1 activity through a different mechanism, such as the inhibition of cellular regulators of L1.

4.3.2 Heavy Metals and Carcinogens

Heavy metals are of great interest as they are well-known carcinogens (1993). However, the mechanism by which they cause cancer is not well defined. Interestingly, multiple reports indicate that a variety of heavy metals stimulate L1 activity in culture (Table 3), which suggests that L1-induced genetic instability may be an underlying mechanism of heavy metal carcinogenesis (reviewed in (Morales et al. 2015b)). The list of heavy metals tested include both soluble and particulate mercury (Habibi et al. 2014; Kale et al. 2005), arsenic trioxide (Karimi et al. 2014a, b), aluminum (Karimi et al. 2014a), copper (Habibi et al. 2013), both soluble and particular cadmium (Kale et al. 2005, 2006), and nickel (Kale et al. 2005).

Although most heavy metals contribute to genetic instability by generating reactive oxygen species (ROS) (Ercal et al. 2001; Klein et al. 1991), data indicate that their effect on L1 is not likely a generalized stress response to metal exposure. Not all metals affect L1 retrotransposition, and the ones that stimulate L1 activity do not share the same mechanism of action. For example, treatment with cobalt ($CoCl_2$) increases L1 mRNA and L1 promoter activity (Habibi et al. 2014), but does not increase L1 retrotransposition (Kale et al. 2005; Habibi et al. 2014). Furthermore, analyses of the new L1 inserts showed that treatment with cobalt lead to a greater proportion of full-length L1 inserts (Kale et al. 2005). In addition, increases in L1 retrotransposition by nickel exposure can be reversed by zinc and magnesium treatment (Kale et al. 2006). In this case, nickel is thought to displace Zn and Mg (which function as enzymatic cofactors) from important cellular proteins, such as DNA repair proteins (Hartwig et al. 1994, 2002a, b; Asmuss et al. 2000). A separate study hypothesized that mercury sulfide (HgS) treatment could lead to an increase in L1 retrotransposition by removing the silencing factors associated with cell senescence (Habibi et al. 2013). Other studies in neuroblastoma cells show that iron and copper only increased LINE-1 retrotransposition in dividing cells, while mercury stimulated activity in both dividing and nondividing cells (Habibi et al. 2013). However, the effect of copper on L1 activity is controversial. One study showed that exposure caused a decrease in L1 silencing in a dividing neuroblastoma cell line and increased L1 retrotransposition. In contrast, copper exposure in a hepatocellular carcinoma cell line decreased L1 retrotransposition (Karimi et al. 2014a, 2015). The complexity of the responses triggered by different heavy metals limits the ability to dissect all of the mechanistic pathways responsible for the effects on the activity of the L1 element. Furthermore, it would not be surprising that each individual heavy metal would have different mechanisms of how they affect L1 activity, which could also be cell type-dependent.

4.3.3 Regulation of L1 Retrotransposition by Circadian Rhythm (Melatonin): In the Jungle the LINE Sleeps at Night

Exposure to light at night disrupts circadian rhythm, which is shown to suppress melatonin production and increase cancer risk (Schernhammer and Schulmeister 2004). Recent studies from the Belancio lab demonstrated that L1 is regulated by the circadian rhythm via a melatonin-mediated mechanism. In this system, tumors established in nude male rats were transfused with human blood collected from healthy adult male donors that was either collected during the day (low melatonin), at night (high melatonin), or at night after exposure to light (reduced melatonin) (Blask et al. 2005). Tumors transfused with blood collected during the day or blood collected at night after light exposure that contained a low concentration of melatonin showed high L1 expression in comparison to tumors transfused with blood collected at night, which contains high concentrations of melatonin (deHaro et al.

2014). Additionally, overexpression of the melatonin receptor MT1 in cultured cells induced a downregulation of L1 retrotransposition which was associated with a decrease in L1 mRNA and ORF1p. Furthermore, the downregulation of ORF1p by overexpression of MT1 receptor is lost when several phosphorylation sites in the ORF1p sequence are mutated. The data suggest that the circadian rhythm regulates L1 expression in a melatonin receptor-dependent manner. Overall, the data suggest that L1 appears to remain inactive at night (i.e., "sleeps"). However, any disruption of the melatonin expression cycle could modify the regulation of L1 expression in tissues and promote genetic instability caused by L1 retrotransposition. Thus, exposure to light or melatonin inhibitors are likely to increase L1 activity. Because the disruption of the circadian rhythm is associated with an elevated cancer risk in night shift workers (Blask et al. 2011; Kochan and Kovalchuk 2015), the upregulation of L1 activity and the accumulation of de novo L1 insertions are predicted to occur in these cancers.

4.3.4 Cellular Reprogramming (iPSC Cells)

Transcriptional silencing by methylation and chromatin remodeling is considered an effective mechanism of global L1 regulation. However, during human development, chromatin is dynamically remodeled through distinct phases of reprogramming and de novo methylation (Hackett and Surani 2013). Thus, it is likely that one of the best moments for L1 retrotransposition to escape control from piRNA or other mechanisms is during development when global demethylation occurs. Not surprisingly, abrogating pathways that regulate endogenous retroelements during development leads to harmful consequences, e.g., the conditional loss of PRMT5 (Kim et al. 2014) and knock-out of Dnmt3L (Bourc'his and Bestor 2004).

Data show that the artificial reprogramming of cells into induced pluripotent stem cells (iPSCs) via overexpression of reprogramming factors involves epigenetic remodeling (Planello et al. 2014). Interestingly, studies show that the artificial reprogramming of cells into iPSCs causes increases in L1 expression and L1 insertions (Klawitter et al. 2016; Wissing et al. 2012; Arokium et al. 2014). An initial report using human embryonic stem cells (hESCs) and iPSCs cells showed significant increases in full-length L1 mRNA and ORF1p expression, which was correlated with an overall hypomethylation of L1 elements (Wissing et al. 2012). A separate study, using deep sequencing of the 3′ ends of L1, also observed increased L1 expression during reprogramming and found potential somatic L1Hs insertions that occurred after reprogramming (Arokium et al. 2014). More recently, Klawitter et al. further verified L1 activity during reprogramming by reporting the detection of de novo somatic L1 inserts by sequencing eight human iPSC lines and three hESC lines (Klawitter et al. 2016). Furthermore, they were able to detect an intronic L1 insertion in the CADPS2 gene that occurred during iPSC cultivation (Klawitter et al. 2016). Although cellular reprogramming would not likely occur in a natural context, these data reinforce the concept that external actions on cells and cellular processes can lead to significant effects on L1 mobilization. For example, the poten-

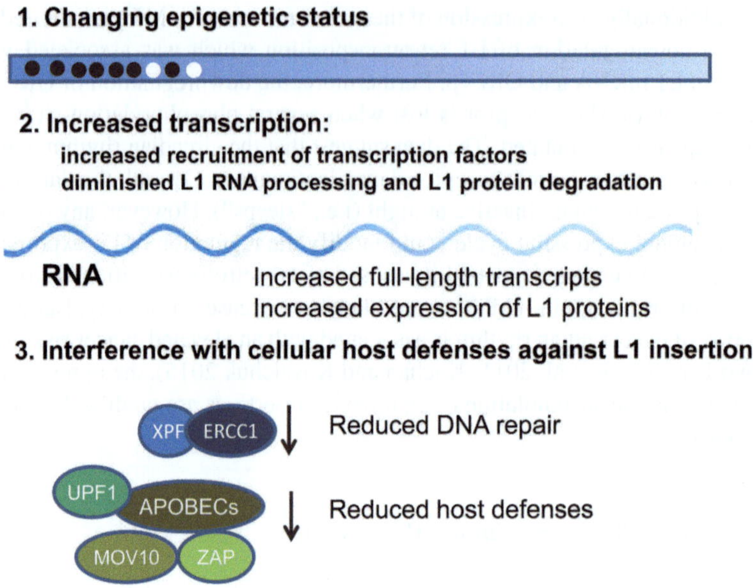

Fig. 5 Proposed mechanisms by which exposure to environmental factors can influence L1 elements. (*1*) Changing epigenetic status: L1 transcription can be affected by environmental factors that modify the epigenetic controls of silent loci. (*2*) Increased amount of full-length L1 transcripts from expressing loci: exposures that reduce L1 RNA processing or promote L1 transcription would directly impact the efficiency full-length L1 RNA production. (*3*) Interference with cellular host defenses against L1 insertion. Compounds inhibiting critical proteins such as the APOBECs would increase L1 activity

tial impact of exposure effects during fertilization and embryonic development is of particular interest in the field of public health.

5 Conclusions

Overall, it is well established that many environmental factors trigger a variety of cellular responses. Thus, it is not surprising that L1 elements also respond to environmental stimuli. TE response to stress is well documented in plants and nonmammalian species, but it is only until recently that studies on the human L1 retrotransposition have become available.

One of the limitations of evaluating xenobiotics or other environmental factors is due to the difficulty of distinguishing between direct and indirect effects. For example, exposure to a compound that affects the ability of the cell to form colonies would cause a decreased number of colonies in the L1 retrotransposition assay. However, the decrease would not be a direct effect of L1 activity, but an indirect effect due to altered cell growth. In addition, some compounds may affect multiple pathways in a cell. For example, cadmium exposure is shown to trigger the DNA damage response, inhibit a variety of enzymatic pathways, as well as induce cellular growth (Waalkes 2000). To complicate matters, because most of the current data are obtained from tissue culture experiments, any negative impact on cell viability caused by the compound tested would affect the experimental outcome. Another limitation is the difficulty in evaluating L1 somatic retrotransposition, creating a void of any exposure studies and L1 in humans.

Although there are multiple complex ways that environmental factors can affect L1, based on the current knowledge we propose three basic mechanisms (Fig. 5). The first mechanism is through the alteration of the epigenetic silencing of the L1 causing expression of normally repressed loci. The next mechanism is through the increase of functional L1 transcripts of expressing loci by directly increasing transcription or diminishing processing/degradation of the L1 RNA. Finally, the third mechanism is through the deregulation of host defense strategies that prevent L1 insertion. In this scenario, the perfect L1-stimulating compound would promote L1 promoter demethylation, increase transcription, and inhibit host defense strategies. It would not be surprising if future studies revealed that the risk of an individual to develop an L1-induced disease is determined by both the presence of "hot" L1s in their genome, as well as by their history of exposure to identified xenobiotics that upregulate L1 activity. Of note, each particular L1 loci in different tissues might respond differently to physiological or environmental triggers, adding a supplemental layer of complexity (Tubio et al. 2014; Scott et al. 2016; Philippe et al. 2016).

References

(1993) Meeting of the IARC working group on beryllium, cadmium, mercury and exposures in the glass manufacturing industry. Scand J Work Environ Health 19: 360–363.

Alisch RS, Garcia-Perez JL, Muotri AR, Gage FH, Moran JV (2006) Unconventional translation of mammalian LINE-1 retrotransposons. Genes Dev 20:210–224

An W, Han JS, Wheelan SJ, Davis ES, Coombes CE, Ye P et al (2006) Active retrotransposition by a synthetic L1 element in mice. Proc Natl Acad Sci U S A 103:18662–18667. doi:10.1073/pnas.0605300103

An W, Han JS, Schrum CM, Maitra A, Koentgen F, Boeke JD (2008) Conditional activation of a single-copy L1 transgene in mice by Cre. Genesis 46:373–383. doi:10.1002/dvg.20407

Anderson P, Kedersha N (2006) RNA granules. J Cell Biol 172:803–808. doi:10.1083/jcb.200512082

Aravin AA, Sachidanandam R, Bourc'his D, Schaefer C, Pezic D, Toth KF et al (2008) A piRNA pathway primed by individual transposons is linked to de novo DNA methylation in mice. Mol Cell 31:785–799

Arokium H, Kamata M, Kim S, Kim N, Liang M, Presson AP et al (2014) Deep sequencing reveals low incidence of endogenous LINE-1 retrotransposition in human induced pluripotent stem cells. PLoS One 9:e108682. doi:10.1371/journal.pone.0108682

Asmuss M, Mullenders LH, Eker A, Hartwig A (2000) Differential effects of toxic metal compounds on the activities of Fpg and XPA, two zinc finger proteins involved in DNA repair. Carcinogenesis 21:2097–2104

Athanikar JN, Badge RM, Moran JV (2004) A YY1-binding site is required for accurate human LINE-1 transcription initiation. Nucleic Acids Res 32:3846–3855

Baillie JK, Barnett MW, Upton KR, Gerhardt DJ, Richmond TA, De SF et al (2011) Somatic retrotransposition alters the genetic landscape of the human brain. Nature. doi:10.1038/nature10531

Banaz-Yasar F, Gedik N, Karahan S, Diaz-Carballo D, Bongartz BM, Ergun S (2012) LINE-1 retrotransposition events regulate gene expression after X-ray irradiation. DNA Cell Biol 31:1458–1467. doi:10.1089/dna.2012.1676

Banerjee G, Gupta N, Tiwari J, Raman G (2005) Ultraviolet-induced transformation of keratinocytes: possible involvement of long interspersed element-1 reverse transcriptase. Photodermatol Photoimmunol Photomed 21:32–39

Beck CR, Collier P, Macfarlane C, Malig M, Kidd JM, Eichler EE et al (2010) LINE-1 retrotransposition activity in human genomes. Cell 141:1159–1170

Becker KG, Swergold GD, Ozato K, Thayer RE (1993) Binding of the ubiquitous nuclear transcription factor YY1 to a cis regulatory sequence in the human LINE-1 transposable element. Hum Mol Genet 2:1697–1702

Belancio VP (2011) Importance of RNA analysis in interpretation of reporter gene expression data. Anal Biochem 417:159–161. doi:10.1016/j.ab.2011.05.035

Belancio VP, Hedges DJ, Deininger P (2006) LINE-1 RNA splicing and influences on mammalian gene expression. Nucleic Acids Res 34:1512–1521

Belancio VP, Deininger PL, Roy-Engel AM (2009) LINE dancing in the human genome: transposable elements and disease. Genome Med 1:97. doi:10.1186/gm97

Belancio VP, Roy-Engel AM, Pochampally RR, Deininger P (2010) Somatic expression of LINE-1 elements in human tissues. Nucleic Acids Res 38:3909–3922

Belgnaoui SM, Gosden RG, Semmes OJ, Haoudi A (2006) Human LINE-1 retrotransposon induces DNA damage and apoptosis in cancer cells. Cancer Cell Int 6:13

Blask DE, Brainard GC, Dauchy RT, Hanifin JP, Davidson LK, Krause JA et al (2005) Melatonin-depleted blood from premenopausal women exposed to light at night stimulates growth of human breast cancer xenografts in nude rats. Cancer Res 65:11174–11184. doi:10.1158/0008-5472.CAN-05-1945

Blask DE, Hill SM, Dauchy RT, Xiang S, Yuan L, Duplessis T et al (2011) Circadian regulation of molecular, dietary, and metabolic signaling mechanisms of human breast cancer growth by the nocturnal melatonin signal and the consequences of its disruption by light at night. J Pineal Res 51:259–269. doi:10.1111/j.1600-079X.2011.00888.x

Boeke JD (1997) LINEs and Alus—the polyA connection. Nat Genet 16:6–7

Bogerd HP, Wiegand HL, Hulme AE, Garcia-Perez JL, O'shea KS, Moran JV et al (2006) Cellular inhibitors of long interspersed element 1 and Alu retrotransposition. Proc Natl Acad Sci U S A 103:8780–8785

Bollati V, Baccarelli A, Hou L, Bonzini M, Fustinoni S, Cavallo D et al (2007) Changes in DNA methylation patterns in subjects exposed to low-dose benzene. Cancer Res 67:876–880. doi:10.1158/0008-5472.CAN-06-2995

Bourc'his D, Bestor TH (2004) Meiotic catastrophe and retrotransposon reactivation in male germ cells lacking Dnmt3L. Nature 431:96–99

Branciforte D, Martin SL (1994) Developmental and cell type specificity of LINE-1 expression in mouse testis: implications for transposition. Mol Cell Biol 14:2584–2592

Brouha B, Schustak J, Badge RM, Lutz-Prigge S, Farley AH, Moran JV et al (2003) Hot L1s account for the bulk of retrotransposition in the human population. Proc Natl Acad Sci U S A 100:5280–5285

Burwinkel B, Kilimann MW (1998) Unequal homologous recombination between LINE-1 elements as a mutational mechanism in human genetic disease. J Mol Biol 277:513–517

Capy P, Gasperi G, Biemont C, Bazin C (2000) Stress and transposable elements: co-evolution or useful parasites? Heredity 85(Pt 2):101–106

Carmell MA, Girard A, van de Kant HJ, Bourc'his D, Bestor TH, de Rooij DG et al (2007) MIWI2 is essential for spermatogenesis and repression of transposons in the mouse male germline. Dev Cell 12:503–514. doi:10.1016/j.devcel.2007.03.001

Casacuberta E, Gonzalez J (2013) The impact of transposable elements in environmental adaptation. Mol Ecol 22:1503–1517. doi:10.1111/mec.12170

Chadha S, Sharma M (2014) Transposable elements as stress adaptive capacitors induce genomic instability in fungal pathogen Magnaporthe oryzae. PLoS One 9:e94415. doi:10.1371/journal.pone.0094415

Chalitchagorn K, Shuangshoti S, Hourpai N, Kongruttanachok N, Tangkijvanich P, Thong-ngam D et al (2004) Distinctive pattern of LINE-1 methylation level in normal tissues and the association with carcinogenesis. Oncogene 23:8841–8846. doi:10.1038/sj.onc.1208137

Chen H, Lilley CE, Yu Q, Lee DV, Chou J, Narvaiza I et al (2006) APOBEC3A is a potent inhibitor of adeno-associated virus and retrotransposons. Curr Biol 16:480–485

Chen JM, Cooper DN, Ferec C, Kehrer-Sawatzki H, Patrinos GP (2010) Genomic rearrangements in inherited disease and cancer. Semin Cancer Biol 20:222–233. doi:10.1016/j.semcancer.2010.05.007

Chiu YL, Witkowska HE, Hall SC, Santiago M, Soros VB, Esnault C et al (2006) High-molecular-mass APOBEC3G complexes restrict Alu retrotransposition. Proc Natl Acad Sci U S A 103:15588–15593

Cho K, Lee YK, Greenhalgh DG (2008) Endogenous retroviruses in systemic response to stress signals. Shock 30:105–116

Christensen SM, Ye J, Eickbush TH (2006) RNA from the 5′ end of the R2 retrotransposon controls R2 protein binding to and cleavage of its DNA target site. Proc Natl Acad Sci U S A 103:17602–17607

Cook PR, Jones CE, Furano AV (2015) Phosphorylation of ORF1p is required for L1 retrotransposition. Proc Natl Acad Sci U S A 112:4298–4303. doi:10.1073/pnas.1416869112

Cordaux R, Batzer MA (2009) The impact of retrotransposons on human genome evolution. Nat Rev Genet 10:691–703. doi:10.1038/nrg2640

Coufal NG, Garcia-Perez JL, Peng GE, Marchetto MC, Muotri AR, Mu Y et al (2011) Ataxia telangiectasia mutated (ATM) modulates long interspersed element-1 (L1) retrotransposition in human neural stem cells. Proc Natl Acad Sci U S A 108:20382–20387. doi:10.1073/pnas.1100273108

Dai L, Taylor MS, O'Donnell KA, Boeke JD (2012) Poly(a) binding protein C1 is essential for efficient L1 retrotransposition and affects L1 RNP formation. Mol Cell Biol 32:4323–4336. doi:10.1128/MCB.06785-11

Dai L, LaCava J, Taylor MS, Boeke JD (2014) Expression and detection of LINE-1 ORF-encoded proteins. Mob Genet Elements 4:e29319. doi:10.4161/mge.29319

de Laat WL, Appeldoorn E, Jaspers NGJ, Hoeijmakers JHJ (1998) DNA Structural elements required for ERCC1-XPF endonuclease activity. J Biol Chem 273:7835–7842

deHaro D, Kines KJ, Sokolowski M, Dauchy RT, Streva VA, Hill SM et al (2014) Regulation of L1 expression and retrotransposition by melatonin and its receptor: implications for cancer risk associated with light exposure at night. Nucleic Acids Res 42:7694–7707. doi:10.1093/nar/gku503

Deininger P, Belancio VP (2016) Detection of LINE-1 RNAs by Northern Blot. Methods Mol Biol 1400:223–236. doi:10.1007/978-1-4939-3372-3_15

Deragon JM, Sinnett D, Labuda D (1990) Reverse transcriptase activity from human embryonal carcinoma cells NTera2D1. Embo J 9:3363–3368

Dewannieux M, Heidmann T (2005) Role of poly(A) tail length in Alu retrotransposition. Genomics 86:378–381

Dittwald P, Gambin T, Szafranski P, Li J, Amato S, Divon MY et al (2013) NAHR-mediated copy-number variants in a clinical population: mechanistic insights into both genomic disorders and Mendelizing traits. Genome Res 23:1395–1409. doi:10.1101/gr.152454.112

Doucet AJ, Hulme AE, Sahinovic E, Kulpa DA, Moldovan JB, Kopera HC et al. (2010) Characterization of LINE-1 ribonucleoprotein particles. PLoS Genet 6. doi:10.1371/journal.pgen.1001150

Doucet AJ, Wilusz JE, Miyoshi T, Liu Y, Moran JV (2015) A 3′ Poly(A) tract is required for LINE-1 retrotransposition. Mol Cell 60:728–741. doi:10.1016/j.molcel.2015.10.012

Doucet-O'Hare TT, Rodic N, Sharma R, Darbari I, Abril G, Choi JA et al (2015) LINE-1 expression and retrotransposition in Barrett's esophagus and esophageal carcinoma. Proc Natl Acad Sci U S A 112:E4894–E4900. doi:10.1073/pnas.1502474112

El Sawy M, Kale SP, Dugan C, Nguyen TQ, Belancio V, Bruch H et al (2005) Nickel stimulates L1 retrotransposition by a post-transcriptional mechanism. J Mol Biol 354:246–257

Ercal N, Gurer-Orhan H, Aykin-Burns N (2001) Toxic metals and oxidative stress part I: mechanisms involved in metal-induced oxidative damage. Curr Top Med Chem 1:529–539

Ergun S, Buschmann C, Heukeshoven J, Dammann K, Schnieders F, Lauke H et al (2004) Cell type-specific expression of LINE-1 open reading frames 1 and 2 in fetal and adult human tissues. J Biol Chem 279:27753–27763

Evrony GD, Cai X, Lee E, Hills LB, Elhosary PC, Lehmann HS et al (2012) Single-neuron sequencing analysis of L1 retrotransposition and somatic mutation in the human brain. Cell 151:483–496. doi:10.1016/j.cell.2012.09.035

Ewing AD, Kazazian HH Jr (2010) High-throughput sequencing reveals extensive variation in human-specific L1 content in individual human genomes. Genome Res 20:1262–1270. doi:10.1101/gr.106419.110

Ewing AD, Gacita A, Wood LD, Ma F, Xing D, Kim MS et al (2015) Widespread somatic L1 retrotransposition occurs early during gastrointestinal cancer evolution. Genome Res 25:1536–1545. doi:10.1101/gr.196238.115

Farkash EA, Prak ET (2006) DNA damage and l1 retrotransposition. J Biomed Biotechnol 2006:37285

Farkash EA, Kao GD, Horman SR, Prak ET (2006) Gamma radiation increases endonuclease-dependent L1 retrotransposition in a cultured cell assay. Nucleic Acids Res 34:1196–1204

Faulkner GJ, Kimura Y, Daub CO, Wani S, Plessy C, Irvine KM et al (2009) The regulated retrotransposon transcriptome of mammalian cells. Nat Genet 41:563–571. doi:10.1038/ng.368

Fedoroff NV (2012) McClintock's challenge in the 21st century. Proc Natl Acad Sci U S A 109:20200–20203. doi:10.1073/pnas.1215482109

Feng Q, Moran JV, Kazazian HH Jr, Boeke JD (1996) Human L1 retrotransposon encodes a conserved endonuclease required for retrotransposition. Cell 87:905–916

Foster PL (2007) Stress-induced mutagenesis in bacteria. Crit Rev Biochem Mol Biol 42:373–397

Garcia-Perez JL, Marchetto MCN, Muotri AR, Coufal NG, Gage FH, O'Shea KS et al (2007) LINE-1 retrotransposition in human embryonic stem cells. Hum Mol Genet 16:1569–1577

Gasior SL, Wakeman TP, Xu B, Deininger PL (2006) The human LINE-1 retrotransposon creates DNA double-strand breaks. J Mol Biol 357:1383–1393

Gasior SL, Roy-Engel AM, Deininger PL (2008) ERCC1/XPF limits L1 retrotransposition. DNA Repair (Amst) 7:983–989

Giorgi G, Marcantonio P, Del RB (2011) LINE-1 retrotransposition in human neuroblastoma cells is affected by oxidative stress. Cell Tissue Res 346:383–391. doi:10.1007/s00441-011-1289-0

Glahn F, Schmidt-Heck W, Zellmer S, Guthke R, Wiese J, Golka K et al (2008) Cadmium, cobalt and lead cause stress response, cell cycle deregulation and increased steroid as well as xenobiotic metabolism in primary normal human bronchial epithelial cells which is coordinated by at least nine transcription factors. Arch Toxicol 82:513–524

Goncalves I, Duret L, Mouchiroud D (2000) Nature and structure of human genes that generate retropseudogenes. Genome Res 10:672–678

Goodier JL, Zhang L, Vetter MR, Kazazian HH Jr (2007) LINE-1 ORF1 protein localizes in stress granules with other RNA-binding proteins, including components of RNA interference RNA-induced silencing complex. Mol Cell Biol 27:6469–6483

Goodier JL, Cheung LE, Kazazian HH Jr (2012) MOV10 RNA helicase is a potent inhibitor of retrotransposition in cells. PLoS Genet 8(10):e1002941

Goodier JL, Pereira GC, Cheung LE, Rose RJ, Kazazian HH Jr (2015) The broad-spectrum antiviral protein ZAP restricts human retrotransposition. PLoS Genet 11:e1005252. doi:10.1371/journal.pgen.1005252

Grandbastien MA (2004) Stress activation and genomic impact of plant retrotransposons. J Soc Biol 198:425–432

Gray NK, Hrabalkova L, Scanlon JP, Smith RW (2015) Poly(A)-binding proteins and mRNA localization: who rules the roost? Biochem Soc Trans 43:1277–1284. doi:10.1042/BST20150171

Gregersen LH, Schueler M, Munschauer M, Mastrobuoni G, Chen W, Kempa S et al (2014) MOV10 Is a 5′ to 3′ RNA helicase contributing to UPF1 mRNA target degradation by translocation along 3′ UTRs. Mol Cell 54:573–585. doi:10.1016/j.molcel.2014.03.017

Gu W, Zhang F, Lupski JR (2008) Mechanisms for human genomic rearrangements. Pathogenetics 1:4. doi:10.1186/1755-8417-1-4

Guleria A, Chandna S (2016) ATM kinase: much more than a DNA damage responsive protein. DNA Repair (Amst) 39:1–20. doi:10.1016/j.dnarep.2015.12.009

Habibi L, Shokrgozar MA, Motamedi M, Akrami SM (2013) Effect of heavy metals on silencing of engineered long interspersed element-1 retrotransposon in nondividing neuroblastoma cell line. Iran Biomed J 17:171–178

Habibi L, Shokrgozar MA, Tabrizi M, Modarressi MH, Akrami SM (2014) Mercury specifically induces LINE-1 activity in a human neuroblastoma cell line. Mutat Res Genet Toxicol Environ Mutagen 759:9–20. doi:10.1016/j.mrgentox.2013.07.015

Hackett JA, Surani MA (2013) DNA methylation dynamics during the mammalian life cycle. Philos Trans R Soc Lond B Biol Sci 368:20110328. doi:10.1098/rstb.2011.0328

Hagan CR, Sheffield RF, Rudin CM (2003) Human Alu element retrotransposition induced by genotoxic stress. Nat Genet 35:219–220

Hamdorf M, Idica A, Zisoulis DG, Gamelin L, Martin C, Sanders KJ et al (2015) miR-128 represses L1 retrotransposition by binding directly to L1 RNA. Nat Struct Mol Biol 22:824–831. doi:10.1038/nsmb.3090

Han K, Konkel MK, Xing J, Wang H, Lee J, Meyer TJ et al (2007) Mobile DNA in Old World monkeys: a glimpse through the rhesus macaque genome. Science 316:238–240

Han K, Lee J, Meyer TJ, Remedios P, Goodwin L, Batzer MA (2008) L1 recombination-associated deletions generate human genomic variation. Proc Natl Acad Sci U S A 105:19366–19371. doi:10.1073/pnas.0807866105

Haoudi A, Semmes OJ, Mason JM, Cannon RE (2004) Retrotransposition-competent human LINE-1 induces apoptosis in cancer cells with intact p53. J Biomed Biotechnol 2004:185–194

Harris CR, Dewan A, Zupnick A, Normart R, Gabriel A, Prives C et al (2009) p53 responsive elements in human retrotransposons. Oncogene 28:3857–3865. doi:10.1038/onc.2009.246

Hartwig A, Mullenders LH, Schlepegrell R, Kasten U, Beyersmann D (1994) Nickel(II) interferes with the incision step in nucleotide excision repair in mammalian cells. Cancer Res 54:4045–4051

Hartwig A, Asmuss M, Ehleben I, Herzer U, Kostelac D, Pelzer A et al (2002a) Interference by toxic metal ions with DNA repair processes and cell cycle control: molecular mechanisms. Environ Health Perspect 110(Suppl 5):797–799

Hartwig A, Asmuss M, Blessing H, Hoffmann S, Jahnke G, Khandelwal S et al (2002b) Interference by toxic metal ions with zinc-dependent proteins involved in maintaining genomic stability. Food Chem Toxicol 40:1179–1184

Hata K, Sakaki Y (1997) Identification of critical CpG sites for repression of L1 transcription by DNA methylation. Gene 189:227–234

Heras SR, Macias S, Plass M, Fernandez N, Cano D, Eyras E et al (2013) The Microprocessor controls the activity of mammalian retrotransposons. Nat Struct Mol Biol. doi:10.1038/nsmb.2658

Horman SR, Svoboda P, Prak ET (2006) The potential regulation of l1 mobility by RNA interference. J Biomed Biotechnol 2006:32713

Horn AV, Klawitter S, Held U, Berger A, Vasudevan AA, Bock A et al (2014) Human LINE-1 restriction by APOBEC3C is deaminase independent and mediated by an ORF1p interaction that affects LINE reverse transcriptase activity. Nucleic Acids Res 42:396–416. doi:10.1093/nar/gkt898

Hossain MB, Vahter M, Concha G, Broberg K (2012) Low-level environmental cadmium exposure is associated with DNA hypomethylation in Argentinean women. Environ Health Perspect 120:879–884. doi:10.1289/ehp.1104600

Huen K, Calafat AM, Bradman A, Yousefi P, Eskenazi B, Holland N (2016) Maternal phthalate exposure during pregnancy is associated with DNA methylation of LINE-1 and Alu repetitive elements in Mexican-American children. Environ Res 148:55–62. doi:10.1016/j.envres.2016.03.025

Intarasunanont P, Navasumrit P, Waraprasit S, Chaisatra K, Suk WA, Mahidol C et al (2012) Effects of arsenic exposure on DNA methylation in cord blood samples from newborn babies and in a human lymphoblast cell line. Environ Health 11:31. doi:10.1186/1476-069X-11-31

Iskow RC, McCabe MT, Mills RE, Torene S, Pittard WS, Neuwald AF et al (2010) Natural mutagenesis of human genomes by endogenous retrotransposons. Cell 141:1253–1261. doi:10.1016/j.cell.2010.05.020

Kaer K, Speek M (2013) Retroelements in human disease. Gene 518:231–241. doi:10.1016/j.gene.2013.01.008

Kale SP, Moore L, Deininger PL, Roy-Engel AM (2005) Heavy metals stimulate human LINE-1 retrotransposition. Int J Env Res Public Health 2:84–90

Kale SP, Carmichael MC, Harris K, Roy-Engel AM (2006) The L1 retrotranspositional stimulation by particulate and soluble cadmium exposure is independent of the generation of DNA breaks. Int J Env Res Public Health 3:121–128

Kano H, Godoy I, Courtney C, Vetter MR, Gerton GL, Ostertag EM et al (2009) L1 retrotransposition occurs mainly in embryogenesis and creates somatic mosaicism. Genes Dev 23:1303–1312. doi:10.1101/gad.1803909

Karimi A, Madjd Z, Habibi L, Akrami SM (2014a) Evaluating the extent of LINE-1 mobility following exposure to heavy metals in HepG2 cells. Biol Trace Elem Res 160:143–151. doi:10.1007/s12011-014-0015-7

Karimi A, Madjd Z, Habibi L, Akrami SM (2014b) Exposure of hepatocellular carcinoma cells to low-level As(2)O(3) causes an extra toxicity pathway via L1 retrotransposition induction. Toxicol Lett 229:111–117. doi:10.1016/j.toxlet.2014.05.025

Karimi A, Majidzadeh A, Madjd Z, Akbari A, Habibi L, Akrami SM (2015) Effect of copper sulfate on expression of endogenous L1 retrotransposons in HepG2 cells (hepatocellular carcinoma). Biol Trace Elem Res 165:131–134. doi:10.1007/s12011-015-0256-0

Kato Y, Kaneda M, Hata K, Kumaki K, Hisano M, Kohara Y et al (2007) Role of the Dnmt3 family in de novo methylation of imprinted and repetitive sequences during male germ cell development in the mouse. Hum Mol Genet 16:2272–2280. doi:10.1093/hmg/ddm179

Kazazian HH, Wong C, Youssoufian H, Scott AF, Phillips DG, Antonarakis SE (1988) Haemophilia A resulting from *de novo* insertion of L1 sequences represents a novel mechanism for mutation in man. Nature 332:164–166

Khan H, Smit A, Boissinot S (2006) Molecular evolution and tempo of amplification of human LINE-1 retrotransposons since the origin of primates. Genome Res 16:78–87

Khazina E, Truffault V, Buttner R, Schmidt S, Coles M, Weichenrieder O (2011) Trimeric structure and flexibility of the L1ORF1 protein in human L1 retrotransposition. Nat Struct Mol Biol 18:1006–1014. doi:10.1038/nsmb.2097

Kim S, Gunesdogan U, Zylicz JJ, Hackett JA, Cougot D, Bao S et al (2014) PRMT5 protects genomic integrity during global DNA demethylation in primordial germ cells and preimplantation embryos. Mol Cell 56:564–579. doi:10.1016/j.molcel.2014.10.003

Kines KJ, Sokolowski M, deHaro DL, Christian CM, Belancio VP (2014) Potential for genomic instability associated with retrotranspositionally-incompetent L1 loci. Nucleic Acids Res 42:10488–10502. doi:10.1093/nar/gku687

Kines KJ, Sokolowski M, deHaro DL, Christian CM, Baddoo M, Smither ME et al (2016) The endonuclease domain of the LINE-1 ORF2 protein can tolerate multiple mutations. Mob DNA 7:8. doi:10.1186/s13100-016-0064-x

Kinomoto M, Kanno T, Shimura M, Ishizaka Y, Kojima A, Kurata T et al (2007) All APOBEC3 family proteins differentially inhibit LINE-1 retrotransposition. Nucleic Acids Res 35:2955–2964. doi:10.1093/nar/gkm181

Kitkumthorn N, Mutirangura A (2011) Long interspersed nuclear element-1 hypomethylation in cancer: biology and clinical applications. Clin Epigenetics 2:315–330. doi:10.1007/s13148-011-0032-8

Kitkumthorn N, Tuangsintanakul T, Rattanatanyong P, Tiwawech D, Mutirangura A (2012) LINE-1 methylation in the peripheral blood mononuclear cells of cancer patients. Clin Chim Acta 413:869–874. doi:10.1016/j.cca.2012.01.024

Klawitter S, Fuchs NV, Upton KR, Munoz-Lopez M, Shukla R, Wang J et al (2016) Reprogramming triggers endogenous L1 and Alu retrotransposition in human induced pluripotent stem cells. Nat Commun 7:10286. doi:10.1038/ncomms10286

Klein CB, Frenkel K, Costa M (1991) The role of oxidative processes in metal carcinogenesis. Chem Res Toxicol 4:592–604

Kloypan C, Srisa-art M, Mutirangura A, Boonla C (2015) LINE-1 hypomethylation induced by reactive oxygen species is mediated via depletion of S-adenosylmethionine. Cell Biochem Funct 33:375–385. doi:10.1002/cbf.3124

Kochan DZ, Kovalchuk O (2015) Circadian disruption and breast cancer: an epigenetic link? Oncotarget 6:16866–16882. doi:10.18632/oncotarget.4343

Kovalenko VM, Bagnyukova TV, Sergienko OV, Bondarenko LB, Shayakhmetova GM, Matvienko AV et al (2007) Epigenetic changes in the rat livers induced by pyrazinamide treatment. Toxicol Appl Pharmacol 225:293–299. doi:10.1016/j.taap.2007.08.011

Kurosaki T, Maquat LE (2016) Nonsense-mediated mRNA decay in humans at a glance. J Cell Sci 129:461–467. doi:10.1242/jcs.181008

Lander ES, Linton LM, Birren B, Nusbaum C, Zody MC, Baldwin J et al (2001) Initial sequencing and analysis of the human genome. Nature 409:860–921

Lavie L, Maldener E, Brouha B, Meese EU, Mayer J (2004) The human L1 promoter: variable transcription initiation sites and a major impact of upstream flanking sequence on promoter activity. Genome Res 14:2253–2260. doi:10.1101/gr.2745804

Lee E, Iskow R, Yang L, Gokcumen O, Haseley P, Luquette LJ III et al (2012a) Landscape of somatic retrotransposition in human cancers. Science 337:967–971. doi:10.1126/science.1222077

Lee J, Mun S, Meyer TJ, Han K (2012b) High levels of sequence diversity in the 5′ UTRs of human-specific L1 elements. Comp Funct Genomics 2012:129416. doi:10.1155/2012/129416

Li X, Zhang J, Jia R, Cheng V, Xu X, Qiao W et al (2013) The MOV10 helicase inhibits LINE-1 mobility. J Biol Chem 288:21148–21160. doi:10.1074/jbc.M113.465856

Lin C, Yang L, Tanasa B, Hutt K, Ju BG, Ohgi K et al (2009) Nuclear receptor-induced chromosomal proximity and DNA breaks underlie specific translocations in cancer. Cell 139:1069–1083. doi:10.1016/j.cell.2009.11.030

Lu KP, Hallberg LM, Tomlinson J, Ramos KS (2000) Benzo(a)pyrene activates L1Md retrotransposon and inhibits DNA repair in vascular smooth muscle cells. Mutat Res 454:35–44

Luan DD, Korman MH, Jakubczak JL, Eickbush TH (1993) Reverse transcription of R2Bm RNA is primed by a nick at the chromosomal target site: a mechanism for non-LTR retrotransposition. Cell 72:595–605

Luteijn MJ, Ketting RF (2013) PIWI-interacting RNAs: from generation to transgenerational epigenetics. Nat Rev Genet 14:523–534. doi:10.1038/nrg3495

Luzhna L, Ilnytskyy Y, Kovalchuk O (2015) Mobilization of LINE-1 in irradiated mammary gland tissue may potentially contribute to low dose radiation-induced genomic instability. Genes Cancer 6:71–81. doi:10.18632/genesandcancer.50

Mao R, Nie H, Cai D, Zhang J, Liu H, Yan R et al (2013) Inhibition of hepatitis B virus replication by the host zinc finger antiviral protein. PLoS Pathog 9:e1003494. doi:10.1371/journal.ppat.1003494

Mathias SL, Scott AF, Kazazian HH Jr, Boeke JD, Gabriel A (1991) Reverse transcriptase encoded by a human transposable element. Science 254:1808–1810

Matlik K, Redik K (2006) Speek M (2006) L1 antisense promoter drives tissue-specific transcription of human genes. J Biomed Biotechnol. doi:10.1155/JBB/2006/71753

Maze I, Feng J, Wilkinson MB, Sun H, Shen L, Nestler EJ (2011) Cocaine dynamically regulates heterochromatin and repetitive element unsilencing in nucleus accumbens. Proc Natl Acad Sci U S A 108:3035–3040. doi:10.1073/pnas.1015483108

McCLINTOCK B (1950) The origin and behavior of mutable loci in maize. Proc Natl Acad Sci U S A 36:344–355

Meister G, Landthaler M, Peters L, Chen PY, Urlaub H, Luhrmann R et al (2005) Identification of novel argonaute-associated proteins. Curr Biol 15:2149–2155. doi:10.1016/j.cub.2005.10.048
Miao M, Zhou X, Li Y, Zhang O, Zhou Z, Li T et al (2014) LINE-1 hypomethylation in spermatozoa is associated with Bisphenol A exposure. Andrology 2:138–144. doi:10.1111/j.2047-2927.2013.00166.x
Miglino N, Roth M, Lardinois D, Sadowski C, Tamm M, Borger P (2012) Cigarette smoke inhibits lung fibroblast proliferation by translational mechanisms. Eur Respir J 39:705–711. doi:10.1183/09031936.00174310
Minakami R, Kurose K, Etoh K, Furuhata Y, Hattori M, Sakaki Y (1992) Identification of an internal cis-element essential for the human L1 transcription and a nuclear factor(s) binding to the element. Nucleic Acids Res 20:3139–3145
Miousse IR, Koturbash I (2015) The fine LINE: methylation drawing the cancer landscape. Biomed Res Int 2015:131547. doi:10.1155/2015/131547
Moldovan JB, Moran JV (2015) The zinc-finger antiviral protein ZAP inhibits LINE and Alu retrotransposition. PLoS Genet 11:e1005121. doi:10.1371/journal.pgen.1005121
Monot C, Kuciak M, Viollet S, Mir AA, Gabus C, Darlix JL et al (2013) The specificity and flexibility of l1 reverse transcription priming at imperfect T-tracts. PLoS Genet 9:e1003499. doi:10.1371/journal.pgen.1003499
Morales JF, Snow ET, Murnane JP (2002) Environmental factors affecting transcription of the human L1 retrotransposon. I. Steroid hormone-like agents. Mutagenesis 17:193–200
Morales JF, Snow ET, Murnane JP (2003) Environmental factors affecting transcription of the human L1 retrotransposon. II. Stressors. Mutagenesis 18:151–158
Morales ME, White TB, Streva VA, DeFreece CB, Hedges DJ, Deininger PL (2015a) The contribution of alu elements to mutagenic DNA double-strand break repair. PLoS Genet 11:e1005016. doi:10.1371/journal.pgen.1005016
Morales ME, Servant G, Ade C, Roy-Engel AM (2015b) Altering genomic integrity: heavy metal exposure promotes transposable element-mediated damage. Biol Trace Elem Res. doi:10.1007/s12011-015-0298-3
Moran JV, Holmes SE, Naas TP, DeBerardinis RJ, Boeke JD, Kazazian HH Jr (1996) High frequency retrotransposition in cultured mammalian cells. Cell 87:917–927
Morrish TA, Gilbert N, Myers JS, Vincent BJ, Stamato TD, Taccioli GE et al (2002) DNA repair mediated by endonuclease-independent LINE-1 retrotransposition. Nat Genet 31:159–165
Moszczynska A, Flack A, Qiu P, Muotri AR, Killinger BA (2015) Neurotoxic methamphetamine doses increase LINE-1 expression in the neurogenic zones of the adult rat brain. Sci Rep 5:14356. doi:10.1038/srep14356
Mourier T, Nielsen LP, Hansen AJ, Willerslev E (2014) Transposable elements in cancer as a by-product of stress-induced evolvability. Front Genet 5:156. doi:10.3389/fgene.2014.00156
Muckenfuss H, Hamdorf M, Held U, Perkovic M, Lower J, Cichutek K et al (2006) APOBEC3 proteins inhibit human LINE-1 retrotransposition. J Biol Chem 281:22161–22172
Muddashetty R, Khanam T, Kondrashov A, Bundman M, Iacoangeli A, Kremerskothen J et al (2002) Poly(a)-binding protein is associated with neuronal BC1 and BC200 ribonucleoprotein particles. J Mol Biol 321:433–445
Muotri AR, Chu VT, Marchetto MC, Deng W, Moran JV, Gage FH (2005) Somatic mosaicism in neuronal precursor cells mediated by L1 retrotransposition. Nature 435:903–910
Muotri AR, Marchetto MC, Coufal NG, Oefner R, Yeo G, Nakashima K et al (2010) L1 retrotransposition in neurons is modulated by MeCP2. Nature 468:443–446. doi:10.1038/nature09544
Nelson WG, Kastan MB (1994) DNA strand breaks: the DNA template alterations that trigger p53-dependent DNA damage response pathways. Mol Cell Biol 14:1815–1823
Okudaira N, Iijima K, Koyama T, Minemoto Y, Kano S, Mimori A et al (2010) Induction of long interspersed nucleotide element-1 (L1) retrotransposition by 6-formylindolo[3,2-b]carbazole (FICZ), a tryptophan photoproduct. Proc Natl Acad Sci U S A 107:18487–18492. doi:10.1073/pnas.1001252107
Okudaira N, Goto M, Yanobu-Takanashi R, Tamura M, An A, Abe Y et al (2011) Involvement of retrotransposition of long interspersed nucleotide element-1 in skin tumorigenesis induced by

7,12-dimethylbenz[a]anthracene and 12-O-tetradecanoylphorbol-13-acetate. Cancer Sci 102:2000–2006. doi:10.1111/j.1349-7006.2011.02060.x

Okudaira N, Ishizaka Y, Nishio H (2014) Retrotransposition of long interspersed element 1 induced by methamphetamine or cocaine. J Biol Chem 289:25476–25485. doi:10.1074/jbc.M114.559419

Okudaira N, Ishizaka Y, Nishio H, Sakagami H (2016) Morphine and fentanyl citrate induce retrotransposition of long interspersed element-1. In Vivo 30:113–118

Ostertag EM, Prak ET, DeBerardinis RJ, Moran JV, Kazazian HH Jr (2000) Determination of L1 retrotransposition kinetics in cultured cells. Nucleic Acids Res 28:1418–1423

Perepelitsa-Belancio V, Deininger PL (2003) RNA truncation by premature polyadenylation attenuates human mobile element activity. Nat Genet 35:363–366

Perrin D, Ballestar E, Fraga MF, Frappart L, Esteller M, Guerin JF et al (2007) Specific hypermethylation of LINE-1 elements during abnormal overgrowth and differentiation of human placenta. Oncogene 26:2518–2524

Philippe C, Vargas-Landin DB, Doucet AJ, van ED, Vera-Otarola J, Kuciak M et al. (2016) Activation of individual L1 retrotransposon instances is restricted to cell-type dependent permissive loci. Elife 5:e13926. doi:10.7554/eLife.13926

Pilsner JR, Hu H, Ettinger A, Sanchez BN, Wright RO, Cantonwine D et al (2009) Influence of prenatal lead exposure on genomic methylation of cord blood DNA. Environ Health Perspect 117:1466–1471. doi:10.1289/ehp.0800497

Pizarro JG, Cristofari G (2016) Post-transcriptional control of LINE-1 retrotransposition by cellular host factors in somatic cells. Front Cell Dev Biol 4:14. doi:10.3389/fcell.2016.00014

Planello AC, Ji J, Sharma V, Singhania R, Mbabaali F, Muller F et al (2014) Aberrant DNA methylation reprogramming during induced pluripotent stem cell generation is dependent on the choice of reprogramming factors. Cell Regen (Lond) 3:4. doi:10.1186/2045-9769-3-4

Puszyk W, Down T, Grimwade D, Chomienne C, Oakey RJ, Solomon E et al (2013) The epigenetic regulator PLZF represses L1 retrotransposition in germ and progenitor cells. Embo J 32:1941–1952. doi:10.1038/emboj.2013.118

Rangwala SH, Kazazian HH Jr (2009) The L1 retrotransposition assay: a retrospective and toolkit. Methods 49:219–226. doi:10.1016/j.ymeth.2009.04.012

Richardson SR, Narvaiza I, Planegger RA, Weitzman MD, Moran JV (2014) APOBEC3A deaminates transiently exposed single-strand DNA during LINE-1 retrotransposition. Elife 3:e02008

Rodic N, Sharma R, Sharma R, Zampella J, Dai L, Taylor MS et al (2014) Long interspersed element-1 protein expression is a hallmark of many human cancers. Am J Pathol 184:1280–1286. doi:10.1016/j.ajpath.2014.01.007

Roman-Gomez J, Jimenez-Velasco A, Agirre X, Cervantes F, Sanchez J, Garate L et al (2005) Promoter hypomethylation of the LINE-1 retrotransposable elements activates sense/antisense transcription and marks the progression of chronic myeloid leukemia. Oncogene 24:7213–7223

Roy-Engel AM (2012) A tale of an A-tail: The lifeline of a SINE. Mob Genet Elements 2:282–286. doi:10.4161/mge.23204

Roy-Engel AM, Salem AH, Oyeniran OO, Deininger L, Hedges DJ, Kilroy GE et al (2002) Active alu element "A-Tails": size does matter. Genome Res 12:1333–1344

Salas LA, Villanueva CM, Tajuddin SM, Amaral AF, Fernandez AF, Moore LE et al (2014) LINE-1 methylation in granulocyte DNA and trihalomethane exposure is associated with bladder cancer risk. Epigenetics 9:1532–1539. doi:10.4161/15592294.2014.983377

Schernhammer E, Schulmeister K (2004) Light at night and cancer risk. Photochem Photobiol 79:316–318

Scott EC, Gardner EJ, Masood A, Chuang NT, Vertino PM, Devine SE (2016) A hot L1 retrotransposon evades somatic repression and initiates human colorectal cancer. Genome Res 26:745–755. doi:10.1101/gr.201814.115

Sebat J, Lakshmi B, Troge J, Alexander J, Young J, Lundin P et al (2004) Large-scale copy number polymorphism in the human genome. Science 305:525–528. doi:10.1126/science.1098918

Servant G, Streva VA, Derbes RS, Wijetunge MI, Neeland M, White TB et al (2016) The nucleotide excision repair pathway limits L1 retrotransposition. Genetics. In press.
Servomaa K, Rytomaa T (1988) Suicidal death of rat chloroleukaemia cells by activation of the long interspersed repetitive DNA element (L1Rn). Cell Tissue Kinet 21:33–43
Servomaa K, Rytomaa T (1990) UV light and ionizing radiations cause programmed death of rat chloroleukaemia cells by inducing retropositions of a mobile DNA element (L1Rn). Int J Radiat Biol 57:331–343
Shomron N, Alberstein M, Reznik M, Ast G (2005) Stress alters the subcellular distribution of hSlu7 and thus modulates alternative splicing. J Cell Sci 118:1151–1159
Smit AF (1999) Interspersed repeats and other mementos of transposable elements in mammalian genomes. Curr Opin Genet Dev 9:657–663
Soifer HS, Zaragoza A, Peyvan M, Behlke MA, Rossi JJ (2005) A potential role for RNA interference in controlling the activity of the human LINE-1 retrotransposon. Nucleic Acids Res 33:846–856
Solyom S, Ewing AD, Rahrmann EP, Doucet TT, Nelson HH, Burns MB et al (2012) Extensive somatic L1 retrotransposition in colorectal tumors. Genome Res. doi:10.1101/gr.145235.112
Speek M (2001) Antisense promoter of human L1 retrotransposon drives transcription of adjacent cellular genes. Mol Cell Biol 21:1973–1985
Stankiewicz P, Lupski JR (2010) Structural variation in the human genome and its role in disease. Annu Rev Med 61:437–455. doi:10.1146/annurev-med-100708-204735
Stapley J, Santure AW, Dennis SR (2015) Transposable elements as agents of rapid adaptation may explain the genetic paradox of invasive species. Mol Ecol 24:2241–2252. doi:10.1111/mec.13089
Startek M, Szafranski P, Gambin T, Campbell IM, Hixson P, Shaw CA et al (2015) Genome-wide analyses of LINE-LINE-mediated nonallelic homologous recombination. Nucleic Acids Res 43:2188–2198. doi:10.1093/nar/gku1394
Streva VA, Jordan VE, Linker S, Hedges DJ, Batzer MA, Deininger PL (2015) Sequencing, identification and mapping of primed L1 elements (SIMPLE) reveals significant variation in full length L1 elements between individuals. BMC Genomics 16:220. doi:10.1186/s12864-015-1374-y
Stribinskis V, Ramos KS (2006) Activation of human long interspersed nuclear element 1 retrotransposition by benzo(a)pyrene, an ubiquitous environmental carcinogen. Cancer Res 66:2616–2620
Suter CM, Martin DI, Ward RL (2004) Hypomethylation of L1 retrotransposons in colorectal cancer and adjacent normal tissue. Int J Colorectal Dis 19:95–101. doi:10.1007/s00384-003-0539-3
Suzuki J, Yamaguchi K, Kajikawa M, Ichiyanagi K, Adachi N, Koyama H et al (2009) Genetic evidence that the non-homologous end-joining repair pathway is involved in LINE retrotransposition. PLoS Genet 5:e1000461. doi:10.1371/journal.pgen.1000461
Swergold GD (1990) Identification, characterization, and cell specificity of a human LINE-1 promoter. Mol Cell Biol 10:6718–6729
Szafranski P, Dharmadhikari AV, Brosens E, Gurha P, Kolodziejska KE, Zhishuo O et al (2013) Small noncoding differentially methylated copy-number variants, including lncRNA genes, cause a lethal lung developmental disorder. Genome Res 23:23–33. doi:10.1101/gr.141887.112
Tanaka A, Nakatani Y, Hamada N, Jinno-Oue A, Shimizu N, Wada S et al (2012) Ionising irradiation alters the dynamics of human long interspersed nuclear elements 1 (LINE1) retrotransposon. Mutagenesis 27:599–607. doi:10.1093/mutage/ges025
Taruscio D, Mantovani A (2004) Factors regulating endogenous retroviral sequences in human and mouse. Cytogenet Genome Res 105:351–362. doi:10.1159/000078208
Taylor MS, LaCava J, Mita P, Molloy KR, Huang CR, Li D et al (2013) Affinity proteomics reveals human host factors implicated in discrete stages of LINE-1 retrotransposition. Cell 155:1034–1048. doi:10.1016/j.cell.2013.10.021
Tchenio T, Casella JF, Heidmann T (2000) Members of the SRY family regulate the human LINE retrotransposons. Nucleic Acids Res 28:411–415
Temtamy SA, Aglan MS, Valencia M, Cocchi G, Pacheco M, Ashour AM et al (2008) Long interspersed nuclear element-1 (LINE1)-mediated deletion of EVC, EVC2, C4orf6, and STK32B in

Ellis-van Creveld syndrome with borderline intelligence. Hum Mutat 29:931–938. doi:10.1002/humu.20778

Teneng I, Stribinskis V, Ramos KS (2007) Context-specific regulation of LINE-1. Genes Cells 12:1101–1110

Teneng I, Montoya-Durango DE, Quertermous JL, Lacy ME, Ramos KS (2011) Reactivation of L1 retrotransposon by benzo(a)pyrene involves complex genetic and epigenetic regulation. Epigenetics 6:355–367

Terasaki N, Goodier JL, Cheung LE, Wang YJ, Kajikawa M, Kazazian HH Jr et al (2013) In vitro screening for compounds that enhance human L1 mobilization. PLoS One 8:e74629. doi:10.1371/journal.pone.0074629

Thayer RE, Singer MF, Fanning TG (1993) Undermethylation of specific LINE-1 sequences in human cells producing a LINE-1-encoded protein. Gene 133:273–277

Thirunavukkarasu C, Choudhury K, Ninichuck AJ, Choudhury I, Eliceiri GL (1988) Effect of ultraviolet light on the expression of genes for human U1 RNA. J Cell Physiol 137:55–64

Thomson T, Lin H (2009) The biogenesis and function of PIWI proteins and piRNAs: progress and prospect. Annu Rev Cell Dev Biol 25:355–376. doi:10.1146/annurev.cellbio.24.110707.175327

Trelogan SA, Martin SL (1995) Tightly regulated, developmentally specific expression of the first open reading frame from LINE-1 during mouse embryogenesis. Proc Natl Acad Sci U S A 92:1520–1524

Tryndyak VP, Kovalchuk O, Muskhelishvili L, Montgomery B, Rodriguez-Juarez R, Melnyk S et al (2007) Epigenetic reprogramming of liver cells in tamoxifen-induced rat hepatocarcinogenesis. Mol Carcinog 46:187–197. doi:10.1002/mc.20263

Tubio JM, Li Y, Ju YS, Martincorena I, Cooke SL, Tojo M et al (2014) Mobile DNA in cancer. Extensive transduction of nonrepetitive DNA mediated by L1 retrotransposition in cancer genomes. Science 345:1251343. doi:10.1126/science.1251343

van den Hurk JA, Meij IC, Seleme MC, Kano H, Nikopoulos K, Hoefsloot LH et al (2007) L1 retrotransposition can occur early in human embryonic development. Hum Mol Genet 16:1587–1592. doi:10.1093/hmg/ddm108

Vilahur N, Bustamante M, Byun HM, Fernandez MF, Santa ML, Basterrechea M et al (2014) Prenatal exposure to mixtures of xenoestrogens and repetitive element DNA methylation changes in human placenta. Environ Int 71:81–87. doi:10.1016/j.envint.2014.06.006

Virani S, Rentschler KM, Nishijo M, Ruangyuttikarn W, Swaddiwudhipong W, Basu N et al (2016) DNA methylation is differentially associated with environmental cadmium exposure based on sex and smoking status. Chemosphere 145:284–290. doi:10.1016/j.chemosphere.2015.10.123

Vissers LE, Bhatt SS, Janssen IM, Xia Z, Lalani SR, Pfundt R et al (2009) Rare pathogenic microdeletions and tandem duplications are microhomology-mediated and stimulated by local genomic architecture. Hum Mol Genet 18:3579–3593. doi:10.1093/hmg/ddp306

Waalkes MP (2000) Cadmium carcinogenesis in review. J Inorg Biochem 79:241–244

Wallace NA, Belancio VP, Faber Z, Deininger P (2010) Feedback inhibition of L1 and alu retrotransposition through altered double strand break repair kinetics. Mob DNA 1:22. doi:10.1186/1759-8753-1-22

Wang SC (2014) PCNA: a silent housekeeper or a potential therapeutic target? Trends Pharmacol Sci 35:178–186. doi:10.1016/j.tips.2014.02.004

Wang H, Xing J, Grover D, Hedges DJ, Han K, Walker JA et al (2005) SVA elements: a hominid-specific retroposon family. J Mol Biol 354:994–1007

Wangsri S, Subbalekha K, Kitkumthorn N, Mutirangura A (2012) Patterns and possible roles of LINE-1 methylation changes in smoke-exposed epithelia. PLoS One 7:e45292. doi:10.1371/journal.pone.0045292

Wei W, Gilbert N, Ooi SL, Lawler JF, Ostertag EM, Kazazian HH et al (2001) Human L1 retrotransposition: cis preference versus trans complementation. Mol Cell Biol 21:1429–1439

Weisenberger DJ, Romano LJ (1999) Cytosine methylation in a CpG sequence leads to enhanced reactivity with Benzo[a]pyrene diol epoxide that correlates with a conformational change. J Biol Chem 274:23948–23955

Wilhelm-Benartzi CS, Houseman EA, Maccani MA, Poage GM, Koestler DC, Langevin SM et al (2012) In utero exposures, infant growth, and DNA methylation of repetitive elements and developmentally related genes in human placenta. Environ Health Perspect 120:296–302. doi:10.1289/ehp.1103927

Wissing S, Munoz-Lopez M, Macia A, Yang Z, Montano M, Collins W et al (2012) Reprogramming somatic cells into iPS cells activates LINE-1 retroelement mobility. Hum Mol Genet 21:208–218. doi:10.1093/hmg/ddr455

Xuan Y, Gong D, Qi J, Han C, Deng H, Gao G (2013) ZAP inhibits murine gammaherpesvirus 68 ORF64 expression and is antagonized by RTA. J Virol 87:2735–2743. doi:10.1128/JVI.03015-12

Yang N, Kazazian HH Jr (2006) L1 retrotransposition is suppressed by endogenously encoded small interfering RNAs in human cultured cells. Nat Struct Mol Biol 13:763–771

Yang N, Zhang L, Zhang Y, Kazazian HH (2003) An important role for RUNX3 in human L1 transcription and retrotransposition. Nucleic Acids Res 31:4929–4940

Yoshida K, Miki Y (2010) The cell death machinery governed by the p53 tumor suppressor in response to DNA damage. Cancer Sci 101:831–835. doi:10.1111/j.1349-7006.2010.01488.x

Yu F, Zingler N, Schumann G, Stratling WH (2001) Methyl-CpG-binding protein 2 represses LINE-1 expression and retrotransposition but not Alu transcription. Nucleic Acids Res 29:4493–4501

Zamudio N, Bourc'his D (2010) Transposable elements in the mammalian germline: a comfortable niche or a deadly trap? Heredity (Edinb) 105:92–104. doi:10.1038/hdy.2010.53

Zeng L, Ma H, Pan S, You J, Zhang G, Yu Z et al (2016) LINE-1 gene hypomethylation and p16 gene hypermethylation in HepG2 cells induced by low-dose and long-term triclosan exposure: The role of hydroxyl group. Toxicol In Vitro 34:35–44. doi:10.1016/j.tiv.2016.03.002

Retrotransposon-Derived Regulatory Regions and Transcripts in Stemness

Zhaohui Su and Guillaume Bourque

Abbreviations

ERV	Endogenous retrovirus
ESC	Embryonic stem cell
HERVH	Human endogenous retrovirus histidine
iPSC	Induced PSC
LINE	Long interspersed nuclear element
lncRNA	Long noncoding RNA
LTR	Long terminal repeat
PSC	Pluripotent stem cell
RABS	Repeat associated binding site
SINE	Short interspersed nuclear element
TE	Transposable element
TF	Transcription factor
TFBS	Transcription factor-binding site

Z. Su
Department of Human Genetics, McGill University, Montréal, QC, Canada

G. Bourque (✉)
Department of Human Genetics, McGill University, Montréal, QC, Canada

McGill University and Génome Québec Innovation Center, Montréal, QC, Canada
e-mail: guil.bourque@mcgill.ca

G. Cristofari (ed.), *Human Retrotransposons in Health and Disease*,
DOI 10.1007/978-3-319-48344-3_8

1 Introduction

TEs have made many important contributions to mammalian gene regulation (Rebollo et al. 2012; Sundaram et al. 2014) and, because TE expansions are often lineage specific, they have helped shape the regulatory landscape of different species (Bourque et al. 2008; Bourque 2009; Jacques et al. 2013). One human cell type that sees species-specific regulation associated with TEs is the pluripotent embryonic stem cells (ESCs) (Kunarso et al. 2010). Retrotransposons in these cells have been shown to affect pluripotency through the introduction of transcription factor (TF)-binding sites (Kunarso et al. 2010). Additionally, because the regulatory machinery tries to shut down retrotransposons (Jacobs et al. 2014; Wolf et al. 2015), TE-targeting repressive mechanisms can also affect neighboring genes.

The discovery of the contributions of TEs to stemness came from studies combining unbiased genome-wide profiling of chromatin and unannotated RNA transcripts in ESCs. Some of the studies that have characterized long noncoding RNAs (lncRNAs) have associated them with TEs (Kelley and Rinn 2012; Kapusta et al. 2013) and have shown that, in some cases, they prevent the degradation of pluripotency-related transcription factor mRNAs (Wang et al. 2013). In pluripotent cells, one key role of these TE-derived lncRNAs is in transcriptional activation through mechanisms such as recruiting pluripotency-related transcription factors (Lu et al. 2014). The combination of some of these different functions relates in a feedback loop that maintains the pluripotent state (Wang et al. 2013). The hallmark retrotransposon in defining the pluripotent state in humans is the human endogenous retrovirus histidine (HERVH), which can serve as a precise marker for naive pluripotency (Lu et al. 2014; Santoni et al. 2012; Wang et al. 2014). HERVH is not the only TE of importance in human stemness and various other LINEs, SINEs, and ERVs have also been shown to be expressed in pluripotent cells (Chen and Carmichael 2009; Fuchs et al. 2013). Ultimately, these studies have changed the way we view the interaction between foreign and host DNA, especially within the transcriptomes of pluripotent cells.

Unlike previous reviews in this field that have focused on describing the broad impact of TEs on gene regulation (Rebollo et al. 2012; Bourque 2009) or on the balance between the activation and the repression of TEs (Rowe and Trono 2011; Robbez-Masson and Rowe 2015), this chapter summarizes recent advances that have shown that retrotransposons have contributed not only novel regulatory regions, but also functional noncoding RNAs and new proteins. We focus on presenting results on human and other mammalian pluripotent stem cells, because of their importance and since that is where many of these recent studies have been performed. Finally, this review also discusses the applications, implications, and questions that are raised from this newfound knowledge.

2 Pluripotency and Stemness

Pluripotent cells are cells that can differentiate into any cell whose origin is one of the three germ layers. The study of pluripotent cells plays an important role in both understanding early development and health care, for example the regrowth of damaged tissues through reprograming healthy tissues. The topic of pluripotent cells and their applications in science and medicine has been extensively reviewed (Nichols and Smith 2012; De Los Angeles et al. 2015; Ji et al. 2016; Romito and Cobellis 2016); we only introduce the basics here. Two important features define pluripotent stem cells (PSCs). First, these cells can continue to be pluripotent given the correct external stimuli. Second, with appropriate changes to these external stimuli, these cells are able to differentiate into any desired progenitor cell with an ectodermal, mesodermal, or endodermal origin, which includes almost all human cell types. PSCs include ESCs, which are acquired from the inner cell mass of the blastocyst in preimplantation embryos during the early stages of pregnancy, and induced pluripotent stem cells (iPSCs), which are produced when a differentiated cell is reprogrammed into a pluripotent cell through the introduction of various combinations of TFs, for instance, using the Yamanaka factors OCT4, SOX2, cMYC, and KLF4 (Takahashi and Yamanaka 2006), or other TFs like NANOG (Mitsui et al. 2003). Reprogramming results in a noticeable increase in the expression of certain TEs (Friedli et al. 2014; Garcia-Perez et al. 2007; Klawitter et al. 2016).

Studies have shown that the transcription profiles are highly similar between iPSCs and ESCs (Wang et al. 2014), though not all iPSC cells in culture express the same transcripts (Santoni et al. 2012). These differences can likely be attributed to factors such as heterogeneity within iPSC cultures as both naive and primed iPSCs are present (Wang et al. 2014). Some PSCs are more dedicated in their fate or "primed" than the fully pluripotent, or "naive" pluripotent cells, which—as we will see later on—can be easily identified through the contrasting contributions of some specific endogenous retroviral transcripts in their transcriptomes (Wang et al. 2014; Weinberger et al. 2016). As the scientific usage of PSCs expands, understanding the environments required to maintain a pluripotent state and to differentiate between different pluripotent cell types becomes exceedingly important.

3 TEs Have Provided Novel Regulatory Elements in ESCs

TF-binding sites (TFBS) are sometimes found in repetitive elements. In particular, it has been shown that many of the TFBS for the regulatory proteins ESR1, TP53, OCT4, SOX2, and CTCF are found in distinctive classes of TE, which are referred to as repeat-associated binding sites (RABS) (Bourque et al. 2008). For example, B2 repeats are rich in CTCF-binding sites, while ERVK repeats are rich in OCT4- and SOX2-binding sites in mouse ESCs. In human, it was also shown that the ERV families LTR10 and MER61 are highly enriched in the tumor-suppressor protein p53 TFBS (Wang et al. 2007). Moreover, certain SINE insertions have produced species-specific binding sites for CTCF (Schmidt et al. 2012). TFBS are predominantly found within the TE untranslated regions, such as long terminal repeats (LTR) of HERVs (Bourque et al. 2008; Jacques et al. 2013; Kunarso et al. 2010) and within LINEs (Kuwabara et al. 2009). This type of contribution of TEs to the regulatory landscape of a genome is summarized in Fig. 1a. Additionally, Table 1 provides a summary of the studies that have demonstrated the roles that TE play in stemness.

In addition to being associated with the binding sites of specific TFs, TEs are also associated with functional genomic features, such as promoters, enhancers, or open chromatin. Several regions containing ERV sequences have transcriptional activities characteristic of enhancer regions, namely, a bidirectional distribution of transcripts around TSSs unlike canonical TSSs with transcripts synthesized mostly downstream of it. In human, LTR7 (flanking HERVH elements) and LTR9 are responsible for the most abundant of these putative enhancer RNAs (Fort et al. 2014), which are highly specific for pluripotent cells. The majority of the long-range chromatin interactions stemming from these LTR enhancers are intrachromosomal and within 100 kb of the LTR enhancer. The affected genes have varied function, from metabolism to chromatin structure. In

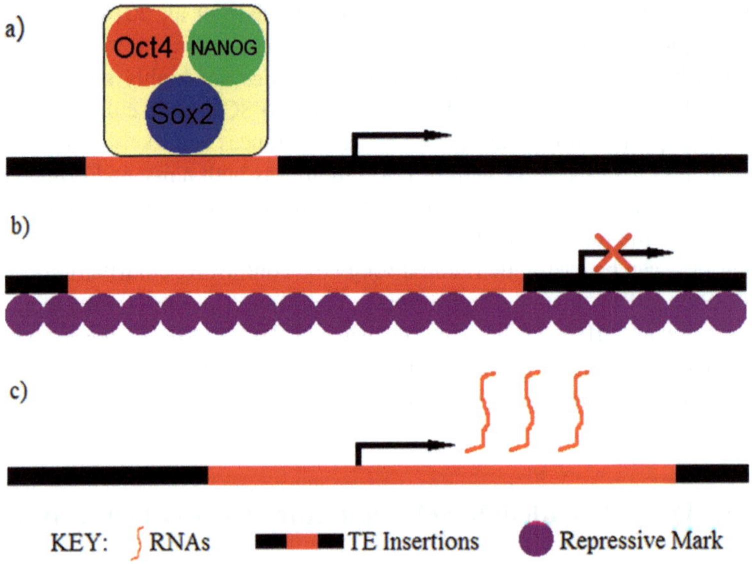

Fig. 1 Insertion of transposable elements (TEs) has affected gene regulation in several ways. (**a**) TEs have transcription factor-binding sites that activate nearby genes. (**b**) TE insertions have been silenced through mechanisms such as DNA methylation or histone modifications such as H3K9me3. This may also silence or otherwise affect expression of nearby genes. (**c**) TEs have added themselves to the host cell's transcriptome—mostly as noncoding RNAs, but some transcripts can be translated into functional protein. These two mechanisms often work together as some insertions both provide binding sites for transcription factors and are themselves transcribed

Table 1 Summary of the roles that TEs play in human pluripotent stem cells

Role of TEs	References
Regulatory elements	
– Binding site for TFs	Bourque et al. (2008), Bourque (2009), Jacques et al. (2013), Kunarso et al. (2010) and Wang et al. (2007)
– Repression	Jacobs et al. (2014), Wolf et al. (2015), Rowe and Trono (2011), Rowe et al. (2013) and Reynolds et al. (2012)
Transcripts	
– Source of lncRNAs	Kelley and Rinn (2012), Kapusta et al. (2013) and Santoni et al. (2012)
– Sources of protein	Fuchs et al. (2013), Qiu et al. (2010), Shi et al. (2015) and Parker and Sheth (2007)
Function	
– Self-regulation	Wang et al. (2013), Lu et al. (2014), Schmidt et al. (2012), Loewer et al. (2010), Guttman et al. (2011) and Qiu et al. (2010)
– Other function	Fuchs et al. (2013) and Shi et al. (2015)

ESCs, several HERVH loci are associated with open chromatin and enriched for active histone marks (H3K4me3), an enrichment not seen in other cell types nor with other repetitive elements (Kelley and Rinn 2012; Santoni et al. 2012).

Another study showed that a large portion of human open chromatin regions, determined via DNase I hypersensitivity, overlap TEs (Jacques et al. 2013). Indeed, 44 % of human open chromatin regions were found to overlap TEs, and even a larger portion (63 %) when considering primate-specific open chromatin regions. The LTR/ERV class of repetitive elements were overrepresented in cell-type-specific regions. The majority of cell-type-specific repeat-associated DNase hypersensitive sites (DHSs) were found in ESCs as well as in cancer cells. For example, 40 % of LTR7 repeats were annotated as enhancers in ESCs but only 2.2 % in other cells. The LTR9B repeat, which was observed to be bound by OCT4 and SOX2, was also overrepresented in open chromatin regions in stem cells. KLF4, another pluripotency marker, joins the two aforementioned TFs among the three most enriched motifs found in ESC-specific open chromatin sites. Though the majority of cell-type-specific DHS-associated repeats were found in ESCs and cancer cells, a few examples, such as LTR2B and MER121, were enriched in the DHSs of differentiated cell types. Altogether, these results highlight the idea that the introduction of certain repetitive sequences into the human genome has provided novel transcription factor-binding sites, many of which are specific to individual cell types of individual species.

4 TEs Have Wired New Genes in the Regulatory Network of ESCs

The lineage specificity of some of these TEs highlights the importance of RABS as a source of non-conserved regulatory elements that may explain some of the phenotypic differences observed between species. The largest contributor of repeat binding sites for OCT4 and NANOG in human is the ERV1 lineage-specific repeat family. As an example, 33 % of LTR9B repeats, a type of ERV1-associated LTR, contain an OCT4-binding site. One study showed that unlike the ubiquitous master gene expression regulator CTCF (Kim et al. 2015), TFs indicative of pluripotency, such as OCT4 and NANOG, have less conserved genome-wide binding profiles, when comparing human and mouse (Kunarso et al. 2010). About 17 % of human CTCF-binding sites have homologs in mouse, but only about 2 % of OCT4 and NANOG do; and this is despite the fact that their DNA-binding sequences themselves are highly conserved between the two species.

The same study showed that following *OCT4* knockdown in ESCs, 137 genes are downregulated in both human and mouse, half of them having OCT4-NANOG-binding sites, but only 15 % of the binding sites being conserved between the two species (Kunarso et al. 2010). The OCT4-binding regions in these human-specific OCT4-regulated genes have about twice as much overlap with repetitive elements than genes regulated by OCT4 in both human and mouse. For example, AEBP2 is a conserved gene regulated by OCT4 but has a human-specific promoter site that overlaps the insertion of a TE. One example of a gene highly expressed in human ESCs but not in mouse ESCs is *SCGB3A2*, a secretoglobin gene. The most likely contributor of this differential regulation is a human-specific ERV1 insertion, which brought new OCT4-binding sites driving *SCGB3A2* expression in human ESCs. The same study identified up to 50 genes which acquired human-specific transcription factor-binding sites originating from human-specific TE insertions, 23 of which are from an ERV1 insertion (Kunarso et al. 2010). That study concluded that many new genes have been rewired into the regulatory network of human embryonic stem cells by TEs.

5 Repression of TEs in ESCs Alters Gene Expression

Because uncontrolled transposon activity and repeated insertions could have a deleterious effect on the host, TEs are typically silenced (Rowe and Trono 2011; Slotkin and Martienssen 2007). But this repression itself can also have an impact on gene regulatory networks. The TRIM28 pathway is prevalent for this silencing in ESCs at a transcriptional level. Through the interaction with sequence-specific Krüppel-associated box zinc finger proteins (KRAB-ZFPs), TRIM28—also known as Kap1—associates with DNA to silence gene expression through the activation and/or recruitment of histone methyltransferases to trimethylate lysine 9 of histone 3, histone deacetylases such as the NuRD complex, DNA methyltransferases, and heterochromatin protein 1 (Rowe et al. 2013). Specifically, through histone methyltransferase activation, TRIM28 silences ERVs in ESCs. The knockdown of TRIM28 in ESCs results in immediate cell death or differentiation and a lack of repressive histone mark along retrotransposon loci, leading to the activation of retrotransposon-based enhancers. The loci that do not undergo TRIM28 silencing can act as activators. One way that TEs are able to escape TRIM28-mediated silencing is through mutations, such that the TRIM28 or the associated KRAB-ZFPs lose the ability to recognize them. One example of this evolutionary arms race is the evolution of ZFP91 to silence a new SVA SINE insertion, or the evolution of ZFP to silence a new L1 LINE insertion (Jacobs et al. 2014). This leads to a scenario where the newly integrated retrotransposons are not silenced because none of the KRAB-ZFP of the human repertoire has (yet) the ability to recognize them. This mechanism does not seem to target the oldest classes of retrotransposons, which are anyway inactive. Intermediately aged TE families are the ones that are the most heavily regulated through this mechanism; while DNA methylation helps silence the most recent retrotransposons (Castro-Diaz et al. 2014).

Importantly, TRIM28-mediated KRAB-ZFP silencing not only affects the targeted TEs, but also silences nearby genes. For example, upon knockdown of *TRIM28*, genes such as ZFP575, Prnp, and Serinc3 are upregulated in ESCs, likely because they are located nearby a TRIM28-regulated ERV (IAP575) (Rowe et al. 2013). As another example, the ectopic expression of ZNF91, which represses some SVA repeats, results in a 70 % inhibition of the expression of genes surrounding the SVA elements, in a 25 kb window (Jacobs et al. 2014). Interestingly, these local side effects have only been observed for ERVs and SVAs but so far not for L1 LINEs. A graphical summary of these side effects is shown in Fig. 1b and additional information is shown in Table 1.

6 Noncoding RNA Sequences are Prevalent in Human Pluripotent Stem Cells

Ever since the human transcriptome started to be analyzed globally in the early 2000s, a large amount of transcripts not matching with the known proteome have been identified. These RNAs, which some have labeled as "dark matter" RNA, can be antisense transcripts, novel protein-coding genes, or alternatively spliced isoforms of known genes (Johnson et al. 2005). Even then, the noncoding RNA can make up somewhere between 50 and 65 % of the transcripts in normal, non-diseased, cells (Kapranov et al. 2010), most of them not being annotated. Their functions and roles in gene regulation (Cabili et al. 2011), diseases (Wapinski and Chang 2011), and human pluripotency (Schmidt et al. 2012; Loewer et al. 2010; Guttman et al. 2011; Fatica and Bozzoni 2014) are just starting to be uncovered. These transcripts tend to localize in the nucleus, be less conserved, be shorter, be expressed at lower levels, and overlap TEs more than known transcripts (Fort et al. 2014). In PSCs, long noncoding RNAs (lncRNAs) affect gene expression on a global scale through trans-acting effects to maintain a pluripotent state while repressing differentiation. The expression of these lncRNAs is often promoted by core pluripotency factors mentioned previously and interacts with chromatin remodelling complexes (Guttman et al. 2011). About 300 of the lncRNAs that Guttman et al. identified are transcribed in human ESC (Tang et al. 2013). Half of them have orthologues in mouse and 78 are expressed significantly higher in ESCs than in any other cell types. Gene ontology analyses show that conserved lncRNAs are associated with processes linked to embryonic development, chromatin segregation, and ribosome biogenesis, while the human-specific lncRNAs are mostly involved in mitosis, embryonic development, and mRNA processing (Tang et al. 2013).

7 Noncoding RNA in Human Pluripotent Stem Cells are Prevalent in TEs

One additional way by which retrotransposons contribute to pluripotency gene regulatory network is through their transcription in a manner that is separate from the retrotransposon cycle (see Fig. 1c). These TE-derived loci produce mostly noncoding RNAs (ncRNAs) though some functional proteins have also been identified (Kelley and Rinn 2012; Kapusta et al. 2013), as reviewed in Robbez-Masson and Rowe (2015). Although the produced ncRNAs may vary in function, one major role is the regulation of pluripotency-related transcription factors (Wang et al. 2013; Lu et al. 2014; Qiu et al. 2010).

Many of the previously discovered functional noncoding transcripts ESCs were shown to overlap with different TEs including those involved in stem cell identity and pluripotency (Kelley and Rinn 2012; Kapusta et al. 2013). TEs may overlap with any part of a transcript, including the exons, TSS, polyadenylation sites, splicing sites, or any combination of the above. TE insertions can also help lncRNAs to form specific secondary structures (Kapusta et al. 2013). In one study, 83.4 % of all known lncRNAs were shown to overlap TEs, while the overlap of protein-coding genes with TEs was 6.2 % (Kelley and Rinn 2012). The promoters of these transcripts are located more often than expected by chance in LTR-containing subfamilies in stem cells but not in any of the differentiated cell types tested (Kelley and Rinn 2012). The level of expression of these transcripts is higher for transcripts that contain an LTR in their promoters, something that is not seen with any known gene. Unlike non-expressed repetitive elements associated with the repressive histone mark H3K9me3 (Karimi et al. 2011), these transcripts are associated with the active histone mark H3K4me3 (Kelley and Rinn 2012). Within the collection of non-annotated RNAs, LINEs and SINEs are depleted despite actually being the most prevalent in both the human genome and lncRNAs, but endogenous retroviruses are significantly enriched, for example, ERV1, ERVL-MaLR, ERVL, ERVK, and most significantly ERVH, one of the most recent endogenous retroviruses (Kelley and Rinn 2012). Kelley and Rinn also showed that seven out of the ten identified lncRNAs involved in pluripotency from a previous study had HERVH elements near their TSS (Loewer et al. 2010). A majority of these LTR-associated transcripts are not canonical full-length retrotransposons as they lack nearby open reading frames (Fort et al. 2014).

The LTR of ERVs can also initiate the formation of chimeric transcripts, and splice donor sites embedded within these retroelements can result into fusions of ERV sequences to canonical gene transcripts (Wang et al. 2014; Göke et al. 2015). This phenomenon is common in cancer and is described in details in Chap. 11 of this book. Over 80 % of these fusion transcripts have not been characterized, while the remaining are split equally between known lncRNAs and known protein-coding genes. ERV families such as MER73, THE1A, HERVH, and HERVK not only generate pure ERV transcripts, but they also link near-identical ERV-derived sequences to a vast array of non-ERV exons, further diversifying the transcriptome of many specific stages of the embryonic development.

These TE-derived transcripts are not only present in the pluripotent stem cells but in fact all throughout early development. XIST, an important conserved embryonic lncRNA involved in X chromosome inactivation in females, also overlaps with retrotransposon sequences (Elisaphenko et al. 2008).

Specific ERVs are transcribed at each developmental stage of preimplantation embryos, from the oocyte to the ESCs, but then most of them are silenced in adult cell types (Göke et al. 2015). The overall expression of ERVs rises from oocyte to zygote and decreases from zygote through the 2-, 4-, and 8-cell stages, the morula and the blastocyst, before a sharp rise in expression after the formation of ESCs where HERVH becomes the most expressed transposon-derived transcript (Göke et al. 2015). Until recently, however, functional roles for these transcripts were unknown. These roles are summarized in Fig. 2 and discussed below.

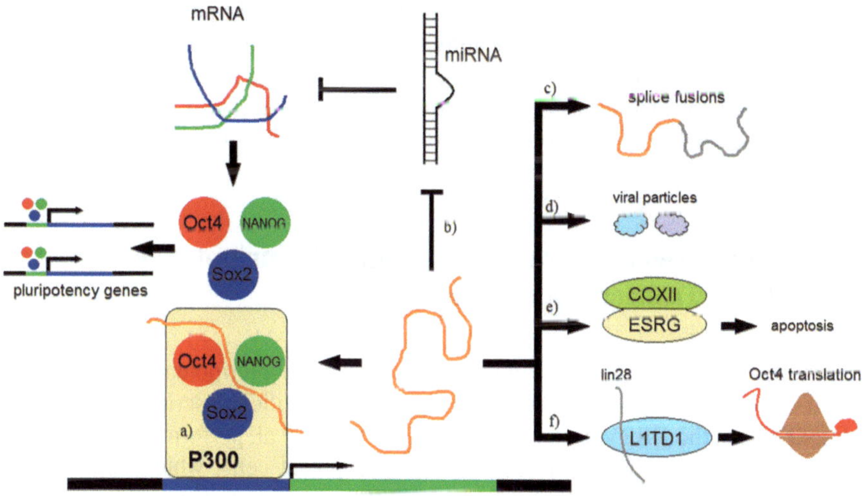

Fig. 2 Various roles of TE-derived transcripts in pluripotency. (**a**) Act as a scaffold for chromatin remodelers and transcription factors. (**b**) Repress repressors of transcription factor mRNA. (**c**) Function in alternative splicing of known genes. (**d**) Form virus-derived proteins. (**e**) Associate with apoptosis factors. (**f**) Upregulate translation of transcription factor mRNA

8 The Most Prevalent Retrotransposon-Derived Transcript in ESCs is HERVH

Through the study of the contribution of TEs and lncRNAs in ESCs, the endogenous retrovirus HERVH rose to prominence. A global annotation of total RNA in many cell types showed that ESCs and iPSCs had a significantly higher number of non-annotated (stem) transcripts (NASTs) compared to differentiated cells, both in human and mouse (Fort et al. 2014). About 40 % of NASTs show interspecies homology, compared to about 80 % for known transcripts. The non-conserved NASTs show high levels of enrichment in TEs. About 80 % of the genomic loci of these NASTs show promoter- or enhancer-related chromatin marks. This study also showed that LTR7 and its associated HERVH element were the most expressed retrotransposon-derived transcripts in ESCs. Variants of the HERVH LTR7 promoter, such as LTR7B and LTR7Y, were also highly expressed. This high level of expression is surprising despite the fact that HERVH and LTR7 families did not have the highest number of copy numbers of TEs in the human genome (Fort et al. 2014). HERVH internal genes (gag, pol, and env) are also significantly more transcribed than neighboring genomic regions at a level comparable to regular protein-coding genes. Estimates of the total contribution of HERVH-associated RNA to the human ESCs transcriptome suggest that they might represent up to 2 % of polyA+ RNA (Santoni et al. 2012). HERVH is not only the most expressed member of the ERV family but also the most expressed retrotransposon (Santoni et al. 2012). Fifty-four of all the HERVH-associated transcripts have been implicated as playing some role in pluripotency (Wang et al. 2014). HERVH-associated transcripts are specific to stem cells and are not seen in healthy differentiated cells or in many malignant cancerous cells (Lu et al. 2014; Santoni et al. 2012; Wang et al. 2014). The decrease of HERVH-associated transcripts is in line with the decrease of SOX2, NANOG, and OCT4 as embryonic stem cell differentiation progresses (Santoni et al. 2012).

9 HERVH-Derived Transcripts Control Pluripotency-Related Transcription Factors

Given the very strong association between HERVH and pluripotency through both stem-cell-specific regulatory regions and stem-cell-specific transcripts, we will now describe what the HERVH loci and transcripts do. Lu et al. performed knockdowns of genes containing HERVH sequences in pluripotent cells, which gave them a fibroblast-like appearance (Lu et al. 2014). Differentiation markers such as GATA6 and RUNX1 were upregulated, while canonical pluripotency markers such as NANOG, SOX2, and OCT4 were downregulated. These results make sense as factors of pluripotency bind to HERVH loci. NANOG binds to 96 % of the LTRs of the 50 most highly expressed HERVH elements. The other pluripotency factors OCT4 and SOX2 bind upstream of the LTR element, at about 1 kb and 2 kb, respectively, away from the 5′ end of the LTR of the HERVH sequence. Another pluripotency marker, KLF4, does not associate with HERVH (Santoni et al. 2012). ShRNA-mediated HERVH knockdown reduces not only HERVH transcript levels, but also the levels of expression of coding and noncoding genes nearby HERVH loci, suggesting that HERVH can act as an enhancer regulating surrounding genes on a long distance. RNA cross-linking and RNA immunoprecipitation assays have further shown that HERVH RNA is bound by the pluripotency transcription factors OCT4 (Lu et al. 2014), and NANOG (Wang et al. 2014), further supporting its property of self-enhancer. HERVH transcripts also associate with coactivators CBP, p300, MED6, and MED12 (Lu et al. 2014) chromatin-modifying complexes (Wang et al. 2014) but not with corepressors ESET, HDAC1, and PRC2 (Lu et al. 2014) (Fig. 2a).

A specific HERVH repeat-associated transcript is lncRNA-ES3, a putative NANOG partner, which, like previously mentioned transcripts, is also specific of ESCs and iPSCs (Schmidt et al. 2012). The lncRNA linc-RoR (long intergenic nonprotein-coding RNA, regulator of reprogramming), one of the lncRNAs that overlap HERVH, is another well-studied example of an upregulated lncRNA in iPSCs whose expression levels decrease after exposure to differentiation signals (Fig. 2b). Linc-RoR is involved in pluripotency as its knockdown reduces the size of iPSC colonies after a return to pluripotency, without effect on the source cells (Loewer et al. 2010). It was later shown that linc-RoR acts like a scaffold that binds to various micro-RNAs (miRNAs) that would normally silence pluripotency factors (Wang et al. 2013). Like other HERVH elements, linc-RoR is present in undifferentiated ESCs. During pluripotency, but not during differentiation, the pluripotency factors are present at the promoter of LincRoR. Moreover, the levels of pluripotency transcription factors and lincRoR decrease simultaneously. This suggests that linc-ROR and the pluripotency factors are co-regulated. Indeed, Linc-RoR shares the same miRNA response elements as the pluripotency transcription factors and becomes the target for these miRNAs, preventing them from targeting the pluripotent transcription factors. However, to ultimately differentiate, the levels of miRNA outclass those of Linc-RoR which turns off both the pluripotency transcription factors and Linc-RoR itself (Wang et al. 2013). This adds to the current knowledge that we have on miRNAs playing an important role in cellular differentiation (Ivey and Srivastava 2010).

Not all HERVH-derived transcripts function as noncoding genes. ESRG is a translated HERVH-derived gene that interacts with COXII, a protein that promotes apoptosis in ESCs (Shi et al. 2015). Though present in the genome of multiple closely related primate species, ESRG may only be translated in humans (Wang et al. 2014). The exons of ESRG contain some of the intronic sequences from the genes in which it originally inserted in addition to two HERVH-derived exons (Fig. 2e).

One question that has been brought up is whether the presence of TEs within lncRNAs came from the fact that TEs are inserted in preexisting lncRNAs, or whether a conglomeration of several TE insertions introduced a transcription start site and a novel lncRNA gene (Kapusta et al. 2013). For some of the most ancestral and conserved lncRNAs that contain less significant portions of nonfunctional TEs, the "lncRNA-first" model is more likely, this is the case for example for the XIST lncRNA. On the other hand, new HERVH-derived lncRNAs are more likely to have been derived from de novo insertions due to their lineage specificity.

10 Presence of HERVH Allows Differentiation of Naive and Primed Pluripotent Cells

Cultures of pluripotent cells are often heterogeneous and include cells with different levels of pluripotency. Some cells are naive, the most pluripotent state, while some are primed, i.e., able to differentiate only into specific cell types (Weinberger et al. 2016). The cellular phenotypes of naive pluripotent cells include higher levels of pluripotency-associated transcription factors, specifically LBP9 and KLF4, DNA hypomethylation, two active X chromosomes in females, and reduced levels of the repressive H3K27me3 histone mark on genes important for development (Nichols and Smith 2009). Surprisingly, the level of HERVH transcripts is able to distinguish distinct pluripotency states within these cell cultures. Using a GFP reporter gene under the control of the LTR7 of some HERVH sequences and selecting cells with the highest levels of GFP allow the isolation of colonies of naive pluripotent cells from an ESC culture (Wang et al. 2014, 2016). In contrast, cells with low GFP levels match a more differentiated state from an epigenetic and morphological point of view. However, this method is still in its infancy and imperfect. This approach of isolating pluripotent stem cells requires a significant number of in vitro stages and may not perfectly match the in vivo environment (Wang et al. 2016). Hence, Wang et al. also pointed out that in vitro-sorted cells and inner cell mass cells from preimplantation embryos differ in their respective transcriptomes. Knowledge that HERVH is specifically transcribed in the most naive pluripotent cells provides an excellent tool to isolate these highly undifferentiated cells from a culture of iPSCs or ESCs and help develop in vitro models of human development.

11 Parallel Examples Exist in Other Mammalian Species

In addition to being important in human ESCs, TE-derived lncRNAs have shown to play roles in mouse pluripotency (Dinger et al. 2008). Mice also contain a specific ERV, the MERVL, which define the two-cell state (Macfarlan et al. 2012). Not only are these elements present in the two-cell stage of early embryonic development, pluripotent cell cultures were shown to contain specific individual cells that contained the two-cell-stage gene expression, namely, lacking pluripotency transcription factors and containing MERVL. These cells are totipotent, meaning they can differentiate not only into the three germ layers but also into the placenta.

In human, the HERVH lncRNAs contain a binding motif for LBP9, a well-characterized pluripotency marker and reprogramming factor in mouse that upregulates NANOG and promotes self-renewal (Martello et al. 2013; Ye et al. 2013). In mouse, this interaction is independent of ERV regulation, but in human, the disruption of LBP9 promotes differentiation through an interaction with ERV loci (Wang et al. 2014). Either HERVH or LBP9 knockdowns are able to modify the transcriptome to a similar outcome. It has been suggested that LBP9, a transcription factor that can differ in function based on its binding partner, has been adopted in primates to guard against transposition and viral infection, and also suppresses the HERVH loci in human ESCs. Some have suggested that this allowed HERVH to be incorporated into the pluripotency interactome but, due to the bivalent nature of LBP9, allowed HERVH to be expressed (Izsvák et al. 2016).

The specific lncRNAs that have been discussed are also found in closely related primate species. The lincROR is found in all apes, while the lncRNA-ES3 is evolutionarily more recent, having been inserted after the marmoset split off from the human-orangutan common ancestor (Kapusta et al. 2013). Their functions in these primate species are not well characterized, but it is quite possible that they serve the same function as in human. Other characterized endogenous retroviruses include BERV-P in cows (Nakagawa et al. 2013), ERVs in rabbits (Heidmann et al. 2009), and ERV-Ls in elephants (Greenwood et al. 2004). As with humans, many of these ERVs have roles in the placenta (Chuong 2013). Endogenous retroviruses have entered the evolution of clades many different times in the past producing a wide range of diversity (Gifford and Tristem 2003). For example, the aforementioned ERVL sequences are present in all placental mammals, entering the evolutionary tree over 100 Ma ago (Gifford and Tristem 2003; Bénit et al. 1999). Since the specific examples of HERVH-derived regulatory sites and lncRNA are so recent in evolutionary terms, it is interesting to think about what parallel mechanisms exist not only in distantly related species that are not well characterized, but also in ancestral species where new insertions such as HERVH were not yet present. Did the transcripts from other—now silenced—retroelements play a role in maintaining pluripotency? Will a novel insertion lead to a new lncRNA that forever changes the gene regulatory network in human pluripotent cells, which may in turn drive evolution and speciation? Answering these questions may lead to a better understanding of lineage-specific developmental processes.

12 Other Proteins Derived from Retrotransposon Sequences

The viral transcripts of some other HERV insertions, like ESRG, are still competent for translation. HERV-K (HML2) is the newest human ERV (Subramanian et al. 2011). Unlike other human ERVs, HERV-K (HML2) still contains retroviral open reading frames that code for retroviral proteins. HERV-K (HML-2) transcription is induced through OCT4 binding, and thus the transcript is present in ESCs and iPSCs and silenced during differentiation. HERV-K elements also contain an LTR promoter, LTR5, which has LTR5Hs, LTR5A, and LTR5b types, the former of which corresponds to the most recent HERV in the human genome (Hanke et al. 2016). This has been shown to code for several viral proteins such as Np9, Gag, and Rec. Np9 protein comes from a mutated form of the ERV where the Rec ORF is lost but a new ORF is created (Armbruester et al. 2002). Rec and Np9 associate with MYC and NOTCH, two proteins involved in pluripotency, respectively (Fuchs et al. 2013). Rec binds to viral RNAs and promotes their nuclear export and translation, and inhibits viral infection in pluripotent cell lines. It has been suggested that this process may lead to innate antiviral responses (Grow et al. 2015). Rec also upregulates *IFITM1*, a gene that codes for a viral restriction factor. These pathways may protect embryos from both external viral infections and transposition of actively jumping retrotransposons. Interestingly, these viral particles not only likely play a role in differentiation or pluripotency maintenance, but they could also be human specific as Rec and Np9 are not encoded by rodent ERVs (Fuchs et al. 2013). It has been suggested that the HERVK could possibly be a better single marker than HERVH to distinguish between human PSCs naive and primed states, as it is also expressed specifically in the most naive pluripotent cells (Robbez-Masson and Rowe 2015). The aforementioned mouse MERVL also produces its own type of viral protein derived from the gag gene (Ribet et al. 2008) (see Fig. 2d and Table 1). Note that the role of ERV envelope proteins in placental development is addressed in Chap. 9 of this book.

13 LINEs and ALUs

Despite the focus of the field on ERV-derived regulation and gene expression in pluripotent cells, other retrotransposons also might contribute to pluripotency with related mechanisms. For example, *L1TD1*, a gene evolved from LINE-1 ORF1, is required for the self-renewal of pluripotent cells and is a marker for undifferentiated human ESCs (Wong et al. 2011). The gene lacks a promoter site for OCT4, but both SOX2 and NANOG bind to its promoter. The L1TD1 protein contains an RNA-binding domain with high homology to its ancestor ORF1 protein. L1TD1 helps maintaining P-bodies, cytoplasmic bodies containing a collection of untranslated RNAs within ribonucleoprotein complexes (Parker and Sheth 2007). L1TD1 associates with another RNA-binding protein, LIN28, through binding to a common RNA and can also complex together with RNA helicase A (Närvä et al. 2012) (Fig. 2f).

Finally, ESCs also express many different Alu elements including subfamilies of different ages, though the most highly expressed Alu RNAs are from the recent AluY and middle-aged AluS subfamilies (Macia et al. 2011). mRNAs containing inverted Alu elements are retained in the nucleus through ADAR1-mediated A-to-I RNA editing and sequestration into nuclear bodies called paraspeckles. The lncRNA NEAT1 helps forming paraspeckles and retaining such mRNAs. However, NEAT1 is not expressed in hESCs. Thus, despite the editing of these mRNA in ESCs, these transcripts are not retained in the nucleus. The example of the Lin-28 mRNA, which contains inverted Alu sequences in its 3′ UTR, and contributes to the maintenance of pluripotency in ESCs, illustrates well this process. Lin-28 mRNA is not retained in the nucleus specifically in ESCs even though the editing is active, and is translated (Chen and Carmichael 2009). Through interactions with the aforementioned L1TD1, the Lin-28 protein promotes OCT4 expression at the level of translation through binding to the OCT4 mRNA. This upregulation of OCT4 coupled with other effects, such as cell cycle modulation and suppression of the differentiation and cell growth by the let-7 miRNA, places Lin-28 as an important contributor in the reprogramming process (Qiu et al. 2010). Like HERVH, L1TD1 is also directly involved in pluripotent stem cell physiology through the upregulation of pluripotency transcription factors.

14 Perspectives

Why are retrotransposons, especially ERVs, involved in the pluripotency networks of various mammalian species? It has been proposed that, among other possibilities, it is perhaps purely by chance that TE-derived elements have been incorporated into this network (Izsvák et al. 2016). The alternative is that there is a tug-of-war between host differentiation pathways and retroviral replication pathways (Schlesinger and Goff 2015). Host organisms have given TEs a chance to escape their silencing through a global loss of repression during stemness, so it makes sense that that's also where they have been co-opted to play important roles. Interestingly, the focus of the roles that have been identified so far is to maintain the pluripotent state, which in turn continues to allow TEs to be unrepressed and attempt to replicate, opposing differentiation.

The metastability of ERV loci has been suggested as a reason why retroviruses belong so well in pluripotency networks (Schlesinger and Goff 2015). The NuRD complex is important in shutting down transcription in the absence of pluripotency factors and is regulated through the opposing action of these factors (Reynolds et al. 2012). NuRD regulates transcription by reducing gene expression levels rather than completely removing it. It has been long observed that there is transcriptional heterogeneity in a few genes within pluripotent cells, even those grown using a standard medium. Even though the majority of cells are in a naive pluripotent state, a fraction of them generally starts differentiating in the population. One mouse study showed that ERVs are among the loci regulated by NuRD and affect pluripotency and differentiation (Ramírez et al. 2006). ERVs, which are prone to variegation and subjected to stochastic silencing mechanisms, may have been recruited to pluripotency networks as facilitators of cell heterogeneity and localized differentiation (Reynolds et al. 2012).

Though much of the focus of this chapter has been on the highly researched ERV-derived functional elements, there has been some work done on the role of other repeats, such as LINEs and SINEs, in pluripotency. The evolution of new retrotransposon function has also been suggested to be caused by rapid genome rearrangements (Schlesinger and Goff 2015). Although these studies have highlighted the preponderant effects of TEs at the genomic level, research still needs to be done to uncover the specific phenotypic changes linked to this rapid evolutionary process.

15 Conclusion and Outlook

The maintenance of pluripotency in human embryonic stem cells relies on different retrotransposons, many of which are lineage specific, to provide binding sites for transcription factors and various functional transcripts. Notably, many of these TE-derived transcripts and proteins are involved in a feedback loop in which they maintain their own expression levels. These transcripts and proteins, which manage to evade highly effective repression processes, can uniquely define a naive pluripotent state and their mechanisms have been shown to be crucial to stemness. Scientists have advanced significantly from the era where much of the DNA sequences and transcripts were labeled as dark matter, to show that these entities are crucial in lineage differentiation and species specificity. Yet, it is remarkable how elements, that are classically considered to be disruptive, can integrate so smoothly into very important developmental networks. Though multiple theories have been proposed in the literature, there is still much speculation on how these retrotransposon-derived functional elements fit into the big picture of evolutionary biology and why they play such an important role in developmental biology. Ultimately, to better understand human stemness, we should characterize the developmental gene networks of different extant organisms and comprehensively assess the impact of lineage-specific retrotransposon insertions.

Acknowledgements This work was supported by a grant from the Canadian Institute for Health Research (CIHR-MOP-115090). GB is supported by the Fonds de Recherche Santé Québec (FRSQ-25348).

References

Armbruester V et al (2002) A novel gene from the human endogenous retrovirus K expressed in transformed cells. Clin Cancer Res 8(6):1800–1807

Bénit L et al (1999) ERV-L elements: a family of endogenous retrovirus-like elements active throughout the evolution of mammals. J Virol 73(4):3301–3308

Bourque G (2009) Transposable elements in gene regulation and in the evolution of vertebrate genomes. Curr Opin Genet Dev 19(6):607–612

Bourque G et al (2008) Evolution of the mammalian transcription factor binding repertoire via transposable elements. Genome Res 18(11):1752–1762

Cabili MN et al (2011) Integrative annotation of human large intergenic noncoding RNAs reveals global properties and specific subclasses. Genes Dev 25(18):1915–1927

Castro-Diaz N et al (2014) Evolutionally dynamic L1 regulation in embryonic stem cells. Genes Dev 28(13):1397–1409

Chen L-L, Carmichael GG (2009) Altered nuclear retention of mRNAs containing inverted repeats in human embryonic stem cells: functional role of a nuclear noncoding RNA. Mol Cell 35(4):467–478

Chuong EB (2013) Retroviruses facilitate the rapid evolution of the mammalian placenta. Bioessays 35(10):853–861

De Los Angeles A et al (2015) Hallmarks of pluripotency. Nature 525(7570):469–478

Dinger ME et al (2008) Long noncoding RNAs in mouse embryonic stem cell pluripotency and differentiation. Genome Res 18(9):1433–1445

Elisaphenko EA et al (2008) A dual origin of the Xist gene from a protein-coding gene and a Set of transposable elements. PLoS One 3(6):e2521

Fatica A, Bozzoni I (2014) Long non-coding RNAs: new players in cell differentiation and development. Nat Rev Genet 15(1):7–21

Fort A et al (2014) Deep transcriptome profiling of mammalian stem cells supports a regulatory role for retrotransposons in pluripotency maintenance. Nat Genet 46(6):558–566

Friedli M et al (2014) Loss of transcriptional control over endogenous retroelements during reprogramming to pluripotency. Genome Res 24(8):1251–1259

Fuchs NV et al (2013) Human endogenous retrovirus K (HML-2) RNA and protein expression is a marker for human embryonic and induced pluripotent stem cells. Retrovirology 10(1):1–6

Garcia-Perez JL et al (2007) LINE-1 retrotransposition in human embryonic stem cells. Hum Mol Genet 16(13):1569–1577

Gifford R, Tristem M (2003) The evolution, distribution and diversity of endogenous retroviruses. Virus Genes 26(3):291–315

Göke J et al (2015) Dynamic transcription of distinct classes of endogenous retroviral elements marks specific populations of early human embryonic cells. Cell Stem Cell 16(2):135–141

Greenwood AD, Englbrecht CC, MacPhee RDE (2004) Characterization of an endogenous retrovirus class in elephants and their relatives. BMC Evol Biol 4(1):1–10

Grow EJ et al (2015) Intrinsic retroviral reactivation in human preimplantation embryos and pluripotent cells. Nature 522(7555):221–225

Guttman M et al (2011) lincRNAs act in the circuitry controlling pluripotency and differentiation. Nature 477(7364):295–300

Hanke K, Hohn O, Bannert N (2016) HERV-K(HML-2), a seemingly silent subtenant—but still waters run deep. APMIS 124(1–2):67–87

Heidmann O et al (2009) Identification of an endogenous retroviral envelope gene with fusogenic activity and placenta-specific expression in the rabbit: a new “syncytin” in a third order of mammals. Retrovirology 6(1):1–11

Ivey KN, Srivastava D (2010) MicroRNAs as regulators of differentiation and cell fate decisions. Cell Stem Cell 7(1):36–41

Izsvák Z et al (2016) Pluripotency and the endogenous retrovirus HERVH: conflict or serendipity? Bioessays 38(1):109–117

Jacobs FMJ et al (2014) An evolutionary arms race between KRAB zinc-finger genes ZNF91/93 and SVA/L1 retrotransposons. Nature 516(7530):242–245

Jacques P-É, Jeyakani J, Bourque G (2013) The majority of primate-specific regulatory sequences Are derived from transposable elements. PLoS Genet 9(5):e1003504

Ji P et al (2016) Induced pluripotent stem cells: generation strategy and epigenetic mystery behind reprogramming. Stem Cells Int 2016:8415010

Johnson JM et al (2005) Dark matter in the genome: evidence of widespread transcription detected by microarray tiling experiments. Trends Genet 21(2):93–102

Kapranov P et al (2010) The majority of total nuclear-encoded non-ribosomal RNA in a human cell is ‘dark matter’ un-annotated RNA. BMC Biol 8(1):1–15

Kapusta A et al (2013) Transposable elements are major contributors to the origin, diversification, and regulation of vertebrate long noncoding RNAs. PLoS Genet 9(4):e1003470

Karimi MM et al (2011) DNA methylation and SETDB1/H3K9me3 regulate predominantly distinct sets of genes, retroelements and chimaeric transcripts in mouse ES cells. Cell Stem Cell 8(6):676–687. doi:10.1016/j.stem.2011.04.004

Kelley D, Rinn J (2012) Transposable elements reveal a stem cell-specific class of long noncoding RNAs. Genome Biol 13(11):R107
Kim S, Yu N-K, Kaang B-K (2015) CTCF as a multifunctional protein in genome regulation and gene expression. Exp Mol Med 47(6):e166
Klawitter S et al (2016) Reprogramming triggers endogenous L1 and Alu retrotransposition in human induced pluripotent stem cells. Nat Commun 7:10286
Kunarso G et al (2010) Transposable elements have rewired the core regulatory network of human embryonic stem cells. Nat Genet 42(7):631–634
Kuwabara T et al (2009) Wnt-mediated activation of NeuroD1 and retro-elements during adult neurogenesis. Nat Neurosci 12(9):1097–1105
Loewer S et al (2010) Large intergenic non-coding RNA-RoR modulates reprogramming of human induced pluripotent stem cells. Nat Genet 42(12):1113–1117
Lu X et al (2014) The retrovirus HERVH is a long noncoding RNA required for human embryonic stem cell identity. Nat Struct Mol Biol 21(4):423–425
Macfarlan TS et al (2012) Embryonic stem cell potency fluctuates with endogenous retrovirus activity. Nature 487(7405):57–63
Macia A et al (2011) Epigenetic control of retrotransposon expression in human embryonic stem cells. Mol Cell Biol 31(2):300–316
Martello G, Bertone P, Smith A (2013) Identification of the missing pluripotency mediator downstream of leukaemia inhibitory factor. EMBO J 32(19):2561–2574
Mitsui K et al (2003) The homeoprotein Nanog is required for maintenance of pluripotency in mouse epiblast and ES cells. Cell 113(5):631–642
Nakagawa S et al (2013) Dynamic evolution of endogenous retrovirus-derived genes expressed in bovine conceptuses during the period of placentation. Genome Biol Evol 5(2):296–306
Närvä E et al (2012) RNA-binding protein L1TD1 interacts with LIN28 via RNA and is required for human embryonic stem cell self-renewal and cancer cell proliferation. Stem Cells 30(3):452–460
Nichols J, Smith A (2009) Naive and primed pluripotent states. Cell Stem Cell 4(6):487–492
Nichols J, Smith A (2012) Pluripotency in the embryo and in culture. Cold Spring Harb Perspect Biol 4(8):a008128
Parker R, Sheth U (2007) P bodies and the control of mRNA translation and degradation. Mol Cell 25(5):635–646
Qiu C et al (2010) Lin28-mediated post-transcriptional regulation of Oct4 expression in human embryonic stem cells. Nucleic Acids Res 38(4):1240–1248
Ramírez MA et al (2006) Transcriptional and post-transcriptional regulation of retrotransposons IAP and MuERV-L affect pluripotency of mice ES cells. Reprod Biol Endocrinol 4(1):1–12
Rebollo R, Romanish MT, Mager DL (2012) Transposable elements: an abundant and natural source of regulatory sequences for host genes. Annu Rev Genet 46(1):21–42
Reynolds N et al (2012) NuRD suppresses pluripotency gene expression to promote transcriptional heterogeneity and lineage commitment. Cell Stem Cell 10(5):583–594
Ribet D et al (2008) Murine endogenous retrovirus MuERV-L is the progenitor of the "Orphan" epsilon viruslike particles of the early mouse embryo. J Virol 82(3):1622–1625
Robbez-Masson L, Rowe HM (2015) Retrotransposons shape species-specific embryonic stem cell gene expression. Retrovirology 12(1):1–12
Romito A, Cobellis G (2016) Pluripotent stem cells: current understanding and future directions. Stem Cells Int 2016:9451492
Rowe HM, Trono D (2011) Dynamic control of endogenous retroviruses during development. Virology 411(2):273–287
Rowe HM et al (2013) TRIM28 repression of retrotransposon-based enhancers is necessary to preserve transcriptional dynamics in embryonic stem cells. Genome Res 23(3):452–461
Santoni FA, Guerra J, Luban J (2012) HERV-H RNA is abundant in human embryonic stem cells and a precise marker for pluripotency. Retrovirology 9(1):1–15
Schlesinger S, Goff SP (2015) Retroviral transcriptional regulation and embryonic stem cells: war and peace. Mol Cell Biol 35(5):770–777

Schmidt D et al (2012) Waves of retrotransposon expansion remodel genome organization and CTCF binding in multiple mammalian lineages. Cell 148(1):335–348

Shi J et al (2015) An ESRG-interacting protein, COXII, is involved in pro-apoptosis of human embryonic stem cells. Biochem Biophys Res Commun 460(2):130–135

Slotkin RK, Martienssen R (2007) Transposable elements and the epigenetic regulation of the genome. Nat Rev Genet 8(4):272–285

Subramanian RP et al (2011) Identification, characterization, and comparative genomic distribution of the HERV-K (HML-2) group of human endogenous retroviruses. Retrovirology 8(1):1–22

Sundaram V et al (2014) Widespread contribution of transposable elements to the innovation of gene regulatory networks. Genome Res 24(12):1963–1976

Takahashi K, Yamanaka S (2006) Induction of pluripotent stem cells from mouse embryonic and adult fibroblast cultures by defined factors. Cell 126(4):663–676

Tang X et al (2013) Systematically profiling and annotating long intergenic non-coding RNAs in human embryonic stem cell. BMC Genomics 14(Suppl 5):S3

Wang T et al (2007) Species-specific endogenous retroviruses shape the transcriptional network of the human tumor suppressor protein p53. Proc Natl Acad Sci 104(47):18613–18618

Wang Y et al (2013) Endogenous miRNA Sponge lincRNA-RoR Regulates Oct4, Nanog, and Sox2 in Human Embryonic Stem Cell Self-Renewal. Dev Cell 25(1):69–80

Wang J et al (2014) Primate-specific endogenous retrovirus-driven transcription defines naive-like stem cells. Nature 516(7531):405–409

Wang J et al (2016) Isolation and cultivation of naive-like human pluripotent stem cells based on HERVH expression. Nat Protoc 11(2):327–346

Wapinski O, Chang HY (2011) Long noncoding RNAs and human disease. Trends Cell Biol 21(6):354–361

Weinberger L et al (2016) Dynamic stem cell states: naive to primed pluripotency in rodents and humans. Nat Rev Mol Cell Biol 17(3):155–169

Wolf G et al (2015) The KRAB zinc finger protein ZFP809 is required to initiate epigenetic silencing of endogenous retroviruses. Genes Dev 29(5):538–554

Wong RC-B et al (2011) L1TD1 is a marker for undifferentiated human embryonic stem cells. PLoS One 6(4):e19355

Ye S et al (2013) Embryonic stem cell self-renewal pathways converge on the transcription factor Tfcp2l1. EMBO J 32(19):2548–2560

Roles of Endogenous Retrovirus-Encoded Syncytins in Human Placentation

Caroline Toudic, Xavier Elisseeff, Adjimon Gatien Lokossou, and Benoit Barbeau

1 Introduction

The human genome sequencing project revealed that transposable elements represent 45 % of the human genome and that, among these, 8 % are LTR retrotransposons (Lander et al. 2001). This class of retrotransposons are composed of human endogenous retroviruses (human ERVs) and attest to the ancient relationship between retroviruses and humans. Indeed, human ERVs are genomic traces of ancient retroviral infections that occurred throughout human evolution. The particularity of these ancestral retroviruses is that they infected germ cells of our ancestors and, in doing so, some of the integrated proviruses have been vertically transmitted from one generation to the other over millions of years (Dewannieux and Heidmann 2013). In our genome, most human ERV sequences are degenerated or were excised from the genome after recombination events between both LTRs, leaving solo LTRs only in the genome (Lander et al. 2001; Dewannieux and Heidmann 2013; de Parseval and Heidmann 2005). However, some human ERVs remarkably possess intact open reading frames (ORFs), in particular in the envelope-coding region (de Parseval and Heidmann 2005). In fact, the search for intact env ORFs in the human genome led to the identification of 16 coding env genes that are transcriptionally active in different tissues including the placenta, where three env transcripts were particularly highly expressed (de Parseval et al. 2003). Additional studies showed

C. Toudic • X. Elisseeff • B. Barbeau (✉)
Département des Sciences Biologiques and Centre de recherche BioMed, Université du Québec à Montréal, 2080 Saint-Urbain, Montréal, Québec, Canada, H2X 3X8
e-mail: barbeau.benoit@uqam.ca

A.G. Lokossou
Département des Sciences Biologiques and Centre de recherche BioMed, Université du Québec à Montréal, 2080 Saint-Urbain, Montréal, Québec, Canada, H2X 3X8

Université d'Abomey Calavi, Cotonou, Benin

G. Cristofari (ed.), *Human Retrotransposons in Health and Disease*,
DOI 10.1007/978-3-319-48344-3_9

that two human ERV envelope proteins, called Syncytin-1 and -2, are expressed in the placenta and exert specific functions in this organ (Mi et al. 2000; Blond et al. 2000; Blaise et al. 2003).

The placenta is a transitory and complex organ with numerous functions during pregnancy. It allows exchanges of nutrients and gas between maternal and fetal circulations but also protects the fetus against pathogens and the maternal immune system. Furthermore, the placenta possesses endocrine functions and is responsible for the release of several hormones and soluble factors that are necessary for the maintenance of pregnancy (Costa 2016). The placenta derives from epithelial cells, called trophoblasts, that differentiate into villous and extravillous cytotrophoblast (CTB) cells. Extravillous cytotrophoblasts (EVCTBs) have an invasive phenotype and penetrate the maternal endometrium, whereas villous cytotrophoblasts (VCTB) are noninvasive and further differentiate into fusogenic CTs that form the syncytiotrophoblast (STB) layer of the placenta (Gerbaud and Pidoux 2015). The STB corresponds to a giant multinucleated structure that covers the chorionic villi and is in direct contact with maternal blood (Carr 1967). The STB functions as a nutrient and hormone exchange barrier and plays a major role in the establishment of the feto-maternal tolerance state by regulating the function of numerous immune cells (Muyan and Boime 1997; Lacroix et al. 2002; Munoz-Suano et al. 2011; Warning et al. 2011). The STB is in a continuous state of renewal and its regeneration depends on the fusion of underlying CTBs. Importantly, Syncytin proteins were shown to be involved in this process, as these proteins possess fusogenic capacities and induce cell fusion upon interaction with specific cellular receptors (Mi et al. 2000; Blond et al. 2000; Blaise et al. 2003; Frendo et al. 2003). Moreover, Syncytin could favor immune tolerance at the placental and systemic levels during pregnancy. Indeed, Syncytin-1 and -2 harbor an active immunosuppressive domain (ISD) and by being incorporated in placental extracellular vesicles, such as exosomes, might act distally on the various immune cell populations (Vargas et al. 2014; Lokossou et al. 2014).

The objective of this chapter is to provide an overview of the known functions of the Syncytin proteins and to provide and update on recent findings over their function in the development and function of the human placenta as well as their potential implication in various placental disorder and cancers.

2 The Human Endogenous Retroviral Proteins Syncytin-1 and Syncytin-2

2.1 Endogenous Retroviruses

Retroviruses are enveloped RNA viruses that exist under infectious forms (termed exogenous retroviruses) or endogenous forms. Endogenous retroviruses (ERVs) are ancient exogenous viruses that infected germ cells of several species ancestors, including humans, and that became endogenized during evolution. Alike retroviral sequences, ERVs can be composed of elements present in all infectious retroviruses: that is, two long terminal repeats (LTRs) that surround three essential coding genes: *gag*, *pol*, and *env* (Stoye 2012). During the life cycle of an infectious retrovirus, the viral RNA genome is reverse transcribed into a double-stranded proviral DNA by the viral reverse transcriptase (RT), which is then integrated by the viral integrase (IN) into the host genomic DNA. Once integrated, these retroviral genomic sequences, called proviruses, can be transcribed by the host machinery into genomic and coding viral RNA in order to produce new infectious viral particles. In the case of ERVs, the initial integration of proviral DNAs in the germline genome led to the Mendelian transmission of these proviruses to the descendants, and some of these retroviral sequences have been maintained throughout evolution (Dewannieux and Heidmann 2013). Interestingly, in some cases, exogenous and endogenous viruses coexist. This is the case in koalas and sheep, in which both infectious and endogenous forms of the KoRV (koala retrovirus) and JSRV (jaagsiekte sheep retrovirus) are present in some of these animal populations (Tarlinton et al. 2006; Varela et al. 2009). In the human lineage, independent and repetitive infection of germ cells by different exogenous retroviruses in the course of time gave rise to the incorporation of several ERV sequences in the human genome. Human ERVs and other retrovirus-like sequences represent up to 8 % of the human genome and are dispersed throughout the chromosomes (Lander et al. 2001). Despite the large copy number of human ERVs, most proviruses are defective and/or noncoding due to accumulation of mutations, deletions, or complete excision of viral genes after homologous recombination between LTRs (Dupressoir et al. 2012). Nevertheless, some human ERVs conserved coding sequences such as the *env* gene and three of these remnant *env* genes are capable of encoding proteins expressed in the placenta, including Syncytin-1 and -2 (de Parseval et al. 2003).

2.2 *ERV and Placentation*

In eutherian mammals, gestation is supported by a fetal-derived organ called placenta. The placenta is essential for nutrient and gas exchange between mother and fetal blood flow. Among the four eutherian clades (Afrotheria, Xenarthra, Euarchontoglires, and Laurasiatheria), different placental structures and invasion properties have been reported (Carter and Enders 2004). From a histological point of view, placentas are classified in four types according to the connection between the chorion and the endometrium: epitheliochorial, synepitheliochorial, endotheliochorial, and hemochorial. The latter type can be further subdivided into three subtypes based on the number of trophoblast cell layers: hemomonochorial, hemodichorial, and hemotrichorial (Carter and Enders 2004; Furukawa et al. 2014). In eutherian mammals, several Syncytin-related genes were successively discovered in three clades, representing all types of placental structures found in eutherian mammals: Syncytin-A and -B in Muroidea (Dupressoir et al. 2005), Syncytin-Ory1 in lagomorphs (Heidmann et al. 2009) and Syncytin-Mar1 in the Rodentia order Sciuromorpha (Redelsperger et al. 2014) (Euarchontoglires clade), Syncytin-Car1 (Cornelis et al. 2012) and Syncytin-Rum1 (Cornelis et al. 2013) (Laurasiatheria clade), and Syncytin-Ten1 (Cornelis et al. 2014) (Afrotheria clade). Although unrelated to the human Syncytins, these Syncytin-like proteins are all expressed in the placenta and harbor fusogenic capacities (Lavialle et al. 2013). In the mouse, two Syncytin proteins, Syncytin-A and -B, are functional homologues of Syncytin-1 and -2 and knockout transgenic mice are unable to support gestation (Dupressoir et al. 2009). In the human placenta, several ERVs are expressed: ERVW-1, ERVFRD-1, ERV-3, ERVV-1, ERVV-2, and ERV-Pb. However, only ERVW-1- and ERVFRD-1-encoded proteins have been shown to be biologically functional (Mi et al. 2000; Blond et al. 1999; Blaise et al. 2003, 2005; Kato et al. 1987; Kjeldbjerg et al. 2008; Esnault et al. 2013).

Initially, three human ERV sequences were found to be expressed in the human placenta, to possess intact ORF in the envelope-coding region and to code for a fusogenic env protein, namely ERV-3 (previously, known as HERV-R), ERVW-1, and ERVFRD-1. In the case of ERV-3, a stop-codon polymorphism that leads to the synthesis of a nonfunctional truncated envelope protein exists in 1 % of the Caucasian population, thus indicating that this ERV-encoded protein is not essential for placental functions (de Parseval and Heidmann 1998). In contrast, ERVW-1 and ERVFRD-1 endogenous retroviral sequences express fully functional envelope glycoproteins with conserved fusogenic properties, renamed Syncytin-1 and Syncytin-2 (Mi et al. 2000; Blond et al. 2000; Blaise et al. 2003). These particular human ERV copies are found on the seventh and the sixth human chromosomes, respectively, and, while the gag and pol genes have been rendered inactive during evolution, both express a glycosylated envelope glycoprotein with features typical of other known retrovirus envelope proteins. Syncytin-1 and -2 are thus encoded by these preserved *env* genes originating from past retroviral infections dating back to 25 and 40 millions of years, respectively (Blaise et al. 2003; Voisset et al. 1999). As these conserved *env* genes are highly expressed in the placenta, their early identification suggested that they could play a role in placental function and/or development (Mi et al. 2000; Blond et al. 2000; Blaise et al. 2003).

2.3 Structural Organization and Functional Domains of Syncytin Proteins

Syncytin proteins demonstrate a structure, which resembles exogenous retroviral envelope proteins. The function of Syncytin proteins in the human placenta is linked to three functional domains: the fusion peptide, the receptor-binding domain, and the immunosuppressive domain (Fig. 1). Syncytin-1 and -2 are two 538-amino acid-long proteins, synthesized as 73–75 kDa precursors (Mi et al. 2000; Blond et al. 2000; Blaise et al. 2003). They both possess a surface (SU) and a transmembrane (TM) subunit and are expressed as homotrimers at the cell plasma membrane

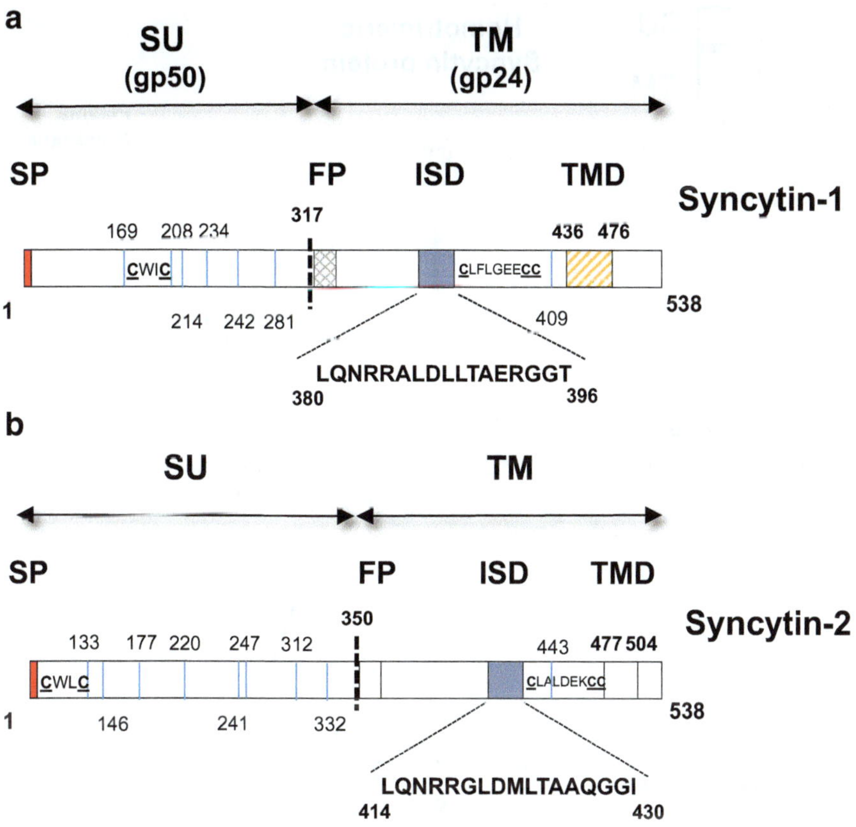

Fig. 1 Structural organization of Syncytin 1 and Syncytin-2. Human Syncytin-1 (**a**) and Syncytin-2 (**b**) are 538 amino acid long and are depicted with their various motifs and domains. The surface (SU) and transmembrane (TM) subunits are indicated with *thick black arrows*. The furin cleavage sites are marked by a *thick discontinuous line* with the position of the site indicated in *bold above*. The signal peptide (SP) is represented as a *red plain box*, the fusion peptide (FP) as a *grey crossed box*, the immunosuppressive domain (ISD) as a *plain blue box*, and the transmembrane domain (TMD) as an *orange-striped box*. Of note, both FP and ISD are included in the ectodomain, which is referred to the segment of the TM subunit, which is exposed extracellularly (amino acids 317–436 for Syncytin-1 and 350–477 for Syncytin-2). The sequence for each ISD is shown in *bold capital letters* with amino acid positions being indicated on each side. Potential sites of glycosylation are indicated by *blue vertical lines* along with their amino acid position. The CX_2C and CX_7C sequences of interaction between the SU and TM subunits are indicated in *small capital letters* inside the SU and TM subunits

(Chang et al. 2004; Chen et al. 2008) (Figs. 1 and 2). The SU subunit is responsible for recognition and binding to the cellular receptor (see below) and thereby possesses a receptor-binding domain, which has been tentatively positioned between amino 1 and 124 for Syncytin-1 (Cheynet et al. 2006). The TM subunit is rather involved in the immunosuppressive and fusogenic functions attributed to retroviral envelope proteins and contains three domains: the fusion peptide (FP) and the

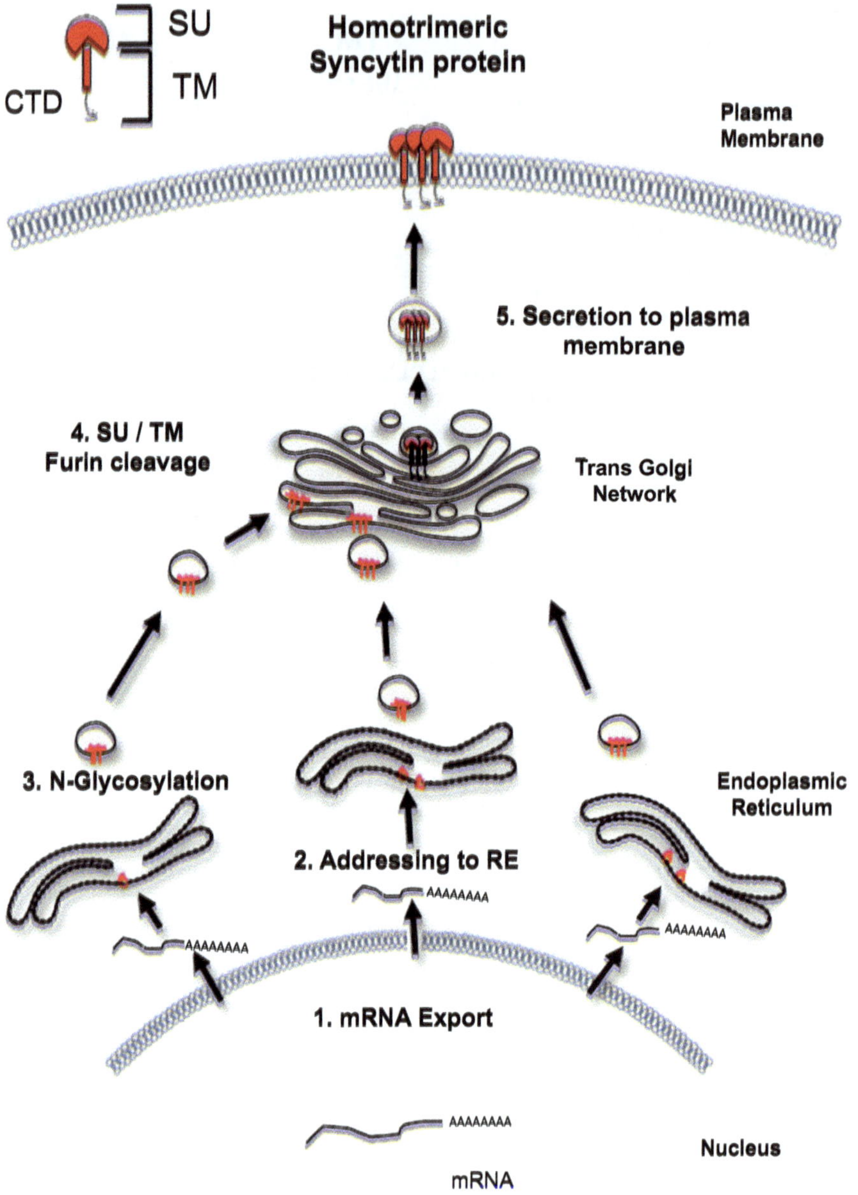

Fig. 2 Processing of Syncytin proteins and their addressing to the plasma membrane. In this figure, Syncytin proteins are represented in *red*. A monomeric protein is composed of the surface (SU) and transmembrane (TM) subunits. The intracellular part of the protein corresponds to the C-terminal domain (CTD) in this figure. Following transcription, Syncytin mRNA are exported in the cytoplasm (*1*) and translated. The presence of a signal peptide addresses the Syncytins to the endoplasmic reticulum (RE) (*2*). In the RE, Syncytins form homotrimeric structures and are addressed to the trans-Golgi network (TGN). During this process, proteins are glycosylated (*3*). The SU and TM moieties are then cleaved by furin in the TGN (*4*). This cleavage is necessary to obtain mature proteins. Finally, Syncytins are routed to the plasma membrane (*5*), where they will be accessible to interact with their specific receptors and induce fusion

immunosuppressive domain (ISD), both being part of the external ectodomain of this subunit, and the transmembrane domain (TMD), anchoring envelope proteins to the plasma membrane (Fig. 1). Both SU and TM subunits are held together by disulfide bonds and interact over two sequences: CX_2C in the SU and CX_7C in the TM.

Similar to exogenous envelope glycoproteins, Syncytins are synthesized as full-length precursors, following a canonical secretory pathway. They are initially addressed to the endoplasmic reticulum where the signal peptide is cleaved and are embedded in the membrane to form homotrimers (Cheynet et al. 2005) (Fig. 2). Syncytin proteins then pass through the trans-Golgi network where the endoprotease Furin cleaves the precursors (at a specific cleavage site matching the consensus sequence RXXR) into two different subunits, after which both subunits are glycosylated and addressed to the plasma membrane. Importantly, all Syncytin-like proteins described to date have the same structural organization, i.e., SU and TM subunits, separated by an RXXR cleavage site; the TM subunits containing an FP motif, an ISD, and a transmembrane domain (Dupressoir et al. 2011; Redelsperger et al. 2014; Cornelis et al. 2012, 2014, 2015).

2.3.1 ERV-W and Syncytin-1

ERV-W is a multicopy family of ERVs that was characterized in 1999, with only one member that has conserved an intact envelope open reading frame with a coding potential (Blond et al. 1999). In 2000, two successive studies characterized this envelope glycoprotein. Mi and coworkers first identified ERV-W transcripts in testis, brain, and placental tissues by Northern blot analyses and localized the only protein-coding human ERV gene, *ERVWE1*, on chromosome 7 (position 7q21–7q22). Finally, in situ hybridization and cell-cell fusion assays showed that Syncytin-1 expression seems restricted to STB and possesses fusogenic properties (Mi et al. 2000). The fusogenic capacity of Syncytin-1 was later confirmed in a subsequent study, which also showed that, despite a stronger Syncytin-1 signal in STB, the protein could also be detected in CTB cells (Blond et al. 2000). Moreover, in this report, SLC1A5 (previously known as ASCT2 or type D mammalian retrovirus receptor) was identified as the cellular receptor of Syncytin-1, thereby mediating fusion upon interaction with this ERV protein (Blond et al. 2000). SCL1A4 (also known as ASCT1) has also been identified as another receptor (Lavillette et al. 2002). Both SCL1A4 and SCL1A5 are amino acid transporters. Further characterization showed that the fusogenicity of Syncytin-1 is similar to that of other retroviral envelope glycoproteins and depends on two heptad repeats, HRA and HRB, localized in the ectodomain of the TM subunit between the FP and TMD regions (Chang et al. 2004).

The *ERVWE1* locus is composed of 5′ and 3′ LTRs surrounding the Env open reading frame. The 5′ LTR region contains the transcription start site, a core promoter with a CAAT box, several transcription factor-binding sites, and two enhancer regions that specifically regulate Syncytin-1 expression in trophoblasts (Cheng et al. 2004; Cheng and Handwerger 2005). Syncytin-1 expression is induced by cAMP in trophoblastic cells and GCM1a is an essential transcription factor, potentially implicating other factors, such as Sp1 and GATA2 and -3 (Prudhomme et al. 2004; Yu et al. 2002; Knerr et al. 2004). Importantly, regulation of *ERVWE1* gene expression depends on DNA methylation as well. Indeed, in the placenta, CpG islands present in the 5′ LTR promoter region are hypomethylated while being hypermethylated in other tissues (Matouskova et al. 2006; Huang et al. 2014a). Also, abnormal increased Syncytin-1 expression is associated with hypomethylation of its promoter and reduced binding of H3K9m3, as opposed to conditions in which its gene is repressed (Yu et al. 2014; Lu et al. 2015; Zhuang et al. 2014; Li et al. 2014).

2.3.2 ERVFRD and Syncytin-2

Syncytin-2 was first described in 2003 as the envelope glycoprotein encoded by the human *ERVFRD-1* locus. The ERVFRD family contains several copies, but only one sequence codes for a functional protein (de Parseval et al. 2003). *ERVFRD-1* is located on chromosome 6 and was acquired around 40 million years ago. The encoded Syncytin-2 protein is placenta specific, is expressed in the syncytiotrophoblast, and demonstrates fusogenic properties (Blaise et al. 2003; Chen et al. 2008; Cui et al. 2016). Syncytin-2 has a similar structure to Syncytin-1, being first synthesized as a 75 kDa precursor (Gp75) and then cleaved into the active TM and SU subunits by furin. Both subunits interact via disulfide bonds and assemble as a homotrimeric structure at the plasma membrane. As expected for a retroviral envelope protein, the SU subunit binds to its receptor, which has been identified as MFSD2a, a lipid membrane transporter (Esnault et al. 2008). The TM subunit harbors typical functional domains, such as the fusion peptide (FP) and the ISD (Fig. 1b). In addition to the FP, the fusogenic capacity of the protein seems to depend on a 54-amino acid-long ectodomain, which has functional similarities with other retroviruses (Renard et al. 2005), and on its C-terminal domain (Chen et al. 2008). In normal placenta, Syncytin-2 levels increase from the first to the third trimester (Chen et al. 2008) and similarly augment during syncytiotrophoblast differentiation (Vargas et al. 2009). Interestingly, in preeclampsia, Syncytin-2 expression is drastically reduced in placental tissue (Chen et al. 2008).

ERVFRD-1 gene expression is activated by cAMP signaling in trophoblastic-like fusogenic cells as shown by pharmacological approaches, such as treatment with forskolin, an activator of adenylyl cyclase, or inhibition of protein tyrosine phosphatase activity (Vargas et al. 2009; Chen et al. 2008). In non-stimulated cells, Syncytin-2 is generally very poorly expressed when compared to Syncytin-1. A number of studies have looked at the promoter region of the Syncytin-2 gene, and, upon testing of the 5′ LTR, revealed that the GCM1 transcription factor was implicated in its modulation but that, most importantly, CREB2 and JunD were upregulating Syncytin-2 expression via the binding of a CRE/AP-1-like motif (Liang et al. 2010; Toufaily et al. 2015). In non-trophoblastic cells, Syncytin-2 expression is also regulated by epigenetic and posttranscriptional mechanisms, involving CpG methylation, H3K9 trimethylation, and alternative splicing (Trejbalova et al. 2011; Liang et al. 2010). Of note, comparable mechanisms of regulation of the inducible Syncytin-2 receptor gene, MFSD2a, seem to be operating in trophoblast cells, including the involvement of identical transcription factors and promoter methylation (Liang et al. 2010).

2.4 *Syncytins and the Formation of the Placental Syncytiotrophoblast Layer*

2.4.1 Different Types of Syncytiotrophoblast Structures in Mammals

In eutherian mammals, different types of placental structures exist and are distinguished through histological observations (Carter and Enders 2004). The human placenta has a hemomonochorial structure, with direct contact of the placental villi with maternal blood, and an outer syncytiotrophoblast layer that overlays the inner cytotrophoblast layer. Hemodi or -trichorial placentas, such as in rabbits and rodents, possess two or three layers of trophoblastic cells, respectively (Metz et al. 1978; Furukawa et al. 2014). The second type of placenta is the endotheliochorial type, which shows less invasion of maternal tissues. In endotheliochorial placentas, an STB layer is present but, instead of being in direct contact with the maternal blood, this cellular structure comes in contact with the endothelium of maternal vessels (Furukawa et al. 2014). Finally, in epitheliochorial placentas, maternal and fetal blood are not in direct contact and are separated by six layers of tissues. In these placentas, there is no STB layer but, in the syndesmochorial type, such as represented in ruminants, cell fusion events occur and give rise to trinucleated or multinucleated cells (Furukawa et al. 2014; Cornelis et al. 2013).

2.4.2 Role of Syncytins in STB formation

In the early steps of their characterization, Syncytin-1 and -2 were both described as human placenta-specific and fusogenic proteins (Mi et al. 2000; Blond et al. 2000; Blaise et al. 2003). The work of Frendo and coworkers soon clearly established that Syncytin-1 had a direct role in STB formation, mediating cytotrophoblast fusion (Frendo et al. 2003). Thus Syncytin-1 was considered as the major STB-forming factor during pregnancy. Syncytin-2 has initially been considered to play a different role from Syncytin-1 in the placenta because of its distinct expression pattern. Indeed, Malassine et al. showed that Syncytin-2 is only expressed in villous CTBs, whereas Syncytin-1 is expressed in villous and extravillous CTBs. Furthermore, Syncytin-2 expression is restricted to cytotrophoblasts, showing a membranous localization in these cells, at the interface between CTB and STB, whereas Syncytin-1 is mostly expressed in the STB layer (Malassine et al. 2007; Blond et al. 2000). However, our team later presented evidence that Syncytin-2 could play a determinant role in STB formation (Vargas et al. 2009). Indeed, the specific siRNA-mediated inhibition of Syncytin-1 and -2 expression in the choriocarcinoma cell line BeWo and in CTBs revealed a more pronounced inhibition of cell fusion when Syncytin-2 was targeted as compared to Syncytin-1 (Vargas et al. 2009). The demonstrated role of MFSD2a in trophoblast fusion and its restricted expression to the STB layer further supported Syncytin-2 as a major contributor of CTB fusion (Esnault et al. 2008; Toufaily et al. 2013; Chen et al. 2006). It is also important to note that the expression of the Syncytin-2 receptor is reduced in severe preeclamptic placentas whereas ASCT2 (SLC1A5) is seemingly not affected (Toufaily et al. 2013). Thus, the interaction between Syncytin-2 and MFSD2a indeed seems to play a critical role in the STB formation. In summary, Syncytin-1 and -2 are retrovirus-derived sequences, captured independently in the course of primate evolution. They exhibit partially overlapping functions, being important for both the formation and renewal of STB, although Syncytin-2 might be more critical for cytotrophoblast fusion.

Other Syncytin proteins have been implicated in fusion processes associated with the formation of multinucleated cell layers in other eutherian mammals. For several years, Heidmann's group has focused on the identification of other Syncytin or Syncytin-like proteins in these mammal species in order to establish a possible link between Syncytin capture and placentation (Dupressoir et al. 2012). After the human Syncytins, the first Syncytin-like *env* gene to be discovered was murine *syncytin-A* and *-B* (Dupressoir et al. 2005). Both genes encode functional fusogenic proteins and were thereby named Syncytin-A and -B. These proteins show the characteristic structure of envelope glycoproteins, are specifically expressed in the placenta (hemotrichorial type), and are localized in the labyrinthine zona, in the two STB layers (Dupressoir et al. 2005). Interestingly, double-knockout mice led to 100 % fetal death, while depletion of either gene impaired the formation of the STB layers (Dupressoir et al. 2009, 2011). Interestingly, Syncytin-Ory1, Syncytin-Mar1, and Syncytin-Ten1 were added to the list of fusogenic proteins with a retroviral origin, implicated in the formation of the STB layer of other orders demonstrating a hemochorial organization (Heidmann et al. 2009; Redelsperger et al. 2014; Cornelis et al. 2014). In the Carnivora order, Syncytin-Car1 was identified as the envelope glycoprotein from a retroviral element acquired more than 60 Ma ago by its Carnivora ancestor. In the feline and canine placentas, which has an endotheliochorial structure, Syncytin-Car1 is expressed at the junctional zone and more specifically in the STB layer close to maternal vessels (Cornelis et al. 2012). In the ruminant order, showing a placenta with a poorly invasive synepitheliochorial structure, the ERV *syncytin-rum1* envelope protein is expressed in binucleated trophoblastic cells before their fusion with maternal mononucleated cells and allows the formation of multinucleated structures, in which the protein is no longer expressed (Cornelis et al. 2013). Finally, in the marsupial order, which is not included in eutherian mammals and therefore presents a very different placentation process, *Syncytin-Opo1* was recently identified as a Syncytin-like gene coding for a fusogenic protein in the syncytial part of the placenta (Cornelis et al. 2015).

2.5 Syncytins, Exosomes, and Immunosuppressive Functions

A number of recent studies have demonstrated that Syncytin proteins are present at the surface of varying types of extracellular vesicles, including exosomes. Exosomes are 30–100 nm extracellular vesicles, which play important roles in intercellular communication (Denzer et al. 2000a; Mathivanan et al. 2010). Exosomes originate in the endosomal compartment following internal invagination of the membrane of early endosomes, which forms intraluminal vesicles (ILVs) resulting in multivesicular bodies. The process of ILV formation requires several factors (ESCRT-dependent and -independent Rab GTPase) (Babst 2011). Exosomes are released from numerous cell types, including immune cells (Denzer et al. 2000b; Raposo et al. 1996), stem cells (Lai et al. 2010; Sahoo et al. 2011), placental cells (Vargas et al. 2014; Luo et al. 2009; Atay et al. 2011), or cancerous cells (Sun and Liu 2014). During pregnancy, placental exosomes are released with other extracellular vesicles from syncytiotrophoblasts and carry specific placental markers, such as the placental alkaline phosphatase (PLAP) enzyme (Mincheva-Nilsson and Baranov 2010). It has been suggested that the number of placental exosomes increases during pregnancy and that these exosomes could have specific functions during gestation, including materno-fetal communication, cell migration, vascular remodelling, and immune functions (Salomon et al. 2014; Mitchell et al. 2015; Hedlund et al. 2009; Stenqvist et al. 2013). Interestingly, Syncytin-1 and -2 are present at the surface of exosomes isolated from culture media of villous cytotrophoblasts. Furthermore, in preeclamptic patients, the level of Syncytin proteins in the placenta, which is reduced compared to unaffected women, is reflected by the change in their abundance on the surface of circulating placental exosomes (Vargas et al. 2014).

The incorporation of Syncytin protein in exosomes bears a very interesting implication over one of their potential functions. Indeed, during pregnancy, an important immune balance state has to be maintained to allow the growth and development of a semi-allogenic fetus. Several lines of evidence have shown that the ISD domain of Syncytins could be involved in this state of immunosuppression. The study of Mangeney et al. first showed that human and murine Syncytin proteins harbor immunosuppressive functions. Results from tumor rejection assays suggested that the human Syncytin-2, together with three other human ERV envelope glycoproteins, namely ERV-3 env, EnvV2, and EnvP(b), and the murine Syncytin-B were immunosuppressive, whereas Syncytin-1 and -A were not (Mangeney et al. 2007). For human Syncytins, recent works from our group and others came to a different conclusion regarding Syncytin-1. Indeed, in vitro analysis of Syncytin-1 immunosuppressive activity using either a Syncytin-1 recombinant ectodomain or a synthetic peptide corresponding to the ISD domain showed that Syncytin-1 inhibits the production and release of Th1 cytokines (TNF-α and IFN-γ) and of the CXCL10 chemokine by human blood cells (Tolosa et al. 2012). A more recent study suggested that Syncytin-1 and -2 could indirectly modify T cell response, through the conditioning of dendritic cells (DCs) (Hummel et al. 2015). In this study, Syncytin-1- and -2-expressing cells were cocultured with T cells or DCs. Syncytin-1 or -2-expressing CHO cells cocultured with T cells promoted the production of pro-inflammatory cytokines (IL-12 and TNF-α), and Syncytin-expressing choriocarcinoma cells stimulated the release of anti-inflammatory cytokines (mostly IL-10) when incubated in the presence of T cells. Furthermore, conditioning DCs with culture media of human ERV-expressing cells, before initiating coculture of DCs with T cells, resulted in reduced T cell activation. Hence, the association of Syncytin proteins to exosomes could lead to a more systemic modulation of the immune response by these placental proteins. Indeed, Holder et al. showed that Syncytin-1-expressing BeWo microvesicles dampen PBMC activation and that this effect was Syncytin-1 dependent (Holder et al. 2012b). Our unpublished data further suggests that exosome-associated Syncytin-2 also downmodulates Th1 response of T lymphocytes (Lokossou et al., unpublished results).

Hence, human ERV envelope proteins, specifically Syncytin-1 and -2, are likely implicated in the state of feto-maternal tolerance prevailing in pregnant women and could be mediated by their presence at the surface of various populations of trophoblast or derived microvesicles. Their implication in maintaining a maternal immune balance might thereby be connected to pregnancy-related disorders associated with reduced expression of Syncytin proteins.

3 Syncytins and Pathologies

3.1 Syncytins and Placental Pathogenesis

3.1.1 Preeclampsia

Preeclampsia (PE) is a life-threatening disease that concerns 5–8 % of pregnant women worldwide and in fact accounts for around 12 % of maternal mortality in certain countries (Duley 2009). Preeclampsia is characterized and often diagnosed by two symptoms that appear near the 20th week of gestation, i.e., recurrent/permanent hypertension (values >140/90 mmHg for several measurements separately taken 4–6 h apart) and proteinuria (>300 mg of proteins in urine in 24 h) (Sibai et al. 2005). The only known cure for PE is delivery, which often leads to early preterm birth and fetal mortality (Duley 2009). Preeclampsia is a multimodal disease, to which is associated several risk factors such as obesity, diabetes, hypertension, and ethnic and genetic factors. Despite intensive research efforts, the cause of this disorder remains unknown (Walker 2000; Sibai et al. 2005).

PE is associated with placental dysfunction and chronic inflammation. At the placental level, PE is characterized by poor vascularization due to poor extravillous trophoblast invasion and moderate remodelling of maternal spiral arteries (Fischer 2015). In addition, preeclamptic placentas demonstrate a reduced syncytiotrophoblast layer. This decrease in the size of the layer is believed to be caused by reduced trophoblast cell fusion associated to a concomitant decrease in Syncytin expression. The initial study of Lee et al. demonstrated a reduced expression of Syncytin-1 in preeclamptic placentas by in situ hybridization and further suggested that Syncytin-1 proteins were abnormally localized at the apical syncytiotrophoblast membrane in preeclamptic placentas instead of being expressed at the syncytiotrophoblast-cytotrophoblast interface (Lee et al. 2001). In subsequent studies, lower Syncytin-1 expression was confirmed in PE placental tissues at the mRNA and protein levels (Chen et al. 2006; Vargas et al. 2011; Zhuang et al. 2014), although one study did not confirm reduced Syncytin-1 levels in PE placentas (Holder et al. 2012a). Of interest, Kudo et al. associated hypoxic conditions with downexpression of Syncytin-1 and its cellular receptor, SLC1A5 (ASCT2), thus leading to decreased cell-cell fusion events at least in the trophoblastic-like BeWo cell line (Kudo et al. 2003). Hypoxic conditions can also alter Syncytin-1 expression in cultured cytotrophoblasts, with a significant decrease in mRNA levels after 12 h of culture under hypoxic conditions (Chen et al. 2006). The reduced Syncytin-1 expression levels under hypoxic conditions might be explained by transcriptional deregulation as hypoxia modifies the PI3K-Akt signaling pathway, leading to the degradation of the glial cell missing-1 (GCM-1) transcription factor (Chiang et al. 2009; Wich et al. 2009). Recently, Huang and coworkers also correlated Syncytin-1 low expression levels with the apoptotic rates observed in trophoblasts in PE placentas. Syncytin-1 expression was inversely correlated with apoptosis-inducing factor (AIF) and Calpain1 in PE placenta. These results were confirmed in BeWo cells using siRNA against Syncytin-1 to mimic the typical expression profile of Syncytin-1 in PE (Huang et al. 2014b).

Although the majority of studies converge toward the importance of Syncytin-1 in cytotrophoblast fusion and that its reduced expression in PE might explain the decrease of syncytialization observed in PE placentas, the role of Syncytin-2 must not be underestimated. Indeed, Chen et al. and Kudaka et al. showed that Syncytin-2 expression was significantly reduced in PE and hypertensive placentas compared to controls (Chen et al. 2008; Kudaka et al. 2008). We have also reported that both Syncytins are reduced at the mRNA and protein levels in PE placentas compared to controls and that the downregulation of Syncytin expression correlated with the severity of PE symptoms (Vargas et al. 2011). Moreover, our results also suggest that Syncytin-2 expression is more affected than Syncytin-1 in PE placentas. An additional interesting aspect of PE relates to its association with increased inflammation and immune reactivity of the maternal immune system against the fetal antigens (Borzychowski et al. 2006; Raghupathy 2013; Taufield et al. 1983; Vlkova et al. 2010). In this respect, the reduction in the levels of Syncytin-1 and -2 at the surface of trophoblasts and derived exosomes could also be linked to an inadequate immune response prevailing in PE patients.

Preeclampsia diagnosis is still dependent on the monitoring of blood pressure and metabolic symptoms and no early diagnosis currently exists. Clinical diagnosis can also be ambiguous as moderate PE can be asymptomatic (Duley 2009). Because Syncytin expression is significantly reduced in PE and reduced abundance is reflected in circulating placental exosomes (Vargas et al. 2014), monitoring exosome-associated Syncytin proteins (and most specifically, Syncytin-2) during the early stages of pregnancy could allow the early identification of pregnant women predisposed to develop severe PE. As depicted in Fig. 3, early testing could be accomplished through the isolation of exosomes from standard blood samples har-

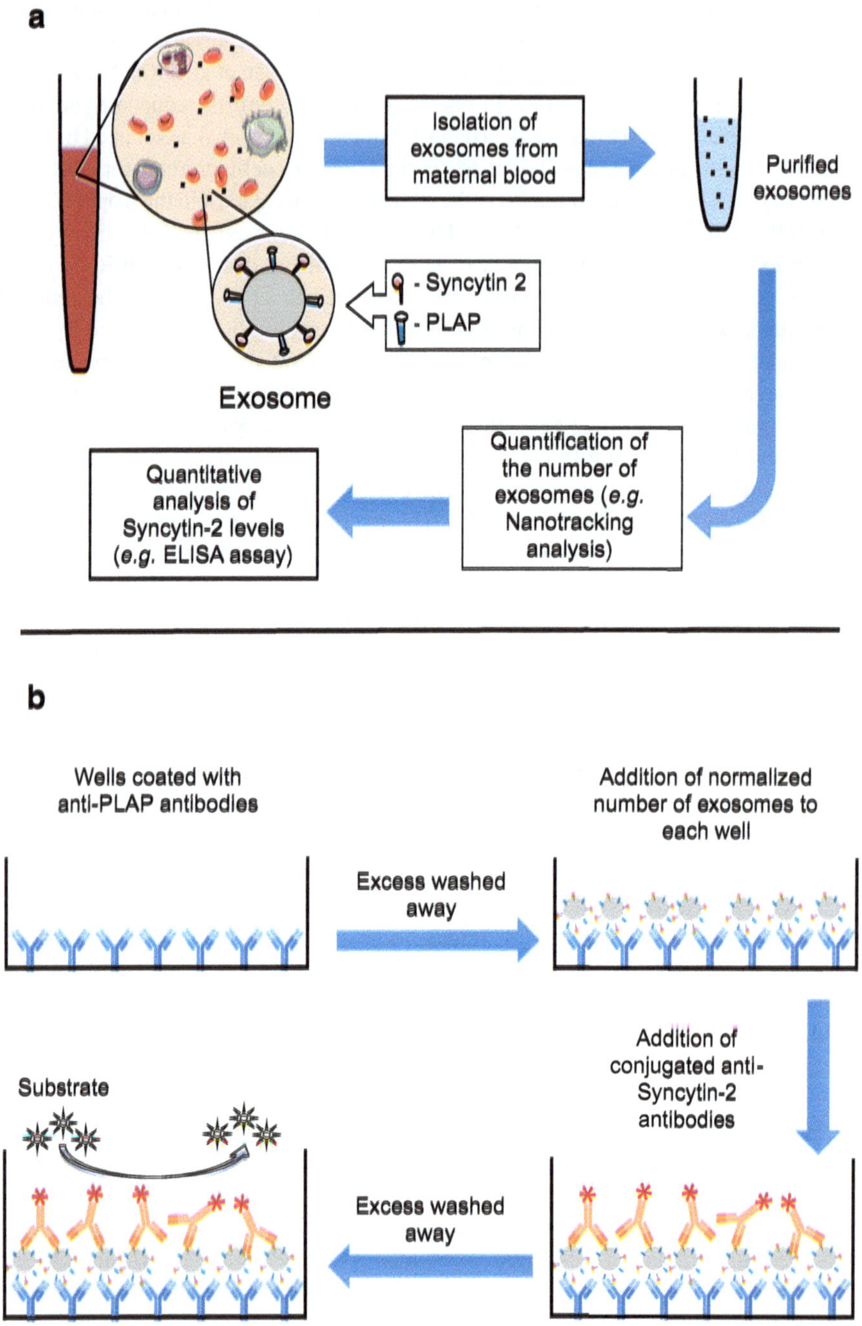

Fig. 3 Isolation of maternal blood-circulating exosomes and quantification of surface Syncytin-2 expression. Isolated exosomes from blood samples can serve as potential biological material for diagnostic kit development. (**a**). Exosomes are isolated from maternal blood harvested as early as

vested during the first trimester and analyzed directly for Syncytin-2 levels using exosomal extracts or with a preceding step involving binding of exosomes to the plate through the placenta-specific PLAP marker with subsequent detection of Syncytin-2 at their surface. Such an assay obviously requires rigorous quantification of exosomes to correctly monitor the intensity of the Syncytin-2 signal. This step of the assay remains challenging and efforts are currently under way to improve the protocols for exosome isolation and quantification.

3.1.2 IUGR and HELLP Syndromes

The hemolysis, elevated liver enzymes, and low platelet (HELLP) syndrome was described by Weinstein in 1982, through the clinical review of 29 patients with severe preeclampsia presenting other symptoms in addition to those associated to PE: thrombocytopenia, hemolysis, and liver dysfunction (Weinstein 1982). The HELLP syndrome develops in less than 1 % of pregnancies but is often associated with PE (10–20 % of PE patients), considerably increasing the risk of mortality for the mother and the fetus (Haram et al. 2014; Dusse et al. 2015). Intrauterine growth restriction (IUGR) is defined by a fetal weight below the 10th percentile of its gestational age and fetal abdominal circumference below the 2.5th percentile (Peleg et al. 1998). IUGR is often associated with both PE and HELLP syndromes.

Because HELLP and IUGR are associated with placental dysfunction and abnormal placentation (Fischer 2015; Khong 2004; Abildgaard and Heimdal 2013), Syncytin expression might also be dysregulated in these complications. Few studies are available regarding the assessment of Syncytin protein levels in the context of IUGR or HELLP, but these reports all point toward a downregulation of Syncytin-1. Knerr et al. studied Syncytin-1 mRNA expression by RT-PCR in placental villi of 30 normal, 16 PE, and 6 HELLP pregnancies. Although the reduction of Syncytin-1 mRNA was not statistically significant compared to controls and PE placentas, a reduced expression of Syncytin-1 in HELLP placentas was observed (Knerr et al. 2002). In another report, Langbein et al. showed a reduced cytotrophoblast fusion index and a highly and significant reduction in Syncytin-1 mRNA levels by qRT-PCR in both cultured villous cytotrophoblasts and tissue extracts from PE- and HELLP-associated IUGR placentas (8.1-fold and 222.7-fold lower in PE-IUGR and HELLP-IUGR cytotrophoblasts and 5.4- and 10.6-fold lower in tissues, respectively) as well as an elevated apoptotic rate of trophoblasts (Langbein et al. 2008). Due to its role in villous CTB fusion, reduced expression level of Syncytin-1 could partly explain placental dysfunction in these pathologies. Finally, Holder et al. have

Fig. 3 (continued) the first trimester. Exosome number can be quantified using various approaches and equal number of exosomes from different samples are lysed, added in 96-well plates and analyzed for Syncytin-2 levels using a standard ELISA assay. (**b**). In this protocol, identical number of blood-derived exosomes are added to anti-PLAP-coated wells and bound exosomes are then directly analyzed for the amount of Syncytin-2 by the addition of HRP-conjugated anti-Syncytin-2 antibodies followed by substrate addition and revelation also studied Syncytin-1 expression in IUGR- and PE-IUGR-associated pregnancies and showed a significant reduction in protein levels in IUGR and a more severe reduction in PE-IUGR placentas (Holder et al. 2012a). Since Syncytin-2 downregulation seems more severe in the case of PE compared to Syncytin-1 (Vargas et al. 2011), it will be important to examine Syncytin-2 expression in HELLP and IUGR pathologies.

3.2 *Syncytins and Cancers*

Although several changes in ERV expression have been described in animal and human cancers, the causality between ERV expression and tumorigenicity has only been demonstrated in certain animal cancers and remains uncertain in humans (Ruprecht et al. 2008). In certain cancers, Syncytin-1 expression has been shown to be increased when compared to normal tissue (Maliniemi et al. 2013; Buslei et al. 2015) and occasionally induced in cancerous cells in which it is normally silenced (Sun et al. 2010). This induced expression can be linked to demethylation of the Syncytin-1 promoter region in ovarian, testis, and endometrial carcinomas (Strissel et al. 2012; Huang et al. 2014a). Furthermore, in urothelial carcinomas, mutations in the *ERVWE1* promoter region were described, leading to increased expression of Syncytin-1 compared to non-tumorigenic tissues. The analysis of urothelial carcinomas showed that 75 % of the tumor cells express high levels of Syncytin-1 and that, in 88 % of tumor cells, the Syncytin-1 3′ LTR region presents two mutations responsible for the upregulation of its expression. Of interest, the 142 T-to-C mutation in the 3′ LTR region was associated with increased binding of the c-Myb transcription factor to the Syncytin-1 promoter region, thus causing *ERVWE1* transcriptional activation (Yu et al. 2014). In contrast, in pancreatic cancers, Syncytin-1 expression was reduced. Indeed, the study of Lu et al. reported constitutive Syncytin-1 expression in normal pancreas tissues, as determined by RT-PCR, Western blot, and immunohistochemistry analyses, while, in pancreatic adenocarcinomas, Syncytin-1 protein levels and *ERVWE1* promoter activity were downregulated (Lu et al. 2015).

Unlike Syncytin-1, very few investigations have been undertaken to study modification in Syncytin-2 expression in tumors, but overexpression of this gene has also been found in certain types of cancer. Indeed, Larsson and coworkers showed that both Syncytin-1 and -2 are expressed in breast cancer carcinomas and breast cancer cell lines (Larsson et al. 2007a).

Based on these previous reports, it has been suggested that Syncytin expression levels could have a prognostic value in various cancers. As an example, in breast cancers, Larsson et al. analyzed Syncytin-1 expression levels in breast cancer-tumor and found that it positively correlated with an absence of cancer recurrence following treatment (Larsson et al. 2007b). One proposed hypothesis is related to previous data showing expression of SLC1A5 (the cellular receptor of Syncytin-1) in noncancerous endothelial cells and fusion between breast cancer cells and endothelial cells (Bjerregaard et al. 2006). By inducing the fusion of cancerous to noncancerous cells, tumor cells could, for example, activate tumor-suppressor-expressing genes and become less malignant (Anderson and Stanbridge 1993). Also, expression of Syncytin-1 in a melanoma cancer cell model has been associated with a reduced invasive phenotype (Mo et al. 2013). Thus, induced Syncytin expression could reduce the extent of invasiveness of tumor cells. However, cell fusion events in cancer have also been shown to increase cancerous cell malignancy (Mohr et al. 2015). The study of Larsen and colleagues illustrates this idea by showing that Syncytin-1 expression correlates with a poor prognosis in rectal carcinomas but not in colonic tumors (Larsen et al. 2009). Thus, the level and impact of Syncytin-1 expression in tumors clearly depend on the cancer type.

4 General Conclusion

Human ERV sequences represent an important component of the human genome and, despite intense effort over recent years, their function and role remain largely unknown. The implication of human ERV-encoded fusogenic Syncytin proteins in the development and function of the human placenta is an outstanding exception to this fact in that it is currently firmly believed that these proteins are strongly altering the surface of placental villi by maintaining the formation of the syncytiotrophoblast layer through induced fusion with underlying cytotrophoblasts. Besides the involvement of Syncytin-1 and -2 in the formation and maintenance of the placenta structure, recent findings also argue that these proteins, through their immunosuppressive domain, could act on an otherwise fatal alloantigen-induced immune response against the fetus and contribute to the state of feto-maternal tolerance existing in pregnant women. Furthermore, the recently reported association of Syncytin proteins with extracellular vesicles, such as exosomes, could extend their immunosuppressive activity by reaching more distal immune cell populations. More studies are needed to understand the functional roles of these proteins and the potential use of murine models, which express their own distinct Syncytin proteins demonstrating shared features to the human Syncytins, will be of interest to further examine functions of Syncytin, either in their cell- or exosome-associated state.

The implication of Syncytin proteins in various obstetrical disorders and cancer has also been of great interest in recent years, and generally tends to be linked to modified expression of these genes at the transcriptional level. Reduced expression of Syncytin proteins in preeclampsia thereby provides a potential explanation for the association of this disorder with altered immune response and abnormal placentation. For the link between Syncytin expression and cancer, further work is needed to determine if these proteins could indeed contribute to the transformation of cells and/or the modulation of their phenotype. The fusogenic potential of these proteins conferred to expressing tumor cells is an interesting hypothesis, which will need further testing. Finally, the association between Syncytin and exosomes provides a potential use of these vesicles as a noninvasive diagnostic tool to assess the abundance of these human ERV proteins and their relationship with the predisposition for certain disorders or with the progression of diseases, such as certain types of cancer.

Bibliography

Abildgaard U, Heimdal K (2013) Pathogenesis of the syndrome of hemolysis, elevated liver enzymes, and low platelet count (HELLP): a review. Eur J Obstet Gynecol Reprod Biol 166:117–123

Anderson MJ, Stanbridge EJ (1993) Tumor suppressor genes studied by cell hybridization and chromosome transfer. Faseb J 7:826–833

Atay S, Gercel-Taylor C, Suttles J, Mor G, Taylor DD (2011) Trophoblast-derived exosomes mediate monocyte recruitment and differentiation. Am J Reprod Immunol 65:65–77

Babst M (2011) MVB vesicle formation: ESCRT-dependent, ESCRT-independent and everything in between. Curr Opin Cell Biol 23:452–457

Bjerregaard B, Holck S, Christensen IJ, Larsson LI (2006) Syncytin is involved in breast cancer-endothelial cell fusions. Cell Mol Life Sci 63:1906–1911

Blaise S, de Parseval N, Benit L, Heidmann T (2003) Genomewide screening for fusogenic human endogenous retrovirus envelopes identifies syncytin 2, a gene conserved on primate evolution. Proc Natl Acad Sci U S A 100:13013–13018

Blaise S, de Parseval N, Heidmann T (2005) Functional characterization of two newly identified Human Endogenous Retrovirus coding envelope genes. Retrovirology 2:19

Blond JL, Besème F, Duret L, Bouton O, Bedin F, Perron H, Mandrand B, Mallet F (1999) Molecular characterization and placental expression of HERV-W, a new human endogenous retrovirus family. J Virol 73:1175–1185

Blond JL, Lavillette D, Cheynet V, Bouton O, Oriol G, Chapel-Fernandes S, Mandrand B, Mallet F, Cosset FL (2000) An envelope glycoprotein of the human endogenous retrovirus HERV-W is expressed in the human placenta and fuses cells expressing the type D mammalian retrovirus receptor. J Virol 74:3321–3329

Borzychowski AM, Sargent IL, Redman CW (2006) Inflammation and pre-eclampsia. Semin Fetal Neonatal Med 11:309–316

Buslei R, Strissel PL, Henke C, Schey R, Lang N, Ruebner M, Stolt CC, Fabry B, Buchfelder M, Strick R (2015) Activation and regulation of endogenous retroviral genes in the human pituitary gland and related endocrine tumours. Neuropathol Appl Neurobiol 41:180–200

Carr MC (1967) Biology of human trophoblast. Calif Med 107:338–343

Carter AM, Enders AC (2004) Comparative aspects of trophoblast development and placentation. Reprod Biol Endocrinol 2:46

Chang C, Chen PT, Chang GD, Huang CJ, Chen H (2004) Functional characterization of the placental fusogenic membrane protein syncytin. Biol Reprod 71:1956–1962

Chen CP, Wang KG, Chen CY, Yu C, Chuang HC, Chen H (2006) Altered placental syncytin and its receptor ASCT2 expression in placental development and pre-eclampsia. BJOG 113:152–158

Chen CP, Chen LF, Yang SR, Chen CY, Ko CC, Chang GD, Chen H (2008) Functional characterization of the human placental fusogenic membrane protein syncytin 2. Biol Reprod 79:815–823

Cheng YH, Handwerger S (2005) A placenta-specific enhancer of the human syncytin gene. Biol Reprod 73:500–509

Cheng YH, Richardson BD, Hubert MA, Handwerger S (2004) Isolation and characterization of the human syncytin gene promoter. Biol Reprod 70:694–701

Cheynet V, Ruggieri A, Oriol G, Blond JL, Boson B, Vachot L, Verrier B, Cosset FL, Mallet F (2005) Synthesis, assembly, and processing of the Env ERVWE1/syncytin human endogenous retroviral envelope. J Virol 79:5585–5593
Cheynet V, Oriol G, Mallet F (2006) Identification of the hASCT2-binding domain of the Env ERVWE1/syncytin-1 fusogenic glycoprotein. Retrovirology 3:41
Chiang M-H, Liang F-Y, Chen CP, Chang CW, Cheong ML, Wang LJ, Liang C-Y, Lin F-Y, Chou CC, Chen H (2009) Mechanism of hypoxia-induced GCM1 degradation. Implications for the pathogenesis of preeclampsia. J Biol Chem 284:17411–17419
Cornelis G, Heidmann O, Bernard-Stoecklin S, Reynaud K, Véron G, Mulot B, Dupressoir A, Heidemann T (2012) Ancestral capture of syncytin-Car1, a fusogenic endogenous retroviral envelope gene involved in placentation and conserved in Carnivora. Proc Natl Acad Sci U S A 109:E432–E441
Cornelis G, Heidmann O, Degrelle S, Vernochet C, Lavialle C, Letzelter C, Bernard-Stoecklin S, Hassanin A, Mulot B, Guillomot M, Hue I, Heidmann T, Dupressoir A (2013) Captured retroviral envelope syncytin gene associated with the unique placental structure of higher ruminants. Proc Natl Acad Sci U S A 110:E828–E837
Cornelis G, Vernochet C, Malicorne S, Souquere S, Tzika A, Goodman S, Catzeflis F, Robinson T, Milinkovitch M, Pierron G, Heidmann O, Dupressoir A, Heidmann T (2014) Retroviral envelope syncytin capture in an ancestrally diverged mammalian clade for placentation in the primitive Afrotherian tenrecs. Proc Natl Acad Sci U S A 111:E4332–E4341
Cornelis G, Vernochet C, Carradec Q, Souquere S, Mulot B, Catzeflis F, Nilsson MA, Menzies BR, Renfree MB, Pierron G, Zeller U, Heidmann O, Dupressoir A, Heidmann T (2015) Retroviral envelope gene captures and syncytin exaptation for placentation in marsupials. Proc Natl Acad Sci U S A 112:E487–E496
Costa MA (2016) The endocrine function of human placenta: an overview. Reprod Biomed Online 32:14–43
Cui L, Wang H, Lu X, Wang R, Zheng R, Li Y, Yang X, Jia WT, Zhao Y, Wang Y, Wang H, Wang YL, Zhu C, Lin HY, Wang H (2016) Effects of individually silenced N-glycosylation sites and non-synonymous single-nucleotide polymorphisms on the fusogenic function of human syncytin-2. Cell Adh Migr 10(1–2):39–55
De Parseval N, Heidmann T (1998) Physiological knockout of the envelope gene of the single-copy ERV-3 human endogenous retrovirus in a fraction of the Caucasian population. J Virol 72:3442–3445
De Parseval N, Heidmann T (2005) Human endogenous retroviruses: from infectious elements to human genes. Cytogenet Genome Res 110:318–332
De Parseval N, Lazar V, Casella JF, Benit L, Heidmann T (2003) Survey of human genes of retroviral origin: identification and transcriptome of the genes with coding capacity for complete envelope proteins. J Virol 77:10414–10422
Denzer K, Kleijmeer MJ, Heijnen HFJ, Stoorvogel W, Geuze HJ (2000a) Exosome: from internal vesicle of the multivesicular body to intercellular signaling device. J Cell Sci 113:3365–3374
Denzer K, VAN Eijk M, Kleijmeer MJ, Jakobson E, DE Groot C, Geuze HJ (2000b) Follicular dendritic cells carry MHC class II-expressing microvesicles at their surface. J Immunol 165:1259–1265
Dewannieux M, Heidmann T (2013) Endogenous retroviruses: acquisition, amplification and taming of genome invaders. Curr Opin Virol 3:646–656
Duley L (2009) The global impact of pre-eclampsia and eclampsia. Semin Perinatol 33:130–137
Dupressoir A, Marceau G, Vernochet C, Benit L, Kanellopoulos C, Sapin V, Heidmann T (2005) Syncytin-A and syncytin-B, two fusogenic placenta-specific murine envelope genes of retroviral origin conserved in Muridae. Proc Natl Acad Sci U S A 102:725–730
Dupressoir A, Vernochet C, Bawa O, Harper F, Pierron G, Opolon P, Heidmann T (2009) Syncytin-A knockout mice demonstrate the critical role in placentation of a fusogenic, endogenous retrovirus-derived, envelope gene. Proc Natl Acad Sci U S A 106:12127–12132
Dupressoir A, Vernochet C, Harper F, Guégan J, Dessen P, Pierron G, Heidmann T (2011) A pair of co-opted retroviral envelope syncytin genes is required for formation of the two-layered murine placental syncytiotrophoblast. Proc Natl Acad Sci U S A 108:E1164–E1173

Dupressoir A, Lavialle C, Heidmann T (2012) From ancestral infectious retroviruses to bona fide cellular genes: role of the captured syncytins in placentation. Placenta 33:663–671

Dusse LM, Alpoim PN, Silva JT, Rios DR, Brandao AH, Cabral AC (2015) Revisiting HELLP syndrome. Clin Chim Acta 451:117–120

Esnault C, Priet S, Ribet D, Vernochet C, Bruls T, Lavialle C, Weissenbach J, Heidmann T (2008) A placenta-specific receptor for the fusogenic, endogenous retrovirus-derived, human syncytin-2. Proc Natl Acad Sci U S A 105:17532–17537

Esnault C, Cornelis G, Heidmann O, Heidmann T (2013) Differential evolutionary fate of an ancestral primate endogenous retrovirus envelope gene, the EnvV syncytin, captured for a function in placentation. PLoS Genet 9:e1003400

Fischer SJ (2015) Why is placentation abnormal in preeclampsia? Am J Obstet Gynecol 213(4 Suppl):S115–S122

Frendo JL, Olivier D, Cheynet V, Blond JL, Bouton O, Vidaud M, Rabreau M, Evain-Brion D, Mallet F (2003) Direct involvement of HERV-W Env glycoprotein in human trophoblast cell fusion and differentiation. Mol Cell Biol 23:3566–3574

Furukawa S, Kuroda Y, Sugiyama A (2014) A comparison of the histological structure of the placenta in experimental animals. J Toxicol Pathol 27:11–18

Gerbaud P, Pidoux G (2015) Review: an overview of molecular events occurring in human trophoblast fusion. Placenta 36(Suppl 1):S35–S42

Haram K, Mortensen JH, Nagy B (2014) Genetic aspects of preeclampsia and the HELLP syndrome. J Pregnancy 2014:1–13

Hedlund M, Stenqvist AC, Nagaeva O, Kjellberg L, Wulff M, Baranov V, Mincheva-Nilsson L (2009) Human placenta expresses and secretes NKG2D ligands via exosomes that down-modulate the cognate receptor expression: evidence for immunosuppressive function. J Immunol 183:340–351

Heidmann O, Vernochet C, Dupressoir A, Heidmann T (2009) Identification of an endogenous retroviral envelope gene with fusogenic activity and placenta-specific expression in the rabbit: a new "syncytin" in a third order of mammals. Retrovirology 6:107

Holder BS, Tower CL, Abrahams VM, Aplin JD (2012a) Syncytin-1 in the human placenta. Placenta 33:460–466

Holder BS, Tower CL, Forbes K, Mulla MJ, Aplin JD, Abrahams VM (2012b) Immune cell activation by trophoblast-derived microvesicles is mediated by syncytin 1. Immunology 136:184–191

Huang Q, Chen H, Li J, Oliver M, Ma X, Byck D, Gao Y, Jiang SW (2014a) Epigenetic and non-epigenetic regulation of syncytin-1 expression in human placenta and cancer tissues. Cell Signal 26:648–656

Huang Q, Chen H, Wang F, Brost BC, Li J, Gao Y, Li Z, Gao Y, Jiang SW (2014b) Reduced syncytin-1 expression in choriocarcinoma BeWo cells activates the calpain1-AIF-mediated apoptosis, implication for preeclampsia. Cell Mol Life Sci 71:3151–3164

Hummel J, Kammerer U, Muller N, Avota E, Schneider-Schaulies S (2015) Human endogenous retrovirus envelope proteins target dendritic cells to suppress T-cell activation. Eur J Immunol 45:1748–1759

Kato N, Pfeifer-Ohlsson S, Kato M, Larsson E, Rydnert J, Ohlsson R, Cohen M (1987) Tissue-specific expression of human provirus ERV3 mRNA in human placenta: two of the three ERV3 mRNAs contain human cellular sequences. J Virol 61:2182–2191

Khong TY (2004) Placental vascular development and neonatal outcome. Semin Neonatol 9:255–263

Kjeldbjerg AL, Villesen P, Aagaard L, Pedersen FS (2008) Gene conversion and purifying selection of a placenta-specific ERV-V envelope gene during simian evolution. BMC Evol Biol 8:266

Knerr I, Beinder E, Rascher W (2002) Syncytin, a novel human endogenous retroviral gene in human placenta: evidence for its dysregulation in preeclampsia and HELLP syndrome. Am J Obstet Gynecol 186:210–213

Knerr I, Huppertz B, Weigel C, Dotsch J, Wich C, Schild RL, Beckmann MW, Rascher W (2004) Endogenous retroviral syncytin: compilation of experimental research on syncytin and its possible role in normal and disturbed human placentogenesis. Mol Hum Reprod 10:581–588

Kudaka W, Oda T, Jinno Y, Yoshimi N, Aoki Y (2008) Cellular localization of placenta-specific human endogenous retrovirus (HERV) transcripts and their possible implication in pregnancy-induced hypertension. Placenta 29:282–289

Kudo Y, Boyd CAR, Sargent IL, Redman CWG (2003) Hypoxia alters expression and function of syncytin and its receptor during trophoblast cell fusion of human placental BeWo cells: implications for impaired trophoblast syncytialisation in pre-eclampsia. Biochim Biophys Acta 1638:63–67

Lacroix MC, Guibourdenche J, Frendo JL, Pidoux G, Evain-Brion D (2002) Placental growth hormones. Endocrine 19:73–79

Lai RC, Arslan F, Lee MM, Sze NS, Choo A, Chen TS, Salto-Tellez M, Timmers L, Lee CN, EL Oakley RM, Pasterkamp G, DE Kleijn DP, Lim SK (2010) Exosome secreted by MSC reduces myocardial ischemia/reperfusion injury. Stem Cell Res 4:214–222

Lander ES, Linton LM, Birren B, Nusbaum C, Zody MC, Baldwin J, Devon K, Dewar K, Doyle M, Fitzhugh W, Funke R, Gage D, Harris K, Heaford A, Howland J, Kann L, Lehoczky J, Levine R, McEwan P, Mckernan K, Meldrim J, Mesirov JP, Miranda C, Morris W, Naylor J, Raymond C, Rosetti M, Santos R, Sheridan A, Sougnez C, Stange-Thomann Y, Stojanovic N, Subramanian A, Wyman D, Rogers J, Sulston J, Ainscough R, Beck S, Bentley D, Burton J, Clee C, Carter N, Coulson A, Deadman R, Deloukas P, Dunham A, Dunham I, Durbin R, French L, Grafham D, Gregory S, Hubbard T, Humphray S, Hunt A, Jones M, Lloyd C, Mcmurray A, Matthews L, Mercer S, Milne S, Mullikin JC, Mungall A, Plumb R, Ross M, Shownkeen R, Sims S, Waterston RH, Wilson RK, Hillier LW, Mcpherson JD, Marra MA, Mardis ER, Fulton LA, Chinwalla AT, Pepin KH, Gish WR, Chissoe SL, Wendl MC, Delehaunty KD, Miner TL, Delehaunty A, Kramer JB, Cook LL, Fulton RS, Johnson DL, Minx PJ, Clifton SW, Hawkins T, Branscomb E, Predki P, Richardson P, Wenning S, Slezak T, Doggett N, Cheng JF, Olsen A, Lucas S, Elkin C, Uberbacher E, Frazier M et al (2001) Initial sequencing and analysis of the human genome. Nature 409:860–921

Langbein M, Strick R, Strissel PL, Vogt N, Parsch H, Beckmann MW, Schild RL (2008) Impaired cytotrophoblast cell–cell fusion is associated with reduced syncytin and increased apoptosis in patients with placental dysfunction. Mol Reprod Dev 75:175–183

Larsen JM, Christensen IJ, Nielsen HJ, Hansen U, Bjerregaard B, Talts JF, Larsson LI (2009) Syncytin immunoreactivity in colorectal cancer: potential prognostic impact. Cancer Lett 280:44–49

Larsson L-I, Bjerregaard B, Wulf-Andersen L, Talts JF (2007a) Syncytin and cancer cell fusions. Sci World J 7:1193–1197

Larsson LI, Holck S, Christensen IJ (2007b) Prognostic role of syncytin expression in breast cancer. Hum Pathol 38:726–731

Lavialle C, Cornelis G, Dupressoir A, Esnault C, Heidmann O, Vernochet C, Heidmann T (2013) Paleovirology of 'syncytins', retroviral env genes exapted for a role in placentation. Philos Trans R Soc Lond B Biol Sci 368:20120507

Lavillette D, Marin M, Ruggieri A, Mallet F, Cosset FL, Kabat D (2002) The envelope glycoprotein of human endogenous retrovirus type W uses a divergent family of amino acid transporters/cell surface receptors. J Virol 76:6442–6452

Lee X, Keith JC Jr, Stumm N, Moutsatsos I, McCoy JM, Crum CP, Genest D, Chin D, Ehrenfels C, Pijnenborg R, Van Assche FA, Mi S (2001) Downregulation of placental syncytin expression and abnormal protein localization in pre-eclampsia. Placenta 22:808–812

Li F, Nellaker C, Sabunciyan S, Yolken RH, Jones-Brando L, Johansson AS, Owe-Larsson B, Karlsson H (2014) Transcriptional derepression of the ERVWE1 locus following influenza A virus infection. J Virol 88:4328–4337

Liang CY, Wang LJ, Chen CP, Chen LF, Chen YH, Chen H (2010) GCM1 regulation of the expression of syncytin 2 and its cognate receptor MFSD2A in human placenta. Biol Reprod 83:387–395

Lokossou AG, Toudic C, Barbeau B (2014) Implication of human endogenous retrovirus envelope proteins in placental functions. Viruses 6:4609–4627

Lu Q, Li J, Senkowski C, Tang Z, Wang J, Huang T, Wang X, Terry K, Brower S, Glasgow W, Chen H, Jiang SW (2015) Promoter hypermethylation and decreased expression of syncytin-1 in pancreatic adenocarcinomas. PLoS One 10:e0134412

Luo SS, Ishibashi O, Ishikawa G, Ishikawa T, Katayama A, Mishima T, Takizawa T, Shigihara T, Goto T, Izumi A, Ohkuchi A, Matsubara S, Takeshita T, Takizawa T (2009) Human villous trophoblasts express and secrete placenta-specific microRNAs into maternal circulation via exosomes. Biol Reprod 81:717–729

Malassine A, Blaise S, Handschuh K, Lalucque H, Dupressoir A, Evain-Brion D, Heidmann T (2007) Expression of the fusogenic HERV-FRD Env glycoprotein (syncytin 2) in human placenta is restricted to villous cytotrophoblastic cells. Placenta 28:185–191

Maliniemi P, Vincendeau M, Mayer J, Frank O, Hahtola S, Karenko L, Carlsson E, Mallet F, Seifarth W, Leib-Mosch C, Ranki A (2013) Expression of human endogenous retrovirus-w including syncytin-1 in cutaneous T-cell lymphoma. PLoS One 8:e76281

Mangeney M, Renard M, Schlecht-Louf G, Bouallaga I, Heidmann O, Letzelter C, Richaud A, Ducos B, Heidmann T (2007) Placental syncytins: genetic disjunction between the fusogenic and immunosuppressive activity of retroviral envelope proteins. Proc Natl Acad Sci U S A 104:20534–20539

Mathivanan S, Ji H, Simpson RJ (2010) Exosomes: extracellular organelles important in intercellular communication. J Proteomics 73:1907–1920

Matouskova M, Blazkova J, Pajer P, Pavlicek A, Hejnar J (2006) CpG methylation suppresses transcriptional activity of human syncytin-1 in non-placental tissues. Exp Cell Res 312:1011–1020

Metz J, Aoki A, Forssmann WG (1978) Studies on the ultrastructure and permeability of the hemotrichorial placenta. I. Intercellular junctions of layer I and tracer administration into the maternal compartment. Cell Tissue Res 192:391–407

Mi S, Lee X, Li X, Veldman G, Finnerty H, Racie L, Lavallie E, Tang X, Edouard P, Howes S, Keith J Jr, McCoy J (2000) Syncytin is a captive retroviral envelope protein involved in human placental morphogenesis. Nature 403(6771):785–789

Mincheva-Nilsson L, Baranov V (2010) The role of placental exosomes in reproduction. Am J Reprod Immunol 63:520–533

Mitchell MD, Peiris HN, Kobayashi M, Koh YQ, Duncombe G, Illanes SE, Rice GE, Salomon C (2015) Placental exosomes in normal and complicated pregnancy. Am J Obstet Gynecol 213:S173–S181

Mo H, Ouyang D, Xu L, Gao Q, He X (2013) Human endogenous retroviral syncytin exerts inhibitory effect on invasive phenotype of B16F10 melanoma cells. Chin J Cancer Res 25:556–564

Mohr M, Zaenker KS, Dittmar T (2015) Fusion in cancer: an explanatory model for aneuploidy, metastasis formation, and drug resistance. Methods Mol Biol 1313:21–40

Munoz-Suano A, Hamilton AB, Betz AG (2011) Gimme shelter: the immune system during pregnancy. Immunol Rev 241:20–38

Muyan M, Boime I (1997) Secretion of chorionic gonadotropin from human trophoblasts. Placenta 18:237–241

Peleg D, Kennedy CM, Hunter SK (1998) Intrauterine growth restriction: identification and management. Am Fam Physician 58(453–60):466–467

Prudhomme S, Oriol G, Mallet F (2004) A retroviral promoter and a cellular enhancer define a bipartite element which controls env ERVWE1 placental expression. J Virol 78:12157–12168

Raghupathy R (2013) Cytokines as key players in the pathophysiology of preeclampsia. Med Princ Pract 22(Suppl 1):8–19

Raposo G, Nijman HW, Stoorvogel W, Liejendekker R, Harding CV, Melief CJ, Geuze HJ (1996) B lymphocytes secrete antigen-presenting vesicles. J Exp Med 183:1161–1172

Redelsperger F, Cornelis G, Vernochet C, Tennant BC, Catzeflis F, Mulot B, Heidmann O, Heidmann T, Dupressoir A (2014) Capture of syncytin-Mar1, a fusogenic endogenous retroviral envelope gene involved in placentation in the Rodentia squirrel-related clade. J Virol 88:7915–7928

Renard M, Varela PF, Letzelter C, Duquerroy S, Rey FA, Heidmann T (2005) Crystal structure of a pivotal domain of human syncytin-2, a 40 million years old endogenous retrovirus fusogenic envelope gene captured by primates. J Mol Biol 352:1029–1034

Ruprecht K, Mayerb J, Sautera M, Roemerc K, Mueller-Lantzscha N (2008) Endogenous retroviruses and cancer. Cell Mol Life Sci 65:3366–3382

Sahoo S, Klychko E, Thorne T, Misener S, Schultz KM, Millay M, Ito A, Liu T, Kamide C, Agrawal H, Perlman H, Qin G, Kishore R, Losordo DW (2011) Exosomes from human CD34(+) stem cells mediate their proangiogenic paracrine activity. Circ Res 109:724–728

Salomon C, Torres MJ, Kobayashi M, Scholz-Romero K, Sobrevia L, Dobierzewska A, Illanes SE, Mitchell MD, Rice GE (2014) A gestational profile of placental exosomes in maternal plasma and their effects on endothelial cell migration. PLoS One 9:e98667

Sibai B, Dekker G, Kupferminc M (2005) Pre-eclampsia. Lancet 365:785–799

Stenqvist AC, Nagaeva O, Baranov V, Mincheva-Nilsson L (2013) Exosomes secreted by human placenta carry functional Fas ligand and TRAIL molecules and convey apoptosis in activated immune cells, suggesting exosome-mediated immune privilege of the fetus. J Immunol 191:5515–5523

Stoye JP (2012) Studies of endogenous retroviruses reveal a continuing evolutionary saga. Nat Rev Microbiol 10:395–406

Strissel PL, Ruebner M, Thiel F, Wachter D, Ekici AB, Wolf F, Thieme F, Ruprecht K, Beckmann MW, Strick R (2012) Reactivation of codogenic endogenous retroviral (ERV) envelope genes in human endometrial carcinoma and prestages: emergence of new molecular targets. Oncotarget 3:1204–1219

Sun Y, Liu J (2014) Potential of cancer cell-derived exosomes in clinical application: a review of recent research advances. Clin Ther 36:863–872

Sun Y, Ouyang DY, Pang W, Tu YQ, Li YY, Shen XM, Tam SC, Yang HY, Zheng YT (2010) Expression of syncytin in leukemia and lymphoma cells. Leuk Res 34:1195–1202

Tarlinton RE, Meers J, Young PR (2006) Retroviral invasion of the koala genome. Nature 442:79–81

Taufield PA, Suthanthiran M, Ales K, Druzin M, Resnick LM, Laragh JH, Stenzel KH, Rubin AL (1983) Maternal-fetal immunity: presence of specific cellular hyporesponsiveness and humoral suppressor activity in normal pregnancy and their absence in preeclampsia. Clin Exp Hypertens B 2:123–131

Tolosa JM, Schjenken JE, Clifton VL, Vargas A, Barbeau B, Lowry P, Maiti K, Smith R (2012) The endogenous retroviral envelope protein syncytin-1 inhibits LPS/PHA-stimulated cytokine responses in human blood and is sorted into placental exosomes. Placenta 33:933–941

Toufaily C, Vargas A, Lemire M, Lafond J, Rassart E, Barbeau B (2013) MFSD2a, the Syncytin-2 receptor, is important for trophoblast fusion. Placenta 34:85–88

Toufaily C, Lokossou AG, Vargas A, Rassart E, Barbeau B (2015) A CRE/AP-1-like motif is essential for induced syncytin-2 expression and fusion in human trophoblast-like model. PLoS One 10:e0121468

Trejbalova K, Blazkova J, Matouskova M, Kucerova D, Pecnova L, Vernerova Z, Heracek J, Hirsch I, Hejnar J (2011) Epigenetic regulation of transcription and splicing of syncytins, fusogenic glycoproteins of retroviral origin. Nucleic Acids Res 39:8728–8739

Varela M, Spencer TE, Palmarini M, Arnaud F (2009) Friendly viruses: the special relationship between endogenous retroviruses and their host. Ann N Y Acad Sci 1178:157–172

Vargas A, Moreau J, Landry S, Lebellego F, Toufaily C, Rassart E, Lafond J, Barbeau B (2009) Syncytin-2 plays an important role in the fusion of human trophoblast cells. J Mol Biol 392:301–318

Vargas A, Toufaily C, Lebellego F, Rassart E, Lafond J, Barbeau B (2011) Reduced expression of both syncytin 1 and syncytin 2 correlates with severity of preeclampsia. Reprod Sci 18:1085–1091

Vargas A, Zhou S, Ethier-Chiasson M, Flipo D, Lafond J, Gilbert C, Barbeau B (2014) Syncytin proteins incorporated in placenta exosomes are important for cell uptake and show variation in abundance in serum exosomes from patients with preeclampsia. FASEB J 28:3703–3719

Vlkova B, Szemes T, Minarik G, Turna J, Celec P (2010) Circulating free fetal nucleic acids in maternal plasma and preeclampsia. Med Hypotheses 74:1030–1032

Voisset C, Blancher A, Perron H, Mandrand B, Mallet F, Paranhos-Baccalà G (1999) Phylogeny of a novel family of human endogenous retrovirus sequences, HERV-W, in humans and other primates. AIDS Res Hum Retroviruses 15:1529–1533

Walker JJ (2000) Pre eclampsia. Lancet 356:1260–1265

Warning JC, McCracken SA, Morris JM (2011) A balancing act: mechanisms by which the fetus avoids rejection by the maternal immune system. Reproduction 141:715–724

Weinstein L (1982) Syndrome of hemolysis, elevated liver enzymes, and low platelet count: a severe consequence of hypertension in pregnancy. Am J Obstet Gynecol 142:159–167

Wich C, Kausler S, Dotsch J, Rascher W, Knerr I (2009) Syncytin-1 and glial cells missing a: hypoxia-induced deregulated gene expression along with disordered cell fusion in primary term human trophoblasts. Gynecol Obstet Invest 68:9–18

Yu C, Shen K, Lin M, Chen P, Lin C, Chang GD, Chen H (2002) GCMa regulates the syncytin-mediated trophoblastic fusion. J Biol Chem 277:50062–50068

Yu H, Liu T, Zhao Z, Chen Y, Zeng J, Liu S, Zhu F (2014) Mutations in 3′-long terminal repeat of HERV-W family in chromosome 7 upregulate syncytin-1 expression in urothelial cell carcinoma of the bladder through interacting with c-Myb. Oncogene 33:3947–3958

Zhuang XW, Li J, Brost BC, Xia XY, Chen HB, Wang CX, Jiang SW (2014) Decreased expression and altered methylation of syncytin-1 gene in human placentas associated with preeclampsia. Curr Pharm Des 20:1796–1802

Alu-Alu Recombinations in Genetic Diseases

Maria E. Morales, Geraldine Servant, Catherine M. Ade, and Prescott Deininger

1 Alu Elements

In 2001 the complete sequencing of the human genome revealed that less than 2 % of its nucleotide sequences code for proteins, while almost 50 % of its mass is composed of repetitive elements (http://genome.ucsc.edu). Alu elements are SINE elements (Short, INterspersed Elements) that make up the largest family of repetitive elements in terms of copy number, with over one million copies comprising an estimated 11 % of the human genome (Lander et al. 2001). Alu elements are specific to primates having emerged more than 65 Ma ago derived from the 7SL RNA gene (Deininger et al. 2003). They have increased to this enormous copy number through an RNA-mediated copy-and-paste process termed retrotransposition or retroposition that utilizes an RNA intermediate to insert a new copy in the genome. One key feature of Alu elements is an internal RNA polymerase III promoter that initiates transcription at the beginning of the Alu to make the RNA intermediate used in their retrotransposition. They also have an A-tail encoded at their 3′ end that is critical for priming reverse transcription at a nicked site in the genome, a process termed target-primed reverse transcription (TPRT) (Luan et al. 1993). Alu elements do not encode any proteins and are completely dependent on L1 elements (the currently active Long INterspersed Element (LINE) in the human genome) for retrotransposition (Dewannieux et al. 2003). Alu RNA binds to the L1-encoded protein (ORF2) that contains an endonuclease domain (to nick an AT-rich genomic region and initiate the insertion process) and a reverse transcriptase activity to copy the RNA into cDNA (Fig. 1). The rest of the insertion process is still poorly understood, but it is hypothesized to involve other proteins such as those used in DNA replication or repair (Pizarro and Cristofari 2016).

M.E. Morales • G. Servant • C.M. Ade • P. Deininger (✉)
Department of Epidemiology and the Tulane Cancer Center, Tulane University Health Sciences Center, New Orleans, LA, USA
e-mail: pdeinin@tulane.edu

G. Cristofari (ed.), *Human Retrotransposons in Health and Disease*,
DOI 10.1007/978-3-319-48344-3_10

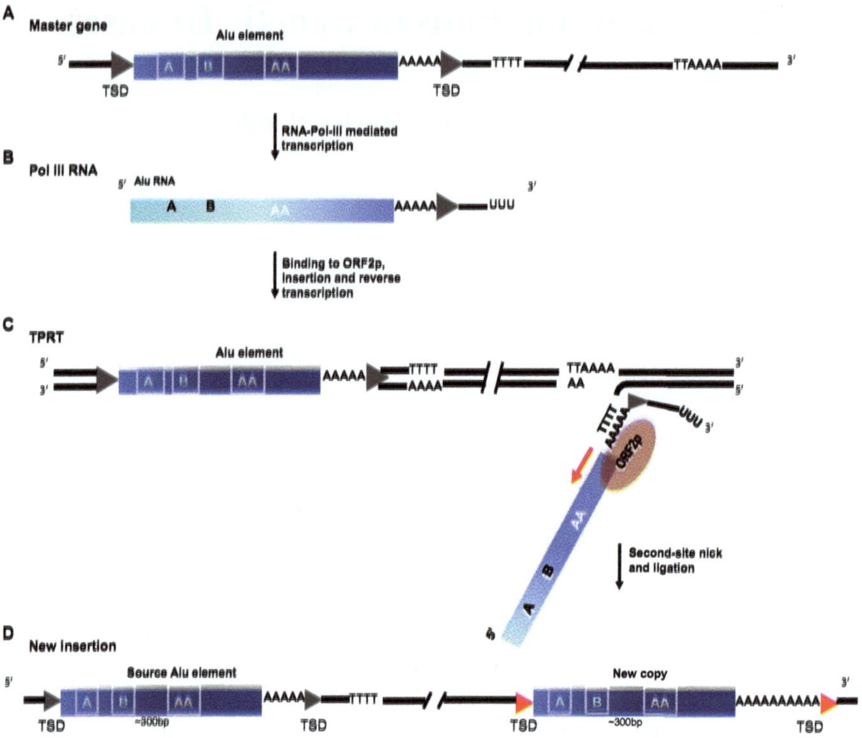

Fig. 1 Alu retrotransposition process by the target-primed reverse transcription (TPRT) mechanism. (**a**) The *top* portion shows a genomic Alu element between two direct repeats formed at the site of insertion (*black arrowheads*). The Alu ends with a long A-run, often referred to as the A-tail, and it also has a smaller A-rich region (indicated by *AA*) separating the two halves of a diverged dimer structure. Alu elements have the internal components of an RNA polymerase III promoter (*boxes A* and *B*), but they do not encode a terminator for RNA polymerase III. They utilize whatever stretch of T nucleotides is found at various distances downstream of the Alu element to terminate transcription. (**b**) A typical Alu transcript is shown below the genomic Alu, showing that it encompasses the entire Alu, including the A-tail, and has a 3′ region that is unique for each locus. (**c**) In the target-primed reverse transcription mechanism, the Alu RNA (*blue*) brings the ORF2p to the genome where its endonuclease activity cleaves at a T-rich consensus sequence. The T-rich region primes reverse transcription by ORF2p on the 3′ A-tail region of the *Alu* element. This creates a cDNA copy of the body of the *Alu* element. A nick occurs by an unknown mechanism on the second strand and second-strand synthesis is primed. The new *Alu* element is then flanked by short direct repeats that are duplicates of the DNA sequence between the first and second nicks (*red arrowheads*)

Alu elements contribute to human genetic instability in multiple ways. There has been an extensive focus on the role of *de novo* Alu insertions, contributing to disease (Ade et al. 2013; Deininger and Batzer 2002; Kaer and Speek 2013). Retrotransposition of Alu elements can potentially affect the expression of genes by providing alternative promoters, gene silencing by transcriptional or RNA interference, and creating aberrant transcripts by introducing cryptic splice sites and polyadenylation signals (Belancio et al. 2009; Britten 1996; Deininger 2011). However the primary focus of this chapter is on the contributions to genetic instability after the insertion process.

These instabilities all rely on the homology between Alu elements and their abundance to stimulate the instability. For the purpose of our discussion on the impact of Alu elements on recombination and related processes of genetic instability after their insertion, there are several key features of Alu elements that may be relevant:

- Their high copy number (Lander et al. 2001) and relative enrichment in the introns of genes (Medstrand et al. 2002) provide many and varied opportunities for homology-dependent interactions.
- Their insertion over a long evolutionary time (see discussion below) has resulted in extensive sequence heterogeneity between individual Alu elements with most having 15–25 % mismatch between Alu pairs (Sen et al. 2006).
- Their insertion process usually causes duplication of the L1 endonuclease cleavage consensus sequence on both ends of the Alu elements, providing more opportunities for future cleavage by the endonuclease.

2 Alu Element Accumulation and Evolution

SINEs are known to accumulate throughout the genomes of eukaryotes. They originated from the retroposition of small RNAs such as 7SL RNA, tRNA, or derivatives that feature an internal RNA polymerase III promoter (Kramerov and Vassetzky 2005). Alu elements are derived from the 7SL RNA gene, an integral part of the signal recognition particle (SRP) involved in protein secretion (Walter and Blobel 1982). Although the details of the origin are not known, it seems likely that a relatively inefficient retrotransposon was formed by a truncated version of the 7SL RNA gene sometime before the primate/rodent evolutionary divergence of primates and rodents. This precursor then evolved into modern B1 repeats in rodents, and into FLAM (free left Alu monomer) and FRAM (free right Alu monomer) sequences in the primate lineage (Kriegs et al. 2007; Quentin 1992). A dimer of FLAM and FRAM eventually took on the highly efficient amplification characteristics of the Alu elements.

Large-scale sequencing studies of primate genomes have provided a great deal of detail on the evolution of Alu elements. Because there is no specific mechanism for removal of Alu insertions (Fig. 1), Alu inserts accumulate sequence variation over time. Different subfamilies of Alu elements have amplified at different stages of primate evolution. This has resulted in a very limited and homogeneous group of subfamilies active in any given species, suggesting a limited number of source, or master, Alu loci (Fig. 1a) (Deininger et al. 1992; Quentin 1992; Shen et al. 1991). The earliest Alu elements were the J subfamily, followed by a very active series of S subfamilies. For the last 20 Ma, Alu amplification has been dominated by a series of younger Y subfamily members. The dominant S subfamilies included Sx, Sq, Sp, and Sc (Batzer et al. 1996). There are several variants on the Y subfamily continuing to amplify and cause human disease, mostly of the Ya5 and Yb8 subfamilies (Konkel et al. 2015; Kaer and Speek 2013). Collectively, the young active subfamilies of Alu elements comprise only about 15 % of the Alu elements present within the human genome. In contrast, the older J and S Alu subfamilies comprise approximately 85 % of all Alu elements in the human genome.

Alu elements are a major contributor to genome variation both through *de novo* insertion and Alu-mediated recombination as evidenced by comparisons between various primate species. Overall, insertions of Alu elements have increased the size of the primate genome by approximately 11 % (Lander et al. 2001). Even the relatively recent divergence between the chimpanzee and human genomes has allowed approximately 5000 new Alu element insertions to be fixed in the human genome (Hedges et al. 2004; Mills et al. 2006; Chimpanzee Sequencing and Analysis Consortium 2005). As discussed in a later section, Alu-related deletions, duplications, and inversions have also made a major impact on these genomes. Thus, mobile element insertions have altered a similar fraction of bases in these genomes to point mutations over that period of time.

Alu elements continue to insert in the modern human lineage as evidenced by their contribution to human genetic diseases (see below). It is estimated that there is about one new Alu insert per 20 human births (Xing et al. 2009). A *de novo* Alu insert is responsible for about 1 in every 1000 new human genetic diseases (Deininger and Batzer 1999). Comparison between two completed human sequenced genomes showed that there were approximately 800 polymorphic Alu elements between those two individuals (Xing et al. 2009). Therefore, Alu element insertions are a driver of genetic diversity between individual humans. Most Alu-related genomic instability events will have no major functional consequence; on the many generations the insertions could simply be lost from the human population gene pool through random fixation, or be deleterious and therefore lost through negative selection. Thus, Alu events causing genomic diversity or disease described above represent only a tiny proportion of the overall genetic instability in the human population caused by such elements.

3 Alu Elements, Recombination, and Disease

Because Alu elements are located throughout the genome and even enriched in genes (Deininger 2011), the spectrum of diseases caused by Alu-related recombination events is very broad (Deininger and Batzer 1999). Many observations of Alu elements causing disease are anecdotal examples from investigators who discover insertions or rearrangements in genetic regions they are studying in depth. However, areas with a high density of Alu elements have been particularly associated with genomic instability. A broad range of genes have been reported in the literature which have undergone Alu/Alu-mediated recombination events leading to genetic defects (Batzer and Deininger 2002; Deininger and Batzer 1999; Boone et al. 2011; Gu et al. 2015, 2016). One of the first to be characterized was the low-density lipoprotein receptor (LDLR). In this gene, several different Alu/Alu-mediated recombination events were identified that created deletions (Lehrman et al. 1985). Many such deletions have been characterized in different populations more recently (Faiz et al. 2013; Nissen et al. 2006). Several other genetic diseases regularly arising from Alu inserts have been identified. A high proportion of the defects found in those genes were caused by Alu-mediated deletions, tandem duplications, and complex rearrangements (Strout et al. 1998; Boone et al. 2014; Deininger and Batzer 1999; Gu et al. 2015; Reddy et al. 2016). There have been seven independent recombination events in the BRCA1 gene that have resulted in breast cancer, three mutations in the MSH2 gene, and one duplication and one deletion in MSH6. MSH2 and MSH6 are essential components of the DNA mismatch repair (MMR) pathway, and mutations cause susceptibility to hereditary nonpolyposis colon cancer (HNPCC) (Brosens et al. 2015). The majority of acute myelogenous leukemia cases that have no visible translocation involve Alu/Alu-mediated duplication events in the MLL gene (Strout et al. 1998). Although interchromosomal translocations are generally not mediated by Alu elements, evidence indicates that many of these cases have additional, smaller rearrangements in the MLL gene that may be Alu mediated (Strout et al. 1998). Recent reports suggest that up to 30 % of mutations in the Fanconi anemia gene (FANCA) may be caused by Alu/Alu recombination events (Callen et al. 2004). More recently, a careful analysis of deletions in the SPAST gene, which leads to spastic paraplegia, has shown that 38 out of 54 CNVs found to cause defects involved in recombination events between two nonallelic Alu elements (Boone et al. 2014). Overall, ~0.3–0.5 % of all human genetic diseases seem to have resulted from Alu-mediated unequal homologous recombination (Deininger and Batzer 1999).

A number of studies have also suggested the possibility that deletion and duplication events resembling nonhomologous end joining rather than nonallelic recombination may occur at a higher frequency in the vicinity of Alu elements relative to other regions (Bailey et al. 2003; Sharp et al. 2007; Boone et al. 2011; Gu et al. 2015). Furthermore, some studies also report an enrichment of CNV breakpoint junctions within Alu elements, implicating an Alu microhomology-mediated repair process (e.g., MMEJ or micro-SSA) in the formation of these CNVs (Boone et al. 2011; Erez et al. 2009; de Smith et al. 2008).

One of the main features that seem to contribute to Alu-related recombination rates is the density of the Alu elements. A recent study (Reddy et al. 2016) revealed that ~48 % of the total NLRP7 genomic structure (25,997 bp) is occupied by Alu sequences, which represent one Alu insertion every 450 bp, and is much higher than average Alu density in the human genome. Another detailed study of CNVs in a gene-rich region of chromosome 17 with a high (30 %) density of Alu elements similarly showing extensive contributions of Alu elements to CNVs and other rearrangements (Gu et al. 2015). They evaluated 39 CNVs affecting several genes in this region. They found that 4 out of 5 deletions, 7 out of 14 tandem duplications, and 6 out of 13 more complex rearrangements studied were Alu/Alu CNV events.

Some of the rearrangements also involved inverted duplications resulting from Alu elements being present in an inverted orientation. In these studies, the Alu/Alu nonallelic recombination (NAR) events were often large, involving tens of thousands of bases between the Alu elements, but were generally smaller than deletions that occurred independent of Alu elements. There are exceptions, however, with deletions extending over a megabase (Boone et al. 2014). In general, it seems that Alu elements are more effective at causing moderate-sized rearrangements. The tendency for Alu elements to recombine with a preference for proximity may also help explain why most interchromosomal translocations do not involve Alu/Alu NAR.

The high density of Alu elements alone may be the primary factor that provides so many more permutations for Alu interactions, as well as closer spacing. However, there may also be unknown factors in these regions that make them particularly recombinogenic. Those same factors may have contributed in some manner to the higher levels of Alu elements in those regions in the first place. There have been suggestions that Alu elements may contain sequences that make them particularly prone to recombination (Rudiger et al. 1995). This was first reported because there seemed to be a preferential region within the Alu element that was enriched for recombination junctions between the nonallelic Alu elements. More recently there has been some suggestion that there may be a sequence, CCNCCNTNNCCNC, that may contribute to increased meiotic recombination that is enriched in the Alu elements that are involved in recombination (McVean 2010). This sequence is not part of the consensus of Alu elements and therefore if it forms it requires some mutation in the Alu elements, making it only present in a subset. At this point, the dominant feature driving Alu/Alu CNVs seems to be the homology and whether there are distinct factors that make some Alu elements specifically more recombinogenic is unclear.

4 Alu Elements, Recombination, and Instability in the Primate Genome

Comparative genomics in primates provides further insight into the role of Alu-mediated recombination events on the structure and evolution of genomes. Studies of the human and chimpanzee genomes show that approximately 5000 deletion events have occurred in both genomes affecting well over 1 million bases (Han et al. 2007; Sen et al. 2006). It has been more difficult to assess the duplication events that are also caused by this type of recombination due to bioinformatics limitations, but it is likely that there is approximately the same number of events, and these events have also been suggested to contribute to approximately 50 chromosomal inversions between the chimpanzee and human genomes (Lee et al. 2008). This represents about 20% of the total inversions detected. Thus, there are extensive, relatively small deletions throughout both the human and chimpanzee genomes that contribute significantly to their genomic differentiation.

The human and chimpanzee genomes have been diverged from one another for much longer time than two individual human genomes (approximately 6 Ma vs. no more than 1 Ma). In a comparison of the human reference genome (Lander et al. 2001) against another human genome (Levy et al. 2007), there were 98 confirmed Alu-recombination-mediated deletions (ARMD) (Sen et al. 2006). These deletions were relatively small compared to the ones described above leading to human disease and on the average only deleted about 1 kb, with the longest being less than 8 kb. The small size of these events may be because smaller deletions are much more common but are often too small to contribute to disease. Alternatively, it is also likely that larger deletions are more likely to be deleterious and be lost from the population faster than the smaller deletions. This same rationale may explain why the much longer L1 elements, but which are also at lower copy number and therefore spaced at greater distances, only showed 9 L1-mediated deletions due to recombination (Sen et al. 2006).

From the disease studies (Deininger and Batzer 1999; Hedges and Deininger 2007; Kaer and Speek 2013) and the whole genome comparisons (Sen et al. 2006), two important variables appear to influence the efficiency of the genetic recombination: the density of the Alu elements in a region and the level of sequence identity between the elements. Although all Alu subfamilies contribute to these events, there is enrichment for the younger subfamilies, particularly AluY, in the recombination events. Also, because Alu elements are enriched in gene-rich regions and within genes (Medstrand et al. 2002), they are more likely to preferentially cause deletions within genic regions.

5 Mechanistic Aspects of Alu-Based Recombination Processes

The primary factor leading to recombination in the human genome is a response to double-strand breaks (DSBs). It is critical to repair DSBs to maintain cellular function. Double-strand breaks can be caused by environmental factors damaging DNA, with ionizing radiation being considered a classical cause of such damage. They can also be caused by other DNA-damaging agents, and chemotherapeutic agents (both genotoxic and Topo2 inhibitors) can also lead to extensive DSBs. However, for normal cells, the most common source of DSBs is during replication. In this case, anything that causes any kind of single-strand nick can turn into a DSB if a replication complex reaches the single-strand nick. It has been estimated that as many 50 DSBs occur per cell division by this mechanism (Vilenchik and Knudson 2003). If not accurately repaired, DSBs can result in cell death, mutations, cancer, and premature aging (Ferguson and Alt 2001).

DSBs can be repaired through homologous recombination (HR), nonhomologous end joining (NHEJ), or single-strand annealing (SSA) (Featherstone and Jackson 1999) (Fig. 2). HR is a mechanism where a homologous DNA sequence is

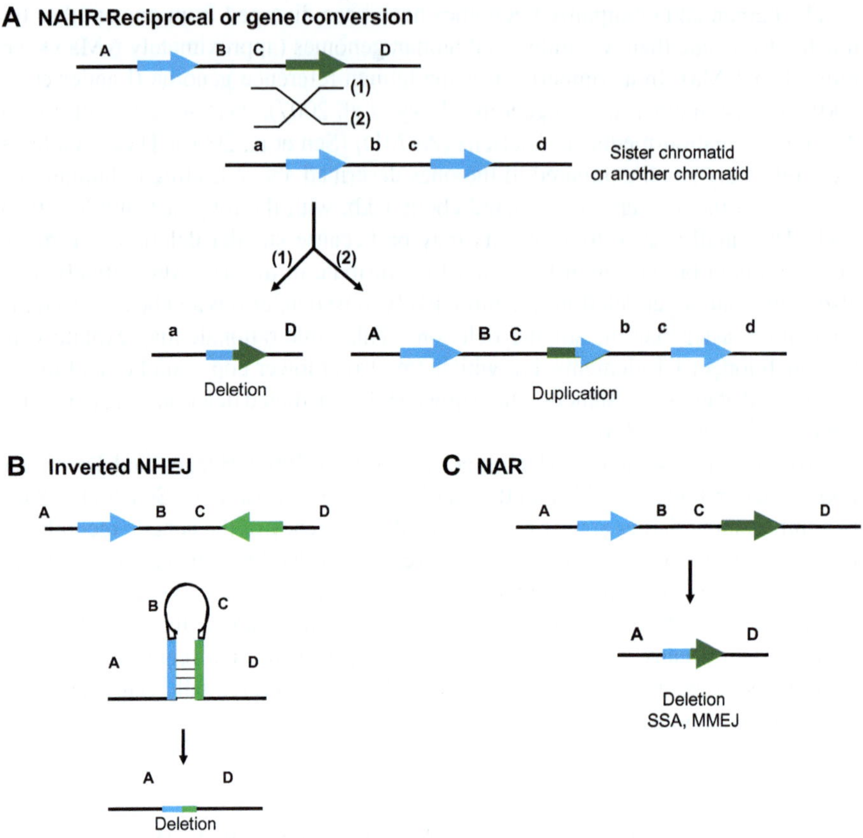

Fig. 2 Genomic rearrangements resulting from recombination between Alu elements. Alu elements are depicted as *blue* or *green arrows* with the orientation indicated by the direction of the *arrowhead. Capital letters above* the *thin horizontal lines* refer to the flanking unique sequences. Homologues on the other strand (can be another chromatid or the homologous chromosome) are also shown. *Thin diagonal lines* refer to a recombination event with the results shown by numbers 1, and 2. (**a**). Nonallelic homologous recombination (NAHR). Recombination between two different chromatids results in deletion and/or duplication. (**b**). Inverted nonhomologous end joining (NHEJ) between inverted repeats results in deletion. (**c**). Nonallelic recombination (NAR) through single-strand annealing (SSA) or microhomology-mediated end joining (MMEJ). Several mechanisms of nonallelic recombination between Alu elements can form a chimeric Alu element from the two flanking elements with the loss of the DNA sequence between them

used as a template to repair a DSB. In mitotic cells, the preferred template is an identical sister chromatid (Johnson and Jasin 2000). HR is highly conserved in prokaryotes and eukaryotes, suggesting that HR is a fundamental biological mechanism. Deficiencies in HR have been associated with cancer development (Kennedy and D'Andrea 2006; Luo et al. 2000). If performed accurately, HR will repair the DNA without a trace. However, the process is often undermined by abundant and/or highly homologous sequences such as Alu elements (Hedges and Deininger 2007; Sen et al. 2006). In these instances, two homologous sequences from different genomic locations recombine in a process called nonallelic homologous recombination (NAHR). NAHR involves the alignment and subsequent crossing over between two sites in the genome that share regions of sequence homology. NAHR can occur both in meiosis (Turner et al. 2008) and, at lower frequencies, in mitosis (Lam and Jeffreys 2006, 2007). NAHR can involve genomic rearrangements between paralogs on homologous chromosomes (interchromosomal) and sister chromatids (interchromatid), and within a chromatid (intrachromatid), which consequently causes deletions and duplications (Gu et al. 2008).

Classical nonhomologous end joining (C-NHEJ) involves the direct ligation of DNA ends and can occur with high fidelity or be commonly associated with small alterations at the junctions (McVey and Lee 2008). Early studies of C-NHEJ-deficient cells identified alternative error-prone mechanisms of end joining, often referred to as alt-NHEJ (Boulton and Jackson 1996; Ma et al. 2003; Yu and Gabriel 2003). In one type of alt-NHEJ, termed microhomology-mediated end joining (MMEJ), repair initiates by resection or unwinding of double-stranded DNA to expose short, single-strand microhomologies on the order of 6–20 nucleotides (Rodgers and McVey 2016). MMEJ is associated with deletions flanking the original DSB site, making this repair pathway particularly error-prone. Alu/Alu recombination events between heterologous elements may arise in some instances from MMEJ where the microhomology happens to be "in register" between the two Alu elements, allowing formation of a single chimeric Alu element (i.e., in register means that the breakpoint contains a complete hybrid Alu, each "half" of which is derived from one of the two Alus that flanked the deleted region prior to its loss) (Elliott et al. 2005). MMEJ is more likely to occur when there is a significant sequence divergence between the two genomic Alu elements. This process has been referred to as "micro-SSA" and occurs more frequently between highly diverged Alu elements (Elliott et al. 2005).

The third pathway is single-strand annealing (SSA). During SSA, the process initiates when a DSB is made between two repeated sequences (e.g., Alu elements) oriented in the same direction. Single-stranded DNA regions are created adjacent to the break that extends to the repeated sequences such that complementary strands can anneal to each other. When enough homology is found, the two single strands of DNA anneal. This annealed intermediate heteroduplex DNA (hDNA) can be processed by digesting away the single-stranded tails and filing in the gaps which results in the deletion of the sequences between the repeat elements as shown in Fig. 3. The resolution of SSA repair is greatly affected by mismatches present in the hDNA. If mismatched DNA bases are present in the annealed hDNA, helicases may be recruited to unwind the mismatched duplex in a process termed heteroduplex rejection (Fig. 3, step 1) (Goldfarb and Alani 2005; Sugawara et al. 2004). Alternatively, molecules that escape heteroduplex rejection are repaired through nonhomologous tail removal (Fig. 3, steps 2–3). There is some data suggesting that the mismatches may then be repaired by the mismatch repair pathway (MMR) or a related pathway (Morales et al. 2015). It is not our goal to review all of the variants of these processes, but we will instead discuss how exposure of Alu elements as single strands can influence the outcome of the recombination events.

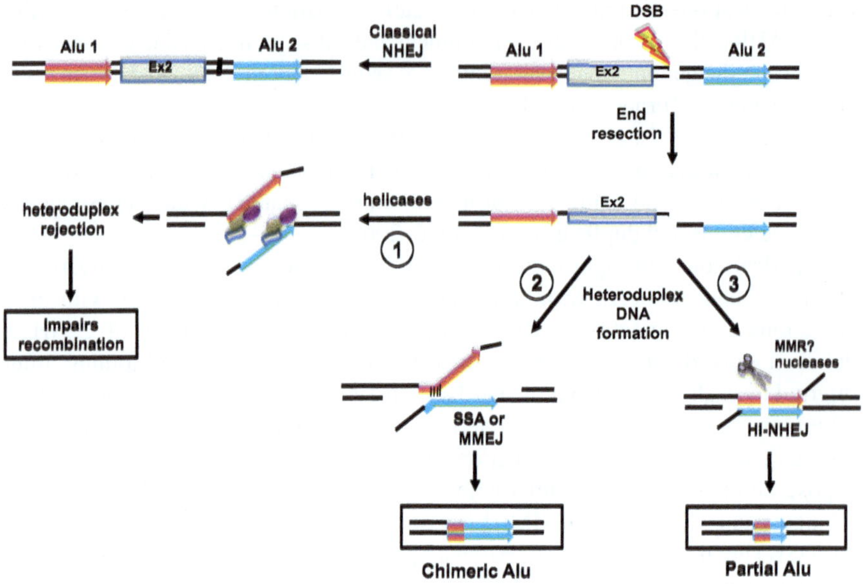

Fig. 3 Model of DSB repair between diverged Alu elements. A double-strand break (lightning bolt) occurs between two diverged Alu elements (*arrows*). While the majority of repair likely occurs through classical NHEJ, a subset of the breaks will undergo end resection followed by Alu heteroduplex DNA formation. Diverged Alu heteroduplexes can be resolved in one of the three independent ways. (*1*) Some divergent Alu heteroduplexes will either undergo helicase-mediated heteroduplex rejection, preventing mutagenic recombination. (*2*) In the absence of heteroduplex rejection, some of the Alu heteroduplex DNA intermediates are repaired either through SSA or MMEJ resulting in the formation of a single chimeric Alu element. (*3*) Some highly divergent (15–25 %) Alu heteroduplexes will be recognized, possibly by mismatch repair (MMR), and destroyed by the creation of nuclease-mediated breaks within the heteroduplex, resulting in a deletion and forming a partial heterologous Alu element. The homeology-influenced NHEJ (HI-NHEJ) repair pathway was proposed by Morales et al. (2015)

6 Orientations of Alu Elements and Different Forms of Rearrangement

The best characterized events in disease are due to unequal recombination between chromosomes because of misalignment of nonallelic Alu elements that are in a direct repeat orientation relative to one another (Fig. 2a, c). These are the dominant forms of recombination discussed in the mechanistic aspects of Alu-based recombination processes section, above. They can result in either a deletion, duplication, or complex rearrangements of the segments between the nonallelic Alu elements (Gu et al. 2015; Morales et al. 2015). In addition, exposed Alu elements flanking a DSB can result in an intrachromosomal interaction of Alu elements that result in repair by SSA and MMEJ (Morales et al. 2015).

It is easy to picture nonallelic homologous recombination occurring between two nearby identical Alu elements in the direct orientation. In fact, it might be somewhat amazing that with so many Alu elements the genome can show any significant stability at all. However, the vast majority of Alu elements near one another share extensive mismatches and will be subjected to heteroduplex rejection. There are a number of lines of evidence that point to the possibility that the interactions of exposed nonallelic Alu elements when they are sequence diverged may trigger alternative outcomes to nonallelic recombination. One of the first direct experiments to demonstrate that diverged Alu/Alu recombination events may be a critical factor in prioritizing DNA DSB repair pathway choice, from either several competing DNA repair pathways or the degree of Alu sequence divergence, was by Morales et al. (2015). The authors developed a novel vector system that, for the first time, allowed a highly flexible measurement of diverged intrachromosomal Alu/Alu recombination events. In this study, the authors investigated Alu/Alu recombination pathway choice between different levels of Alu divergence (e.g., 0–30 %). When greater levels of Alu sequence divergence (15–30 %) were tested in HEK293 cells, the authors observed a significant increase in variable-length NHEJ repair with junctions around the two Alu elements. This study also showed that highly diverged Alu elements may still be able to form an Alu heteroduplex DNA (hDNA) as shown in Fig. 3, step 3. With greater degrees of mismatch within the Alu heteroduplex (≥15 %), the authors proposed that heteroduplex rejection is less effective relative to a mismatch-related DNA cleavage event. Instead, the Alu hDNA may be preferentially subject to DNA cleavage, which may undergo NHEJ or microhomology-mediated end joining (Fig. 4). This process was proposed to be dependent on the MMR pathway to create the breaks in the two Alu elements (Fig. 3, step 3). Because this localized NHEJ process is dependent on the presence of homeologous Alu hDNA, the authors referred to the influence of Alu elements as homeology-influenced nonhomologous end joining (HI-NHEJ). This observation helps explain anecdotal observations that

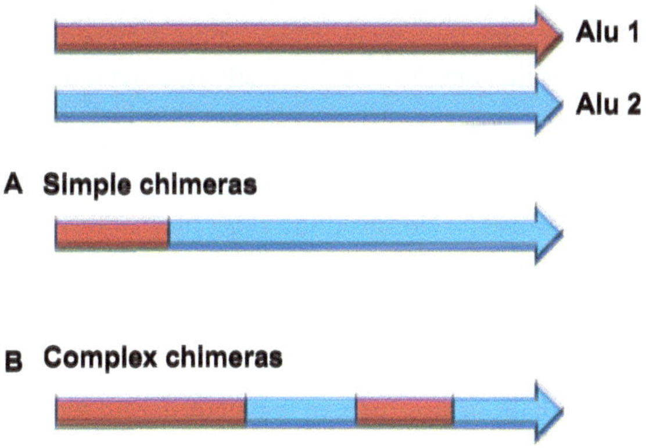

Fig. 4 Distribution of single and complex chimeras. Alu/Alu recombination can generate two types of outcomes: simple chimera or complex chimera. (**a**) A simple chimera shows only one recombination junction (*left* portion of the Alu sequence chimera derived from Alu1 and *right* portion derived from Alu2). (**b**) A complex chimera shows multiple shifts from sequences related to one Alu or the other across the length of the chimera. This is more likely the result of patchy repair processes rather than multiple crossovers

a number of NHEJ events seem to be concentrated in the vicinity of Alu elements (reviewed in (Hedges and Deininger 2007)).

The breakpoint signatures of an SSA product derived from two *direct* homeologous Alu elements have been of particular interest. In our recent study, modest levels of sequence divergence between Alu elements (≤10 %) showed that 88 % of the Alu/Alu recombination events contained a single-crossover region within the Alu element. The remaining events (12 %) contained sequence patches from each Alu element, which represents the formation of complex chimeric Alus (Morales et al. 2015). This paper found that both strands are equivalent targets for mismatch repair processes. These characteristics often result in short stretches of repair on both DNA strands that has the ability to create chimeras with patches of each repeated sequence (Fig. 4). Formation of these complex chimeras seems to reflect two different resolutions of SSA repair. The first one is complex chimeras, which shows a patchwork of sequence variation from different regions of the two different direct Alu sequences (Fig. 4b). Complex chimeras have been observed when solving the heteroduplex formed by two homeologous sequences in yeast (Sugawara et al. 2004). Complex chimeras have also been seen in Alu/Alu recombinations that cause disease. The most recent reports were in the SPAST gene (8 %) (Boone et al. 2014), VHL (7 %) (Franke et al. 2009), and EPCAM (11 %) (Kuiper et al. 2011). They have also been seen in an elegant vector system to measure rare Alu/Alu recombination between chromosomes (Elliott et al. 2005).

Simple Alu recombination junctions (e.g., an intact Alu element with 5′ portion derived from Alu1 and the 3′ portion derived from Alu2) make up the majority of the Alu NAR breakpoint junctions in the genome (Fig. 4a) (Boone et al. 2014; Gu et al. 2015; Morales et al. 2015). Various studies have observed that naturally occurring Alu NAR preferentially occurs near the 5′ region of Alu elements (Morales et al. 2015; Boone et al. 2014; Gu et al. 2015). In one of our recently published studies, we analyzed 100 reported diseases causing Alu NAR events, and found that 53 % of the breakpoint junctions preferred recombination within the first 100 bp (e.g., 5′ end) of the Alu element (Morales et al. 2015). Several hypotheses have been suggested explaining the 5′ preference, including a "chi-like" hotspot for recombination (Rudiger et al. 1995) and the increased accessibility around the transcription factor-binding sites for the RNA polymerase III promoter to recombination proteins (Hedges and Deininger 2007). However, the exact mechanism of why recombination happens at the 5′ end is poorly understood.

An alternative form of Alu-influenced instability occurs when Alu elements are arranged in an inverted orientation near one another (Gebow et al. 2000; Lobachev et al. 2000; Stenger et al. 2001; Gu et al. 2015). These particular inverted Alu elements appear to build hairpin structures in the DNA during replication, which can cause DSBs and excision of inverted Alu elements from the human genome (Fig. 2b) (Lobachev et al. 2000; Voineagu et al. 2008). Moreover, these Alu-mediated hairpin structures appear to cause replication stalling and subsequent collapsing of the replication fork, which can also lead to DSBs and/or intra- or intermolecular template switching (Voineagu et al. 2008). A second possibility for Alu elements oriented in an inverted orientation is for them to contribute to inverting the sequences in between the Alu elements. Gu et al. (2015) found that 25 % of the Alu/Alu-mediated CNVs found in chromosome 17p13.3 lead to an inversion event. Several examples of these types of rearrangements have also been found in the human and chimpanzee genomes (Lee et al. 2008).

7 Methods for Studying Alu-Based Instabilities

Much of our understanding about the contributions of Alu elements comes from anecdotal studies that detect Alu-related instabilities in the course of studying diseases. The ability to detect various types of Alu-related recombination events is dependent on the methods applied in the analysis. For instance, prior to the use of PCR for detecting genetic defects, many studies utilized Southern blots, which were able to robustly detect the larger Alu-related events. However, as PCR became the dominant diagnostic tool, most assays were carried out in a way that had difficulty detecting large deletions. Thus, in recent years, most studies have been strongly biased against finding the larger rearrangements that might be associated with Alu elements.

A number of approaches have been developed to detect CNVs in various genes (Boone et al. 2014; de Smith et al. 2008; Gu et al. 2015; Vissers et al. 2009), some of which have been specifically utilized to detect Alu-related CNVs. The most common approach is to utilize a high-density comparative genomic hybridization (aCGH) microarray to measure CNVs, followed by PCR to identify the specific breakpoint junctions and rearrangements (Gu et al. 2015; Lucito et al. 2003). These studies have been highly effective, and detected 584 Alu-mediated hot spots in a series of genes associated with autism spectrum disorder, as well as an Alu-rich region around the SPAST gene associated with spastic parapalegia (Boone et al. 2011, 2014). These types of studies provide outstanding information on Alu-related duplication and deletion events that may not represent the average behavior of Alu elements in the genome.

Whole-genome analyses for comparative genomics and other next-generation sequencing studies for sequence variation also have tremendous potential to detect Alu-related instabilities. There are, however, major limitations to these studies. For instance, these approaches have been shown to extensively detect Alu-related deletions (Han et al. 2007); the same investigators found it difficult to establish an approach for detecting duplications.

Similarly, paired-end NGS studies have the potential for characterizing most types of genome variations. However, the design of appropriate pipelines to detect Alu-related deletions and duplications is difficult and rarely used. Thus, the vast majority of NGS studies have missed Alu-related CNVs. For instance, with paired-end reads spanning 300–400 base fragments, few reads will have one read on either side of an Alu element. Therefore, these NGS strategies will have difficulty detecting a deletion or duplication between two Alu elements. An approach such as mate pair sequencing (Zeitouni et al. 2010), which allows sequencing the ends of longer fragments, would detect such events better. Mate paired sequencing is rarely performed, and will only be sensitive to those that change the fragment size sufficiently to clearly show discordance vs. the size of the original fragment size. Thus, depending on the preparation and the analytical pipeline, each approach creates biases.

All of the methods described above have the limitation that they are looking at events that occurred at an earlier point and have been under varying levels of selection that may bias the types and sizes of events detected. Furthermore, although it is easy to assess that Alu elements must have played some role in any of the NAR-related events, one can only model the mechanisms that contributed to them. Although several genomic studies have suggested that either direct Alu repeats (Bailey et al. 2003) or inverted Alu repeats (Cook et al. 2013; Konkel and Batzer 2010) may lead to an enrichment of NHEJ-like repair events in the vicinity of Alu elements, these observations are relatively indirect and do not allow testing of mechanisms.

A complementary approach for observing natural Alu-related CNVs detected through the methods above is presented by reporter-gene approaches (Elliott et al. 2005; Gebow et al. 2000; Morales et al. 2015). Although reporter gene approaches do not utilize Alu elements in their natural settings, they do allow complicated manipulations to test specific characteristics of Alu elements. These approaches can be used to test the influence of genes that may influence the Alu-related recombination events in cells where specific genes have been manipulated. Thus, reporter gene approaches have the capability of providing important mechanistic insights, as well as the ability to directly test theories developed from observations of potential Alu-related rearrangements in genomes.

Three previous investigators devised vector systems specifically to study Alu/Alu recombination. The first used the human thymidine kinase gene (HSVtk) as a reporter, with the Alu elements placed in introns of the gene (Gebow et al. 2000). The Gebow system worked reasonably well, but was integrated randomly into chromosomes and limited to studies in tk-negative cells. More recently, an additional system placed two portions of a reporter gene on different chromosomes that could be linked by Alu/Alu recombination (Elliott et al. 2005). Although this system detected Alu recombination events, trans-chromosomal Alu recombination events are rare. Therefore, this system does not mimic the most common forms of Alu/Alu recombination that occur in a more proximal, intrachromosomal manner in vivo. The third Alu recombination system was described in Morales et al. (2015). This vector system has the benefits of the first system and is able to use almost any pair of reporter genes, with one allowing selection against accumulation of background, as well as allowing its use at a specific chromosomal site using the FLP/In system that will allow direct comparisons between recombination vectors with variations, such as mismatch, orientation, and spacing between Alus. This group found that highly diverged Alu elements have a higher rate of NAR in mammalian cells than expected by processing the Alu heteroduplex DNA through HI-NHEJ. This study has laid the foundation for future work on Alu NAR utilizing a unique tool to gain a better understanding of this complex and highly regulated process in maintaining genome integrity in mammalian cells.

References

Ade C, Roy-Engel AM, Deininger PL (2013) Alu elements: an intrinsic source of human genome instability. Curr Opin Virol 3:639–645

Bailey JA, Liu G, Eichler EE (2003) An Alu transposition model for the origin and expansion of human segmental duplications. Am J Hum Genet 73:823–834

Batzer MA, Deininger PL (2002) Alu repeats and human genomic diversity. Nat Rev Genet 3:370–379

Batzer MA, Deininger PL, Hellmann-Blumberg U, Jurka J, Labuda D, Rubin CM, Schmid CW, Zietkiewicz E, Zuckerkandl E (1996) Standardized nomenclature for Alu repeats. J Mol Evol 42:3–6

Belancio VP, Deininger PL, Roy-Engel AM (2009) LINE dancing in the human genome: transposable elements and disease. Genome Med 1:97

Boone PM, Liu P, Zhang F, Carvalho CM, Towne CF, Batish SD, Lupski JR (2011) Alu-specific microhomology-mediated deletion of the final exon of SPAST in three unrelated subjects with hereditary spastic paraplegia. Genet Med 13:582–592

Boone PM, Yuan B, Campbell IM, Scull JC, Withers MA, Baggett BC, Beck CR, Shaw CJ, Stankiewicz P, Moretti P, Goodwin WE, Hein N, Fink JK, Seong MW, Seo SH, Park SS, Karbassi ID, Batish SD, Ordonez-Ugalde A, Quintans B, Sobrido MJ, Stemmler S, Lupski JR (2014) The Alu-rich genomic architecture of SPAST predisposes to diverse and functionally distinct disease-associated CNV alleles. Am J Hum Genet 95:143–161

Boulton SJ, Jackson SP (1996) Saccharomyces cerevisiae Ku70 potentiates illegitimate DNA double-strand break repair and serves as a barrier to error-prone DNA repair pathways. EMBO J 15:5093–5103

Britten RJ (1996) DNA sequence insertion and evolutionary variation in gene regulation. Proc Natl Acad Sci U S A 93:9374–9377

Brosens LA, Offerhaus GJ, Giardiello FM (2015) Hereditary colorectal cancer: genetics and screening. Surg Clin North Am 95:1067–1080

Callen E, Tischkowitz MD, Creus A, Marcos R, Bueren JA, Casado JA, Mathew CG, Surralles J (2004) Quantitative PCR analysis reveals a high incidence of large intragenic deletions in the FANCA gene in Spanish Fanconi anemia patients. Cytogenet Genome Res 104:341–345

Chimpanzee Sequencing and Analysis Consortium (2005) Initial sequence of the chimpanzee genome and comparison with the human genome. Nature 437:69–87

Cook GW, Konkel MK, Walker JA, Bourgeois MG, Fullerton ML, Fussell JT, Herbold HD, Batzer MA (2013) A comparison of 100 human genes using an alu element-based instability model. PLoS One 8:e65188

de Smith AJ, Walters RG, Coin LJ, Steinfeld I, Yakhini Z, Sladek R, Froguel P, Blakemore AI (2008) Small deletion variants have stable breakpoints commonly associated with alu elements. PLoS One 3:e3104

Deininger P (2011) Alu elements: know the SINEs. Genome Biol 12:236

Deininger PL, Batzer MA (1999) Alu repeats and human disease. Mol Genet Metab 67:183–193

Deininger PL, Batzer MA (2002) Mammalian retroelements. Genome Res 12:1455–1465

Deininger PL, Batzer MA, Hutchison CA III, Edgell MH (1992) Master genes in mammalian repetitive DNA amplification. Trends Genet 8:307–311

Deininger PL, Moran JV, Batzer MA, Kazazian HH Jr (2003) Mobile elements and mammalian genome evolution. Curr Opin Genet Dev 13:651–658

Dewannieux M, Esnault C, Heidmann T (2003) LINE-mediated retrotransposition of marked Alu sequences. Nat Genet 35:41–48

Elliott B, Richardson C, Jasin M (2005) Chromosomal translocation mechanisms at intronic alu elements in mammalian cells. Mol Cell 17:885–894

Erez A, Patel AJ, Wang X, Xia Z, Bhatt SS, Craigen W, Cheung SW, Lewis RA, Fang P, Davenport SL, Stankiewicz P, Lalani SR (2009) Alu-specific microhomology-mediated deletions in CDKL5 in females with early-onset seizure disorder. Neurogenetics 10:363–369

Faiz F, Allcock RJ, Hooper AJ, van Bockxmeer FM (2013) Detection of variations and identifying genomic breakpoints for large deletions in the LDLR by Ion Torrent semiconductor sequencing. Atherosclerosis 230:249–255

Featherstone C, Jackson SP (1999) DNA double-strand break repair. Curr Biol 9:R759–R761

Ferguson DO, Alt FW (2001) DNA double strand break repair and chromosomal translocation: lessons from animal models. Oncogene 20:5572–5579

Franke G, Bausch B, Hoffmann MM, Cybulla M, Wilhelm C, Kohlhase J, Scherer G, Neumann HP (2009) Alu-Alu recombination underlies the vast majority of large VHL germline deletions: Molecular characterization and genotype-phenotype correlations in VHL patients. Hum Mutat 30:776–786

Gebow D, Miselis N, Liber HL (2000) Homologous and nonhomologous recombination resulting in deletion: effects of p53 status, microhomology, and repetitive DNA length and orientation. Mol Cell Biol 20:4028–4035

Goldfarb T, Alani E (2005) Distinct roles for the Saccharomyces cerevisiae mismatch repair proteins in heteroduplex rejection, mismatch repair and nonhomologous tail removal. Genetics 169:563–574

Gu W, Zhang F, Lupski JR (2008) Mechanisms for human genomic rearrangements. Pathogenetics 1:4

Gu S, Yuan B, Campbell IM, Beck CR, Carvalho CM, Nagamani SC, Erez A, Patel A, Bacino CA, Shaw CA, Stankiewicz P, Cheung SW, Bi W, Lupski JR (2015) Alu-mediated diverse and complex pathogenic copy-number variants within human chromosome 17 at p13.3. Hum Mol Genet 24:4061–4077

Gu S, Posey JE, Yuan B, Carvalho CM, Luk HM, Erikson K, Lo IF, Leung GK, Pickering CR, Chung BH, Lupski JR (2016) Mechanisms for the generation of two quadruplications associated with split-hand malformation. Hum Mutat 37:160–164

Han K, Lee J, Meyer TJ, Wang J, Sen SK, Srikanta D, Liang P, Batzer MA (2007) Alu recombination-mediated structural deletions in the chimpanzee genome. PLoS Genet 3:1939–1949

Hedges DJ, Deininger PL (2007) Inviting instability: transposable elements, double-strand breaks, and the maintenance of genome integrity. Mutat Res 616:46–59

Hedges DJ, Callinan PA, Cordaux R, Xing J, Barnes E, Batzer MA (2004) Differential alu mobilization and polymorphism among the human and chimpanzee lineages. Genome Res 14:1068–1075

Johnson RD, Jasin M (2000) Sister chromatid gene conversion is a prominent double-strand break repair pathway in mammalian cells. EMBO J 19:3398–3407

Kaer K, Speek M (2013) Retroelements in human disease. Gene 518:231–241

Kennedy RD, D'Andrea AD (2006) DNA repair pathways in clinical practice: lessons from pediatric cancer susceptibility syndromes. J Clin Oncol 24:3799–3808

Konkel MK, Batzer MA (2010) A mobile threat to genome stability: the impact of non-LTR retrotransposons upon the human genome. Semin Cancer Biol 20:211–221

Konkel MK, Walker JA, Hotard AB, Ranck MC, Fontenot CC, Storer J, Stewart C, Marth GT, Batzer MA (2015) Sequence analysis and characterization of active human Alu subfamilies based on the 1000 genomes pilot project. Genome Biol Evol 7:2608–2622

Kramerov DA, Vassetzky NS (2005) Short retroposons in eukaryotic genomes. Int Rev Cytol 247:165–221

Kriegs JO, Churakov G, Jurka J, Brosius J, Schmitz J (2007) Evolutionary history of 7SL RNA-derived SINEs in Supraprimates. Trends Genet 23:158–161

Kuiper RP, Vissers LE, Venkatachalam R, Bodmer D, Hoenselaar E, Goossens M, Haufe A, Kamping E, Niessen RC, Hogervorst FB, Gille JJ, Redeker B, Tops CM, van Gijn ME, van den Ouweland AM, Rahner N, Steinke V, Kahl P, Holinski-Feder E, Morak M, Kloor M, Stemmler S, Betz B, Hutter P, Bunyan DJ, Syngal S, Culver JO, Graham T, Chan TL, Nagtegaal ID, van Krieken JH, Schackert HK, Hoogerbrugge N, van Kessel AG, Ligtenberg MJ (2011) Recurrence and variability of germline EPCAM deletions in Lynch syndrome. Hum Mutat 32:407–414

Lam KW, Jeffreys AJ (2006) Processes of copy-number change in human DNA: the dynamics of {alpha}-globin gene deletion. Proc Natl Acad Sci U S A 103:8921–8927

Lam KW, Jeffreys AJ (2007) Processes of de novo duplication of human alpha-globin genes. Proc Natl Acad Sci U S A 104:10950–10955

Lander ES, Linton LM, Birren B, Nusbaum C, Zody MC, Baldwin J, Devon K, Dewar K, Doyle M, FitzHugh W, Funke R, Gage D, Harris K, Heaford A, Howland J, Kann L, Lehoczky J, LeVine R, McEwan P, McKernan K, Meldrim J, Mesirov JP, Miranda C, Morris W, Naylor J, Raymond C, Rosetti M, Santos R, Sheridan A, Sougnez C, Stange-Thomann Y, Stojanovic N, Subramanian A, Wyman D, Rogers J, Sulston J, Ainscough R, Beck S, Bentley D, Burton J, Clee C, Carter N, Coulson A, Deadman R, Deloukas P, Dunham A, Dunham I, Durbin R, French L, Grafham D, Gregory S, Hubbard T, Humphray S, Hunt A, Jones M, Lloyd C, McMurray A, Matthews L, Mercer S, Milne S, Mullikin JC, Mungall A, Plumb R, Ross M, Shownkeen R, Sims S, Waterston RH, Wilson RK, Hillier LW, McPherson JD, Marra MA, Mardis ER, Fulton LA, Chinwalla AT, Pepin KH, Gish WR, Chissoe SL, Wendl MC, Delehaunty KD, Miner TL, Delehaunty A, Kramer JB, Cook LL, Fulton RS, Johnson DL, Minx PJ, Clifton SW, Hawkins T, Branscomb E, Predki P, Richardson P, Wenning S, Slezak T, Doggett N, Cheng JF, Olsen A, Lucas S, Elkin C, Uberbacher E, Frazier M, Gibbs RA, Muzny DM, Scherer SE, Bouck JB, Sodergren EJ, Worley KC, Rives CM, Gorrell JH, Metzker ML, Naylor SL, Kucherlapati RS, Nelson DL, Weinstock GM, Sakaki Y, Fujiyama A, Hattori M, Yada T, Toyoda A, Itoh T, Kawagoe C, Watanabe H, Totoki Y, Taylor T, Weissenbach J, Heilig R, Saurin W, Artiguenave F, Brottier P, Bruls T, Pelletier E, Robert C, Wincker P, Smith DR, Doucette-Stamm L, Rubenfield M, Weinstock K, Lee HM, Dubois J, Rosenthal A, Platzer M, Nyakatura G, Taudien S, Rump A, Yang H, Yu J, Wang J, Huang G, Gu J, Hood L, Rowen L, Madan A, Qin S, Davis RW, Federspiel NA, Abola AP, Proctor MJ, Myers RM, Schmutz J, Dickson M, Grimwood J, Cox DR, Olson MV, Kaul R, Raymond C, Shimizu N, Kawasaki K, Minoshima S, Evans GA, Athanasiou M, Schultz R, Roe BA, Chen F, Pan H, Ramser J, Lehrach H, Reinhardt R, McCombie WR, de la Bastide M, Dedhia N, Blocker H, Hornischer K, Nordsiek G, Agarwala R, Aravind L, Bailey JA, Bateman A, Batzoglou S, Birney E, Bork P, Brown DG, Burge CB, Cerutti L, Chen HC, Church D, Clamp M, Copley RR, Doerks T, Eddy SR, Eichler EE, Furey TS, Galagan J, Gilbert JG, Harmon C, Hayashizaki Y, Haussler D, Hermjakob H, Hokamp K, Jang W, Johnson LS, Jones TA, Kasif S, Kaspryzk A, Kennedy S, Kent WJ, Kitts P, Koonin EV, Korf I, Kulp D, Lancet D, Lowe TM, McLysaght A, Mikkelsen T, Moran JV, Mulder N, Pollara VJ, Ponting CP, Schuler G, Schultz J, Slater G, Smit AF, Stupka E, Szustakowki J, Thierry-Mieg D, Thierry-Mieg J, Wagner L, Wallis J, Wheeler R, Williams A, Wolf YI, Wolfe KH, Yang SP, Yeh RF, Collins F, Guyer MS, Peterson J, Felsenfeld A, Wetterstrand KA, Patrinos A, Morgan MJ, de Jong P, Catanese JJ, Osoegawa K, Shizuya H, Choi S, Chen YJ (2001) Initial sequencing and analysis of the human genome. Nature 409:860–921

Lee J, Han K, Meyer TJ, Kim HS, Batzer MA (2008) Chromosomal inversions between human and chimpanzee lineages caused by retrotransposons. PLoS One 3:e4047

Lehrman MA, Schneider WJ, Sudhof TC, Brown MS, Goldstein JL, Russell DW (1985) Mutation in LDL receptor: Alu-Alu recombination deletes exons encoding transmembrane and cytoplasmic domains. Science 227:140–146

Levy S, Sutton G, Ng PC, Feuk L, Halpern AL, Walenz BP, Axelrod N, Huang J, Kirkness EF, Denisov G, Lin Y, MacDonald JR, Pang AW, Shago M, Stockwell TB, Tsiamouri A, Bafna V, Bansal V, Kravitz SA, Busam DA, Beeson KY, McIntosh TC, Remington KA, Abril JF, Gill J, Borman J, Rogers YH, Frazier ME, Scherer SW, Strausberg RL, Venter JC (2007) The diploid genome sequence of an individual human. PLoS Biol 5:e254

Lobachev KS, Stenger JE, Kozyreva OG, Jurka J, Gordenin DA, Resnick MA (2000) Inverted Alu repeats unstable in yeast are excluded from the human genome. EMBO J 19:3822–3830

Luan DD, Korman MH, Jakubczak JL, Eickbush TH (1993) Reverse transcription of R2Bm RNA is primed by a nick at the chromosomal target site: a mechanism for non-LTR retrotransposition. Cell 72:595–605

Lucito R, Healy J, Alexander J, Reiner A, Esposito D, Chi M, Rodgers L, Brady A, Sebat J, Troge J, West JA, Rostan S, Nguyen KC, Powers S, Ye KQ, Olshen A, Venkatraman E, Norton L,

Wigler M (2003) Representational oligonucleotide microarray analysis: a high-resolution method to detect genome copy number variation. Genome Res 13:2291–2305
Luo G, Santoro IM, McDaniel LD, Nishijima I, Mills M, Youssoufian H, Vogel H, Schultz RA, Bradley A (2000) Cancer predisposition caused by elevated mitotic recombination in Bloom mice. Nat Genet 26:424–429
Ma JL, Kim EM, Haber JE, Lee SE (2003) Yeast Mre11 and Rad1 proteins define a Ku-independent mechanism to repair double-strand breaks lacking overlapping end sequences. Mol Cell Biol 23:8820–8828
McVean G (2010) What drives recombination hotspots to repeat DNA in humans? Philos Trans R Soc Lond B Biol Sci 365:1213–1218
McVey M, Lee SE (2008) MMEJ repair of double-strand breaks (director's cut): deleted sequences and alternative endings. Trends Genet 24:529–538
Medstrand P, van de Lagemaat LN, Mager DL (2002) Retroelement distributions in the human genome: variations associated with age and proximity to genes. Genome Res 12:1483–1495
Mills RE, Bennett EA, Iskow RC, Luttig CT, Tsui C, Pittard WS, Devine SE (2006) Recently mobilized transposons in the human and chimpanzee genomes. Am J Hum Genet 78:671–679
Morales ME, White TB, Streva VA, DeFreece CB, Hedges DJ, Deininger PL (2015) The contribution of alu elements to mutagenic DNA double-strand break repair. PLoS Genet 11:e1005016
Nissen PH, Damgaard D, Stenderup A, Nielsen GG, Larsen ML, Faergeman O (2006) Genomic characterization of five deletions in the LDL receptor gene in Danish Familial Hypercholesterolemic subjects. BMC Med Genet 7:55
Pizarro JG, Cristofari G (2016) Post-transcriptional control of LINE-1 retrotransposition by cellular host factors in somatic cells. Front Cell Dev Biol 4:14
Quentin Y (1992) Fusion of a free left Alu monomer and a free right Alu monomer at the origin of the Alu family in the primate genomes. Nucleic Acids Res 20:487–493
Reddy R, Nguyen NM, Sarrabay G, Rezaei M, Rivas MC, Kavasoglu A, Berkil H, Elshafey A, Nunez KP, Dreyfus H, Philippe M, Hadipour Z, Durmaz A, Eaton EE, Schubert B, Ulker V, Hadipour F, Ahmadpour F, Touitou I, Fardaei M, Slim R (2016) The genomic architecture of NLRP7 is Alu rich and predisposes to disease-associated large deletions. Eur J Hum Genet 24(10):1445–1452
Rodgers K, McVey M (2016) Error-prone repair of DNA double-strand breaks. J Cell Physiol 231:15–24
Rudiger NS, Gregersen N, Kielland-Brandt MC (1995) One short well conserved region of Alu-sequences is involved in human gene rearrangements and has homology with prokaryotic chi. Nucleic Acids Res 23:256–260
Sen SK, Han K, Wang J, Lee J, Wang H, Callinan PA, Dyer M, Cordaux R, Liang P, Batzer MA (2006) Human genomic deletions mediated by recombination between Alu elements. Am J Hum Genet 79:41–53
Sharp AJ, Itsara A, Cheng Z, Alkan C, Schwartz S, Eichler EE (2007) Optimal design of oligonucleotide microarrays for measurement of DNA copy-number. Hum Mol Genet 16:2770–2779
Shen MR, Batzer MA, Deininger PL (1991) Evolution of the master Alu gene(s). J Mol Evol 33:311–320
Stenger JE, Lobachev KS, Gordenin D, Darden TA, Jurka J, Resnick MA (2001) Biased distribution of inverted and direct Alus in the human genome: implications for insertion, exclusion, and genome stability. Genome Res 11:12–27
Strout MP, Marcucci G, Bloomfield CD, Caligiuri MA (1998) The partial tandem duplication of ALL1 (MLL) is consistently generated by Alu-mediated homologous recombination in acute myeloid leukemia. Proc Natl Acad Sci U S A 95:2390–2395
Sugawara N, Goldfarb T, Studamire B, Alani E, Haber JE (2004) Heteroduplex rejection during single-strand annealing requires Sgs1 helicase and mismatch repair proteins Msh2 and Msh6 but not Pms1. Proc Natl Acad Sci U S A 101:9315–9320
Turner DJ, Miretti M, Rajan D, Fiegler H, Carter NP, Blayney ML, Beck S, Hurles ME (2008) Germline rates of de novo meiotic deletions and duplications causing several genomic disorders. Nat Genet 40:90–95

Vilenchik MM, Knudson AG (2003) Endogenous DNA double-strand breaks: production, fidelity of repair, and induction of cancer. Proc Natl Acad Sci U S A 100:12871–12876

Vissers LE, Bhatt SS, Janssen IM, Xia Z, Lalani SR, Pfundt R, Derwinska K, de Vries BB, Gilissen C, Hoischen A, Nesteruk M, Wisniowiecka-Kowalnik B, Smyk M, Brunner HG, Cheung SW, van Kessel AG, Veltman JA, Stankiewicz P (2009) Rare pathogenic microdeletions and tandem duplications are microhomology-mediated and stimulated by local genomic architecture. Hum Mol Genet 18:3579–3593

Voineagu I, Narayanan V, Lobachev KS, Mirkin SM (2008) Replication stalling at unstable inverted repeats: interplay between DNA hairpins and fork stabilizing proteins. Proc Natl Acad Sci U S A 105:9936–9941

Walter P, Blobel G (1982) Signal recognition particle contains a 7S RNA essential for protein translocation across the endoplasmic reticulum. Nature 299:691–698

Xing J, Zhang Y, Han K, Salem AH, Sen SK, Huff CD, Zhou Q, Kirkness EF, Levy S, Batzer MA, Jorde LB (2009) Mobile elements create structural variation: analysis of a complete human genome. Genome Res 19:1516–1526

Yu X, Gabriel A (2003) Ku-dependent and Ku-independent end-joining pathways lead to chromosomal rearrangements during double-strand break repair in Saccharomyces cerevisiae. Genetics 163:843–856

Zeitouni B, Boeva V, Janoueix-Lerosey I, Loeillet S, Legoix-ne P, Nicolas A, Delattre O, Barillot E (2010) SVDetect: a tool to identify genomic structural variations from paired-end and mate-pair sequencing data. Bioinformatics 26:1895–1896

Retrotransposon-Driven Transcription and Cancer

Cristina Tufarelli and Richard M. Badge

1 Introduction

Retrotransposons (RTNs), also known as Type 1 or RNA transposons, are repetitive sequences that comprise about 43 % of the human genome (Lander et al. 2001). RTNs are divided into virus-like Long Terminal Repeat (LTR), and non-LTR containing elements. Whilst mobilization of LTR elements in humans is limited, several subclasses of non-LTR elements remain actively mobile (Mills et al. 2007). "Long INterspersed Element 1" (LINE-1s) are the only known autonomous class of non-LTR RTNs, accounting for about 17 % of the human genome, nearly eight times the amount of protein coding sequences (Lander et al. 2001). LINE-1s carry their own promoters and the information to copy and paste themselves to different locations in the genome; they are also responsible for the mobilization of nonautonomous Short Interspersed Nuclear Element (SINE) retrotransposons such as Alu and SVA (Dewannieux et al. 2003; Raiz et al. 2012), as well as some cellular coding and non-coding RNAs (Garcia-Perez et al. 2007).

RTNs owe their name to their ability to move to different locations within the genome using a "copy-and-paste" mechanism. This requires the elements to be transcribed into an RNA intermediate that is then reverse transcribed to cDNA for insertion at a new site within the genome (Wicker et al. 2007). RTN mobilization can thus promote genetic variation and genome reorganization (Ewing and Kazazian 2010), and not surprisingly can also contribute to disease states, through mutagenic insertion (Beck et al. 2011; Belancio et al. 2008; Tubio et al. 2014). In fact, recently

C. Tufarelli (✉)
School of Medicine, University of Nottingham, Royal Derby Hospital,
Uttoxeter Road, Derby DE22 3DT, UK
e-mail: cristina.tufarelli@nottingham.ac.uk

R.M. Badge
Department of Genetics, University of Leicester, University Road, Leicester LE1 7RH, UK

G. Cristofari (ed.), *Human Retrotransposons in Health and Disease*,
DOI 10.1007/978-3-319-48344-3_11

it has become clear that, at least in epithelial cancers, activation of RTNs correlates with their mobilization and genomic rearrangements (Lee et al. 2012; Solyom et al. 2012; Rangasamy et al. 2015; Kassiotis 2014). The impact of retrotransposon mobilization is discussed in chapters “Retrotransposon Contribution to Genomic Plasticity,” “The Mobilisation of Processed Transcripts in Germline and Somatic Tissues,” “Neuronal Genome Plasticity: Retrotransposons, Environment and Disease,” and “Activity of Retrotransposons in Stem Cells and Differentiated Cells” of this book. However, in addition to the effects of mobilization, RTNs can contribute to regulating host genome activity independently of retrotransposition (Gifford et al. 2013; Goodier and Kazazian 2008). The data supporting these effects in mammals provide evidence that during the course of evolution some RTNs have developed retrotransposition-independent regulatory functions that can be advantageous (Donnelly et al. 1999; Lowe and Haussler 2012) or detrimental (Cruickshanks et al. 2013; Wolff et al. 2010; Wilkins 2010) to their hosts. Most notable is the contribution of endogenous retroviruses or ERVs (LTR containing RTNs) to placentation, suggesting that RTNs have been pivotal in the evolution of placental mammals (Dupressoir et al. 2012) (see also chapter “Roles of Endogenous Retrovirus-Encoded Syncytins in Human Placentation” on syncytins). Instances of RTNs acquiring regulatory functions in a normal physiological context have also been described for non-LTR RTNs; for example, conserved LINE-2 fragments have been shown to act as T-cell-specific silencers (Donnelly et al. 1999) and LINE-1s have been proposed to play a role in X-inactivation (Lyon 2006; Chow et al. 2010). In the context of cancer, retrotransposition-independent effects have been ascribed to both ERVs and LINEs (Rangasamy et al. 2015; Kassiotis 2014; Wilkins 2010). Some of these effects occur without active transcription from the promoters of the elements, for example, premature transcription termination caused by the usage of an RTN polyA site, or RTN exonization caused by usage of cryptic splice sites within the elements (Kaer et al. 2011; Cowley and Oakey 2013).

Activation of transcription from RTNs promoters is a necessary step for retrotransposition; nevertheless, transcription activation per se can have functional consequences regardless of mobilization (Fig. 1). It was proposed that RTNs have frequently acted as alternative promoters for protein coding genes (Faulkner et al. 2009) and that they can contribute to tissue-specific expression (Matlik et al. 2006; Peaston et al. 2004). Remarkably, up to 30% of all RNA transcripts have been found to initiate within RTNs (Faulkner et al. 2009). These studies support a functional role for RTNs in genome regulation, suggesting that RTNs have coevolved with the host genome, which has developed mechanisms to control their activity whilst exploiting RTNs for regulatory functions (Gifford et al. 2013; Goodier and Kazazian 2008; Goke and Ng 2016). Given that aberrant transcription from RTNs is a well-recognized hallmark of cancer, it is possible that activation of RTN promoters could have direct roles, independent of mobilization, in the transition from normalcy to malignancy, by altering the genome’s transcriptional and epigenetic profiles. Evidence supporting this hypothesis is explored within this chapter. Retrotransposon-derived regulatory regions and transcripts in the context of stem cells and pluripotency are reviewed in chapter “Retrotransposon-Derived Regulatory Regions and Transcripts in Stemness.”

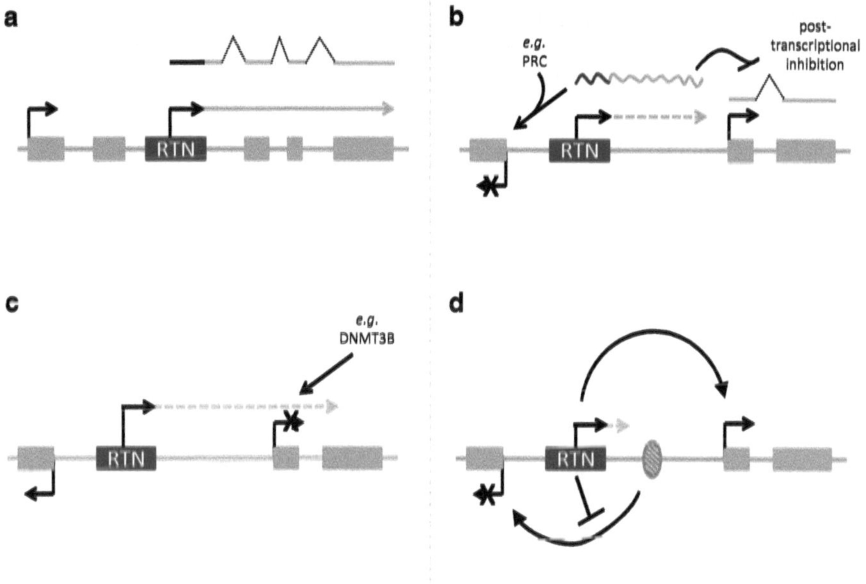

Fig. 1 Consequences of activation of RTN promoters in cancer. (**a**) Activation of the promoters of RTNs (*black box*) can act to drive chimeric transcripts (*black* and *gray lines*) that are alternative isoforms of protein coding genes (*gray boxes*), some of which may have oncogenic potential (e.g., Wolff et al. 2010; Faulkner et al. 2009; Hur et al. 2013). (**b**) RTNs can drive expression of non-coding chimeric RNAs (*black* and *gray wavy line*; e.g., Faulkner et al. 2009; Lu et al. 2014) that, similar to the HOTAIR lncRNA (Gupta et al. 2010), can act as a scaffold for chromatin remodeling complexes (e.g., PRC), resulting in silencing of regions that are complementary to them. Alternatively, non-coding RDTs can be further processed into small and micro RNAs to induce posttranscriptional inhibition of complementary RNAs (e.g., Ohms and Rangasamy 2014; Yang and Kazazian 2006). (**c**) In a more speculative scenario, as has been shown for gene body regions (Baubec et al. 2015), transcription reading through at host gene promoters could drive their silencing by inducing recruitment at these promoters of DNA methyltransferases like DNMT3B. (**d**) Activation of some RTNs may be associated with their activity as enhancers for other genes (*curved arrow*; e.g., Fort et al. 2014) or as boundary elements that could either prevent interaction of host genes with their enhancers (*gray oval*) or allow host gene expression by preventing the spreading of flanking heterochromatin (not shown; e.g., Lunyak et al. 2007). Note that for simplicity in each panel an RTN located either within (intragenic) or between (intergenic) genes is illustrated with only one promoter active. Most RTNs have bidirectional promoters and the effects described can be ascribed to both promoter activities and brought about both by intergenic and intragenic RTNs. ^ = splicing; *bent arrows* = promoters

2 RTN Promoters' Activity

It has long been recognized that RTN promoters—generally silenced and heavily methylated in normal somatic cells—are usually hypomethylated and aberrantly active in cancer. The observed association between decreased levels of DNA methylation and RTN activity has led to the notion that it is the decrease in methylation level that triggers RTN activation in cancer (Eden et al. 2003; Gaudet et al. 2003; Symer et al. 2002). It has been proposed that DNA methylation initially evolved as a mean to repress endogenous transposable elements (Yoder et al. 1997; Jones and Baylin 2007), and indeed loss of methylation is associated with RTN transcriptional activation (Walsh et al. 1998). Decreased methylation at transposable element sequences occurs naturally in the early embryo, where increased transcription from RTN promoters may be essential for correct development (Peaston et al. 2004; Macfarlan et al. 2012). However, whether hypomethylation is a cause or a consequence of RTN activation remains to be established. In colorectal cancer, the study of LINE-1 methylation has produced controversy, with divergent findings reported, for example, with regard to the association of LINE-1 methylation level and tumor stage (Ogino et al. 2008; Sunami et al. 2011; Matsunoki et al. 2012; Murata et al. 2013). Though it remains to be proven whether methylation of RTNs directly regulates their promoters or is rather a reflection of the chromosomal context (Sproul et al. 2011, 2012), it has been established that transcriptional activity of RTN promoters is increased in cancer (Criscione et al. 2014). Intriguingly, whilst accumulation of mutations during the course of evolution has rendered the majority of RTNs immobile, many non-mobile RTNs have retained strong promoters. For example, of the 500,000 LINE-1s present in the human genome only 80–100 elements are retrotransposition proficient (RC-L1s) (Brouha et al. 2003); by contrast, as many as 7000 L1s contain intact promoters from which transcription can be initiated (Khan et al. 2006).

Considering that there are more RTN promoters than actively mobile elements, it is feasible to propose that the effects of aberrant transcription from RTN promoters maybe as relevant to cancer as the effects of retrotransposition itself. Many RTN promoters are bidirectional and this property allows them to not only drive sense transcription of the element itself, but also on the antisense strand (Speek 2001; Domansky et al. 2000; Dunn et al. 2006; Cruickshanks and Tufarelli 2009). Both activities can extend transcription past the RTN into adjacent sequences giving rise to RNA transcripts comprising part repetitive element and part flanking genomic sequence, referred to hereafter as RTN-driven transcripts (RDTs). RDTs can possess protein coding potential, or be non-coding, or can simply be the by-product of aberrant transcription. Nevertheless, their presence demonstrates that RTN promoters are active in cancer; as described in detail below, such aberrant transcription could in some instances contribute to cancer development or progression as a consequence of the action of the transcripts generated or of the transcription event itself.

2.1 RTNs as Promoters of Protein Coding Transcripts

RTNs' ability to drive expression of protein coding genes was reported, even before the advent of high-throughput sequencing technologies. For example, expression of human β1,3-galactosyltransferase 5 in the colon, was shown to be driven by a

HERV-LTR (Dunn et al. 2003). Similarly, studies in mouse mature oocytes and cleavage stage embryos demonstrated that expression of a subset of host genes is driven and regulated by RTNs (Peaston et al. 2004). The propensity of RTNs to act as promoters and first exons for host genes in normal and disease situations was later confirmed by genome-wide studies (Faulkner et al. 2009). In cancer, examples of both LTR and non-LTR RTNs driving coding chimeric transcripts with potential roles in malignancy have been reported. An upstream LTR2 has been found to drive aberrant transcription, in diffuse large B-cell lymphoma, of an alternative isoform of the brain-specific fatty acid-binding protein gene FABP7 (Lock et al. 2014); similarly, upregulation of interferon regulatory factor 5 (IRF5) in Hodgkin lymphoma is driven by aberrant activation of an endogenous retroviral LTR upstream of IRF5 (Babaian et al. 2015). Promoters of LINE-1 elements have also been shown to drive transcription of protein coding chimeric transcripts in cancer. Activation of the antisense promoter of an intronic LINE-1 within the receptor tyrosine kinase *cMET* gene causes expression of a truncated isoform of cMET (L1-MET) with oncogenic properties both in chronic myeloid leukemia (Roman-Gomez et al. 2005) and in bladder cancer (Wolff et al. 2010). Expression of L1-MET was also observed in colon cancer alongside that of other chimeric transcripts coding oncogenic isoforms of the RAB3A interacting protein (RAB3IP) and the Cholinergic Receptor Muscarinic 3 (CHRM3), also driven by activation of intronic LINE-1 promoters (Hur et al. 2013).

These reports highlight the potential contribution of RTNs to cancer progression by inducing either ectopic expression of host genes or the production of truncated isoforms of cellular genes with oncogenic properties (Fig. 1a). It is worth noting that at the moment all the examples of coding chimeric transcripts are limited to those including sequences of known proteins; further research combining high-throughput transcriptomics and proteomics approaches with the development of appropriate bioinformatic tools will be required to uncover whether novel coding transcripts can arise from activation of RTNs within region expected to be non-coding, such as gene desert regions.

2.2 *RTN-Driven Non-coding RNA Transcription*

An increasing body of evidence indicates that, in addition to their contribution to coding transcripts, RTNs can act as regulators of gene expression (Elbarbary et al. 2016). It has been proposed that initiation of many cancers is triggered by epigenetic alterations brought about by activation of RTN promoters (Wilkins 2010). RTN promoters can drive transcription of antisense and non-coding RNAs; both non-coding and antisense-RNAs are able to alter the epigenetic profile (e.g., DNA methylation) of the regions through which they are transcribed, or to which they are complementary, causing changes in the way in which genes are expressed (Guttman and Rinn 2012; Werner 2013). The contribution of RTN promoters to transcription initiation of non-coding RNAs in stem cells had been previously reported (Faulkner et al. 2009). This is epitomized by the recent finding that subfamily H of human endogenous retroviruses (HERVH) gives rise to nuclear non-coding RNAs necessary for human embryonic stem cell identity through their interaction with coactivators and pluripotency factors such as OCT4 (Lu et al. 2014). Recent work also suggests that RTN-driven non-coding RNAs predominantly derive from LTR/ERVs promoters rather than from LINE-1 and Alu (Kapusta et al. 2013). However, antisense transcription from LINE-1 elements on the X chromosome has been observed and is proposed to play a role in silencing of genes that are prone to escape X-inactivation (Chow et al. 2010). Given that the majority of long non-coding RNAs tend to be retained within the nucleus, and are of considerably lower abundance than coding transcripts, it is possible that the contribution of RTNs to non-coding RNAs is currently underestimated; indeed, most reports have performed their analyses on polyadenylated RNA which is predominantly cytosolic (Djebali et al. 2012). In support of this, a recent study has shown that in mouse stem cells many previously unknown nuclear non-coding RNAs derived from ERVs can only be found when analyzing the nuclear RNA fraction (Fort et al. 2014).

Non-coding and antisense-RNAs as well as the process of transcription can have functional roles in health and in disease situations (Baubec et al. 2015; Guttman and Rinn 2012; Magistri et al. 2012; Tufarelli et al. 2003; Yu et al. 2008). A number of long non-coding RNAs (lncRNAs), i.e., those longer than 200 nt, have been shown to have genome regulatory functions (Rinn and Chang 2012). lncRNAs are deregulated in a variety of cancer types (Gutschner and Diederichs 2012) and evidence is accumulating suggesting that they can play key roles in tumorigenesis (Cheetham et al. 2013). An early example of a long non-coding RNA in cancer is that of HOTAIR, an RNA that binds to polycomb repressive complex 2 (PRC2), and whose overexpression in breast cancer leads to increased invasiveness and metastasis (Gupta et al. 2010). This example highlights the ability of some lncRNAs to act as scaffolds for chromatin-remodeling complexes and regulate the complexes' spatial and temporal function (Khalil et al. 2009; Koziol and Rinn 2010). In addition to function as scaffolds, lncRNAs can exert other molecular functions—e.g., act as signals, decoys, and guides—which given the modular structure of RNAs can be exercised in different combinations by individual lncRNAs (Rinn and Chang 2012; Wang and Chang 2011). Not do only RTNs contribute heavily to lncRNA sequences (Ganesh and Svoboda 2016), but more importantly they provide promoters for many lncRNAs, as the majority of naturally occurring antisense transcripts and a significant number of non-coding RNAs have been found to initiate at RTNs (Faulkner et al. 2009; Conley et al. 2008). It is therefore possible that activation of RTN transcription in cancer can also contribute to cancer development or progression by deregulating lncRNAs (Fig. 1b).

In addition to lncRNAs, RTNs can give rise to short RNAs (including siRNAs and miRNAs), some of which play roles in regulating RTN expression, via RNA interference pathways (Yang and Kazazian 2006). It has been proposed that RNA interference systems that evolved to silence RTNs have been co-opted to give rise to miRNA pathways shown to regulate host gene expression, and also that several of the miRNAs acting on host gene are derived from RTNs (Piriyapongsa et al. 2007).

Interestingly, recent work indicates that expression of LINE-1 in human cancers may contribute to downregulation of several miRNAs that may be important in maintaining genome integrity (Ohms and Rangasamy 2014), suggesting an additional way in which RTNs can contribute to cancer progression (Fig. 1b).

Finally, it is worth noting that, in the last few years, it has been established that the process of transcription can alter the chromatin of read-through regions; transcribing RNA polymerase II interacts with the histone methyltransferase SETD2 causing trimethylation at lysine 36 of histone H3 (H3K36me3), which in turn recruits the de novo DNA methyltransferase DNMT3B, leading to methylation of the read-through region (Baubec et al. 2015). It is therefore possible that transcription initiated at aberrantly active RTN promoters in cancer could trigger some of the instances of tumor suppressor gene promoter methylation observed in cancer (Fig. 1c). Recently, expression of a long RNA driven by the antisense promoter of a LINE-1 element (LCT13) has been linked to epigenetic silencing of a known tumor suppressor gene (*TFPI-2*) in breast and colon cancer (Cruickshanks et al. 2013). The mechanism underlying this event has yet to be elucidated and the observations are compatible with both of the scenarios depicted in Fig. 1b, c. Nevertheless this example further supports the hypothesis that transcription driven by LINE-1 promoters can extend over long distances and contribute to epigenetic remodeling of the cancer cell genome (Tufarelli et al. 2013).

2.3 Other Regulatory Functions of RTN Promoters

RTN promoter activity can be associated with other regulatory functions (e.g., enhancer and boundary activities) besides those directly attributable to the effects of transcription described above. The potential for RTNs to regulate gene expression has been further supported by the observation that at least 20 % of the known regulatory sequences in the human genome have been exapted from RTNs (Lowe and Haussler 2012). Intriguingly, intragenic enhancers have been found to act as alternative promoters (Kowalczyk et al. 2012), and promoters of intergenic long non-coding RNAs can show enhancer properties (Zentner et al. 2011). Indeed, a large number of functional enhancers have been found to drive transcription of non-coding enhancer RNAs, highlighting the similarities between enhancers and promoters (Li et al. 2016). It is therefore possible that some intergenic RTNs that act as alternative promoters for host genes or intergenic RTNs that initiate transcription of long non-coding RNAs might have enhancer activity. Non-LTR SINEs, for example, have been shown to act as enhancers in the developing brain to regulate expression of fibroblast growth factor 8 (Fgf8) and special AT-rich sequence-binding protein 2 (Satb2) (Sasaki et al. 2008; Tashiro et al. 2011), creating a precedent for RTNs in this role. Genome-wide approaches have demonstrated that studying the nuclear RNA fraction allowed the identification of many LTR-RTN derived transcripts, several of which co-localize with enhancer activity (Fort et al. 2014).

Transcription events have also been found to be associated with boundary activities. In the mouse, transcription from a SINE B2 within the growth hormone locus is required to prevent flanking heterochromatin from silencing the five genes within the locus in a developmentally specific manner (Lunyak et al. 2007). More recently, a cross species comparative analysis of sites bound by the CCCTC-binding factor (CTCF), a DNA-binding protein found at chromatin boundary regions, has revealed that RTNs often flank highly conserved CTCF sites (Schmidt et al. 2012).

Taken together, these observations suggest that transcription from some RTN promoters in cancer maybe an indicator of their activity as regulatory elements (e.g., enhancers and/or boundaries; Fig. 1d) and the RNA is just a by-product of such activity. In this way, RTNs could contribute to cancer transformation by altering the structural organization of chromatin domains within cells, with consequent deregulation of host gene expression.

3 Final Remarks

The findings summarized above support the concept that activation of RTN promoters can contribute to cancer in ways that go beyond retrotransposition and thus can also be exerted by immobilized elements, including ancient RTN subfamilies (Fig. 2).

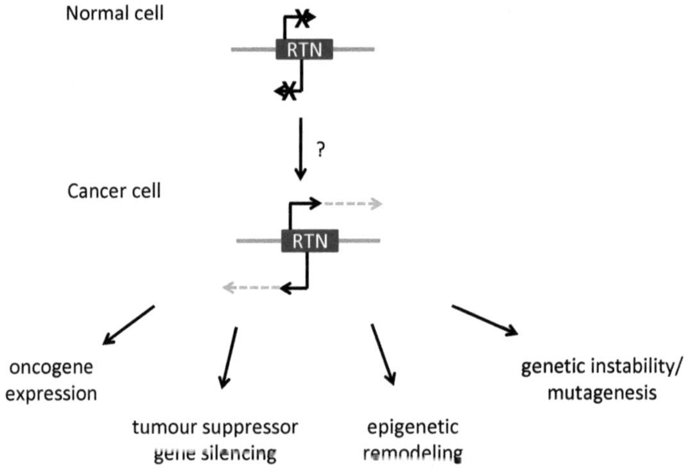

Fig. 2 Biological implications of RTN promoters activation in cancer—in normal cells RTN (*black box*) promoters (*bent arrows*) are silenced (*crossed out bent arrows*) by epigenetic means (Goodier and Kazazian 2008). Activation of promoters of RTNs is associated with cancer but the mechanisms leading to this have not yet been fully elucidated as shown by the "?". In addition to genome instability and mutagenesis associated with retrotransposition (not discussed in this chapter), activation of RTN promoters can lead to cancer in ways that are independent of retrotransposition, including the effects depicted in Fig. 1 that can promote oncogene expression, tumor suppressor gene silencing, and epigenetic remodeling

Activation of RTN promoters can play important roles in regulating genome function at those stages in which RTNs are active (germline, early embryo stem cells/placentation, and cancer (Yoder et al. 1997; Kazazian 2004; Ostertag and Kazazian 2001)). Given that transcription necessarily precedes retrotransposition, these effects can also play important roles in the early stages of cancer. Understanding the molecular pathways leading to RTN activation in cancer remains a largely unanswered question. Although hypomethylation has been proposed as a cause of RTN activation, it is unclear how this comes about and in which instances this is the trigger of activation, or simply reflects the methylation of the tissue and/or cell of origin of the cancer, or is just a consequence of activation. Intriguingly, the majority of cancer-inducing agents—e.g., irradiation, UV exposure, and drugs—cause RTN activation (Banerjee et al. 2005; Farkash et al. 2006; Terasaki et al. 2013). RTN activation is also observed following viral infection (Karijolich et al. 2015) and in ageing (Zane et al. 2014), both associated with an increased risk of cancer. It is possible that different subsets of RTNs will act in different cancer types and at different stages of cancer development. Given the large number of RTNs, it will be important to identify the individual elements whose activation drives cancer (drivers) from those that become active as a consequence of cancer (passengers). However, due to their repetitive nature, until recently it has been challenging to study individual elements at their natural integration sites.

Rapid advances in sequencing technologies now make it possible to gather information about virtually every nucleotide in the genome. Indeed, a newly published study used GeneBank ESTs databases and long reads RNA-seq data obtained using Pacific Bioscience technology to characterize a large number of transcripts initiating at LINE-1s (Criscione et al. 2016). Ongoing work to improve both sequencing technologies and the bioinformatic analysis tools for repetitive elements will enable a thorough investigation of individual RTN insertions. In this way it should be possible to determine which RTNs play functional roles in individual cancers and at which stages of cancer progression. Recent progresses in measuring the transcriptional activity of individual RTN copies (Philippe et al. 2016; Scott et al. 2016) will help profiling RTN promoters that become active in cancer, and will be fundamental to establish which ones play functional roles, and in studying the mechanisms that lead to their activation. This will bring insights into how RTN promoters are regulated, the factors that control their activity, and the functional consequences of this activity. This knowledge will be pivotal in establishing which RDTs or components of the molecular machinery involved in their activation and/or functional outcomes warrant further investigation as potential targets for therapy. Finally, it is worth noting that regardless of their functional roles, RDTs have the potential to distinguish normal from cancerous tissues (Cruickshanks et al. 2013; Wolff et al. 2010; Hur et al. 2013), suggesting they may be a rich source of novel cancer biomarkers to complement those already available. As discrete, locus-specific transcripts, they may also be appropriate "liquid biopsy" targets, with potential to improve the early detection of cancer.

Acknowledgments Work in CT lab is supported by the Derby Colorectal Cancer Fund, in RB lab by the University of Leicester.

References

Babaian A, Romanish MT, Gagnier L, Kuo LY, Karimi MM, Steidl C, Mager DL (2015) Onco-exaptation of an endogenous retroviral LTR drives IRF5 expression in Hodgkin lymphoma. Oncogene 35(19):2542–2546. doi:10.1038/onc.2015.308

Banerjee G, Gupta N, Tiwari J, Raman G (2005) Ultraviolet-induced transformation of keratinocytes: possible involvement of long interspersed element-1 reverse transcriptase. Photodermatol Photoimmunol Photomed 21(1):32–39. doi:10.1111/j.1600-0781.2005.00136.x

Baubec T, Colombo DF, Wirbelauer C, Schmidt J, Burger L, Krebs AR, Akalin A, Schubeler D (2015) Genomic profiling of DNA methyltransferases reveals a role for DNMT3B in genic methylation. Nature 520(7546):243–247. doi:10.1038/nature14176

Beck CR, Garcia-Perez JL, Badge RM, Moran JV (2011) LINE-1 elements in structural variation and disease. Annu Rev Genomics Hum Genet 12:187–215. doi:10.1146/annurev-genom-082509-141802

Belancio VP, Hedges DJ, Deininger P (2008) Mammalian non-LTR retrotransposons: for better or worse, in sickness and in health. Genome Res 18(3):343–358. doi:10.1101/gr.5558208

Brouha B, Schustak J, Badge RM, Lutz-Prigge S, Farley AH, Moran JV, Kazazian HH Jr (2003) Hot L1s account for the bulk of retrotransposition in the human population. Proc Natl Acad Sci U S A 100(9):5280–5285. doi:10.1073/pnas.0831042100

Cheetham SW, Gruhl F, Mattick JS, Dinger ME (2013) Long noncoding RNAs and the genetics of cancer. Br J Cancer 108(12):2419–2425. doi:10.1038/bjc.2013.233

Chow JC, Ciaudo C, Fazzari MJ, Mise N, Servant N, Glass JL, Attreed M, Avner P, Wutz A, Barillot E et al (2010) LINE-1 activity in facultative heterochromatin formation during X chromosome inactivation. Cell 141(6):956–969. doi:10.1016/j.cell.2010.04.042

Conley AB, Miller WJ, Jordan IK (2008) Human cis natural antisense transcripts initiated by transposable elements. Trends Genet 24(2):53–56

Cowley M, Oakey RJ (2013) Transposable elements re-wire and fine-tune the transcriptome. PLoS Genet 9(1):e1003234. doi:10.1371/journal.pgen.1003234

Criscione SW, Zhang Y, Thompson W, Sedivy JM, Neretti N (2014) Transcriptional landscape of repetitive elements in normal and cancer human cells. BMC Genomics 15:583. doi:10.1186/1471-2164-15-583

Criscione SW, Theodosakis N, Micevic G, Cornish TC, Burns KH, Neretti N, Rodic N (2016) Genome-wide characterization of human L1 antisense promoter-driven transcripts. BMC Genomics 17:463. doi:10.1186/s12864-016-2800-5

Cruickshanks HA, Tufarelli C (2009) Isolation of cancer-specific chimeric transcripts induced by hypomethylation of the LINE-1 antisense promoter. Genomics 94(6):397–406. doi:10.1016/j.ygeno.2009.08.013

Cruickshanks HA, Vafadar-Isfahani N, Dunican DS, Lee A, Sproul D, Lund JN, Meehan RR, Tufarelli C (2013) Expression of a large LINE-1-driven antisense RNA is linked to epigenetic silencing of the metastasis suppressor gene TFPI-2 in cancer. Nucleic Acids Res 41(14):6857–6869. doi:10.1093/nar/gkt438

Dewannieux M, Esnault C, Heidmann T (2003) LINE-mediated retrotransposition of marked Alu sequences. Nat Genet 35(1):41–48. doi:10.1038/ng1223

Djebali S, Davis CA, Merkel A, Dobin A, Lassmann T, Mortazavi A, Tanzer A, Lagarde J, Lin W, Schlesinger F et al (2012) Landscape of transcription in human cells. Nature 489(7414):101–108. doi:10.1038/nature11233

Domansky AN, Kopantzev EP, Snezhkov EV, Lebedev YB, Leib-Mosch C, Sverdlov ED (2000) Solitary HERV-K LTRs possess bi-directional promoter activity and contain a negative regulatory element in the U5 region. FEBS Lett 472(2–3):191–195

Donnelly SR, Hawkins TE, Moss SE (1999) A conserved nuclear element with a role in mammalian gene regulation. Hum Mol Genet 8(9):1723–1728

Dunn CA, Medstrand P, Mager DL (2003) An endogenous retroviral long terminal repeat is the dominant promoter for human beta1,3-galactosyltransferase 5 in the colon. Proc Natl Acad Sci U S A 100(22):12841–12846. doi:10.1073/pnas.2134464100

Dunn CA, Romanish MT, Gutierrez LE, van de Lagemaat LN, Mager DL (2006) Transcription of two human genes from a bidirectional endogenous retrovirus promoter. Gene 366(2):335–342. doi:10.1016/j.gene.2005.09.003

Dupressoir A, Lavialle C, Heidmann T (2012) From ancestral infectious retroviruses to bona fide cellular genes: role of the captured syncytins in placentation. Placenta 33(9):663–671. doi:10.1016/j.placenta.2012.05.005

Eden A, Gaudet F, Waghmare A, Jaenisch R (2003) Chromosomal instability and tumors promoted by DNA hypomethylation. Science 300(5618):455

Elbarbary RA, Lucas BA, Maquat LE (2016) Retrotransposons as regulators of gene expression. Science 351(6274):aac7247. doi:10.1126/science.aac7247

Ewing AD, Kazazian HH Jr (2010) High-throughput sequencing reveals extensive variation in human-specific L1 content in individual human genomes. Genome Res 20(9):1262–1270. doi:10.1101/gr.106419.110

Farkash EA, Kao GD, Horman SR, Prak ET (2006) Gamma radiation increases endonuclease-dependent L1 retrotransposition in a cultured cell assay. Nucleic Acids Res 34(4):1196–1204. doi:10.1093/nar/gkj522

Faulkner GJ, Kimura Y, Daub CO, Wani S, Plessy C, Irvine KM, Schroder K, Cloonan N, Steptoe AL, Lassmann T et al (2009) The regulated retrotransposon transcriptome of mammalian cells. Nat Genet 41(5):563–571. doi:10.1038/ng.368

Fort A, Hashimoto K, Yamada D, Salimullah M, Keya CA, Saxena A, Bonetti A, Voineagu I, Bertin N, Kratz A et al (2014) Deep transcriptome profiling of mammalian stem cells supports a regulatory role for retrotransposons in pluripotency maintenance. Nat Genet 46(6):558–566. doi:10.1038/ng.2965

Ganesh S, Svoboda P (2016) Retrotransposon-associated long non-coding RNAs in mice and men. Pflugers Arch 468(6):1049–1060. doi:10.1007/s00424-016-1818-5

Garcia-Perez JL, Doucet AJ, Bucheton A, Moran JV, Gilbert N (2007) Distinct mechanisms for trans-mediated mobilization of cellular RNAs by the LINE-1 reverse transcriptase. Genome Res 17(5):602–611. doi:10.1101/gr.5870107

Gaudet F, Hodgson JG, Eden A, Jackson-Grusby L, Dausman J, Gray JW, Leonhardt H, Jaenisch R (2003) Induction of tumors in mice by genomic hypomethylation. Science 300(5618): 489–492

Gifford WD, Pfaff SL, Macfarlan TS (2013) Transposable elements as genetic regulatory substrates in early development. Trends Cell Biol 23(5):218–226. doi:10.1016/j.tcb.2013.01.001

Goke J, Ng HH (2016) CTRL+INSERT: retrotransposons and their contribution to regulation and innovation of the transcriptome. EMBO Rep 17(8):1131–1144. doi:10.15252/embr.201642743

Goodier JL, Kazazian HH Jr (2008) Retrotransposons revisited: the restraint and rehabilitation of parasites. Cell 135(1):23–35. doi:10.1016/j.cell.2008.09.022

Gupta RA, Shah N, Wang KC, Kim J, Horlings HM, Wong DJ, Tsai MC, Hung T, Argani P, Rinn JL et al (2010) Long non-coding RNA HOTAIR reprograms chromatin state to promote cancer metastasis. Nature 464(7291):1071–1076. doi:10.1038/nature08975

Gutschner T, Diederichs S (2012) The hallmarks of cancer: a long non-coding RNA point of view. RNA Biol 9(6):703–719. doi:10.4161/rna.20481

Guttman M, Rinn JL (2012) Modular regulatory principles of large non-coding RNAs. Nature 482(7385):339–346. doi:10.1038/nature10887

Hur K, Cejas P, Feliu J, Moreno-Rubio J, Burgos E, Boland CR, Goel A (2013) Hypomethylation of long interspersed nuclear element-1 (LINE-1) leads to activation of proto-oncogenes in human colorectal cancer metastasis. Gut 63(4):635–646. doi:10.1136/gutjnl-2012-304219

Jones PA, Baylin SB (2007) The epigenomics of cancer. Cell 128(4):683–692. doi:10.1016/j.cell.2007.01.029

Kaer K, Branovets J, Hallikma A, Nigumann P, Speek M (2011) Intronic L1 retrotransposons and nested genes cause transcriptional interference by inducing intron retention, exonization and cryptic polyadenylation. PLoS One 6(10):e26099. doi:10.1371/journal.pone.0026099

Kapusta A, Kronenberg Z, Lynch VJ, Zhuo X, Ramsay L, Bourque G, Yandell M, Feschotte C (2013) Transposable elements are major contributors to the origin, diversification, and regulation of vertebrate long noncoding RNAs. PLoS Genet 9(4):e1003470. doi:10.1371/journal.pgen.1003470

Karijolich J, Abernathy E, Glaunsinger BA (2015) Infection-induced retrotransposon-derived noncoding RNAs enhance herpesviral gene expression via the NF-kappaB pathway. PLoS Pathog 11(11):e1005260. doi:10.1371/journal.ppat.1005260

Kassiotis G (2014) Endogenous retroviruses and the development of cancer. J Immunol 192(4):1343–1349. doi:10.4049/jimmunol.1302972

Kazazian HH Jr (2004) Mobile elements: drivers of genome evolution. Science 303(5664):1626–1632. doi:10.1126/science.1089670

Khalil AM, Guttman M, Huarte M, Garber M, Raj A, Rivea Morales D, Thomas K, Presser A, Bernstein BE, van Oudenaarden A et al (2009) Many human large intergenic noncoding RNAs associate with chromatin-modifying complexes and affect gene expression. Proc Natl Acad Sci U S A 106(28):11667–11672. doi:10.1073/pnas.0904715106

Khan H, Smit A, Boissinot S (2006) Molecular evolution and tempo of amplification of human LINE-1 retrotransposons since the origin of primates. Genome Res 16(1):78–87. doi:10.1101/gr.4001406

Kowalczyk MS, Hughes JR, Garrick D, Lynch MD, Sharpe JA, Sloane-Stanley JA, McGowan SJ, De Gobbi M, Hosseini M, Vernimmen D et al (2012) Intragenic enhancers act as alternative promoters. Mol Cell 45(4):447–458. doi:10.1016/j.molcel.2011.12.021

Koziol MJ, Rinn JL (2010) RNA traffic control of chromatin complexes. Curr Opin Genet Dev 20(2):142–148. doi:10.1016/j.gde.2010.03.003

Lander ES, Linton LM, Birren B, Nusbaum C, Zody MC, Baldwin J, Devon K, Dewar K, Doyle M, FitzHugh W et al (2001) Initial sequencing and analysis of the human genome. Nature 409(6822):860–921. doi:10.1038/35057062

Lee E, Iskow R, Yang L, Gokcumen O, Haseley P, Luquette LJ III, Lohr JG, Harris CC, Ding L, Wilson RK et al (2012) Landscape of somatic retrotransposition in human cancers. Science 337(6097):967–971. doi:10.1126/science.1222077

Li W, Notani D, Rosenfeld MG (2016) Enhancers as non-coding RNA transcription units: recent insights and future perspectives. Nat Rev Genet 17(4):207–223. doi:10.1038/nrg.2016.4

Lock FE, Rebollo R, Miceli-Royer K, Gagnier L, Kuah S, Babaian A, Sistiaga-Poveda M, Lai CB, Nemirovsky O, Serrano I et al (2014) Distinct isoform of FABP7 revealed by screening for retroelement-activated genes in diffuse large B-cell lymphoma. Proc Natl Acad Sci U S A 111(34):E3534–E3543. doi:10.1073/pnas.1405507111

Lowe CB, Haussler D (2012) 29 mammalian genomes reveal novel exaptations of mobile elements for likely regulatory functions in the human genome. PLoS One 7(8):e43128. doi:10.1371/journal.pone.0043128

Lu X, Sachs F, Ramsay L, Jacques PE, Goke J, Bourque G, Ng HH (2014) The retrovirus HERVH is a long noncoding RNA required for human embryonic stem cell identity. Nat Struct Mol Biol 21(4):423–425. doi:10.1038/nsmb.2799

Lunyak VV, Prefontaine GG, Nunez E, Cramer T, Ju BG, Ohgi KA, Hutt K, Roy R, Garcia-Diaz A, Zhu X et al (2007) Developmentally regulated activation of a SINE B2 repeat as a domain boundary in organogenesis. Science 317(5835):248–251. doi:10.1126/science.1140871

Lyon MF (2006) Do LINEs have a role in X-chromosome inactivation? J Biomed Biotechnol 2006(1):59746. doi:10.1155/JBB/2006/59746

Macfarlan TS, Gifford WD, Driscoll S, Lettieri K, Rowe HM, Bonanomi D, Firth A, Singer O, Trono D, Pfaff SL (2012) Embryonic stem cell potency fluctuates with endogenous retrovirus activity. Nature 487(7405):57–63. doi:10.1038/nature11244

Magistri M, Faghihi MA, St Laurent G III, Wahlestedt C (2012) Regulation of chromatin structure by long noncoding RNAs: focus on natural antisense transcripts. Trends Genet 28(8):389–396. doi:10.1016/j.tig.2012.03.013

Matlik K, Redik K, Speek M (2006) L1 antisense promoter drives tissue-specific transcription of human genes. J Biomed Biotechnol 2006(1):71753

Matsunoki A, Kawakami K, Kotake M, Kaneko M, Kitamura H, Ooi A, Watanabe G, Minamoto T (2012) LINE-1 methylation shows little intra-patient heterogeneity in primary and synchronous metastatic colorectal cancer. BMC Cancer 12:574. doi:10.1186/1471-2407-12-574

Mills RE, Bennett EA, Iskow RC, Devine SE (2007) Which transposable elements are active in the human genome? Trends Genet 23(4):183–191. doi:10.1016/j.tig.2007.02.006

Murata A, Baba Y, Watanabe M, Shigaki H, Miyake K, Ishimoto T, Iwatsuki M, Iwagami S, Sakamoto Y, Miyamoto Y et al (2013) Methylation levels of LINE-1 in primary lesion and matched metastatic lesions of colorectal cancer. Br J Cancer 109(2):408–415. doi:10.1038/bjc.2013.289

Ogino S, Nosho K, Kirkner GJ, Kawasaki T, Chan AT, Schernhammer ES, Giovannucci EL, Fuchs CS (2008) A cohort study of tumoral LINE-1 hypomethylation and prognosis in colon cancer. J Natl Cancer Inst 100(23):1734–1738. doi:10.1093/jnci/djn359

Ohms S, Rangasamy D (2014) Silencing of LINE-1 retrotransposons contributes to variation in small noncoding RNA expression in human cancer cells. Oncotarget 5(12):4103–4117. doi:10.18632/oncotarget.1822

Ostertag EM, Kazazian HH Jr (2001) Biology of mammalian L1 retrotransposons. Annu Rev Genet 35:501–538. doi:10.1146/annurev.genet.35.102401.091032

Peaston AE, Evsikov AV, Graber JH, de Vries WN, Holbrook AE, Solter D, Knowles BB (2004) Retrotransposons regulate host genes in mouse oocytes and preimplantation embryos. Dev Cell 7(4):597–606. doi:10.1016/j.devcel.2004.09.004

Philippe C, Vargas-Landin DB, Doucet AJ, van Essen D, Vera-Otarola J, Kuciak M, Corbin A, Nigumann P, Cristofari G (2016) Activation of individual L1 retrotransposon instances is restricted to cell-type dependent permissive loci. Elife 5:e13926. doi:10.7554/eLife.13926

Piriyapongsa J, Marino-Ramirez L, Jordan IK (2007) Origin and evolution of human microRNAs from transposable elements. Genetics 176(2):1323–1337. doi:10.1534/genetics.107.072553

Raiz J, Damert A, Chira S, Held U, Klawitter S, Hamdorf M, Lower J, Stratling WH, Lower R, Schumann GG (2012) The non-autonomous retrotransposon SVA is trans-mobilized by the human LINE-1 protein machinery. Nucleic Acids Res 40(4):1666–1683. doi:10.1093/nar/gkr863

Rangasamy D, Lenka N, Ohms S, Dahlstrom JE, Blackburn AC, Board PG (2015) Activation of LINE-1 retrotransposon increases the risk of epithelial-mesenchymal transition and metastasis in epithelial cancer. Curr Mol Med 15(7):588–597

Rinn JL, Chang HY (2012) Genome regulation by long noncoding RNAs. Annu Rev Biochem 81:145–166. doi:10.1146/annurev-biochem-051410-092902

Roman-Gomez J, Jimenez-Velasco A, Agirre X, Cervantes F, Sanchez J, Garate L, Barrios M, Castillejo JA, Navarro G, Colomer D et al (2005) Promoter hypomethylation of the LINE-1 retrotransposable elements activates sense/antisense transcription and marks the progression of chronic myeloid leukemia. Oncogene 24(48):7213–7223. doi:10.1038/sj.onc.1208866

Sasaki T, Nishihara H, Hirakawa M, Fujimura K, Tanaka M, Kokubo N, Kimura-Yoshida C, Matsuo I, Sumiyama K, Saitou N et al (2008) Possible involvement of SINEs in mammalian-specific brain formation. Proc Natl Acad Sci U S A 105(11):4220–4225. doi:10.1073/pnas.0709398105

Schmidt D, Schwalie PC, Wilson MD, Ballester B, Goncalves A, Kutter C, Brown GD, Marshall A, Flicek P, Odom DT (2012) Waves of retrotransposon expansion remodel genome organization and CTCF binding in multiple mammalian lineages. Cell 148(1–2):335–348. doi:10.1016/j.cell.2011.11.058

Scott EC, Gardner EJ, Masood A, Chuang NT, Vertino PM, Devine SE (2016) A hot L1 retrotransposon evades somatic repression and initiates human colorectal cancer. Genome Res 26(6):745–755. doi:10.1101/gr.201814.115

Solyom S, Ewing AD, Rahrmann EP, Doucet TT, Nelson HH, Burns MB, Harris RS, Sigmon DF, Casella A, Erlanger B et al (2012) Extensive somatic L1 retrotransposition in colorectal tumors. Genome Res 22(12):2328–2338. doi:10.1101/gr.145235.112

Speek M (2001) Antisense promoter of human L1 retrotransposon drives transcription of adjacent cellular genes. Mol Cell Biol 21(6):1973–1985

Sproul D, Nestor C, Culley J, Dickson JH, Dixon JM, Harrison DJ, Meehan RR, Sims AH, Ramsahoye BH (2011) Transcriptionally repressed genes become aberrantly methylated and distinguish tumors of different lineages in breast cancer. Proc Natl Acad Sci U S A 108(11):4364–4369. doi:10.1073/pnas.1013224108

Sproul D, Kitchen RR, Nestor CE, Dixon JM, Sims AH, Harrison DJ, Ramsahoye BH, Meehan RR (2012) Tissue of origin determines cancer-associated CpG island promoter hypermethylation patterns. Genome Biol 13(10):R84. doi:10.1186/gb-2012-13-10-r84

Sunami E, de Maat M, Vu A, Turner RR, Hoon DS (2011) LINE-1 hypomethylation during primary colon cancer progression. PLoS One 6(4):e18884. doi:10.1371/journal.pone.0018884

Symer DE, Connelly C, Szak ST, Caputo EM, Cost GJ, Parmigiani G, Boeke JD (2002) Human l1 retrotransposition is associated with genetic instability in vivo. Cell 110(3):327–338

Tashiro K, Teissier A, Kobayashi N, Nakanishi A, Sasaki T, Yan K, Tarabykin V, Vigier L, Sumiyama K, Hirakawa M et al (2011) A mammalian conserved element derived from SINE displays enhancer properties recapitulating Satb2 expression in early-born callosal projection neurons. PLoS One 6(12):e28497. doi:10.1371/journal.pone.0028497

Terasaki N, Goodier JL, Cheung LE, Wang YJ, Kajikawa M, Kazazian HH Jr, Okada N (2013) In vitro screening for compounds that enhance human L1 mobilization. PLoS One 8(9):e74629. doi:10.1371/journal.pone.0074629

Tubio JM, Li Y, Ju YS, Martincorena I, Cooke SL, Tojo M, Gundem G, Pipinikas CP, Zamora J, Raine K et al (2014) Mobile DNA in cancer. Extensive transduction of nonrepetitive DNA mediated by L1 retrotransposition in cancer genomes. Science 345(6196):1251343. doi:10.1126/science.1251343

Tufarelli C, Stanley JA, Garrick D, Sharpe JA, Ayyub H, Wood WG, Higgs DR (2003) Transcription of antisense RNA leading to gene silencing and methylation as a novel cause of human genetic disease. Nat Genet 34(2):157–165

Tufarelli C, Cruickshanks HA, Meehan RR (2013) LINE-1 activation and epigenetic silencing of suppressor genes in cancer: causally related events? Mob Genet Elements 3(5):e26832. doi:10.4161/mge.26832

Walsh CP, Chaillet JR, Bestor TH (1998) Transcription of IAP endogenous retroviruses is constrained by cytosine methylation. Nat Genet 20(2):116–117. doi:10.1038/2413

Wang KC, Chang HY (2011) Molecular mechanisms of long noncoding RNAs. Mol Cell 43(6):904–914. doi:10.1016/j.molcel.2011.08.018

Werner A (2013) Biological functions of natural antisense transcripts. BMC Biol 11:31. doi:10.1186/1741-7007-11-31

Wicker T, Sabot F, Hua-Van A, Bennetzen JL, Capy P, Chalhoub B, Flavell A, Leroy P, Morgante M, Panaud O et al (2007) A unified classification system for eukaryotic transposable elements. Nat Rev Genet 8(12):973–982. doi:10.1038/nrg2165

Wilkins AS (2010) The enemy within: an epigenetic role of retrotransposons in cancer initiation. Bioessays 32(10):856–865. doi:10.1002/bies.201000008

Wolff EM, Byun HM, Han HF, Sharma S, Nichols PW, Siegmund KD, Yang AS, Jones PA, Liang G (2010) Hypomethylation of a LINE-1 promoter activates an alternate transcript of the MET oncogene in bladders with cancer. PLoS Genet 6(4):e1000917. doi:10.1371/journal.pgen.1000917

Yang N, Kazazian HH Jr (2006) L1 retrotransposition is suppressed by endogenously encoded small interfering RNAs in human cultured cells. Nat Struct Mol Biol 13(9):763–771. doi:10.1038/nsmb1141

Yoder JA, Walsh CP, Bestor TH (1997) Cytosine methylation and the ecology of intragenomic parasites. Trends Genet 13(8):335–340

Yu W, Gius D, Onyango P, Muldoon-Jacobs K, Karp J, Feinberg AP, Cui H (2008) Epigenetic silencing of tumour suppressor gene p15 by its antisense RNA. Nature 451(7175):202–206
Zane L, Sharma V, Misteli T (2014) Common features of chromatin in aging and cancer: cause or coincidence? Trends Cell Biol 24(11):686–694. doi:10.1016/j.tcb.2014.07.001
Zentner GE, Tesar PJ, Scacheri PC (2011) Epigenetic signatures distinguish multiple classes of enhancers with distinct cellular functions. Genome Res 21(8):1273–1283. doi:10.1101/gr.122382.111

LINE-1 Retrotransposons as Neoplastic Biomarkers

Nemanja Rodic

1 Introduction

Knowledge is an unending adventure at the edge of uncertainty.

Jacob Bronowski

Thus far, Koch's postulates have not been satisfied on the question of a putative etiologic role of any human endogenous retroelements in human tumorigenesis. The above quote illustrates the uncertainty of precisely how, or even if, endogenous retroelements play an etiologic role in tumorigenesis. These difficult-to-study genetic elements have been suspected to play at least a partial role in tumorigenesis for the following two principal reasons. First, germline insertions of most well-studied endogenous retrotransposon, Long INterspesed Element-1 (LINE-1 or L1), cause many nonneoplastic heritable genetic disorders in humans (Belancio et al. 2008). Second, increased activity of endogenous retroelements has been observed in human cancers (Ting et al. 2011). More recent studies on LINE-1 antisense promoter-driven transcription in both human health and disease highlight novel and varied aspects of endogenous retrotransposons life cycle (Denli et al. 2015). Therefore, it is entirely possible that endogenous retrotransposons play a putative etiologic role in tumorigenesis.

In an authoritative review on the mechanisms of neoplasia development in humans, Vogelstein and colleagues remind the readers that much of the primary nuclear genome is occupied by repetitive elements (Vogelstein et al. 2013). Selected features of such so-called genomic dark matter will be, at least in part, demystified through my exposition. What follows is a principled discussion on varied aspects of the role of endogenous retrotransposons in tumorigenesis. First, two most well-studied features of LINE-1 in cancers will be defined: measurement of LINE-1 promoter methylation

N. Rodic (✉)
Yale Dermatopathology Laboratory, Yale School of Medicine,
15 York Street, LMP5031, P.O.Box 208059, New Haven, CT 06520-8059, USA
e-mail: Nemanja.Rodic@yale.edu

G. Cristofari (ed.), *Human Retrotransposons in Health and Disease*,
DOI 10.1007/978-3-319-48344-3_12

by PCR assay, a proxy for genome-wide DNA methylation level, and measurement of LINE-1-encoded proteins (ORF1p≫ORF2p), usually as determined by immunohistochemistry. Next, using a current WHO anatomic classification of human neoplasias as a conceptual framework, I will discuss salient features of LINE-1 retrotransposon biology in selected cancers derived from different somatic organs. Finally, I will conclude the exposition with a brief discussion of most notable mechanistic studies implicating LINE-1 retrotransposons in human tumorigenesis.

As of January 2016, there are 932 primary published reports focusing, at least in part, on detection of some feature of LINE-1 biology in human neoplasms (Fig. 1). Remarkably, almost 13 % of all these reports are collective work of only three research groups: (1) the group led by Drs. Shuji Ogino and Charles S. Fuchs, both of Harvard Medical School; (2) collaborators of the former group, Drs. Yoshifumi Baba and Hideo Baba, both of Kumamoto University; and (3) work of Dr. Prescott Deininger, of Tulane University (Fig. 2). Also notable is the pioneering work by Drs. Fanning and Singer on LINE-1 expression in cancers. These two remarkable authors have, in a series of early original articles, established that LINE-1 is an independent genetic element (Fanning and Singer 1987), which is both transcribed and translated at increased levels in breast and germ cell cancers (Skowronski et al. 1988; Leibold et al. 1990; Bratthauer and Fanning 1992; Asch et al. 1996). Finally, Miki and colleagues were the first to document LINE-1 retrotransposition in cancer (Miki et al. 1992). Specifically, these authors identified LINE-1-derived sequence in the last exon of APC gene of colorectal carcinoma but not in adjoining normal tissue. This particular LINE-1 insertion gives us arguably the most compelling record, to date, for unequivocal deleterious genetic variant caused by LINE-1 retrotransposition.

The vast majority of LINE-1 studies in cancer rely on detection of LINE-1 promoter methylation by PCR assay. Nearly all such studies require pretreatment of target analytes, i.e., genomic DNAs, with bisulfite reagent, which in turn allows the identification of unmodified cytosine moieties and methylated cytosine bases. The analyte in such studies is heterogeneous, composed of unknowable and variable proportions of both lesional cancer cells and intercalated contaminating normal cells. The second—less used—method to assay LINE-1-encoded proteins in cancer specimens is immunohistochemistry. Here, the analyte is individual cancer cell, which allows evaluation of both LINE-1 protein expression levels and cellular localization in lesional cells. Comparably fewer studies rely on DNA sequencing technology, a laborious and imperfect high-complexity assay, to identify somatic LINE-1 retrotransposition events in cancer.

2 LINE-1 in Prostate Cancer

To date, studies of LINE-1 in prostate cancer support the notion that some features of LINE-1 biology, such as LINE-1 hypomethylation and translation of LINE-1-encoded proteins, are detectable in prostate cancers. For instance, a retrospective study of 737 men showed that relative decrease in LINE-1 methylation in initial biopsy is associated with prostate cancer on re-biopsy, but only in patients with

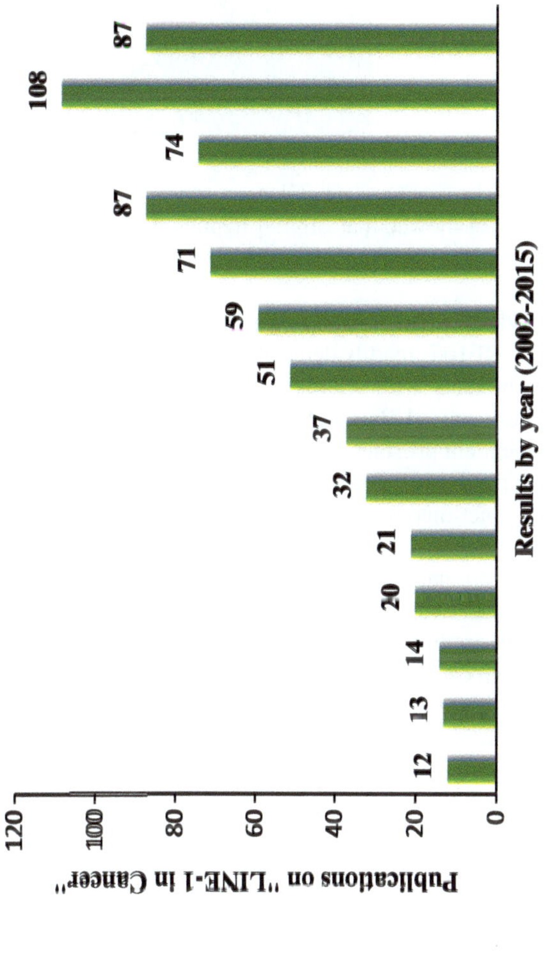

Fig. 1 Pictorial representation of published reports on LINE-1 retrotransposons in human cancers in the period from 2002 to 2015. Results were obtained by searching the National Center for Biotechnology Information PubMed database using keywords “LINE-1 AND cancer”

Fig. 2 Wordle representation of most published authors on LINE-1 retrotransposons in human cancers to date. Image was obtained by analyzing the names of all PubMed authors (keywords: "LINE-1 AND cancer") using free wordle software at http://worditout.com/word-cloud/make-a-new-one

higher pathologic stage prostate cancer (Gleason score >4+3) (Zelic et al. 2015). Another study of 703 men showed that LINE-1 hypomethylation in matched peripheral blood specimens is not associated with subsequent risk of prostate cancer (Barry et al. 2015).

An intriguing in vitro study of human prostate cancer cell lines showed that endogenous LINE-1 ORF1p is readily detectable in nuclei of human prostate cancer cell lines, where it interacts with androgen receptor (Lu et al. 2013). Transient downregulation of endogenous LINE-1 ORF1p attenuated growth of both androgen-dependent and -independent prostate cancer cell lines, as well as caused decrease in anchorage-dependent proliferation (Lu et al. 2013) (Table 1). The authors concluded that LINE-1 ORF1p functions as a novel androgen receptor co-activator and promotes the growth of human prostatic carcinoma cells (Lu et al. 2013). Another in vitro study of two human prostate cancer cell lines showed that exogenous LINE-1 ORF2p interacts with androgen receptor and promotes double-stranded break formation, which are in turn required for generation of *TMPRSS2:ERGb* and *TMPRSS2:ETV1b* translocations (Lin et al. 2009). Both studies deserve further follow-up inquiries to attribute a definite and easily observable phenotype to exogenous LINE-1 ORF2p overexpression (Table 1).

A small study of six precancerous prostate lesions, termed prostatic intraepithelial neoplasias (PINs), revealed that 5/6 PINs express LINE-1 ORF2p, suggesting that LINE-1-encoded protein expression occurs early during tumorigenesis and may be useful as an early biomarker of incipient prostate cancer (De Luca et al. 2015). Another small in vitro study of 14 human prostate cancers revealed that cancer cells with *MPRSS2:ETV1b* translocation show a relative loss of methylation (~30 %) in genomic loci in and around LINE-1 retrotransposon sequences compared to translocation-negative cohort (~50 %) (Kim et al. 2011). Because study design involved deep sequencing and genomic localization of at least some individual reads, the findings support the notion that attenuation of LINE-1 methylation is roughly equally distributed within the primary nuclear genome (Kim et al. 2011).

There are two research articles documenting LINE-1 retrotransposition in prostate cancer. First, work by Lee and colleagues revealed that LINE-1 retrotransposition is a frequent event in prostate cancers (present in 6/7 tumor specimens studied)

Table 1 Phenotypes attributed to LINE-1 retrotransposons in empirical cancer studies

	References
Due to exogenous LINE-1 retrotransposons	
Cell culture	
ORF2p promotes translocations in cultured cancer cells	Lin et al. (2009)
ORF1p promotes breast cancer cell growth	Yang et al. (2013)
Due to endogenous LINE-1 retrotransposons	
Cell culture	
siRNA/RNAi knockdown causes decreased cancer cell line growth	Lu et al. (2013), Sciamanna et al. (2005)
siRNA knockdown causes telomere shortening	Aschacher et al. (2016)
Human tissue studies	
Insertional mutagenesis into MCC intron associated with haploinsufficiency	Shukla et al. (2013)

(Lee et al. 2012). However, on average, there were only four LINE-1 retrotransposition events detected per tumor. The genes near LINE-1 retrotransposition were notable for two features. They were both mutated at a higher rate than background mutational rate and expressed at comparably lower levels relative to normal colorectal tissue. More recent collaborative work by Tubio and colleagues showed that ~75 % of 60 prostate cancers studied supported LINE-1 retrotransposition (Tubio et al. 2014). However, similar to the preceding study, the vast majority of LINE-1 permissive cancers were marked by a single LINE-1 retrotransposition event per tumor.

Taken together, there is some evidence that LINE-1-encoded proteins are expressed early in prostate cancer, and that LINE-1 retrotransposons may play a role in the growth of some prostate cancer cells.

3 LINE-1 in Bladder Cancer

Studies of LINE-1 retrotransposons in bladder cancer are preliminary and few. For instance, there are no studies documenting LINE-1 retrotransposition in bladder cancer. A retrospective study of 548 bladder cancer cases showed modest synergy between LINE-1 hypomethylation in patients' peripheral leukocytes and estimated historical trihalomethanes water source levels, a known risk factor for bladder cancer tumorigenesis (Salas et al. 2014). A retrospective study of 952 bladder cancer cases showed modest tendency for bladder cancer risk in both cases with marked LINE-1 hypomethylation and in a subgroup of patient with five single-nucleotide variants of phosphatidylethanolamine *N*-methyltransferase (PEMT) gene, a transferase enzyme that is involved in biosynthesis of phosphatidylcholine—a putative bladder cancer risk factor (Tajuddin et al. 2014).

A small study of 16 bladder cancers showed that LINE-1 promoter methylation in bladder cancer is decreased compared to cultured urothelial cells (Kreimer et al. 2013). A moderate ($p = +0.63$) positive tendency for concurrent relatively higher expression of LINE-1 transcripts by qRT-PCR in hypomethylated LINE-1 bladder cancers was also noted (Kreimer et al. 2013). This peculiar dissociation between degrees of LINE-1 methylation, a feature of transcriptional potential, and steady-state levels of LINE-1 transcripts in cancer suggests that the cellular machinery might limit LINE-1 transcript accumulation.

A study of 50 bladder cancers noted attenuation of LINE-1 promoter methylation in bladder cancer (average level, 61.36 %) compared to matched normal specimens, including serum (80.47 %), buffy coat (79.36 %), and buccal cell DNA (77.09 %) (van Bemmel et al. 2012). In addition, a moderate association between LINE-1 hypomethylation and higher tumor stage and higher grade was identified ($p = -0.56$ and -0.52, respectively) (van Bemmel et al. 2012). Using DNA from circulating nucleated cells, mostly leukocytes, a retrospective case-controlled study of 285 bladder cancer patients revealed an association between LINE-1 hypomethylation and increased risk of developing bladder cancer (odds ratio, 1.8) (Wilhelm et al. 2010). The association was more pronounced in women than in men (odds ratio, 2.48) (Wilhelm et al. 2010). Of note, in control cohort comprised of patients without diagnosis of bladder cancer, reduced LINE-1 methylation was associated with higher levels of arsenic in blood (Wilhelm et al. 2010).

Taken together, while there is some evidence that the epigenome of bladder cancers is marked by relative LINE-1 hypomethylation, there is no evidence that LINE-1 plays an etiologic role in tumorigenesis of bladder cancers in human.

4 LINE-1 in Renal Cell Carcinoma

There are only a few notable clinical reports focusing on LINE-1 methylation in renal cell carcinoma. A recent study documents that LINE-1 retrotransposition occurs rarely in most renal cell carcinomas (Helman et al. 2014). First notable study found that LINE-1 methylation was lower in renal cell carcinomas from adults compared with neoplasms from younger patients (71.1 % vs. 76.7 %) (Malouf et al. 2013). Second retrospective study notes that comparably higher LINE-1 methylation levels are detected in peripheral leukocytes of patients with renal cell carcinoma compared to healthy controls (Liao et al. 2011). The authors suggest that a risk of developing renal cell cancer may be associated with higher LINE-1 methylation.

Taken together, there is insufficient evidence on what, if any, role does LINE-1 methylation plays in renal cell carcinoma tumorigenesis.

5 LINE-1 in Hepatocellular Carcinoma

Similar to studies of LINE-1 in genitourinary tract, LINE-1 clinical studies in hepatocellular carcinoma model are few. However, a single study by Shukla et al. provided us with one of the most far-reaching evidences for a causative role of LINE-1 in tumorigenesis (Shukla et al. 2013) (Table 1).

First notable study is that of 208 hepatocellular carcinomas (Harada et al. 2015). The authors found that greater degree of LINE-1 hypomethylation is associated with higher recurrence rate, but there was no effect on overall survival (Harada et al. 2015). Interestingly, a more pronounced tendency for tumoral LINE-1 hypomethylation and higher recurrence rate was observed within a subgroup of hepatocellular carcinomas that did not harbor concurrent HBV or HCV infection (Harada et al. 2015).

Sometimes LINE-1 retrotransposons are co-opted during tumorigenic process to provide additional amino acid sequences for alternative chimeric proteins. Studying HBV genomic integration sites Lau et al. discovered that newly integrated HBV near LINE-1 sequences causes HBV-driven HBV-LINE-1 chimeric transcripts that are expressed in up to 30 % of hepatocellular carcinomas (Lau et al. 2014). Of note, cases with detectable HBV-LINE-1 chimeric transcripts had shorter overall survival (Lau et al. 2014).

On the other hand, LINE-1 hypomethylation is common in hepatocellular carcinomas. In a study of 71 hepatocellular carcinomas, a ~20 % decrease in genome-wide LINE-1 methylation was detected in nearly 90 % of cases (Gao et al. 2014). LINE-1 hypomethylation in this cohort was also associated with shorter overall survival tumor (Gao et al. 2014).

In a hallmark study of 19 hepatocellular carcinomas by Shukla et al., the authors reported a handful of cancer-specific LINE-1 retrotransposition events (Shukla et al. 2013). Of note, in three cases, a heterozygous germline LINE-1 insertion into an intron of *MCC*, a known tumor-suppressor gene, was associated with marked haploinsufficiency of both *MCC* mRNA and protein (Shukla et al. 2013). The findings suggest that LINE-1 insert caused *MCC* downregulation. Albeit limited in scope, this study provides the most complete empirical mechanistic evidence to date on how exactly LINE-1 activation in cancer can contribute to tumorigenesis.

A unique prospective study of 305 hepatocellular carcinomas revealed that LINE-1 hypomethylation is detectable in matched pre-diagnostic blood specimens (Wu et al. 2012b). These remarkable results suggest that global hypomethylation may be a useful biomarker of hepatocellular carcinoma susceptibility.

Taken together, LINE-1 hypomethylation is a feature of many hepatocellular carcinomas, but it remains uncertain how these epigenetic abnormalities arise. LINE-1 retrotransposons contribute rarely to hepatocellular carcinoma tumorigenesis.

6 LINE-1 in Pancreatic Carcinomas

There are only a few notable studies of the putative role of LINE-1 retrotransposons in pancreatic carcinomas. My own recent study of 20 pancreatic ductal carcinomas revealed that LINE-1 insertions occur at a comparable rate to single-nucleotide mutations (Rodic et al. 2015). However, we found no easily interpretable mechanistic answers on how LINE-1 inserts possibly affect either the epigenome or the transcriptome of pancreatic ductal carcinomas. Most importantly, we detected no tendency for LINE-1 to insert into or near genes already implicated in pancreatic cancer development (Rodic et al. 2015; Ewing et al. 2015). However, genome-wide changes in LINE-1 methylation levels are readily detectable in many pancreatic cancer cases. A retrospective study of peripheral leukocytes from 559 pancreatic ductal carcinoma patients revealed LINE-1 hypermethylation in cancers compared to healthy controls (Neale et al. 2014; Ewing et al. 2015). The authors noted that selection bias might have affected their findings. In contrast to LINE-1 hypermethylation in peripheral leukocytes, expression of LINE-1-encoded ORF1p is a feature of most pancreatic ductal adenocarcinomas (Rodic et al. 2014). However, it appears that while transcription from LINE-1 promoters and translation of LINE-1-encoded proteins are readily detected in many pancreatic ductal carcinomas, LINE-1 retrotransposition is a comparably rare event in this type of carcinomas.

A study by Stefanoli et al. involving 56 pancreatic neuroendocrine carcinomas revealed cancer-specific LINE-1 hypomethylation compared to that seen in healthy controls (Stefanoli et al. 2014). Due to cleverly executed study design, the authors discerned that LINE-1 hypomethylation is not associated with either copy number alterations or gene-specific methylation changes (Stefanoli et al. 2014). Further, LINE-1 hypomethylation was associated with higher pathologic stage and were found to be independent significant predictors of outcome (Stefanoli et al. 2014).

Taken together, LINE-1 methylation changes may be an early feature of pancreatic cancers, but there is preliminary evidence that endogenous LINE-1 retrotransposition does not affect pancreatic cancer tumorigenesis.

7 LINE-1 in Colorectal Carcinomas

Studies of LINE-1 retrotransposons in colorectal cancers are many; perhaps colorectal cancer is the most well-studied cancer type to date when it comes to LINE-1 retrotransposons, and was the first reported case of somatic retrotransposition in humans (Miki et al. 1992).

Focusing on 129 metastatic colorectal carcinoma cases, Lou et al. elucidated that LINE-1 hypomethylation correlates to shorter disease-free survival (Lou et al. 2014). Of interest, the degree of LINE-1 hypomethylation is more pronounced in older patients (>65 years) (Lou et al. 2014). A study of 219 colorectal carcinomas by Benard et al. discerned an association between LINE-1 hypomethylation and shorter overall survival (Benard et al. 2013). Of note, attenuation of the H3K27me3 histone mark, an epigenetic biomarker of global gene repression, further adversely affected overall survival in the LINE-1 hypomethylated subgroup, indicating that LINE-1 hypomethylation is a likely marker of genome-wide epigenetic changes (Benard et al. 2013).

Colorectal carcinomas are a heterogeneous group of disorders, some marked by genetic hypermutability of microsatellites (microsatellite instability or MSI) that results from impaired DNA mismatch repair. A subgroup analysis of 1336 colorectal cancers by Inamura et al. revealed an association between cancers with MSI high and LINE-1 hypomethylation (Inamura et al. 2015). Another study of a comparable cohort of 1211 colorectal carcinomas by the same authors focused on studying the tumoral LINE-1 methylation level and microsatellite instability in relation to colorectal cancer prognosis. Focusing on the MSI high subgroup, the authors discovered that LINE-1 hypomethylation is associated with a shorter cancer-specific survival (Inamura et al. 2014).

Because colorectal carcinomas are posited to form via a multistep tumorigenic process, LINE-1 hypomethylation has been studied in many early pre-cancerous proliferations and adjoining normal colonic tissue. In essence, colorectal cancer is thought to originate from some benign proliferations, so-called adenomas. A study of 1386 colorectal lesions by Naito et al. revealed that LINE-1 hypomethylation is a feature of some adenomas, revealing that LINE-1 hypomethylation occurs in many benign adenomas (Naito et al. 2014). A similar study of 158 colorectal neoplasms of varied microscopic appearance by Konda et al. revealed that LINE-1 hypomethylation is more pronounced in selected colorectal carcinomas with unique macroscopic features (Konda et al. 2014). A unique study of 40 colorectal carcinomas with adjacent adenomas by Yamada et al. detected LINE-1 hypomethylation in normal colon associated with multiple adenomas, suggesting a cancerization "field defect" effect (Yamada et al. 2014). However, in a study of 77 adenomas by Quintanilla et al., LINE-1 hypomethylation was not present in normal colon associated with adenomas (Quintanilla et al. 2014).

Several studies addressed intratumoral heterogeneity of LINE-1 methylation levels in colorectal carcinomas. In a study of 68 metastatic colorectal carcinomas by Murata et al., LINE-1 methylation levels in matched primary and metastatic specimens were similar (Murata et al. 2013). Prognostic impact of LINE-1 hypomethylation on overall survival was not detected (Murata et al. 2013). A study of 48 colorectal carcinomas with matched primary and metastatic specimens by Matsunoki et al. showed that LINE-1 methylation shows little intra-patient tumor heterogeneity (Matsunoki et al. 2012).

An important retrospective study of 1244 patients with colorectal carcinoma diagnosis by Ogino et al. established that LINE-1 hypomethylation associated with cases with positive family history of colorectal carcinoma (Ogino et al. 2013).

A study of 281 colorectal adenoma cases by Jung et al. found that LINE-1 hypomethylation in peripheral leukocytes is associated with increased risk of colorectal carcinoma (Jung et al. 2013). Interestingly, this study also provides evidence that LINE-1 hypomethylation is in turn associated with decreased plasma folate levels (Jung et al. 2013). A study of 509 colorectal patients by Walters et al. revealed an association between LINE-1 hypermethylation in peripheral leukocytes in cases with colorectal cancer compared to healthy controls (Walters et al. 2013). In a study of 343 colorectal carcinomas by Antelo et al., LINE-1 hypomethylation was most pronounced in early-onset cases (<50 years old at presentation) compared to LINE-1 methylation in other clinical cancer subtypes and normal colonic mucosa (Antelo et al. 2012). Considering the entire cohort, the authors detected shorter overall survival in patients with LINE-1 hypomethylation (Antelo et al. 2012). A recent study of ten colorectal carcinomas by De Luca et al. showed that LINE-1 ORF2p is expressed in all ten cancer specimens by immunohistochemistry (De Luca et al. 2015).

More recent studies by Solyom and Ewing showed insertions in adjacent normal tissue, colonic adenoma, and colorectal carcinoma (Ewing et al. 2015). In addition, TCGA-based pancancer studies by Lee and colleagues (Lee et al. 2012), Helman and colleagues (Helman et al. 2014), and Tubio and colleagues (Tubio et al. 2014) all report LINE-1 retrotransposition in colorectal carcinomas. Similar to several other visceral neoplasms, including head and neck carcinomas, lung carcinomas, and uterine carcinomas, LINE-1 retrotransposition in colorectal carcinomas appears to be biphasic. Specifically, LINE-1 retrotransposition is not identified in the vast majority of the above-mentioned neoplasms, but, in rare cases, marked LINE-1 retrotranposition (>50 inserts per tumor) is seen in each of the above-mentioned neoplasms.

Taken together, there is moderate evidence that LINE-1 hypomethylation occurs early during colorectal carcinoma tumorigenesis. There is an assertion that serum folate levels may be inversely related to cancer-specific LINE-1 hypomethylation. LINE-1 retrotransposition may drive tumorigenesis in rare cases of colorectal carcinomas.

8 LINE-1 in Esophageal Carcinoma

Many organ-specific cancers are commonly divided into histopathologic variants, based on microscopic appearance of cancer cells. A study of 502 esophageal cancers by Baba et al. revealed that basiloid esophageal cancers showed greater degree of LINE-1 hypomethylation than squamous esophageal cancers (Baba et al. 2015).

A study of 140 esophageal cancer cases by Hashimoto et al. revealed that LINE-1 hypomethylation is present in esophageal cancers compared to normal mucosa (Hoshimoto et al. 2015). Furthermore, greater degree of LINE-1 hypomethylation was noted in higher pathologic stage cancers (Hoshimoto et al. 2015). Another study of 125 esophageal carcinomas by Li et al. also detected LINE-1 hypomethylation in esophageal cancers compared to LINE-1 methylation levels in normal mucosa (Li et al. 2014).

A study of 109 esophageal carcinomas by Shigaki et al. discovered an association between LINE-1 hypomethylation in noncancerous esophageal mucosae of esophageal carcinoma cases and smoking history (Shigaki et al. 2012). Importantly, there was an additional association amongst LINE-1 hypomethylation and smoking duration as well as number of cigarettes smoked per day (Shigaki et al. 2012).

Taken together there is moderate epidemiologic evidence documenting LINE-1 hypomethylation in esophageal carcinomas. There are no mechanistic studies implicating putative role of LINE-1 hypomethylation in tumorigenesis of esophageal carcinomas. LINE-1 retrotransposition has been reported in premalignant Barett's esophagus and esophageal carcinomas (Doucet-O'Hare et al. 2015; Paterson et al. 2015), but the role of LINE-1 insertions in esophageal carcinomas remains unclear.

9 LINE-1 in Gastric Cancers

A study of 24 gastric cancers by Dauksa et al. reported LINE-1 hypomethylation in peripheral leukocytes of cancer cases compared to LINE-1 methylation levels of healthy controls (Dauksa et al. 2014). The authors suggest that LINE-1 hypomethylation may be a biomarker and/or predictive of gastric cancer (Dauksa et al. 2014).

A study of 88 gastric cancers by Yang et al. reported LINE-1 hypomethylation of gastric cancers relative to LINE-1 methylation levels in both gastric adenomas and normal gastric mucosa (Yang et al. 2014). A study of 87 gastric carcinoma cases by Kosumi et al. determined LINE-1 hypomethylation in noncancerous gastric mucosa of gastric cancer patients relative to LINE-1 methylation levels in healthy mucosa (Kosumi et al. 2015). The authors note that this feature of LINE-1 abnormality folds under the term "field cancerization" and as such LINE-1 hypomethylation may arise from exposure to a putative injurious environment.

An interesting study by Wang et al. characterizes expression of a gene, GCRG213p, which shares 88 % protein sequence homology with endonuclease C-terminal portion of LINE-1 ORF2p (Wang et al. 2013). Studying a series of 175 gastric carcinomas, the authors discover that GCRG213p is readily detectable by western blotting analysis and immunohistochemistry in gastric carcinomas, but not in normal mucosa (Wang et al. 2013). Highest GCRG213p was observed in well-differentiated histopathologic appearance and late-age onset of gastric carcinomas (Wang et al. 2013). This report highlights notable genetic pleiotropy of LINE-1 retrotransposon biology in cancer.

A study of 203 gastric carcinomas by Shigaki et al. reported an association between LINE-1 hypomethylation and shorter overall survival (Shigaki et al. 2013).

A study of 198 gastric cancers by Bae et al. looked at LINE-1 methylation levels in early, the most incipient pre-stage of gastric cancers, so-called intestinal metaplasia (Bae et al. 2012). The authors detect LINE-1 hypomethylation even in such most early stage of gastric precancer (Bae et al. 2012). Of note, it is thought that only a few intestinal metaplasias progress to gastric adenoma stage, and, in turn, even fewer gastric adenomas progress to gastric carcinomas.

Taken together, there is moderate evidence that LINE-1 hypomethylation is a biomarker of early gastric neoplasias. LINE-1 retrotransposition in gastric carcinomas has been reported (Ewing et al. 2015), but there is no empirical evidence that LINE-1 retrotransposons contribute to tumorigenesis of gastric carcinomas.

10 LINE-1 in Malignant Melanomas

Studies of LINE-1 retrotransposons in malignant melanomas are few but interesting. One such notable study is a report by De Araujo et al. of 69 melanomas (De Araujo et al. 2015). The authors discovered no difference in LINE-1 methylation in peripheral leukocytes of most melanoma cases compared to LINE-1 methylation levels in healthy controls (De Araujo et al. 2015). However, hereditary cases of melanoma carrying germline CDKN2A mutations showed hypermethylation of LINE-1 in peripheral leukocytes. This important finding suggests that LINE-1 hypomethylation is dispensable for tumorigenesis of some malignant melanomas.

Another notable study is that by Pergoli et al. of 167 malignant melanomas (Pergoli et al. 2014). The study revealed no tendency for LINE-1 methylation changes in peripheral leukocytes of malignant melanoma cases compared to LINE-1 methylation levels in healthy controls (Pergoli et al. 2014). A study of 180 melanomas by Hyland et al. also did not detect any significant association between LINE-1 methylation in peripheral leukocytes and risk of malignant melanomas (Hyland et al. 2013). Study of 133 melanomas and 56 peripheral leukocytes specimens from melanoma patients by Hashimoto et al. revealed that tumoral LINE-1 hypomethylation denotes shorter overall prognosis (Hoshimoto et al. 2012). In addition, the authors also report LINE-1 hypomethylation in peripheral leukocytes of high-pathologic-stage melanoma patients (Hoshimoto et al. 2012).

Taken together, there is some evidence that LINE-1 hypomethylation occurs in most malignant melanomas. There are no mechanistic studies of LINE-1 retrotransposons in malignant melanomas.

11 LINE-1 in Breast Cancer

Cleverly designed studies of LINE-1 retrotransposons in breast cancer have provided useful insights into the timing of LINE-1 hypomethylation throughout tumorigenesis. Of particular importance is a study by Delgano-Cruzata et al. of 333 cancer-free females' family members of the New York site of the Breast Cancer Family Registry (Delgado-Cruzata et al. 2014). The authors discovered that LINE-1 hypomethylation is detectable in individuals with three or more first-degree relatives with breast cancer compared to women with only one first-degree relative (Delgado-Cruzata et al. 2014). Of note, hypomethylation of other repetitive sequences, such as Alu and Sat2, also showed a similar association (Delgado-Cruzata et al. 2014). A study of 274 breast cancers by Park et al. identified an association between LINE-1 hypomethylation and several early aberrant molecular features: negative ER status, ERBB2 (HER2) amplification, and p53 overexpression (Park et al. 2014). As in the preceding study, the authors also reported Alu hypomethylation in the same cohort (Park et al. 2014).

Singularly, the most insightful study of this entire chapter is an analysis by Deroo et al. of pre-diagnostic LINE-1 methylation levels in the sister study (Deroo et al. 2014). Here the authors obtained blood specimens from 50,884 females aged 35–74 years who were not diagnosed with breast cancer at the time of blood draw (Deroo et al. 2014). Some females developed breast cancer in the course of the study and two main findings are notable. First, LINE-1 hypomethylation of pre-diagnostic peripheral leukocytes was associated with breast cancer risk (Deroo et al. 2014). Second, the authors also detected a dose-dependent relationship between the degree of LINE-1 hypomethylation and the level of breast cancer risk (Deroo et al. 2014). Another sister study by Wu examined DNA methylation levels in 282 breast cancer cases and 347 unaffected sisters (Wu et al. 2012a). While methylation of tandemly repeating satellite DNA, Sat2, was associated with the risk of breask cancer, no association was detected between breast cancer risk and LINE-1 methylation (Wu et al. 2012a).

Studying a singular breast cancer cell line Yang et al. reported that exogenous LINE-1 ORF1p functions as a novel HGF/ETS-1 signaling pathway co-activator (Yang et al. 2013). Authors also noted that exogenous ORF1p promotes breast cancer cell line growth (Yang et al. 2013). The study is notable because, to my mind, this is one of only a few reported instances of a measurable empirical phenotype following LINE-1, albeit exogenous, expression (Table 1).

Several of the following studies focused on varied other aspects of LINE-1 biology in breast cancers. A study by Cruickshanks et al. identified transcription of novel chimeric transcripts, termed LCT13, in selected breast cancers (Cruickshanks et al. 2013). The authors propose that LCT13 expression likely mediates epigenetic silencing of the metastasis-suppressor gene *TFPI-2* (Cruickshanks et al. 2013). L1-mediated chimeric transcripts are the topic of chapter "Retrotransposon-Driven Transcription and Cancer" in this book.

Pioneering studies by Fanning and colleagues discovered that a majority (~90 %) of breast carcinomas support LINE-1 ORF1p expression by immunohistochemistry (Bratthauer et al. 1994), raising the possibility that LINE-1 expression contributes to the tumorigenesis of breast cancers. A study of 95 breast carcinomas by Chen et al. revealed that cancers with both nuclear LINE-1 ORF1p and ORF2p expression are associated with shorter overall survival (Chen et al. 2012). Of note, roughly 30 % of all breast cancers displayed such phenotype (Chen et al. 2012). A study by van Hoesel et al. examined LINE-1 methylation in 395 breast cancers (van Hoesel et al. 2012). Subgroup analysis showed that LINE-1 hypomethylation is associated with shorter overall survival, but only in younger patients (<55 years old) (van Hoesel et al. 2012). A study of 441 breast cancers by Harris showed that LINE-1 ORF1p expression is detectable in the cytoplasm of tumor cells in approximately 82 % of breast cancers (Harris et al. 2010). As shown in the van Hoesel study, nuclear localization of the LINE-1 ORF1p protein by immunohistochemistry is associated with shorter overall survival (Harris et al. 2010).

Taken together there is evidence that LINE-1 ORF1p nuclear expression is a biomarker of aggressive breast cancers. There is preliminary evidence that LINE-1 hypomethylation is a risk factor for breast cancer. LINE-1 retrotransposition in breast cancers has been noted (Helman et al. 2014), but there is no empirical evidence that LINE-1 insertions can drive the tumorigenesis of these malignant neoplasms.

12 LINE-1 in Lung Cancers

A hallmark study by Iskow et al. reported 9 LINE-1 retrotransposition events in 20 lung cancers (Iskow et al. 2010). Of interest, cases that were LINE-1 retrotransposition permissive also showed hypomethylation of selected single-copy CpG islands suggesting again that global DNA hypomethylation in cancer underlies LINE-1 hypomethylation (Iskow et al. 2010). A study by Saito et al. examined LINE-1 methylation in 364 lung cancers (Saito et al. 2010). The authors find a link between LINE-1 hypomethylation and shorter overall survival, but only in early-pathologic-stage lung cancers (Saito et al. 2010). A study of 211 lung adenocarcinomas by Ikeda et al. showed that LINE-1 hypomethylation is associated with several clinical and pathologic features such as shorter disease-free survival, higher pathologic stage, and presence of lymphovascular invasion (Ikeda et al. 2013).

Taken together, there is only preliminary evidence that LINE-1 hypomethylation occurs in lung cancers. Very little is known about putative effect of LINE-1 hypomethylation on lung cancer tumorigenesis. Some lung carcinomas are one of the most permissive types of neoplasms for somatic LINE-1 retrotransposition (Helman et al. 2014). Future studies focusing on the putative role of LINE-1 retrotransposition in the tumorigenesis of lung cancers are warranted.

13 Perspectives

LINE-1 retrotransposons remain peculiar genetic elements with a wide array of potential biological ramifications: their transcription may cause downregulation of neighboring transcripts, endonuclease activity of LINE-1 ORF2p may cause staggered double-stranded DNA breaks, and completion of LINE-1 retrotransposon life cycle can result in insertional mutagenesis. Each feature of LINE-1 retrotransposon life cycle can occur unchecked and therefore studies of LINE-1 in cancer are not one topic, but many.

The study of LINE-1 retrotransposons as a putative mediator or indicator of mutagenesis/tumorigenesis is confounded by technological difficulties. The analyte is often hard to study for two principal reasons. First, LINE-1 hypomethylation denotes only limited decrease in LINE-1 methylation levels, such that LINE-1 hypomethylation in cancer tissues is usually approximately 70–80 % of that seen in healthy tissue. Alternatively, because methylation-specific PCR primer pairs used in many of the above studies could plausibly anneal preferentially only to a subgroup of LINE-1 sequences, there could be more marked LINE-1 hypomethylation effect, limited only to a subgroup of LINE-1 sequences, which are difficult to measure. Second, clinical specimens are often heterogeneous, comprised of unknown proportion of lesional cancer cells with accompanying and varied assortment of contaminating somatic cells, a mixture of stromal cells—fibroblasts, endothelial cells, and circulating hematolymphoid cells. Many studies of LINE-1 hypomethylation are affected by this type of pre-analytical bias, whereby each LINE-1 methylation measure is a composite average value of LINE-1 methylation obtained from some lesional cells and variable amount of contaminating non-lesional cells. To possibly improve performance characteristics one could consider a change in tissue procurement: dissociating tissue and separating lesional cells by flow cytometry prior to analyses. Finally, when evaluating LINE-1 protein expression by immunohistochemistry or immunofluorescence, one should develop empirical standard, either from transfected cells or specimen types that support high level of LINE-1 ORF1p expression such as placenta or germ cell tumors (Rodic et al. 2014). Low levels of LINE-1-encoded protein expression versus background nonspecific staining can be difficult to discern (Doucet-O'Hare et al. 2015) even with the use of technical controls.

What studies are then needed to discern possible effect(s) of LINE-1 retrotransposons on tumorigenesis? First, an animal model is desperately needed. Despite years of work, current codon-optimized LINE-1 ORFeus transgenic mice display limited skin color variegation phenotype, with no propensity for tumor formation (O'Donnell et al. 2013). As an alternative, one should focus on identifying an observable phenotype, ascribed preferably to endogenous LINE-1 retrotransposons, for instance in cultured cells. Towards that end, works by Yang et al. (2013) and Sciamanna et al. (2005) should be revisited and studied in greater detail; both bodies of work show cancer cell growth inhibition following siRNA-mediated inhibition of endogenous LINE-1 retrotransposons (Table 1). A recent study by Aschacher et al. described how endogenous LINE-1 helps maintain telomere length (Aschacher et al. 2016). This study also deserves further inquiry (Table 1).

14 Conclusions

LINE-1 hypomethylation marks many, both benign and malignant, human neoplasms. The magnitude of hypomethylation is mild, usually ~70–80 % of that seen in normal tissue. On the other hand, LINE-1 retrotransposition is a seemingly rare event in cancer. With rare exceptions, LINE-1 retrotransposition likely does not drive tumorigenesis of most human cancers. Future studies should focus on mechanisms of LINE-1 hypomethylation in cancer.

References

Antelo M, Balaguer F, Shia J, Shen Y, Hur K, Moreira L, Cuatrecasas M, Bujanda L, Giraldez MD, Takahashi M, Cabanne A, Barugel ME, Arnold M, Roca EL, Andreu M, Castellvi-Bel S, Llor X, Jover R, Castells A, Boland CR, Goel A (2012) A high degree of LINE-1 hypomethylation is a unique feature of early-onset colorectal cancer. PLoS One 7:e45357

Asch HL, Eliacin E, Fanning TG, Connolly JL, Bratthauer G, Asch BB (1996) Comparative expression of the LINE-1 p40 protein in human breast carcinomas and normal breast tissues. Oncol Res 8:239–247

Aschacher T, Wolf B, Enzmann F, Kienzl P, Messner B, Sampl S, Svoboda M, Mechtcheriakova D, Holzmann K, Bergmann M (2016) LINE-1 induces hTERT and ensures telomere maintenance in tumour cell lines. Oncogene 35:94–104

Baba Y, Ishimoto T, Harada K, Kosumi K, Murata A, Miyake K, Hiyoshi Y, Kurashige J, Iwatsuki M, Iwagami S, Miyamoto Y, Sakamoto Y, Yoshida N, Oki E, Iyama K, Watanabe M, Baba H (2015) Molecular characteristics of basaloid squamous cell carcinoma of the esophagus: analysis of KRAS, BRAF, and PIK3CA mutations and LINE-1 methylation. Ann Surg Oncol 22:3659–3665

Bae JM, Shin SH, Kwon HJ, Park SY, Kook MC, Kim YW, Cho NY, Kim N, Kim TY, Kim D, Kang GH (2012) ALU and LINE-1 hypomethylations in multistep gastric carcinogenesis and their prognostic implications. Int J Cancer 131:1323–1331

Barry KH, Moore LE, Liao LM, Huang WY, Andreotti G, Poulin M, Berndt SI (2015) Prospective study of DNA methylation at LINE-1 and Alu in peripheral blood and the risk of prostate cancer. Prostate 75:1718–1725

Belancio VP, Hedges DJ, Deininger P (2008) Mammalian non-LTR retrotransposons: for better or worse, in sickness and in health. Genome Res 18:343–358

Benard A, Van De Velde CJ, Lessard L, Putter H, Takeshima L, Kuppen PJ, Hoon DS (2013) Epigenetic status of LINE-1 predicts clinical outcome in early-stage rectal cancer. Br J Cancer 109:3073–3083

Bratthauer GL, Cardiff RD, Fanning TG (1994) Expression of LINE-1 retrotransposons in human breast cancer. Cancer 73:2333–2336

Bratthauer GL, Fanning TG (1992) Active LINE-1 retrotransposons in human testicular cancer. Oncogene 7:507–510

Chen L, Dahlstrom JE, Chandra A, Board P, Rangasamy D (2012) Prognostic value of LINE-1 retrotransposon expression and its subcellular localization in breast cancer. Breast Cancer Res Treat 136:129–142

Cruickshanks HA, Vafadar-Isfahani N, Dunican DS, Lee A, Sproul D, Lund JN, Meehan RR, Tufarelli C (2013) Expression of a large LINE-1-driven antisense RNA is linked to epigenetic silencing of the metastasis suppressor gene TFPI-2 in cancer. Nucleic Acids Res 41:6857–6869

Dauksa A, Gulbinas A, Endzinas Z, Oldenburg J, El-Maarri O (2014) DNA methylation at selected CpG sites in peripheral blood leukocytes is predictive of gastric cancer. Anticancer Res 34:5381–5388

De Araujo ES, Kashiwabara AY, Achatz MI, Moredo LF, De Sa BC, Duprat JP, Rosenberg C, Carraro DM, Krepischi AC (2015) LINE-1 hypermethylation in peripheral blood of cutaneous melanoma patients is associated with metastasis. Melanoma Res 25:173–177

De Luca C, Guadagni F, Sinibaldi-Vallebona P, Sentinelli S, Gallucci M, Hoffmann A, Schumann GG, Spadafora C, Sciamanna I (2015) Enhanced expression of LINE-1-encoded ORF2 protein in early stages of colon and prostate transformation. Oncotarget 7:4048–4061

Delgado-Cruzata L, Wu HC, Liao Y, Santella RM, Terry MB (2014) Differences in DNA methylation by extent of breast cancer family history in unaffected women. Epigenetics 9:243–248

Denli AM, Narvaiza I, Kerman BE, Pena M, Benner C, Marchetto MC, Diedrich JK, Aslanian A, Ma J, Moresco JJ, Moore L, Hunter T, Saghatelian A, Gage FH (2015) Primate-specific ORF0 contributes to retrotransposon-mediated diversity. Cell 163:583–593

Deroo LA, Bolick SC, Xu Z, Umbach DM, Shore D, Weinberg CR, Sandler DP, Taylor JA (2014) Global DNA methylation and one-carbon metabolism gene polymorphisms and the risk of breast cancer in the Sister Study. Carcinogenesis 35:333–338

Doucet-O'Hare TT, Rodic N, Sharma R, Darbari I, Abril G, Choi JA, Ahn Young J, Cheng Y, Anders RA, Burns KH, Meltzer SJ, Kazazian HH Jr (2015) LINE-1 expression and retrotransposition in Barrett's esophagus and esophageal carcinoma. Proc Natl Acad Sci U S A 112:E4894–E4900

Ewing AD, Gacita A, Wood LD, Ma F, Xing D, Kim MS, Manda SS, Abril G, Pereira G, Makohon-Moore A, Looijenga LH, Gillis AJ, Hruban RH, Anders RA, Romans KE, Pandey A, Iacobuzio-Donahue CA, Vogelstein B, Kinzler KW, Kazazian HH Jr, Solyom S (2015) Widespread somatic L1 retrotransposition occurs early during gastrointestinal cancer evolution. Genome Res 25:1536–1545

Fanning TG, Singer MF (1987) LINE-1: a mammalian transposable element. Biochim Biophys Acta 910:203–212

Gao XD, Qu JH, Chang XJ, Lu YY, Bai WL, Wang H, Xu ZX, An LJ, Wang CP, Zeng Z, Yang YP (2014) Hypomethylation of long interspersed nuclear element-1 promoter is associated with poor outcomes for curative resected hepatocellular carcinoma. Liver Int 34:136–146

Harada K, Baba Y, Ishimoto T, Chikamoto A, Kosumi K, Hayashi H, Nitta H, Hashimoto D, Beppu T, Baba H (2015) LINE-1 methylation level and patient prognosis in a database of 208 hepatocellular carcinomas. Ann Surg Oncol 22:1280–1287

Harris CR, Normart R, Yang Q, Stevenson E, Haffty BG, Ganesan S, Cordon-Cardo C, Levine AJ, Tang LH (2010) Association of nuclear localization of a long interspersed nuclear element-1 protein in breast tumors with poor prognostic outcomes. Genes Cancer 1:115–124

Helman E, Lawrence MS, Stewart C, Sougnez C, Getz G, Meyerson M (2014) Somatic retrotransposition in human cancer revealed by whole-genome and exome sequencing. Genome Res 24:1053–1063

Hoshimoto S, Kuo CT, Chong KK, Takeshima TL, Takei Y, Li MW, Huang SK, Sim MS, Morton DL, Hoon DS (2012) AIM1 and LINE-1 epigenetic aberrations in tumor and serum relate to melanoma progression and disease outcome. J Invest Dermatol 132:1689–1697

Hoshimoto S, Takeuchi H, Ono S, Sim MS, Huynh JL, Huang SK, Marzese DM, Kitagawa Y, Hoon DS (2015) Genome-wide hypomethylation and specific tumor-related gene hypermethylation are associated with esophageal squamous cell carcinoma outcome. J Thorac Oncol 10:509–517

Hyland PL, Burke LS, Pfeiffer RM, Mirabello L, Tucker MA, Goldstein AM, Yang XR (2013) LINE-1 methylation in peripheral blood and the risk of melanoma in melanoma-prone families with and without CDKN2A mutations. Melanoma Res 23:55–60

Ikeda K, Shiraishi K, Eguchi A, Shibata H, Yoshimoto K, Mori T, Baba Y, Baba H, Suzuki M (2013) Long interspersed nucleotide element 1 hypomethylation is associated with poor prognosis of lung adenocarcinoma. Ann Thorac Surg 96:1790–1794

Inamura K, Yamauchi M, Nishihara R, Kim SA, Mima K, Sukawa Y, Li T, Yasunari M, Zhang X, Wu K, Meyerhardt JA, Fuchs CS, Harris CC, Qian ZR, Ogino S (2015) Prognostic significance and molecular features of signet-ring cell and mucinous components in colorectal carcinoma. Ann Surg Oncol 22:1226–1235

Inamura K, Yamauchi M, Nishihara R, Lochhead P, Qian ZR, Kuchiba A, Kim SA, Mima K, Sukawa Y, Jung S, Zhang X, Wu K, Cho E, Chan AT, Meyerhardt JA, Harris CC, Fuchs CS, Ogino S (2014) Tumor LINE-1 methylation level and microsatellite instability in relation to colorectal cancer prognosis. J Natl Cancer Inst 106:dju195

Iskow RC, McCabe MT, Mills RE, Torene S, Pittard WS, Neuwald AF, Van Meir EG, Vertino PM, Devine SE (2010) Natural mutagenesis of human genomes by endogenous retrotransposons. Cell 141:1253–1261

Jung AY, Botma A, Lute C, Blom HJ, Ueland PM, Kvalheim G, Midttun O, Nagengast F, Steegenga W, Kampman E (2013) Plasma B vitamins and LINE-1 DNA methylation in leukocytes of patients with a history of colorectal adenomas. Mol Nutr Food Res 57:698–708

Kim JH, Dhanasekaran SM, Prensner JR, Cao X, Robinson D, Kalyana-Sundaram S, Huang C, Shankar S, Jing X, Iyer M, Hu M, Sam L, Grasso C, Maher CA, Palanisamy N, Mehra R, Kominsky HD, Siddiqui J, Yu J, Qin ZS, Chinnaiyan AM (2011) Deep sequencing reveals distinct patterns of DNA methylation in prostate cancer. Genome Res 21:1028–1041

Konda K, Konishi K, Yamochi T, Ito YM, Nozawa H, Tojo M, Shinmura K, Kogo M, Katagiri A, Kubota Y, Muramoto T, Yano Y, Kobayashi Y, Kihara T, Tagawa T, Makino R, Takimoto M, Imawari M, Yoshida H (2014) Distinct molecular features of different macroscopic subtypes of colorectal neoplasms. PLoS One 9:e103822

Kosumi K, Baba Y, Ishimoto T, Harada K, Miyake K, Izumi D, Tokunaga R, Murata A, Eto K, Sugihara H, Shigaki H, Iwagami S, Sakamoto Y, Miyamoto Y, Yoshida N, Watanabe M,

Komohara Y, Takeya M, Baba H (2015) Relationship between LINE-1 hypomethylation and Helicobacter pylori infection in gastric mucosae. Med Oncol 32:117
Kreimer U, Schulz WA, Koch A, Niegisch G, Goering W (2013) HERV-K and LINE-1 DNA Methylation and Reexpression in Urothelial Carcinoma. Front Oncol 3:255
Lau CC, Sun T, Ching AK, He M, Li JW, Wong AM, Co NN, Chan AW, Li PS, Lung RW, Tong JH, Lai PB, Chan HL, To KF, Chan TF, Wong N (2014) Viral-human chimeric transcript predisposes risk to liver cancer development and progression. Cancer Cell 25:335–349
Lee E, Iskow R, Yang L, Gokcumen O, Haseley P, Luquette LJ III, Lohr JG, Harris CC, Ding L, Wilson RK, Wheeler DA, Gibbs RA, Kucherlapati R, Lee C, Kharchenko PV, Park PJ, Cancer Genome Atlas Research Network (2012) Landscape of somatic retrotransposition in human cancers. Science 337:967–971
Leibold DM, Swergold GD, Singer MF, Thayer RE, Dombroski BA, Fanning TG (1990) Translation of LINE-1 DNA elements in vitro and in human cells. Proc Natl Acad Sci U S A 87:6990–6994
Li RN, Yu FJ, Wu CC, Chen YK, Yu CC, Chou SH, Lee JY, Cheng YJ, Wu MT, Wu IC (2014) Methylation status of retinoic acid receptor beta2 promoter and global DNA in esophageal squamous cell carcinoma. J Surg Oncol 109:623–627
Liao LM, Brennan P, Van Bemmel DM, Zaridze D, Matveev V, Janout V, Kollarova H, Bencko V, Navratilova M, Szeszenia-Dabrowska N, Mates D, Rothman N, Boffetta P, Chow WH, Moore LE (2011) LINE-1 methylation levels in leukocyte DNA and risk of renal cell cancer. PLoS One 6:e27361
Lin C, Yang L, Tanasa B, Hutt K, Ju BG, Ohgi K, Zhang J, Rose DW, Fu XD, Glass CK, Rosenfeld MG (2009) Nuclear receptor-induced chromosomal proximity and DNA breaks underlie specific translocations in cancer. Cell 139:1069–1083
Lou YT, Chen CW, Fan YC, Chang WC, Lu CY, Wu IC, Hsu WH, Huang CW, Wang JY (2014) LINE-1 methylation status correlates significantly to post-therapeutic recurrence in stage III Colon cancer patients receiving FOLFOX-4 adjuvant chemotherapy. PLoS One 10:e0123973
Lu Y, Feng F, Yang Y, Gao X, Cui J, Zhang C, Zhang F, Xu Z, Qv J, Wang C, Zeng Z, Zhu Y, Yang Y (2013) LINE-1 ORF-1p functions as a novel androgen receptor co-activator and promotes the growth of human prostatic carcinoma cells. Cell Signal 25:479–489
Malouf GG, Monzon FA, Couturier J, Molinie V, Escudier B, Camparo P, Su X, Yao H, Tamboli P, Lopez-Terrada D, Picken M, Garcia M, Multani AS, Pathak S, Wood CG, Tannir NM (2013) Genomic heterogeneity of translocation renal cell carcinoma. Clin Cancer Res 19:4673–4684
Matsunoki A, Kawakami K, Kotake M, Kaneko M, Kitamura H, Ooi A, Watanabe G, Minamoto T (2012) LINE-1 methylation shows little intra-patient heterogeneity in primary and synchronous metastatic colorectal cancer. BMC Cancer 12:574
Miki Y, Nishisho I, Horii A, Miyoshi Y, Utsunomiya J, Kinzler KW, Vogelstein B, Nakamura Y (1992) Disruption of the APC gene by a retrotransposal insertion of L1 sequence in a colon cancer. Cancer Res 52:643–645
Murata A, Baba Y, Watanabe M, Shigaki H, Miyake K, Ishimoto T, Iwatsuki M, Iwagami S, Sakamoto Y, Miyamoto Y, Yoshida N, Nosho K, Baba H (2013) Methylation levels of LINE-1 in primary lesion and matched metastatic lesions of colorectal cancer. Br J Cancer 109:408–415
Naito T, Nosho K, Ito M, Igarashi H, Mitsuhashi K, Yoshii S, Aoki H, Nomura M, Sukawa Y, Yamamoto E, Adachi Y, Takahashi H, Hosokawa M, Fujita M, Takenouchi T, Maruyama R, Suzuki H, Baba Y, Imai K, Yamamoto H, Ogino S, Shinomura Y (2014) IGF2 differentially methylated region hypomethylation in relation to pathological and molecular features of serrated lesions. World J Gastroenterol 20:10050–10061
Neale RE, Clark PJ, Fawcett J, Fritschi L, Nagler BN, Risch HA, Walters RJ, Crawford WJ, Webb PM, Whiteman DC, Buchanan DD (2014) Association between hypermethylation of DNA repetitive elements in white blood cell DNA and pancreatic cancer. Cancer Epidemiol 38:576–582
O'Donnell KA, An W, Schrum CT, Wheelan SJ, Boeke JD (2013) Controlled insertional mutagenesis using a LINE-1 (ORFeus) gene-trap mouse model. Proc Natl Acad Sci U S A 110:E2706–E2713

Ogino S, Nishihara R, Lochhead P, Imamura Y, Kuchiba A, Morikawa T, Yamauchi M, Liao X, Qian ZR, Sun R, Sato K, Kirkner GJ, Wang M, Spiegelman D, Meyerhardt JA, Schernhammer ES, Chan AT, Giovannucci E, Fuchs CS (2013) Prospective study of family history and colorectal cancer risk by tumor LINE-1 methylation level. J Natl Cancer Inst 105:130–140
Park SY, Seo AN, Jung HY, Gwak JM, Jung N, Cho NY, Kang GH (2014) Alu and LINE-1 hypomethylation is associated with HER2 enriched subtype of breast cancer. PLoS One 9:e100429
Paterson AL, Weaver JM, Eldridge MD, Tavare S, Fitzgerald RC, Edwards PA, Consortium, O. C. (2015) Mobile element insertions are frequent in oesophageal adenocarcinomas and can mislead paired-end sequencing analysis. BMC Genomics 16:473
Pergoli L, Favero C, Pfeiffer RM, Tarantini L, Calista D, Cavalleri T, Angelici L, Consonni D, Bertazzi PA, Pesatori AC, Landi MT, Bollati V (2014) Blood DNA methylation, nevi number, and the risk of melanoma. Melanoma Res 24:480–487
Quintanilla I, Lopez-Ceron M, Jimeno M, Cuatrecasas M, Munoz J, Moreira L, Carballal S, Leoz ML, Camps J, Castells A, Pellise M, Balaguer F (2014) LINE-1 hypomethylation is neither present in rectal aberrant crypt foci nor associated with field defect in sporadic colorectal neoplasia. Clin Epigenetics 6:24
Rodic N, Sharma R, Sharma R, Zampella J, Dai L, Taylor MS, Hruban RH, Iacobuzio-Donahue CA, Maitra A, Torbenson MS, Goggins M, Shih Ie M, Duffield AS, Montgomery EA, Gabrielson E, Netto GJ, Lotan TL, De Marzo AM, Westra W, Binder ZA, Orr BA, Gallia GL, Eberhart CG, Boeke JD, Harris CR, Burns KH (2014) Long interspersed element-1 protein expression is a hallmark of many human cancers. Am J Pathol 184:1280–1286
Rodic N, Steranka JP, Makohon-Moore A, Moyer A, Shen P, Sharma R, Kohutek ZA, Huang CR, Ahn D, Mita P, Taylor MS, Barker NJ, Hruban RH, Iacobuzio-Donahue CA, Boeke JD, Burns KH (2015) Retrotransposon insertions in the clonal evolution of pancreatic ductal adenocarcinoma. Nat Med 21:1060–1064
Saito K, Kawakami K, Matsumoto I, Oda M, Watanabe G, Minamoto T (2010) Long interspersed nuclear element 1 hypomethylation is a marker of poor prognosis in stage IA non-small cell lung cancer. Clin Cancer Res 16:2418–2426
Salas LA, Villanueva CM, Tajuddin SM, Amaral AF, Fernandez AF, Moore LE, Carrato A, Tardon A, Serra C, Garcia-Closas R, Basagana X, Rothman N, Silverman DT, Cantor KP, Kogevinas M, Real FX, Fraga MF, Malats N (2014) LINE-1 methylation in granulocyte DNA and trihalomethane exposure is associated with bladder cancer risk. Epigenetics 9:1532–1539
Sciamanna I, Landriscina M, Pittoggi C, Quirino M, Mearelli C, Beraldi R, Mattei E, Serafino A, Cassano A, Sinibaldi-Vallebona P, Garaci E, Barone C, Spadafora C (2005) Inhibition of endogenous reverse transcriptase antagonizes human tumor growth. Oncogene 24:3923–3931
Shigaki H, Baba Y, Watanabe M, Iwagami S, Miyake K, Ishimoto T, Iwatsuki M, Baba H (2012) LINE-1 hypomethylation in noncancerous esophageal mucosae is associated with smoking history. Ann Surg Oncol 19:4238–4243
Shigaki H, Baba Y, Watanabe M, Murata A, Iwagami S, Miyake K, Ishimoto T, Iwatsuki M, Baba H (2013) LINE-1 hypomethylation in gastric cancer, detected by bisulfite pyrosequencing, is associated with poor prognosis. Gastric Cancer 16:480–487
Shukla R, Upton KR, Munoz-Lopez M, Gerhardt DJ, Fisher ME, Nguyen T, Brennan PM, Baillie JK, Collino A, Ghisletti S, Sinha S, Iannelli F, Radaelli E, Dos Santos A, Rapoud D, Guettier C, Samuel D, Natoli G, Carninci P, Ciccarelli FD, Garcia-Perez JL, Faivre J, Faulkner GJ (2013) Endogenous retrotransposition activates oncogenic pathways in hepatocellular carcinoma. Cell 153:101–111
Skowronski J, Fanning TG, Singer MF (1988) Unit-length line-1 transcripts in human teratocarcinoma cells. Mol Cell Biol 8:1385–1397
Stefanoli M, La Rosa S, Sahnane N, Romualdi C, Pastorino R, Marando A, Capella C, Sessa F, Furlan D (2014) Prognostic relevance of aberrant DNA methylation in g1 and g2 pancreatic neuroendocrine tumors. Neuroendocrinology 100:26–34
Tajuddin SM, Amaral AF, Fernandez AF, Chanock S, Silverman DT, Tardon A, Carrato A, Garcia-Closas M, Jackson BP, Torano EG, Marquez M, Urdinguio RG, Garcia-Closas R, Rothman N,

Kogevinas M, Real FX, Fraga MF, Malats N, Spanish Bladder Cancer, EPICURO Study Investigators (2014) LINE-1 methylation in leukocyte DNA, interaction with phosphatidylethanolamine N-methyltransferase variants and bladder cancer risk. Br J Cancer 110:2123–2130

Ting DT, Lipson D, Paul S, Brannigan BW, Akhavanfard S, Coffman EJ, Contino G, Deshpande V, Iafrate AJ, Letovsky S, Rivera MN, Bardeesy N, Maheswaran S, Haber DA (2011) Aberrant overexpression of satellite repeats in pancreatic and other epithelial cancers. Science 331:593–596

Tubio JM, Li Y, Ju YS, Martincorena I, Cooke SL, Tojo M, Gundem G, Pipinikas CP, Zamora J, Raine K, Menzies A, Roman-Garcia P, Fullam A, Gerstung M, Shlien A, Tarpey PS, Papaemmanuil E, Knappskog S, Van Loo P, Ramakrishna M, Davies HR, Marshall J, Wedge DC, Teague JW, Butler AP, Nik-Zainal S, Alexandrov L, Behjati S, Yates LR, Bolli N, Mudie L, Hardy C, Martin S, McLaren S, O'Meara S, Anderson E, Maddison M, Gamble S, Group IBC, Group IBC, Group IPC, Foster C, Warren AY, Whitaker H, Brewer D, Eeles R, Cooper C, Neal D, Lynch AG, Visakorpi T, Isaacs WB, Van't Veer L, Caldas C, Desmedt C, Sotiriou C, Aparicio S, Foekens JA, Eyfjord JE, Lakhani SR, Thomas G, Myklebost O, Span PN, Borresen-Dale AL, Richardson AL, Van De Vijver M, Vincent-Salomon A, Van Den Eynden GG, Flanagan AM, Futreal PA, Janes SM, Bova GS, Stratton MR, McDermott U, Campbell PJ (2014) Mobile DNA in cancer. Extensive transduction of nonrepetitive DNA mediated by L1 retrotransposition in cancer genomes. Science 345:1251343

Van Bemmel D, Lenz P, Liao LM, Baris D, Sternberg LR, Warner A, Johnson A, Jones M, Kida M, Schwenn M, Schned AR, Silverman DT, Rothman N, Moore LE (2012) Correlation of LINE-1 methylation levels in patient-matched buffy coat, serum, buccal cell, and bladder tumor tissue DNA samples. Cancer Epidemiol Biomarkers Prev 21:1143–1148

Van Hoesel AQ, Van De Velde CJ, Kuppen PJ, Liefers GJ, Putter H, Sato Y, Elashoff DA, Turner RR, Shamonki JM, De Kruijf EM, Van Nes JG, Giuliano AE, Hoon DS (2012) Hypomethylation of LINE-1 in primary tumor has poor prognosis in young breast cancer patients: a retrospective cohort study. Breast Cancer Res Treat 134:1103–1114

Vogelstein B, Papadopoulos N, Velculescu VE, Zhou S, Diaz LA Jr, Kinzler KW (2013) Cancer genome landscapes. Science 339:1546–1558

Walters RJ, Williamson EJ, English DR, Young JP, Rosty C, Clendenning M, Walsh MD, Parry S, Ahnen DJ, Baron JA, Win AK, Giles GG, Hopper JL, Jenkins MA, Buchanan DD (2013) Association between hypermethylation of DNA repetitive elements in white blood cell DNA and early-onset colorectal cancer. Epigenetics 8:748–755

Wang G, Gao J, Huang H, Tian Y, Xue L, Wang W, You W, Lian H, Duan X, Wu B, Wang M (2013) Expression of a LINE-1 endonuclease variant in gastric cancer: its association with clinicopathological parameters. BMC Cancer 13:265

Wilhelm CS, Kelsey KT, Butler R, Plaza S, Gagne L, Zens MS, Andrew AS, Morris S, Nelson HH, Schned AR, Karagas MR, Marsit CJ (2010) Implications of LINE1 methylation for bladder cancer risk in women. Clin Cancer Res 16:1682–1689

Wu HC, Delgado-Cruzata L, Flom JD, Perrin M, Liao Y, Ferris JS, Santella RM, Terry MB (2012a) Repetitive element DNA methylation levels in white blood cell DNA from sisters discordant for breast cancer from the New York site of the Breast Cancer Family Registry. Carcinogenesis 33:1946–1952

Wu HC, Wang Q, Yang HI, Tsai WY, Chen CJ, Santella RM (2012b) Global DNA methylation levels in white blood cells as a biomarker for hepatocellular carcinoma risk: a nested case-control study. Carcinogenesis 33:1340–1345

Yamada A, Minamiguchi S, Sakai Y, Horimatsu T, Muto M, Chiba T, Boland CR, Goel A (2014) Colorectal advanced neoplasms occur through dual carcinogenesis pathways in individuals with coexisting serrated polyps. PLoS One 9:e98059

Yang M, Kim HS, Cho MY (2014) Different methylation profiles between intestinal and diffuse sporadic gastric carcinogenesis. Clin Res Hepatol Gastroenterol 38:613–620

Yang Q, Feng F, Zhang F, Wang C, Lu Y, Gao X, Zhu Y, Yang Y (2013) LINE-1 ORF-1p functions as a novel HGF/ETS-1 signaling pathway co-activator and promotes the growth of MDA-MB-231 cell. Cell Signal 25:2652–2660

Zelic R, Fiano V, Zugna D, Grasso C, Delsedime L, Daniele L, Galliano D, Pettersson A, Gillio-Tos A, Merletti F, Richiardi L (2015) Global Hypomethylation (LINE-1) and Gene-Specific Hypermethylation (GSTP1) on initial negative prostate biopsy as markers of prostate cancer on a rebiopsy. Clin Cancer Res 22:984–992

Contribution of Retrotransposable Elements to Aging

Jill A. Kreiling, Brian C. Jones, Jason G. Wood, Marco De Cecco, Steven W. Criscione, Nicola Neretti, Stephen L. Helfand, and John M. Sedivy

1 Introduction

Aging is characterized by a failure within many cells and organs of the normal homeostatic mechanisms. It is a major risk factor for numerous disorders, including diabetes, hypertension, cardiac disease, osteoarthritis, neurodegeneration, and cancer. Slowing the rate of aging offers an opportunity to prevent, or at least delay, the onset and extent of these disorders, as well as the possibility of extending healthy human life span. Despite the biological complexity that underlies aging, it has repeatedly proven possible to extend the life span of model organisms through modifications of specific physiological systems such as chromatin maintenance, intermediary metabolism, or insulin signaling (Kenyon et al. 1993; Rogina et al. 2000; Clancy et al. 2001; Tatar et al. 2001; Giannakou et al. 2004; Hwangbo et al. 2004; Kapahi et al. 2004; Oberdoerffer and Sinclair 2007; Dang et al. 2009; Sinclair and Oberdoerffer 2009; Feser et al. 2010; Greer et al. 2010; Kenyon 2010; Feser and Tyler 2011; Maures et al. 2011; Han and Brunet 2012; Ni et al. 2012).

A critical aspect of aging is the degradation of fundamental biological structures such as chromatin (Oberdoerffer and Sinclair 2007). In somatic cells, stability of the genome and epigenome is essential for the maintenance of proper gene expression and silencing. Chromatin remodeling, including changes within regions of constitutive heterochromatin that were previously thought to retain repressive characteristics throughout the life of the cell, has emerged as an exciting area in the molecular genetics of aging. Chromatin maintenance, especially

Jill A. Kreiling and Brian C. Jones are co-first authors.

J.A. Kreiling • B.C. Jones • J.G. Wood • M. De Cecco • S.W. Criscione • N. Neretti • S.L. Helfand (✉)• J.M. Sedivy (✉)
Department of Molecular Biology, Cell Biology and Biochemistry, Brown University, Providence, RI 02912, USA
e-mail: stephen_helfand@brown.edu; john_sedivy@brown.edu

G. Cristofari (ed.), *Human Retrotransposons in Health and Disease*,
DOI 10.1007/978-3-319-48344-3_13

that of heterochromatin, has been shown to change with age in yeast, nematodes, flies, mice, and human cell culture, with far-reaching consequences for gene expression and cellular physiology (Kim et al. 1996; Smeal et al. 1996; Dang et al. 2009; Feser et al. 2010; Wood et al. 2010; Feser and Tyler 2011; De Cecco et al. 2013a; Jiang et al. 2013; Sedivy et al. 2013; Wood et al. 2016).

Age-related changes in chromatin states can alter gene transcription, resulting in the expression of genes that are normally silenced (or vice versa), with consequent deleterious effects on cellular physiology (Elgin and Grewal 2003; Berger 2007; Grewal and Jia 2007; Sedivy et al. 2008; Dang et al. 2009; Feser and Tyler 2011; Han and Brunet 2012). The observed loss of silencing in heterochromatic regions with age includes the increased transcription of genes native to heterochromatin, but also transcription and potential mobility of transposable elements (TEs), which make up the majority of transcripts emanating from heterochromatic regions. The ability of transposable elements not just to express themselves, but to mobilize to new genomic locations within individual somatic cells, adds an additional layer of peril to the potential consequences of the loss of heterochromatin silencing with age.

The contributions that loss of silencing of TEs in somatic cells makes toward the inevitable decline in organismal health with age are just beginning to be explored in detail. The recently discovered ability of TEs to promote aging is expected to open a new area of inquiry, with the potential of providing novel insights into the molecular mechanisms underlying the aging process, while simultaneously offering the promise of novel therapeutic interventions for the preservation of a healthier life span.

2 Remodeling of Chromatin During Aging

2.1 Yeast

The link between chromatin and aging has been well interrogated in invertebrate model systems. Early work in the budding yeast *S. cerevisiae* demonstrated a loss of silencing with age in heterochromatic regions of the genome, including telomeres, the mating type loci, and rDNA (Kim et al. 1996; Smeal et al. 1996; Kennedy et al. 1997). More recently, a number of studies have examined the specific chromatin changes that take place as yeasts age. Histone H4K16 acetylation levels increase with age, and Sir2 (which deacetylates H4K16ac) levels drop (Dang et al. 2009). Furthermore, subtelomeric heterochromatic regions lose both histones and silencing as cells age. Another study confirmed the observation of general histone loss with age, and also showed that increasing histone supply genetically is sufficient to extend yeast replicative life span (Feser et al. 2010). This age-related loss of histones is also associated with a breakdown in proper gene regulation, with normally silent genes becoming transcribed with age upon nucleosome loss or rearrangement (Hu et al. 2014). This is accompanied by a general increase in genomic instability, with DNA strand breaks, mitochondrial-nuclear DNA transfer, chromosomal alterations and translocations, and retrotransposition all increasing during yeast aging (Hu et al. 2014). Additionally, manipulating chromatin by deleting the ISWI family chromatin remodeling gene *ISW2* also leads to an extension in life span in a manner mimicking calorie restriction (Dang et al. 2014).

2.2 Nematodes

Results observed in yeast have also been extended to metazoan invertebrate model systems. In *C. elegans* a number of studies have shown links between chromatin structure and life span. Disrupting the ASH-2 complex, which contains a histone H3K4 methyltransferase activity, causes an increase in life span (Greer et al. 2010, 2011). Disruption of the H3K4 demethylases has also been reported to extend life span in several studies. RNAi knockdown or null mutations of the H3K4me3 demethylase RBR-2 as well as the H3K4me1/2 demethylases LSD-1 and SPR-5 extend life span (Lee et al. 2003; McColl et al. 2008; Ni et al. 2012; Alvares et al. 2014). Manipulation of the heterochromatic H3K27me3 mark, which is associated with Polycomb group complex silencing, also showed effects on life span. Two independent studies demonstrated that disrupting the H3K27me3 demethylase UTX-1 leads to increased levels of H3K27me3 accumulation in the genome, as well as increased life span (Jin et al. 2011; Maures et al. 2011). Similarly to yeast, knocking down expression of the ISWI complex member *athp-2* led to an increase in life span (Hu et al. 2014).

2.3 Fruit Flies

In addition to *C. elegans*, *D. melanogaster* has also been a useful model to investigate the association between chromatin structure and organismal life span. The characteristic enrichments of the constitutive heterochromatin mark H3K9me3 and the heterochromatin protein HP1 are lost from pericentric heterochromatin with age in flies (Wood et al. 2010). In conjunction with this observation, heterochromatic silencing of reporter genes in these same regions was lost with age in multiple tissues in the fly (Jiang et al. 2013). Overexpression of HP1 in flies is able to extend life span, suggesting the importance of maintaining proper heterochromatin structure with age (Larson et al. 2012). A study examining aging fly muscle showed an accumulation of γH2AX, a histone variant associated with DNA strand breaks, in old flies (Jeon et al. 2015). Knockdown of HP1 accelerated γH2AX accumulation and also shortened life span (Jeon et al. 2015). Aging effects are however not limited to heterochromatin. Histone acetylation levels also change with age on multiple residues, including an increase of H4K12ac, and mutation of the H4K12 acetyltransferase *Chameau* confers extended life span (Peleg et al. 2016).

2.4 Mammals

Recent studies show that large regions of the genome undergo significant reorganization in cellular senescence and in aged mammalian tissues. Cellular senescence is an irreversible cell cycle arrest that is triggered by replicative exhaustion, DNA damage, oncogene activation, or oxidative stress. Although low in numbers, senescent cells are found in aged tissues and have been shown to contribute to aging phenotypes (Baker et al. 2016). During the onset of senescence large segments of euchromatin become more closed and accumulate heterochromatic marks (Fig. 1) (Kreiling et al. 2011; De Cecco et al. 2013a; Chen et al. 2015; Criscione et al. 2016). A key feature of senescent cells is the formation of senescence-associated heterochromatin foci (SAHF) containing specific heterochromatin signatures (Narita et al. 2003; Zhang et al. 2007; Chandra et al. 2012). In contrast, regions of constitutive heterochromatin, such as lamin-associated domains (LADs) and centromeres, assume more open characteristics, as exemplified by the senescence-associated distention of satellites (SADS) (De Cecco et al. 2013a; Sadaie et al. 2013; Swanson et al. 2013). In addition, genes associated with the senescence-associated secretory phenotype (SASP) take on epigenetic signatures not found in non-senescent cells (Rai and Adams 2013; Chen et al. 2015).

Genome-wide changes in chromatin structure also occur in chronologically aged cells in vivo, with a closing of euchromatic regions and an accumulation of heterochromatic marks, leading to an overall reduction in mRNA expression (Sarg et al. 2002; Shumaker et al. 2006; O'Sullivan et al. 2010; Kreiling et al. 2011; De Cecco et al. 2013b). A corresponding opening of constitutive heterochromatin (De Cecco et al. 2013b) suggests an overall decompaction of the highly heterochromatic regions known to contain large numbers of retrotransposable elements (RTEs). Taken together,

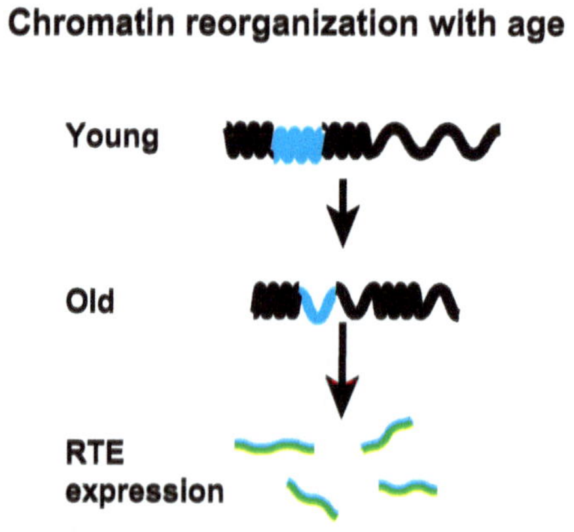

Fig. 1 Age-associated chromatin reorganization. In young cells chromatin is organized into regions of tightly packed heterochromatin (*left*) and relatively open euchromatin (*right*). As cells age, some regions of heterochromatin open up and other regions of euchromatin become more condensed. As a result, repressed genes (such as RTEs, indicated in *blue*) in heterochromatic regions become susceptible to transcription

evidence points to large-scale changes in genome organization, with some regions becoming more closed and others more open, with the latter leading to an increase in the expression of RTEs (O'Sullivan and Karlseder 2012; Sedivy et al. 2013).

The loss of constitutive heterochromatin is correlated with a loss of DNA methylation and histone modifications associated with constitutive heterochromatin. Genome-wide methylation patterns change during cellular senescence and with age in the mammalian genome, with specific regions gaining methylation and others losing methylation (Cruickshanks et al. 2013a; Day et al. 2013; Hanzelmann et al. 2015). In young cells the repetitive regions of the genome show highest levels of DNA methylation, and these regions of hypermethylation become hypomethylated with age (Avrahami et al. 2015; Fernandez et al. 2015; Sun and Yi 2015). This global loss of methylation is coupled with a genome-wide reduction in the H3K9me3 histone modification (Scaffidi and Misteli 2006; Shumaker et al. 2006; O'Sullivan et al. 2010; Zhang et al. 2015), which is associated with repressive heterochromatin and is believed to be actively involved in the repression of RTEs (Scaffidi and Misteli 2006). As discussed below, these heterochromatic marks are involved in silencing RTEs and their loss may contribute to the derepression of these elements.

It is also important to note that results obtained in model organisms are not always completely consistent. For instance, in flies the disruption of *lid*, a LSD-1 H3K4 demethylase analog, shortens life span, in contrast to results observed in *C. elegans* (Li et al. 2010). Disruption of the H3K27 methyltransferase E(Z) in flies leads to reduced levels of H3K27me3 and increased life span (Siebold et al. 2010), in contrast to worms where higher levels of H3K27me3 were associated with long life span (Jin et al. 2011; Maures et al. 2011). Disruption of RBR-2 in worms can have differential effects depending on which allele is used (Greer et al. 2010; Alvares et al. 2014). Nevertheless, although there undoubtedly are tissue-specific and even organism-specific mechanistic details that remain to be worked out, considerable evidence has accumulated for a strong association between chromatin structure, especially that of heterochromatin, and the regulation of longevity in multiple model systems.

2.5 Changes in the 3D Structure of Chromosomes

Aging cells display dramatic alterations in chromatin accessibility, histone modifications, DNA methylation, and nuclear lamina associations. These changes in chromatin architecture were hypothesized to extend even to the 3D structure of the chromosomes. The first hint that chromosome structure may be altered in aging cells came from studies of fibroblasts from patients with the Hutchinson-Gilford progeria syndrome (HGPS) (McCord et al. 2013). HPGS is a premature aging disease that is caused by mutations in the lamin A gene (*LMNA*) that result in disruption of interactions between chromatin and the nuclear lamina. In cell culture HGPS patient skin fibroblasts display misshapen nuclei and a loss of the peripheral heterochromatin compartment (Goldman et al. 2004). When HGPS skin fibroblasts were examined by Hi-C, a method to investigate the three-dimensional architecture of the genome, a breakdown of the compartmentalization of active and inactive chromatin domains was observed (McCord et al. 2013). The alterations are likely caused by the disruption of nuclear lamina-chromatin interactions which normally function to restrict the inactive heterochromatin compartment to the nuclear periphery (Guelen et al. 2008).

The 3D structure of chromosomes has also been explored using Hi-C in oncogene-induced and replicative cellular senescence, which have some overlapping but also distinct features. Oncogene-induced senescence (OIS) is believed to be induced by a DNA damage response that is caused by replication stress (Hills and Diffley 2014), whereas replicative cellular senescence is caused by a DNA damage response due to the progressive shortening and deprotection of telomeres. SAHF are typically observed in OIS (Narita et al. 2003), whereas in many models of replicative senescence SAHF formation is weaker or sometimes not present (Kosar et al. 2011). In OIS regions with heterochromatic histone marks as well as LADs display loss of local interactions and gain of long-range interactions (Chandra et al. 2015). This reorganization is consistent with the presence of SAHF in OIS, since heterochromatic regions could cluster spatially over long distances to form the SAHFs (Chandra et al. 2015). The alterations in 3D chromosome structure observed in OIS are however relatively modest in comparison to the global loss of chromosome compartmentalization found in HPGS.

The alterations in chromosome structure observed during replicative senescence are more extensive than in OIS, but also not as drastic as in HPGS. In replicative senescence chromosomes displayed a global loss of long-range and increase of short-range interactions (Criscione et al. 2016). Chromosome painting experiments additionally showed that these alterations were associated with a decrease in the absolute chromosome volume in senescent cells. In replicative senescence the chromosome compartment organization remained mostly unchanged; however, a subset of compartments switched from active to repressive domains (and vice versa). Similar to the compartment switching observed during cellular differentiation (Dixon et al. 2015), compartment switching in replicative senescence also led to correlated changes in gene expression. Interestingly, similarities to cellular differentiation events were noted in studies of both OIS and replicative senescence (Chandra et al. 2015; Criscione et al. 2016). These observations highlight that cellular senescence is a programmed response to DNA damage that results in the remodeling of chromatin as well as large-scale changes in chromosome architecture, although these processes also include some distinct features that are dependent on the senescence-inducing stimuli.

3 Control of TEs and Their Activation with Aging

A significant fraction of eukaryotic genomes are comprised of repetitive sequences. Among the several types of repetitive sequences, noncoding tandem repeats (satellites, telomeres) and TEs are the most abundant. The TEs can be subdivided into two major groups, the DNA transposons and the retrotransposons (RTEs) (Huang et al. 2012). Many species, including the model organisms *C. elegans* and *D. melanogaster* discussed in this chapter, harbor active elements of both classes. The most prominent TEs in the mammalian genome are the RTEs. There are three major families of RTEs: the long terminal repeat (LTR) RTEs, which include retroviruses; the long interspersed nuclear elements (LINEs); and the short interspersed nuclear elements (SINEs). LTR RTEs and LINEs encode a reverse transcriptase and other proteins required for retrotransposition, and hence intact elements can mobilize autonomously, whereas the SINEs are noncoding and exploit the machinery encoded by LINEs to transpose. It is believed that only the LINE L1 remains capable of autonomous retrotransposition in the human genome, whereas both LINE and LTR elements can mobilize in the mouse genome.

3.1 TEs Are Silenced by RNAi Pathways

Largely conserved across species from plants to animals, RNA interference (RNAi) pathways employ small RNAs (smRNAs) to regulate protein-coding genes as well as endogenous proviral sequences such as TEs (Shabalina and Koonin 2008). TEs are repressed by RNAi at two levels: posttranscriptional regulation by targeting mRNA, and transcriptional regulation by the recruitment of repressive heterochromatic marks to silence the target genes. smRNA pathways known to regulate TEs in animals include the microRNA (miRNA) pathway, the short interfering RNA (siRNA) pathway, and the Piwi-interacting RNA (piRNA) pathways. While these pathways are known for their roles in silencing TEs, they are largely tissue specific with the siRNA and miRNA pathways being active in all tissues while the piRNA pathway is predominantly active in the gonads (Slotkin and Martienssen 2007; Ghildiyal and Zamore 2009; Heras et al. 2013, 2014; Hamdorf et al. 2015).

Each pathway differs somewhat in its effector proteins, manner of smRNA biogenesis, and modes of silencing. The siRNA pathway employs 21 nt siRNAs derived from the cleavage of long double-stranded (dsRNA) substrates by the protein Dicer (Yang and Kazazian 2006; Brennecke et al. 2007; Czech et al. 2008; Ghildiyal et al. 2008; Kawamura et al. 2008). These siRNAs are loaded onto an argonaute (AGO) effector protein, thereby forming an RNA-induced silencing complex (RISC), which then uses its siRNA to target and cleave homologous mRNAs in the cytoplasm. The RISC can also move to the nucleus where it recruits chromatin-modifying enzymes promoting the formation of heterochromatin at the site of TE transcription (Slotkin and Martienssen 2007; Fagegaltier et al. 2009).

The piRNA pathway operates through a mechanism whereby large genomic regions consisting of intact as well as fragmented TEs, called piRNA clusters, are transcribed into large single-stranded RNA precursors that are then processed into smaller 23–29 nt piRNAs. piRNAs are also loaded onto pathway-specific Piwi clade argonaute proteins, thus forming piRNA-RISCs (Brennecke et al. 2007; Ghildiyal and Zamore 2009). Similar to the siRNA pathway, these piRNA-RISCs are able to target TE transcripts for silencing either through catalytic cleavage or heterochromatization (Aravin et al. 2007,

2008; Carmell et al. 2007; Di Giacomo et al. 2013). These smRNA pathways have been shown to be critical in preventing the genomic damage caused by the reactivation of TEs. Evidence is also growing that the ability of these pathways to perform their vital functions of suppressing TEs, in both somatic and reproductive tissues, may be closely linked with aging phenotypes.

3.2 Disruption of RNAi Pathways Correlates with Aging Phenotypes

The role of RNAi in regulating TEs in metazoans has been well characterized in multiple model organisms. In *Drosophila*, mutations in genes of either the siRNA or the piRNA pathways have consistently been associated with a dramatic upregulation of TE transcript levels (Vagin et al. 2006; Rozhkov et al. 2013). This correlates with both an increase in transposition and a change in the heterochromatic marks associated with TEs (Fagegaltier et al. 2009; Gu and Elgin 2013; Perrat et al. 2013). Mutants in the siRNA genes *Dcr-2* and *Ago-2* have dramatically shortened life spans, and this correlates with significant reactivation of TEs (Czech et al. 2008; Ghildiyal et al. 2008; Lim et al. 2011; Li et al. 2013). Interestingly, while TEs have been shown to reactivate with age across multiple species, the transcript levels of RNAi genes that regulate TEs are not known to decline with age and in fact remain constant (Li et al. 2013; Abe et al. 2014). However, the spectrum of available smRNAs that are loaded onto RISCs is known to change with age (Abe et al. 2014). In addition, mutation of known modifiers of RNAi efficacy has been shown to modulate TE activity and life span (Savva et al. 2013). Multiple age-associated diseases are also associated with TE reactivation. For example, macular degeneration in mice and human cell culture has been shown to be dependent upon RNAi machinery where RNAi mutants exhibited increased levels of Alu RNA resulting in RNA toxicity (Kaneko et al. 2011; Tarallo et al. 2012; Gelfand et al. 2015). This suggests that while RNAi proteins themselves may remain relatively constant with age, the many dynamic components and partners of RNAi as well as the overall activity of RNAi may not be as stable. Hence, inhibition or enhancement of RNAi silencing of TEs would be expected to negatively or positively impact life span, respectively. A better understanding of the dynamics of RNAi TE silencing may allow us to control TE reactivation with age.

3.2.1 RNA Editing

In the siRNA pathway, dsRNAs serve as the substrates from which RNAi proteins produce and employ siRNAs in silencing TEs (Ghildiyal and Zamore 2009). dsRNAs in general have also long been known to be substrates for dsRNA-modifying enzymes such as ADAR proteins. These enzymes bind to dsRNAs and are able to convert adenosine bases to inosines, a base analog of guanine (Savva et al. 2012).

This A-to-I editing results in a base pair mismatch between the resulting inosine and the thymine that previously paired with the edited adenosine. The capacity of ADAR proteins to edit dsRNAs has been shown to confer new properties on their substrates, including modified secondary structures, altered stability, nuclear retention, and even novel protein-coding functions (Chen et al. 2008; Jepson et al. 2011; Rieder et al. 2013).

Since dsRNAs are the substrates for siRNA formation, ADAR may also be able to edit these RNAs and thereby modulate the RNAi pathway. The Dicer proteins that catalyze the endonucleolytic cleavage of their dsRNA targets often require a high degree of base pair complementarity, a property that is impaired by RNA editing (Scadden and Smith 2001; Wang et al. 2005; Carpenter et al. 2009; Heale et al. 2009). Hence, ADAR could indirectly inhibit the ability of the siRNA pathway to silence TEs by impairing the access of Dicer to its dsRNA substrates. In fact, it was recently shown that a dsRNA trigger necessary for the silencing of a TE in *Drosophila* was a target of ADAR, and ADAR mutants showed reduced levels of TE transcripts, suggesting enhanced TE silencing (Savva et al. 2013). These mutants also showed altered levels of heterochromatic marks, including HP1 and H3K9me3, and dramatic changes in position effect variegation, a phenotype in *Drosophila* known to be dependent on heterochromatin boundaries. Finally, ADAR mutants showed a dramatic extension of life span. These results suggest that RNA editing may abrogate the TE silencing effects of RNAi and that disrupting genes that impede RNAi, such as ADAR, may enhance TE silencing and thereby extend organismal life span.

3.2.2 RISC Complex Misloading

Argonautes are the main effector proteins that perform RNAi silencing, and the siRNA, piRNA, and miRNA pathways all employ such proteins (Ghildiyal and Zamore 2009). The argonaute proteins act in concert with their respective smRNAs to mediate silencing. The miRNA pathway utilizes miRNAs (21–22 nt long) that often imperfectly base pair with their targets upon association with an argonaute protein. This miRNA-RISC then prevents translation of the target mRNA by one of the two methods: stalling or blocking ribosome access, or cleavage of the target mRNA (Ghildiyal and Zamore 2009). Mammals have four argonautes (AGO1–4), and while only AGO2 is catalytically active, all four argonautes are able to bind smRNAs and facilitate inhibition of translation (Liu 2004; Meister et al. 2004; Wilson and Doudna 2013). Interestingly, human AGO2 can accept both miRNAs and siRNAs (Hutvagner and Zamore 2002; Martinez et al. 2002).

In contrast, in *Drosophila* Ago1 is almost exclusively loaded with miRNAs while Ago2 is loaded mostly with siRNAs (Forstemann et al. 2007). However, recent work has shown that miRNAs and siRNAs compete for loading onto Ago2, the argonaute responsible for TE silencing in flies (Abe et al. 2014). In both flies and mammals, siRNAs are specifically 2′-*O*-methylated at their 3′ termini (Ghildiyal and Zamore 2009). In flies miRNAs were found to be increasingly inappropriately methylated with age, allowing them to be loaded onto Ago2, and thus reducing siRNA access (Abe et al. 2014). This study did not examine the effect of this RISC misloading on the ability of siRNAs to silence TEs. However, this is an interesting possibility, especially in mammals where siRNAs and miRNAs share AGO2 for silencing, and this competition could functionally impact TE silencing.

3.3 The Role of the piRNA Pathway in Aging

3.3.1 piRNA Deficiencies in Aging Gonads

The piRNA pathway has long been known to be a guardian of genomic integrity in the germline. These longer smRNAs (23-29 nt) associate with three Piwi clade argonaute proteins and, similar to siRNAs, are 2′-*O*-methylated (Brennecke et al. 2007; Ghildiyal and Zamore 2009). An exonuclease, known as Nibbler, regulates the length of miRNAs, siRNAs, and piRNAs in both somatic and gonadal tissues (Feltzin et al. 2015). Nibbler is responsible for the appropriate trimming of the 3′ termini of these diverse classes of smRNAs (Liu et al. 2011; Feltzin et al. 2015). In flies, Nibbler mutants showed age-associated accumulation of brain damage and physiological effects such as loss of climbing ability (Abe et al. 2014; Feltzin et al. 2015). Another study showed an association between increased Nibbler activity and TE reactivation (Wang et al. 2016). piRNA length was also shown to increase in aged ovaries and this correlated with lower piRNA abundance, suggesting a disruption of piRNA biogenesis. piRNA pathway mutants display increased TE reactivation and a decline or complete loss in fertility (Wang et al. 2016), and both of these phenotypes are also observed in aging animals. Aging is also known to directly affect the fertility of mammals (Ge et al. 2015) and may be related to reactivation of TEs. In support of this, studies in mice where L1 elements were transgenically overexpressed show increased embryonic lethality suggesting that TEs directly contribute to infertility (Malki et al. 2014). It is possible that reproductive output is reduced with age due to an increase in piRNA trimming, resulting in aberrant piRNA biogenesis and increased TE reactivation

3.3.2 The piRNA Pathway in Somatic Tissues

Recent evidence has begun to suggest that the piRNA pathway may also be active in tissues outside of the gonad. piRNAs and their argonautes have been found in healthy somatic tissues of flies, mice, macaques, and humans (Lee et al. 2011; Yan et al. 2011; Perrat et al. 2013; Jones et al. 2016). In addition, multiple studies have documented reactivation of piRNA pathway machinery in various types of cancer (Ross et al. 2014). However, it is not yet known why these piRNA components are expressed in these situations. It is interesting to note that multiple studies have also shown reactivation of TEs in cancer (Chenais 2013; Doucet-O'Hare et al. 2015; Ewing et al. 2015; Rodic et al. 2015) (see also chapters "Retrotransposon Contribution to Genomic Plasticity" and "LINE-1 Retrotransposons as Neoplastic Biomarkers"). One possibility is that the piRNA pathway, arguably the premier genomic defense against TEs, is activated as a compensatory response to TE derepression in cancerous or aging somatic tissues. Our knowledge of piRNA pathway activity in somatic tissues is very incomplete, and experiments determining a mechanistic cause for its presence and the role it serves in the soma have yet to be performed.

3.4 *TEs in the Mammalian Genome*

Approximately 50 % of mammalian genomes are comprised of repetitive sequences (de Koning et al. 2011). Over evolutionary time most resident RTEs have acquired multiple mutations and are no longer active; however a small fraction retain the ability to transpose (Levin and Moran 2011; Sookdeo et al. 2013). In response, cells have evolved mechanisms to keep these elements tightly repressed. The front-line defense against RTEs is transcriptional silencing (Fig. 2). In mouse embryonic stem

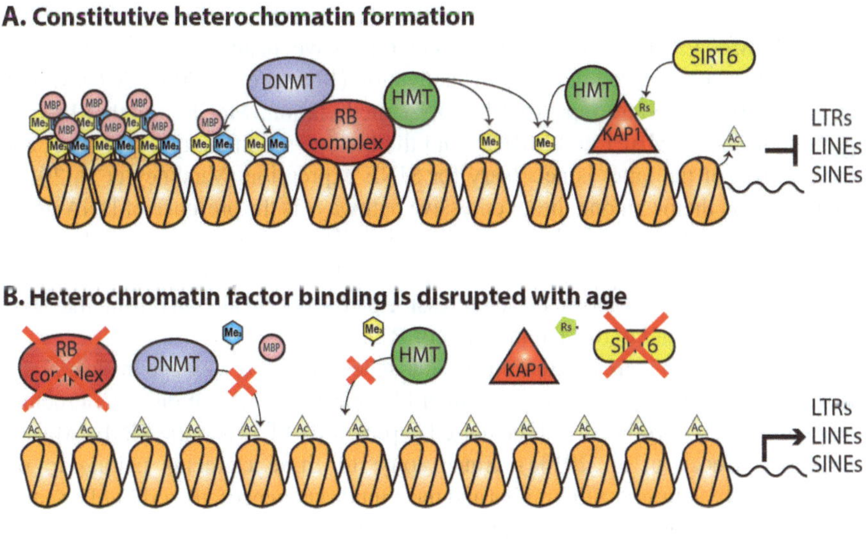

Fig. 2 RTEs in the genome are repressed by heterochromatin. (**a**) Multiple pathways are involved in the establishment and maintenance of heterochromatin. In many regions of the genome, these domains of heterochromatin encompass RTEs and are instrumental in their silencing. The RB complex recruits several histone methyltransferases (HMTs) that methylate specific lysine residues on histones H3 and H4. Additional HMTs are recruited to the site of heterochromatin formation as part of a Kruppel-associated box-associated protein 1 (KAP1)-dependent pathway that requires ribosylation (Rs) by the sirtuin SIRT6. Together, these mechanisms maintain DNA methylation at cytosine residues by methyltransferases (DNMTs). (**b**). These processes are disrupted in aging cells resulting in the relaxation of heterochromatin that in turn allows the expression of RTEs. RB complex: a complex containing the retinoblastoma protein, elongation factor 2, histone deacetylases 1 and 2, methyl CpG-binding protein 2, and the nucleosomal and remodeling deacetylase complex; Me_3: methyl group; Ac: acetyl group; MBP: methylation-binding protein

(ES) cells, LTR RTEs are silenced through multiple mechanisms including DNA methylation by the DNA methyltransferases 1 and 3L (DNMT1 and DNMT3L) (Slotkin and Martienssen 2007). In addition, ES cells use a Kruppel-associated box-associated protein 1 (KAP1)-dependent pathway that results in the tri-methylation of histone H3 on lysine 9 (H3K9me3) by the methyltransferase ESET (Matsui et al. 2010; Rowe et al. 2010). Recruitment of ESET to LTR RTEs requires the deposition of the histone variant H3.3 (Elsasser et al. 2015).

LINE RTEs are also silenced through multiple mechanisms in mammalian cells including DNA methylation and histone modification. Mouse embryonic fibroblasts (MEFs) regulate the expression of the LINE L1 in part through a pathway involving the SIRT6-mediated ribosylation of KAP1 (Van Meter et al. 2014). In addition, MEFs and human cancer cell lines require the recruitment of the EF2/RB complex along with the histone deacetylases 1 and 2 (HDAC1 and HDAC2), the methyl CpG-binding protein 2 (MeCP2), and the nucleosomal and remodeling deacetylase (NuRD) complex to silence L1 expression (Montoya-Durango et al. 2009, 2016; Teneng et al. 2011). In human and mouse neural tissue some L1s become transiently activated during neural progenitor cell differentiation, and this process has been hypothesized to drive variation in neuronal genomes (Muotri et al. 2005; Erwin et al. 2014). Expression of L1s in neural stem cells (NSC) is repressed by SOX2, HDAC1, MeCP2, DNA methylation, and repressive histone modifications, and these factors are reduced during NSC activation (Muotri et al. 2005, 2010; Coufal et al. 2009). SINEs are repressed by DNA methylation, MeCP2, methyl-binding proteins 1 and 2 (MBP1 and MBP2), and the histone modification H3K9me3, and the removal of the latter is necessary for SINE expression (Varshney et al. 2015). The common theme among these repressive pathways is the presence of DNA methylation and the H3K9me3 histone modification. These repressive heterochromatic marks are used by the cell to silence RTEs in an effort to maintain genome integrity. However, as discussed above, these repressive pathways are altered during the aging process and can lead to the derepression of RTEs.

RTE expression increases during cellular senescence and with age in several different mouse tissues. In senescent human fibroblasts the relaxation of heterochromatic regions is correlated with increased expression of L1s and the SINEs Alu and SVA (De Cecco et al. 2013a). Since some of these elements belong to the evolutionarily youngest subfamilies and have intact sequences, they should be capable of transposition (De Cecco et al. 2013a). Indeed, increased genomic copy numbers of L1Hs were observed in senescent cells.

In mouse, members of the LINE (L1), SINE (B1 and B2), and LTR (MusD) families were found to increase expression with age in liver and skeletal muscle (De Cecco et al. 2013b). Interestingly, there appears to be variability between tissues as this increase was more pronounced in muscle than in liver. The transcription of L1s, the largest family of potentially active retrotransposons, was also analyzed in mouse liver by RNA-seq using a bioinformatic pipeline recently developed for the analysis of repetitive sequences in high-throughput DNA sequencing data (Criscione et al. 2014). Many of the L1 subfamilies in the mouse genome were found to increase their expression in liver samples from old animals (De Cecco et al. 2013b).

Expression of RTE mRNA is only the first step that eventually may lead to actual transposition, and several cellular defense mechanisms are known to be active downstream of heterochromatinization. In addition, many elements in the genome have acquired mutations rendering them incapable of transposition. However, current evidence suggests that at least a subset of the derepressed elements are capable of transposition during cellular senescence as well as aging of several mouse tissues (De Cecco et al. 2013a, b).

4 Consequences of Age-Associated TE Activation

4.1 Chimeric Transcripts

In this section we explore the links between RTE activity and changes in the transcriptome. We refer to this process as *transcriptional instability*, and discuss here the different forms it can take and its potential role in aging. The reader can also refer to chapters "Retrotransposon-Derived Regulatory Regions and Transcripts in Stemness" and "Retrotransposon-Driven Transcription and Cancer" for a discussion on RTE-induced transcriptome changes in the context of pluripotent cells and cancer, respectively. Transcriptional noise, defined as increased cell-to-cell variation in gene expression, has been described in the aging mouse heart (Bahar et al. 2006). Dysregulation of alternative splicing has been found in cellular senescence (Cao et al. 2011), in the aging brain (Mazin et al. 2013) and neurodegeneration (Tollervey et al. 2011), and in blood leukocytes (Harries et al. 2011). It has been argued that these changes may be of particular relevance in postmitotic cells and tissues (Warren et al. 2007). Although a direct link between transcriptional instability and RTE activity has not yet been demonstrated in aging, the ability of RTEs to affect the transcriptome is well known in other contexts. First and foremost, over the course of evolution RTEs have rewired the core regulatory network of the mammalian genome (Kunarso et al. 2010). This demonstrates their ability to influence the transcriptome by either disrupting regulatory elements or contributing new ones. For example, Alu elements harbor binding sites for nuclear hormone receptors and can compete or act as promoters for nearby genes (Polak and Domany 2006; Deininger 2011). Their presence in introns can result in alternative or aberrant splicing (Lev-Maor et al. 2008) that can lead to disease (Ganguly et al. 2003).

RTEs, including many transposition-incompetent elements, retain intact promoter sequences that are capable of driving transcription (Faulkner et al. 2009). L1s contain both sense and antisense promoters (ASPs) that can transcribe into adjacent regions to produce chimeric transcripts (Speek 2001; Cruickshanks and Tufarelli 2009). The sense promoter can promote transcription of downstream genes (Abyzov et al. 2013), and L1-ASP transcription of upstream genes has also been found (Speek 2001; Nigumann et al. 2002). Transformed cancer cell lines and prostate tumors display significant upregulation of L1 RNA expression (Criscione et al. 2014). The marked increased in L1 promoter activity in cancer cells has been linked to a variety of aberrant L1 chimeric transcripts. In colorectal cancer, hypomethylation of L1s leads to activation of the methylation-silenced MET and RAB3IP proto-oncogenes (Hur et al. 2014). A truncated isoform of the oncogene c-MET can be driven from an alternative promoter by hypomethylation of an intronic L1-ASP (Roman-Gomez et al. 2005; Weber et al. 2010; Wolff et al. 2010). Conversely, an L1-ASP-driven RNA can silence the metastasis-suppressor gene TFPI-2 (Cruickshanks et al. 2013b). This suggests that activation of L1-ASPs might lead to epigenetic silencing of tumor-suppressor genes, potentially by similar mechanisms as those described for antisense RNAs in development or several diseases (Tufarelli et al. 2003; Matzke and Birchler 2005; Yu et al. 2008; Taft et al. 2010). Hence, it is evident that RTEs are capable of interfering with the transcriptional machinery at multiple levels and could contribute a similar role to cellular dysfunction during aging.

4.2 Characterizing the Transposition Landscape in Aging Cells

TE sequences posed a great challenge for the initial sequencing and assembly of reference genomes. Their analysis has lagged far behind that of non-repetitive sequences, and even the most recent draft of the human genome (GRCh38) contained major updates of TE annotations. Short-read sequencing strategies, such as Illumina's HiSeq, provide an additional challenge: it is not possible to unambiguously assign the genomic locations of many reads originating from repetitive elements. To enable the comprehensive documentation of all existing and novel RTE insertions in the genomes of human somatic cells we would ideally require long reads spanning the entire RTE and flanking sequences on both sides, sufficient coverage of the genome to make statistically significant calls, and low costs to make the profiling of many tissues and ages economically feasible. Recent advances in long-read high-throughput sequencing platforms, including Pacific Biosystems Single Molecule, Real-Time (SMRT) Sequencing, and Oxford Nanopore MinION, will likely aid in discovery of new transposition events; however, these technologies are still costly and yield low coverage.

To further complicate studying RTE mobility during cellular senescence or aging of tissues, many new insertions are likely to be "private," i.e., occurring in an individual cell after it has ceased dividing. This is likely from theoretical considerations, because many cells in the adult organism are postmitotic. Thus characterization of the transposition landscape in aging cells is complicated by the fact that the landscape is likely to be unique for each individual cell. Two approaches have been used to overcome this obstacle: greatly enriching for RTEs before sequencing, or sequencing single-cell genomes. The principle of RTE enrichment is simply to reduce the genomic space that is sequenced in order to increase the coverage and the sensitivity of detection. The caveats of enrichment are that there is selection bias (it requires prior knowledge of active transposons) and enrichment cannot predict transposition frequency (distinguish between equivalent activity in all cells and many hits in some cells). Nevertheless, enrichment methods can provide high coverage, and have been successful in demonstrating the presence of novel events in different biological contexts by several groups (Ewing and Kazazian 2010; Huang et al. 2010; Baillie et al. 2011; Solyom et al. 2012; Shukla et al. 2013).

An attractive alternative approach is high-throughput sequencing of single-cell genomes. Single-cell sequencing was first used to identify copy number variants (CNVs) in single cells from tumors (Navin et al. 2011). Single-cell sequencing was also used to examine retrotransposition frequency in the postmortem adult brain (Evrony et al. 2012, 2015; Upton et al. 2015). While these studies clearly identified novel somatic retrotransposition events in the adult human brain, they differed on the frequency of transposition. One group (Upton et al. 2015) reported a frequency of approximately 14 new retrotranspositions per hippocampal neuron and approximately 11 per cortical neuron, while another group (Evrony et al. 2012, 2015) found that somatic retrotranspositions were relatively infrequent. The reasons for these differences, which may be technical in nature, are currently under discussion (Upton et al. 2015; Evrony et al. 2016). Hence, more work is necessary to document the retrotransposition frequency in the adult brain, and in particular to address the bio informatic challenges of detecting novel transposition events in single-cell high-throughput sequencing data.

4.3 Transposable Elements and Autoimmunity

Studies of the negative effects of RTEs have largely focused on the damage caused by the transposition process to the genomes of their hosts. While many transposition events are abortive, they often cause DNA double-strand breaks and can promote a variety of illegitimate recombination events, such as chromosomal rearrangements (Farkash and Luning Prak 2006). Recent work has shed light on a new dimension of this host-pathogen relationship: an interesting link between RTEs and the development of autoimmune disease (Bhoj and Chen 2008).

Arguably, the primal form of infection is the parasitism of nucleic acids (Stetson 2009). Across billions of years of host-pathogen interactions many antiviral defense mechanisms have evolved and became implemented with various degrees of success (Hannon 2002; Kawai and Akira 2006; Pichlmair and Reis e Sousa 2007). Many of these response networks are centered on the detection of nucleic acids. To discriminate self from non-self, antiviral sensors must detect potentially hazardous invading nucleic acids among the copious amounts of host-derived DNA and RNA. However, overactivation or other failures of these antiviral systems can result in hyperstimulation of the immune system and autoimmune responses (Banchereau and Pascual 2006). The type I interferon (IFN-I) response is in particular important for establishing an antiviral state; however chronic IFN-I signaling can lead to hyperimmune activation and inflammation (Wilson et al. 2013).

Study of one specific human autoimmune disease, Aicardi–Goutieres syndrome (AGS), has provided important insights into the contribution of endogenous RTEs to the development of autoimmunity (Stetson et al. 2008). The *TREX1* gene encodes a 3′ exonuclease that degrades perceived invading DNA, including the cDNAs from endogenous RTEs. Unless eliminated, these DNA fragments accumulate in the cytosol and activate the IFN-stimulatory DNA (ISD) response and innate immune signaling. *TREX1* was found to be mutated in AGS, and the accumulation of RTE-derived cDNAs was associated the hyperactivation of the IFN-I pathway. In addition to *TREX1*, mutations in the RNAseH2 enzyme also cause AGS, suggesting that accumulation of RNA-DNA hybrids derived from endogenous RTEs contributes to the chronic pro-inflammatory state (Bhoj and Chen 2008).

In a fascinating parallel, chronic inflammation was proposed many years ago to play a major role in exacerbating the aging process, referred to as the "inflammaging" theory of aging (Franceschi and Campisi 2014). Inflammaging appears to be significant risk factor for the morbidity and mortality of the elderly, as most, if not all, age-related diseases share an inflammatory component. However, the etiology of inflammaging remains largely unknown. Thus, both aging and RTEs have been independently associated with chronic IFN-I responses, and aging itself has been associated with RTE activation (De Cecco et al. 2013b). It is thus tempting to speculate a direct connection between these factors, and a causal role between the age-associated expression of RTEs and chronic IFN-I activation.

Autoimmune inflammation that may be caused by the accumulation of RTE-derived single-stranded DNA (Yang et al. 2007) can be treated with reverse transcriptase inhibitors (Beck-Engeser et al. 2011). Several different nucleoside reverse transcriptase inhibitors (NRTIs), developed to treat HIV, have been tested against the reverse transcriptase enzyme encoded by L1 elements, with varying degrees of success (Jones et al. 2008; Dai et al. 2011). Note that some of these compounds might also exert their effects indirectly by inhibiting inflammation, independently of reverse transcriptase inhibition (Fowler et al. 2014). It is thus important to consider the possibility that interventions designed specifically against RTE activities may be effective against autoimmune disorders and perhaps other age-related diseases.

5 Conclusions

In this chapter we have summarized recent discoveries documenting age-related changes in chromatin and transposable element activity. What is the significance of these changes to our understanding of aging and for the prospect of developing new interventions to ameliorate the decline of organismal function with age?

Studies in yeast, nematodes, and fruit flies have demonstrated a strong link between the loss of a "youthful" chromatin state and aging. The salient characteristic of youthfulness in this context we believe is the effective partitioning and maintenance of euchromatic and heterochromatic domains of the genome. An important (albeit not only) consequence of the loss of this chromatin homeostasis is a failure to maintain the effective repression of TE activity. The evidence linking TE activity to aging, though less abundant, is steadily growing. Studies in yeast, flies, mice, and human cell culture show that compromising the cellular TE surveillance mechanisms can result in cellular damage, age-associated diseases, and shortened life span (Czech et al. 2008; Ghildiyal et al. 2008; Wallace et al. 2008; Kaneko et al. 2011; Lim et al. 2011; Maxwell et al. 2011; Li et al. 2013; Jeon et al. 2015; Wood et al. 2016). Evidence is also emerging that augmenting the surveillance mechanisms that maintain TE repression improves cellular physiology and may extend healthy life span (Savva et al. 2013; Wood et al. 2016).

Activation of TEs in the germline has been postulated to drive evolution and create genomic diversity. We believe that the sporadic activation of RTEs in somatic cells is unlikely to be beneficial. Instead, RTE activation is more likely to result in a variety of deleterious effects, such as dysregulation of gene expression, transcriptional noise, chronic activation of an antiviral state, insertional mutagenesis, DNA damage, and genome instability (Fig. 3). The emerging understanding of the potential role of RTEs to promote these rather serious consequences has led us (and others) to envision (Li et al. 2013; Sedivy et al. 2013; Volkman and Stetson 2014) that drugs targeting RTEs, such as NRTIs, or more indirect interventions, such as improving repressive heterochromatin or bolstering some other defense mechanisms, may provide new and novel therapeutic modalities to treat diseases of aging and extend healthy life span.

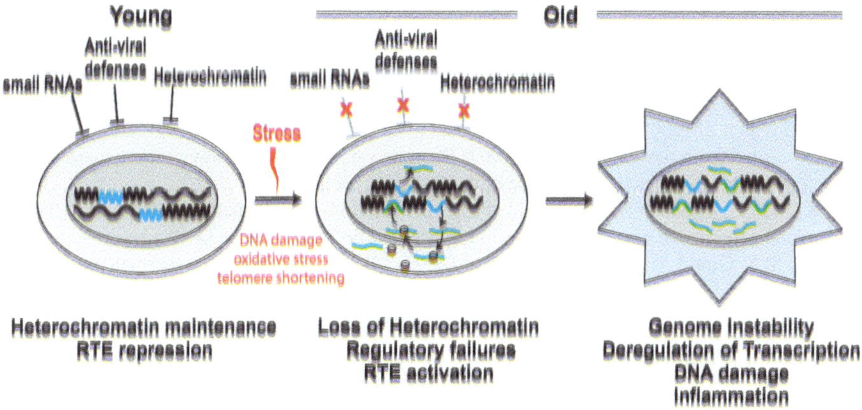

Fig. 3 Retrotransposition theory of aging. RTEs are epigenetically silenced in young somatic cells by their incorporation into constitutive heterochromatin, and additionally targeted by RNAi pathways and a variety of antiviral surveillance systems. Due to an accumulation of macromolecular damage and loss of homeostatic capacity, caused by a variety of extrinsic as well as intrinsic stresses, these cellular defense mechanisms become weakened with age. One consequence of this decline is the activation of dormant RTEs. The age-related increase in RTE expression and mobilization in turn causes further damage, and thus promotes the dysregulation of cellular physiology, loss of tissue function, and ultimately many of the deleterious aspects of aging

Acknowledgments The authors gratefully acknowledge the following sources of financial support: J.A.K., NIH K01 AG039410; B.C.J., NIH T32 AG041688, F31 AG047736; M.D.C., AFAR postdoctoral fellowship; S.W.C., NIH T32 GM007601, F31 AG050365; N.N., NIH R56 AG050582; S.L.H., NIH R01 AG024353, NIH P01 AG051449 and J.M.S., NIH R37 AG016694, NIH P01 AG051449.

References

Abe M, Naqvi A, Hendriks GJ et al (2014) Impact of age-associated increase in 2′-O-methylation of miRNAs on aging and neurodegeneration in Drosophila. Genes Dev 28:44–57

Abyzov A, Iskow R, Gokcumen O et al (2013) Analysis of variable retroduplications in human populations suggests coupling of retrotransposition to cell division. Genome Res 23:2042–2052

Alvares SM, Mayberry GA, Joyner EY et al (2014) H3K4 demethylase activities repress proliferative and postmitotic aging. Aging Cell 13:245–253

Aravin AA, Sachidanandam R, Bourc'his D et al (2008) A piRNA pathway primed by individual transposons is linked to de novo DNA methylation in mice. Mol Cell 31:785–799

Aravin AA, Sachidanandam R, Girard A et al (2007) Developmentally regulated piRNA clusters implicate MILI in transposon control. Science 316:744–747

Avrahami D, Li C, Zhang J et al (2015) Aging-dependent demethylation of regulatory elements correlates with chromatin state and improved beta cell function. Cell Metab 22:619–632

Bahar R, Hartmann CH, Rodriguez KA et al (2006) Increased cell-to-cell variation in gene expression in ageing mouse heart. Nature 441:1011–1014

Baillie JK, Barnett MW, Upton KR et al (2011) Somatic retrotransposition alters the genetic landscape of the human brain. Nature 479:534–537

Baker DJ, Childs BG, Durik M et al (2016) Naturally occurring p16(Ink4a)-positive cells shorten healthy lifespan. Nature 530:184–189

Banchereau J, Pascual V (2006) Type I interferon in systemic lupus erythematosus and other autoimmune diseases. Immunity 25:383–392

Beck-Engeser GB, Eilat D, Wabl M (2011) An autoimmune disease prevented by anti-retroviral drugs. Retrovirology 8:91

Berger SL (2007) The complex language of chromatin regulation during transcription. Nature 447:407–412

Bhoj VG, Chen ZJ (2008) Linking retroelements to autoimmunity. Cell 134:569–571

Brennecke J, Aravin AA, Stark A et al (2007) Discrete small RNA-generating loci as master regulators of transposon activity in Drosophila. Cell 128:1089–1103

Cao K, Blair CD, Faddah DA et al (2011) Progerin and telomere dysfunction collaborate to trigger cellular senescence in normal human fibroblasts. J Clin Invest 121:2833–2844

Carmell MA, Girard A, van de Kant HJ et al (2007) MIWI2 is essential for spermatogenesis and repression of transposons in the mouse male germline. Dev Cell 12:503–514

Carpenter JA, Keegan LP, Wilfert L et al (2009) Evidence for ADAR-induced hypermutation of the Drosophila sigma virus (Rhabdoviridae). BMC Genet 10:75

Chandra T, Ewels PA, Schoenfelder S et al (2015) Global reorganization of the nuclear landscape in senescent cells. Cell Rep 10:471–483

Chandra T, Kirschner K, Thuret JY et al (2012) Independence of repressive histone marks and chromatin compaction during senescent heterochromatic layer formation. Mol Cell 47:203–214

Chen H, Ruiz PD, McKimpson WM et al (2015) MacroH2A1 and ATM play opposing roles in paracrine senescence and the senescence-associated Secretory phenotype. Mol Cell 59:719–731

Chen LL, DeCerbo JN, Carmichael GG (2008) Alu element-mediated gene silencing. EMBO J 27:1694–1705

Chenais B (2013) Transposable elements and human cancer: a causal relationship? Biochim Biophys Acta 1835:28–35
Clancy DJ, Gems D, Harshman LG et al (2001) Extension of life-span by loss of CHICO, a Drosophila insulin receptor substrate protein. Science 292:104–106
Coufal NG, Garcia-Perez JL, Peng GE et al (2009) L1 retrotransposition in human neural progenitor cells. Nature 460:1127–1131
Criscione SW, De Cecco M, Siranosian B et al (2016) Reorganization of chromosome architecture in replicative cellular senescence. Sci Adv 2:e1500882
Criscione SW, Zhang Y, Thompson W et al (2014) Transcriptional landscape of repetitive elements in normal and cancer human cells. BMC Genomics 15:583
Cruickshanks HA, McBryan T, Nelson DM et al (2013a) Senescent cells harbour features of the cancer epigenome. Nat Cell Biol 15:1495–1506
Cruickshanks HA, Tufarelli C (2009) Isolation of cancer-specific chimeric transcripts induced by hypomethylation of the LINE-1 antisense promoter. Genomics 94:397–406
Cruickshanks HA, Vafadar-Isfahani N, Dunican DS et al (2013b) Expression of a large LINE-1-driven antisense RNA is linked to epigenetic silencing of the metastasis suppressor gene TFPI-2 in cancer. Nucleic Acids Res 41:6857–6869
Czech B, Malone CD, Zhou R et al (2008) An endogenous small interfering RNA pathway in Drosophila. Nature 453:798–802
Dai L, Huang Q, Boeke JD (2011) Effect of reverse transcriptase inhibitors on LINE-1 and Ty1 reverse transcriptase activities and on LINE-1 retrotransposition. BMC Biochem 12:18
Dang W, Steffen KK, Perry R et al (2009) Histone H4 lysine 16 acetylation regulates cellular lifespan. Nature 459:802–807
Dang W, Sutphin GL, Dorsey JA et al (2014) Inactivation of yeast Isw2 chromatin remodeling enzyme mimics longevity effect of calorie restriction via induction of genotoxic stress response. Cell Metab 19:952–966
Day K, Waite LL, Thalacker-Mercer A et al (2013) Differential DNA methylation with age displays both common and dynamic features across human tissues that are influenced by CpG landscape. Genome Biol 14:R102
De Cecco M, Criscione SW, Peckham EJ et al (2013a) Genomes of replicatively senescent cells undergo global epigenetic changes leading to gene silencing and activation of transposable elements. Aging Cell 12:247–256
De Cecco M, Criscione SW, Peterson AL et al (2013b) Transposable elements become active and mobile in the genomes of aging mammalian somatic tissues. Aging (Albany NY) 5:867–883
de Koning AP, Gu W, Castoe TA et al (2011) Repetitive elements may comprise over two-thirds of the human genome. PLoS Genet 7:e1002384
Deininger P (2011) Alu elements: know the SINEs. Genome Biol 12:236
Di Giacomo M, Comazzetto S, Saini H et al (2013) Multiple epigenetic mechanisms and the piRNA pathway enforce LINE1 silencing during adult spermatogenesis. Mol Cell 50:601–608
Dixon JR, Jung I, Selvaraj S et al (2015) Chromatin architecture reorganization during stem cell differentiation. Nature 518:331–336
Doucet-O'Hare TT, Rodic N, Sharma R et al (2015) LINE-1 expression and retrotransposition in Barrett's esophagus and esophageal carcinoma. Proc Natl Acad Sci U S A 112:E4894–E4900
Elgin SC, Grewal SI (2003) Heterochromatin: silence is golden. Curr Biol 13:R895–R898
Elsasser SJ, Noh KM, Diaz N et al (2015) Histone H3.3 is required for endogenous retroviral element silencing in embryonic stem cells. Nature 522:240–244
Erwin JA, Marchetto MC, Gage FH (2014) Mobile DNA elements in the generation of diversity and complexity in the brain. Nat Rev Neurosci 15:497–506
Evrony GD, Cai X, Lee E et al (2012) Single-neuron sequencing analysis of L1 retrotransposition and somatic mutation in the human brain. Cell 151:483–496
Evrony GD, Lee E, Mehta BK et al (2015) Cell lineage analysis in human brain using endogenous retroelements. Neuron 85:49–59
Evrony GD, Lee E, Park PJ et al (2016) Resolving rates of mutation in the brain using single-neuron genomics. Elife 5:e12966

Ewing AD, Gacita A, Wood LD et al (2015) Widespread somatic L1 retrotransposition occurs early during gastrointestinal cancer evolution. Genome Res 25:1536–1545
Ewing AD, Kazazian HH Jr (2010) High-throughput sequencing reveals extensive variation in human-specific L1 content in individual human genomes. Genome Res 20:1262–1270
Fagegaltier D, Bouge AL, Berry B et al (2009) The endogenous siRNA pathway is involved in heterochromatin formation in Drosophila. Proc Natl Acad Sci U S A 106:21258–21263
Farkash EA, Luning Prak ET (2006) DNA damage and L1 retrotransposition. J Biomed Biotechnol 2006:37285
Faulkner GJ, Kimura Y, Daub CO et al (2009) The regulated retrotransposon transcriptome of mammalian cells. Nat Genet 41:563–571
Feltzin VL, Khaladkar M, Abe M et al (2015) The exonuclease Nibbler regulates age-associated traits and modulates piRNA length in Drosophila. Aging Cell 14:443–452
Fernandez AF, Bayon GF, Urdinguio RG et al (2015) H3K4me1 marks DNA regions hypomethylated during aging in human stem and differentiated cells. Genome Res 25:27–40
Feser J, Truong D, Das C et al (2010) Elevated histone expression promotes life span extension. Mol Cell 39:724–735
Feser J, Tyler J (2011) Chromatin structure as a mediator of aging. FEBS Lett 585:2041–2048
Forstemann K, Horwich MD, Wee L et al (2007) Drosophila microRNAs are sorted into functionally distinct argonaute complexes after production by dicer-1. Cell 130:287–297
Fowler BJ, Gelfand BD, Kim Y et al (2014) Nucleoside reverse transcriptase inhibitors possess intrinsic anti-inflammatory activity. Science 346:1000–1003
Franceschi C, Campisi J (2014) Chronic inflammation (inflammaging) and its potential contribution to age-associated diseases. J Gerontol A Biol Sci Med Sci 69(Suppl 1):S4–S9
Ganguly A, Dunbar T, Chen P et al (2003) Exon skipping caused by an intronic insertion of a young Alu Yb9 element leads to severe hemophilia A. Hum Genet 113:348–352
Ge ZJ, Schatten H, Zhang CL et al (2015) Oocyte ageing and epigenetics. Reproduction 149:R103–R114
Gelfand BD, Wright CB, Kim Y et al (2015) Iron toxicity in the retina requires Alu RNA and the NLRP3 inflammasome. Cell Rep 11:1686–1693
Ghildiyal M, Seitz H, Horwich MD et al (2008) Endogenous siRNAs derived from transposons and mRNAs in Drosophila somatic cells. Science 320:1077–1081
Ghildiyal M, Zamore PD (2009) Small silencing RNAs: an expanding universe. Nat Rev Genet 10:94–108
Giannakou ME, Goss M, Junger MA et al (2004) Long-lived Drosophila with overexpressed dFOXO in adult fat body. Science 305:361
Goldman RD, Shumaker DK, Erdos MR et al (2004) Accumulation of mutant lamin A causes progressive changes in nuclear architecture in Hutchinson-Gilford progeria syndrome. Proc Natl Acad Sci U S A 101:8963–8968
Greer EL, Maures TJ, Hauswirth AG et al (2010) Members of the H3K4 trimethylation complex regulate lifespan in a germline-dependent manner in C. elegans. Nature 466:383–387
Greer EL, Maures TJ, Ucar D et al (2011) Transgenerational epigenetic inheritance of longevity in *Caenorhabditis elegans*. Nature 479:365–371
Grewal SI, Jia S (2007) Heterochromatin revisited. Nat Rev Genet 8:35–46
Gu T, Elgin SC (2013) Maternal depletion of Piwi, a component of the RNAi system, impacts heterochromatin formation in Drosophila. PLoS Genet 9:e1003780
Guelen L, Pagie L, Brasset E et al (2008) Domain organization of human chromosomes revealed by mapping of nuclear lamina interactions. Nature 453:948–951
Hamdorf M, Idica A, Zisoulis DG et al (2015) miR-128 represses L1 retrotransposition by binding directly to L1 RNA. Nat Struct Mol Biol 22:824–831
Han S, Brunet A (2012) Histone methylation makes its mark on longevity. Trends Cell Biol 22:42–49
Hannon GJ (2002) RNA interference. Nature 418:244–251

Hanzelmann S, Beier F, Gusmao EG et al (2015) Replicative senescence is associated with nuclear reorganization and with DNA methylation at specific transcription factor binding sites. Clin Epigenetics 7:19

Harries LW, Hernandez D, Henley W et al (2011) Human aging is characterized by focused changes in gene expression and deregulation of alternative splicing. Aging Cell 10:868–878

Heale BS, Keegan LP, McGurk L et al (2009) Editing independent effects of ADARs on the miRNA/siRNA pathways. EMBO J 28:3145–3156

Heras SR, Macias S, Caceres JF et al (2014) Control of mammalian retrotransposons by cellular RNA processing activities. Mob Genet Elements 4:e28439

Heras SR, Macias S, Plass M et al (2013) The Microprocessor controls the activity of mammalian retrotransposons. Nat Struct Mol Biol 20:1173–1181

Hills SA, Diffley JF (2014) DNA replication and oncogene-induced replicative stress. Curr Biol 24:R435–R444

Hu Z, Chen K, Xia Z et al (2014) Nucleosome loss leads to global transcriptional up-regulation and genomic instability during yeast aging. Genes Dev 28:396–408

Huang CR, Burns KH, Boeke JD (2012) Active transposition in genomes. Annu Rev Genet 46:651–675

Huang CR, Schneider AM, Lu Y et al (2010) Mobile interspersed repeats are major structural variants in the human genome. Cell 141:1171–1182

Hur K, Cejas P, Feliu J et al (2014) Hypomethylation of long interspersed nuclear element-1 (LINE-1) leads to activation of proto-oncogenes in human colorectal cancer metastasis. Gut 63:635–646

Hutvagner G, Zamore PD (2002) A microRNA in a multiple-turnover RNAi enzyme complex. Science 297:2056–2060

Hwangbo DS, Gershman B, Tu MP et al (2004) Drosophila dFOXO controls lifespan and regulates insulin signalling in brain and fat body. Nature 429:562–566

Jeon HJ, Kim YS, Park JS et al (2015) Age-related change in gammaH2AX of Drosophila muscle: its significance as a marker for muscle damage and longevity. Biogerontology 16:503–516

Jepson JE, Savva YA, Yokose C et al (2011) Engineered alterations in RNA editing modulate complex behavior in Drosophila: regulatory diversity of adenosine deaminase acting on RNA (ADAR) targets. J Biol Chem 286:8325–8337

Jiang N, Du G, Tobias E et al (2013) Dietary and genetic effects on age-related loss of gene silencing reveal epigenetic plasticity of chromatin repression during aging. Aging 5:813–824

Jin C, Li J, Green CD et al (2011) Histone demethylase UTX-1 regulates C. elegans life span by targeting the insulin/IGF-1 signaling pathway. Cell Metab 14:161–172

Jones RB, Garrison KE, Wong JC et al (2008) Nucleoside analogue reverse transcriptase inhibitors differentially inhibit human LINE-1 retrotransposition. PLoS One 3:e1547

Jones BC, Wood JG, Chang C et al (2016) A somatic piRNA pathway in the Drosophila fat body ensures metabolic homeostasis and normal lifespan. Nat Com (in press)

Kaneko H, Dridi S, Tarallo V et al (2011) DICER1 deficit induces Alu RNA toxicity in age-related macular degeneration. Nature 471:325–330

Kapahi P, Zid BM, Harper T et al (2004) Regulation of lifespan in Drosophila by modulation of genes in the TOR signaling pathway. Curr Biol 14:885–890

Kawai T, Akira S (2006) Innate immune recognition of viral infection. Nat Immunol 7:131–137

Kawamura Y, Saito K, Kin T et al (2008) Drosophila endogenous small RNAs bind to Argonaute 2 in somatic cells. Nature 453:793–797

Kennedy BK, Gotta M, Sinclair DA et al (1997) Redistribution of silencing proteins from telomeres to the nucleolus is associated with extension of life span in S. cerevisiae. Cell 89:381–391

Kenyon C, Chang J, Gensch E et al (1993) A C. elegans mutant that lives twice as long as wild type. Nature 366:461–464

Kenyon CJ (2010) The genetics of ageing. Nature 464:504–512

Kim S, Villeponteau B, Jazwinski SM (1996) Effect of replicative age on transcriptional silencing near telomeres in *Saccharomyces cerevisiae*. Biochem Biophys Res Commun 219:370–376

Kosar M, Bartkova J, Hubackova S et al (2011) Senescence-associated heterochromatin foci are dispensable for cellular senescence, occur in a cell type- and insult-dependent manner and follow expression of p16(ink4a). Cell Cycle 10:457–468
Kreiling JA, Tamamori-Adachi M, Sexton AN et al (2011) Age-associated increase in heterochromatic marks in murine and primate tissues. Aging Cell 10:292–304
Kunarso G, Chia NY, Jeyakani J et al (2010) Transposable elements have rewired the core regulatory network of human embryonic stem cells. Nat Genet 42:631–634
Larson K, Yan SJ, Tsurumi A et al (2012) Heterochromatin formation promotes longevity and represses ribosomal RNA synthesis. PLoS Genet 8:e1002473
Lee EJ, Banerjee S, Zhou H et al (2011) Identification of piRNAs in the central nervous system. RNA 17:1090–1099
Lee SS, Kennedy S, Tolonen AC et al (2003) DAF-16 target genes that control C. elegans life-span and metabolism. Science 300:644–647
Lev-Maor G, Ram O, Kim E et al (2008) Intronic Alus influence alternative splicing. PLoS Genet 4:e1000204
Levin HL, Moran JV (2011) Dynamic interactions between transposable elements and their hosts. Nat Rev Genet 12:615–627
Li L, Greer C, Eisenman RN et al (2010) Essential functions of the histone demethylase lid. PLoS Genet 6:e1001221
Li W, Prazak L, Chatterjee N et al (2013) Activation of transposable elements during aging and neuronal decline in Drosophila. Nat Neurosci 16:529–531
Lim DH, Oh CT, Lee L et al (2011) The endogenous siRNA pathway in Drosophila impacts stress resistance and lifespan by regulating metabolic homeostasis. FEBS Lett 585:3079–3085
Liu J (2004) Argonaute2 is the catalytic engine of mammalian RNAi. Science 305:1437–1441
Liu N, Abe M, Sabin LR et al (2011) The exoribonuclease Nibbler controls 3′ end processing of microRNAs in Drosophila. Curr Biol 21:1888–1893
Malki S, van der Heijden GW, O'Donnell KA et al (2014) A role for retrotransposon LINE-1 in fetal oocyte attrition in mice. Dev Cell 29:521–533
Martinez J, Patkaniowska A, Urlaub H et al (2002) Single-stranded antisense siRNAs guide target RNA cleavage in RNAi. Cell 110:563–574
Matsui T, Leung D, Miyashita H et al (2010) Proviral silencing in embryonic stem cells requires the histone methyltransferase ESET. Nature 464:927–931
Matzke MA, Birchler JA (2005) RNAi-mediated pathways in the nucleus. Nat Rev Genet 6:24–35
Maures TJ, Greer EL, Hauswirth AG et al (2011) The H3K27 demethylase UTX-1 regulates C. elegans lifespan in a germline-independent, insulin-dependent manner. Aging Cell 10:980–990
Maxwell PH, Burhans WC, Curcio MJ (2011) Retrotransposition is associated with genome instability during chronological aging. Proc Natl Acad Sci U S A 108:20376–20381
Mazin P, Xiong J, Liu X et al (2013) Widespread splicing changes in human brain development and aging. Mol Syst Biol 9:633
McColl G, Killilea DW, Hubbard AE et al (2008) Pharmacogenetic analysis of lithium-induced delayed aging in Caenorhabditis elegans. J Biol Chem 283:350–357
McCord RP, Nazario-Toole A, Zhang H et al (2013) Correlated alterations in genome organization, histone methylation, and DNA-lamin A/C interactions in Hutchinson-Gilford progeria syndrome. Genome Res 23:260–269
Meister G, Landthaler M, Patkaniowska A et al (2004) Human Argonaute2 mediates RNA cleavage targeted by miRNAs and siRNAs. Mol Cell 15:185–197
Montoya-Durango DE, Liu Y, Teneng I et al (2009) Epigenetic control of mammalian LINE-1 retrotransposon by retinoblastoma proteins. Mutat Res 665:20–28
Montoya-Durango DE, Ramos KA, Bojang P et al (2016) LINE-1 silencing by retinoblastoma proteins is effected through the nucleosomal and remodeling deacetylase multiprotein complex. BMC Cancer 16:38

Muotri AR, Chu VT, Marchetto MC et al (2005) Somatic mosaicism in neuronal precursor cells mediated by L1 retrotransposition. Nature 435:903–910
Muotri AR, Marchetto MC, Coufal NG et al (2010) L1 retrotransposition in neurons is modulated by MeCP2. Nature 468:443–446
Narita M, Nunez S, Heard E et al (2003) Rb-mediated heterochromatin formation and silencing of E2F target genes during cellular senescence. Cell 113:703–716
Navin N, Kendall J, Troge J et al (2011) Tumour evolution inferred by single-cell sequencing. Nature 472:90–94
Ni Z, Ebata A, Alipanahiramandi E et al (2012) Two SET domain containing genes link epigenetic changes and aging in Caenorhabditis elegans. Aging Cell 11:315–325
Nigumann P, Redik K, Matlik K et al (2002) Many human genes are transcribed from the antisense promoter of L1 retrotransposon. Genomics 79:628–634
O'Sullivan RJ, Karlseder J (2012) The great unravelling: chromatin as a modulator of the aging process. Trends Biochem Sci 37:466–476
O'Sullivan RJ, Kubicek S, Schreiber SL et al (2010) Reduced histone biosynthesis and chromatin changes arising from a damage signal at telomeres. Nat Struct Mol Biol 17:1218–1225
Oberdoerffer P, Sinclair DA (2007) The role of nuclear architecture in genomic instability and ageing. Nat Rev Mol Cell Biol 8:692–702
Peleg S, Feller C, Forne I et al (2016) Life span extension by targeting a link between metabolism and histone acetylation in Drosophila. EMBO Rep 17:455–469
Perrat PN, DasGupta S, Wang J et al (2013) Transposition-driven genomic heterogeneity in the Drosophila brain. Science 340:91–95
Pichlmair A, Reis e Sousa C (2007) Innate recognition of viruses. Immunity 27:370–383
Polak P, Domany E (2006) Alu elements contain many binding sites for transcription factors and may play a role in regulation of developmental processes. BMC Genomics 7:133
Rai TS, Adams PD (2013) Lessons from senescence: chromatin maintenance in non-proliferating cells. Biochim Biophys Acta 1819:322–331
Rieder LE, Staber CJ, Hoopengardner B et al (2013) Tertiary structural elements determine the extent and specificity of messenger RNA editing. Nat Commun 4:2232
Rodic N, Steranka JP, Makohon-Moore A et al (2015) Retrotransposon insertions in the clonal evolution of pancreatic ductal adenocarcinoma. Nat Med 21:1060–1064
Rogina B, Reenan RA, Nilsen SP et al (2000) Extended life-span conferred by cotransporter gene mutations in Drosophila. Science 290:2137–2140
Roman-Gomez J, Jimenez-Velasco A, Agirre X et al (2005) Promoter hypomethylation of the LINE-1 retrotransposable elements activates sense/antisense transcription and marks the progression of chronic myeloid leukemia. Oncogene 24:7213–7223
Ross RJ, Weiner MM, Lin H (2014) PIWI proteins and PIWI-interacting RNAs in the soma. Nature 505:353–359
Rowe HM, Jakobsson J, Mesnard D et al (2010) KAP1 controls endogenous retroviruses in embryonic stem cells. Nature 463:237–240
Rozhkov NV, Hammell M, Hannon GJ (2013) Multiple roles for Piwi in silencing Drosophila transposons. Genes Dev 27:400–412
Sadaie M, Salama R, Carroll T et al (2013) Redistribution of the Lamin B1 genomic binding profile affects rearrangement of heterochromatic domains and SAHF formation during senescence. Genes Dev 27:1800–1808
Sarg B, Koutzamani E, Helliger W et al (2002) Postsynthetic trimethylation of histone H4 at lysine 20 in mammalian tissues is associated with aging. J Biol Chem 277:39195–39201
Savva YA, Jepson JE, Chang YJ et al (2013) RNA editing regulates transposon-mediated heterochromatic gene silencing. Nat Commun 4:2745
Savva YA, Rieder LE, Reenan RA (2012) The ADAR protein family. Genome Biol 13:252
Scadden AD, Smith CW (2001) RNAi is antagonized by A→I hyper-editing. EMBO Rep 2:1107–1111
Scaffidi P, Misteli T (2006) Lamin A-dependent nuclear defects in human aging. Science 312:1059–1063

Sedivy JM, Banumathy G, Adams PD (2008) Aging by epigenetics—a consequence of chromatin damage? Exp Cell Res 314:1909–1917
Sedivy JM, Kreiling JA, Neretti N et al (2013) Death by transposition—the enemy within? Bioessays 35:1035–1043
Shabalina SA, Koonin EV (2008) Origins and evolution of eukaryotic RNA interference. Trends Ecol Evol 23:578–587
Shukla R, Upton KR, Munoz-Lopez M et al (2013) Endogenous retrotransposition activates oncogenic pathways in hepatocellular carcinoma. Cell 153:101–111
Shumaker DK, Dechat T, Kohlmaier A et al (2006) Mutant nuclear lamin A leads to progressive alterations of epigenetic control in premature aging. Proc Natl Acad Sci U S A 103:8703–8708
Siebold AP, Banerjee R, Tie F et al (2010) Polycomb repressive complex 2 and Trithorax modulate Drosophila longevity and stress resistance. Proc Natl Acad Sci U S A 107:169–174
Sinclair DA, Oberdoerffer P (2009) The ageing epigenome: damaged beyond repair? Ageing Res Rev 8:189–198
Slotkin RK, Martienssen R (2007) Transposable elements and the epigenetic regulation of the genome. Nat Rev Genet 8:272–285
Smeal T, Claus J, Kennedy B et al (1996) Loss of transcriptional silencing causes sterility in old mother cells of S. cerevisiae. Cell 84:633–642
Solyom S, Ewing AD, Rahrmann EP et al (2012) Extensive somatic L1 retrotransposition in colorectal tumors. Genome Res 22:2328–2338
Sookdeo A, Hepp CM, McClure MA et al (2013) Revisiting the evolution of mouse LINE-1 in the genomic era. Mob DNA 4:3
Speek M (2001) Antisense promoter of human L1 retrotransposon drives transcription of adjacent cellular genes. Mol Cell Biol 21:1973–1985
Stetson DB (2009) Connections between antiviral defense and autoimmunity. Curr Opin Immunol 21:244–250
Stetson DB, Ko JS, Heidmann T et al (2008) Trex1 prevents cell-intrinsic initiation of autoimmunity. Cell 134:587–598
Sun D, Yi SV (2015) Impacts of chromatin states and long-range genomic segments on aging and DNA methylation. PLoS One 10:e0128517
Swanson EC, Manning B, Zhang H et al (2013) Higher-order unfolding of satellite heterochromatin is a consistent and early event in cell senescence. J Cell Biol 203:929–942
Taft RJ, Pang KC, Mercer TR et al (2010) Non-coding RNAs: regulators of disease. J Pathol 220:126–139
Tarallo V, Hirano Y, Gelfand BD et al (2012) DICER1 loss and Alu RNA induce age-related macular degeneration via the NLRP3 inflammasome and MyD88. Cell 149:847–859
Tatar M, Kopelman A, Epstein D et al (2001) A mutant Drosophila insulin receptor homolog that extends life-span and impairs neuroendocrine function. Science 292:107–110
Teneng I, Montoya-Durango DE, Quertermous JL et al (2011) Reactivation of L1 retrotransposon by benzo(a)pyrene involves complex genetic and epigenetic regulation. Epigenetics 6:355–367
Tollervey JR, Wang Z, Hortobagyi T et al (2011) Analysis of alternative splicing associated with aging and neurodegeneration in the human brain. Genome Res 21:1572–1582
Tufarelli C, Stanley JA, Garrick D et al (2003) Transcription of antisense RNA leading to gene silencing and methylation as a novel cause of human genetic disease. Nat Genet 34:157–165
Upton KR, Gerhardt DJ, Jesuadian JS et al (2015) Ubiquitous L1 mosaicism in hippocampal neurons. Cell 161:228–239
Vagin VV, Sigova A, Li C et al (2006) A distinct small RNA pathway silences selfish genetic elements in the germline. Science 313:320–324
Van Meter M, Kashyap M, Rezazadeh S et al (2014) SIRT6 represses LINE1 retrotransposons by ribosylating KAP1 but this repression fails with stress and age. Nat Commun 5:5011
Varshney D, Vavrova-Anderson J, Oler AJ et al (2015) SINE transcription by RNA polymerase III is suppressed by histone methylation but not by DNA methylation. Nat Commun 6:6569

Volkman HE, Stetson DB (2014) The enemy within: endogenous retroelements and autoimmune disease. Nat Immunol 15:415–422
Wallace NA, Belancio VP, Deininger PL (2008) L1 mobile element expression causes multiple types of toxicity. Gene 419:75–81
Wang H, Ma Z, Niu K et al (2016) Antagonistic roles of Nibbler and Hen1 in modulating piRNA 3′ ends in Drosophila. Development 143:530–539
Wang Q, Zhang Z, Blackwell K et al (2005) Vigilins bind to promiscuously A-to-I-edited RNAs and are involved in the formation of heterochromatin. Curr Biol 15:384–391
Warren LA, Rossi DJ, Schiebinger GR et al (2007) Transcriptional instability is not a universal attribute of aging. Aging Cell 6:775–782
Weber B, Kimhi S, Howard G et al (2010) Demethylation of a LINE-1 antisense promoter in the cMet locus impairs Met signalling through induction of illegitimate transcription. Oncogene 29:5775–5784
Wilson EB, Yamada DH, Elsaesser H et al (2013) Blockade of chronic type I interferon signaling to control persistent LCMV infection. Science 340:202–207
Wilson RC, Doudna JA (2013) Molecular mechanisms of RNA interference. Annu Rev Biophys 42:217–239
Wolff EM, Byun HM, Han HF et al (2010) Hypomethylation of a LINE-1 promoter activates an alternate transcript of the MET oncogene in bladders with cancer. PLoS Genet 6:e1000917
Wood JG, Hillenmeyer S, Lawrence C et al (2010) Chromatin remodeling in the aging genome of Drosophila. Aging Cell 9:971–978
Wood JG, Jones BC, Jiang N et al. (2016) Chromatin-modifying genetic interventions suppress age-associated transposable element activation and extend life span in Drosophila. Proc Natl Acad Sci USA, 113:11277-11282
Yan Z, Hu HY, Jiang X et al (2011) Widespread expression of piRNA-like molecules in somatic tissues. Nucleic Acids Res 39:6596–6607
Yang N, Kazazian HH Jr (2006) L1 retrotransposition is suppressed by endogenously encoded small interfering RNAs in human cultured cells. Nat Struct Mol Biol 13:763–771
Yang YG, Lindahl T, Barnes DE (2007) Trex1 exonuclease degrades ssDNA to prevent chronic checkpoint activation and autoimmune disease. Cell 131:873–886
Yu W, Gius D, Onyango P et al (2008) Epigenetic silencing of tumour suppressor gene p15 by its antisense RNA. Nature 451:202–206
Zhang R, Chen W, Adams PD (2007) Molecular dissection of formation of senescence-associated heterochromatin foci. Mol Cell Biol 27:2343–2358
Zhang W, Li J, Suzuki K et al (2015) Aging stem cells. A Werner syndrome stem cell model unveils heterochromatin alterations as a driver of human aging. Science 348:1160–1163

Index

G. Cristofari (ed.), *Human Retrotransposons in Health and Disease*,
DOI 10.1007/978-3-319-48344-3

S

T

MIX
Papier aus verantwortungsvollen Quellen
Paper from responsible sources
FSC® C105338

If you have any concerns about our products,
you can contact us on
ProductSafety@springernature.com

In case Publisher is established outside the EU,
the EU authorized representative is:
Springer Nature Customer Service Center GmbH
Europaplatz 3, 69115 Heidelberg, Germany

Printed by Libri Plureos GmbH
in Hamburg, Germany